Materials Design for Energy Harvesting and Sensor Applications

Materials Design for Energy Harvesting and Sensor Applications

Edited by

Hiroki Kurita

Yu Shi

Tianzhuo Zhan

WILEY VCH

Editors

Prof. Hiroki Kurita
Tohoku University
Sendai, JA

Prof. Yu Shi
University of Leeds
Woodhouse Lane, Leeds
UK

Prof. Tianzhuo Zhan
Beihang University
37 Xueyuan Road, Haidian District, Beijing,
P.R. China

Cover Design: Wiley
Cover Image: Courtesy of the Editors

Typesetting: Lumina Datamatics, Madurai, India

Library of Congress Card No.: applied for

British Library Cataloguing-in-Publication Data:
A catalogue record for this book is available from the British Library.

Bibliographic information published by the Deutsche Nationalbibliothek
The Deutsche Nationalbibliothek lists this publication in the Deutsche Nationalbibliografie; detailed bibliographic data are available on the Internet at <http://dnb.d-nb.de>.

Print ISBN: 9783527353217
ePDF ISBN: 9783527844968
ePub ISBN: 9783527844975
oBook ISBN: 9783527844982

Printed and bound by
CPI Group (UK) Ltd, Croydon, CR0 4YY

C9783527353217_240726

To my colleagues and friends in the JSPS Core-to-Core Program "Establishing an International Research Center for Multi-Energy Harvesting Materials and Devices to Realize the IoT Society,"

To my students, whose curiosity and determination drive innovation forward.

To the pioneering researchers who laid the foundations of this field, upon whose shoulders we continue to build.

And to my wife, Stéphanie, and my son, Seiya-Lancelot, whose love, patience, and courage sustain every journey I take. whose collaboration and trust made this vision possible.

—Hiroki Kurita

Contents

Preface

Energy harvesting and self-powered sensing technologies are entering one of the most dynamic phases in their history. Over the past two decades, the rapid proliferation of distributed electronics—Internet of Things sensor networks, wearable and implantable health-monitoring systems, autonomous infrastructures, and smart manufacturing—has reshaped how we think about power generation at small scales. As these systems increasingly demand long-term, maintenance-free operation, researchers have been compelled to rethink materials, architectures, and energy-conversion principles from the ground up.

This textbook emerges from that momentum. It is built on a simple but profound belief: future energy systems will rely not on larger batteries but on smarter materials—materials capable of harvesting, converting, storing, and responding to the energy that already surrounds us. The chapters collected here span physics, chemistry, mechanics, materials engineering, and device integration, reflecting the inherently interdisciplinary nature of this field.

A defining characteristic of modern energy harvesting is its diversity. Vibration, impact, heat flow, mechanical contact, electromagnetic induction, magnetic-field variation, and even quantum-level spin interactions can each be sources of usable power when coupled with the right material platform. Accordingly, this book covers an exceptionally wide range of transduction mechanisms. Thermoelectric systems are presented from the perspective of both microscopic electronic structure—such as Dirac-electron dispersions—and macroscopic film–substrate engineering. Piezoelectric harvesters are explored from their classical monolithic forms to their modern manifestations within carbon fiber reinforced polymer laminates, bistable composite shells, energy-adaptive sandwich structures, and ultra-flexible polymer nanofiber networks for biomedical use. Magnetostrictive materials, once confined to niche applications, are revisited through Fe–Co-based composites, clad plates, nitrided alloys, additively manufactured lattices, and advanced magneto-mechanical modeling.

Mechanical energy harvesting is no longer limited to linear vibration devices. Several chapters demonstrate the power of nonlinearity—flutter-induced limit-cycle oscillations, bistable dynamics, Duffing-type responses, and digitally controlled synchronized switch harvesting on inductor circuits capable of autonomous, self-powered operation. The book also explores multifield coupling through magnetoelectric laminates and multiscale composite optimization, revealing strategies to amplify electromechanical responses far beyond traditional monolithic limits.

Thermal energy harvesting is similarly revisited from multiple vantage points, including thermoelectric polymers, stretchable thermoelectric generators, origami-inspired architectures, silicon-micromachined micro-thermoelectric generators, and thin-film Bi_2Te_3 devices fabricated through molecular beam epitaxy, pulsed laser deposition, solvothermal growth, or screen printing. Multiple chapters demonstrate that thermal–mechanical compliance—once seen as a limitation—can be transformed into a design advantage when materials are engineered to conform to curved, dynamic, or biological surfaces.

In parallel, the book introduces emerging materials platforms that promise to broaden the future landscape of energy harvesters and sensors. Two-dimensional materials, framework structures (Metal–Organic Frameworks/Covalent Organic Frameworks), hybrid perovskites, MXenes (Mn+1XnTx (transition metal carbides / nitrides / carbonitrides)), electrets, hydrogels, organic–inorganic hybrids, and lead-free ceramics each contribute new options for flexibility, multifunctionality, biocompatibility, sustainability, or low-temperature processing. Their inclusion reflects a critical trend: the field is evolving beyond performance alone to embrace environmental responsibility, circularity, and scalable manufacturability.

Another recurring theme across the chapters is the convergence of materials innovation and computational design. Multiscale homogenization, finite element modeling, domain-switching simulations, microstructural orientation optimization, and thermo–electro–magneto–mechanical coupling analyses now play central roles in guiding material development. These tools allow researchers to predict device behavior before fabrication, reducing cost and accelerating discovery. Such approaches highlight the growing need to integrate data-driven design, digital engineering, and virtual prototyping into the development cycle of smart materials.

Sustainability, circular design, and responsible deployment form the final axis of this book. As self-powered sensors become ubiquitous, their environmental footprint grows in parallel. The chapters dedicated to circular material strategies, waste-to-energy concepts, and policy–business integration remind us that next-generation devices must be designed not only for performance but also for recyclability, safe disposal, and ethical use. Concepts such as modular architectures, product-as-a-service models, digital product passports, and life-cycle-informed design represent the new frontier where materials science meets societal responsibility.

Taken together, the chapters in this volume offer more than a snapshot of current technologies—they form a roadmap toward a future in which energy-harvesting and sensing systems seamlessly integrate into human environments, operate autonomously for years, and are crafted from materials carefully chosen for both performance and planetary health.

We hope that this textbook serves multiple purposes:

- as a foundation for students entering the multidisciplinary world of energy harvesting

- as a reference for researchers seeking to understand emerging materials and design frameworks
- as inspiration for new ideas that transcend conventional boundaries of mechanics, electronics, and materials science
- and as a reminder that technological innovation must advance hand-in-hand with sustainable thinking

Finally, we express our deepest gratitude to all contributing authors for their expertise, their generosity, and their commitment to this shared vision. Their work represents the cutting edge of a rapidly expanding scientific frontier, and their insights will undoubtedly influence the development of self-powered systems for years to come.

We would also like to express our sincere appreciation to Ms. Dominika Demlova for creating the cover image for this book. Her ability to translate complex scientific concepts into a visually compelling design greatly enriches this volume's identity. Her aesthetic sense and professionalism have added a unique and inspiring dimension to the final publication.

December 2025

Prof. Hiroki Kurita, Prof. Yu Shi, and Prof. Tianzhuo Zhan

1

Recent Progress in Theory of Thermoelectric Effect: Focusing on Types I, II, and III Dirac Systems and Film-substrate Systems

Hiroyasu Matsuura[1] and Tomonari Mizoguchi[2]

[1] *National Institute of Advanced Industrial Science and Technology (AIST), Tsukuba, Ibaraki, Japan*
[2] *Department of Physics, Institute of Pure and Applied Sciences, University of Tsukuba, Tsukuba, Ibaraki, Japan*

1.1 Introduction

The thermoelectric effect, which is a phenomenon converting the temperature gradient to the electric voltage, has been used in special applications such as the exploration satellites, while recently the thermoelectric effect has been focused on not only the special applications but also the green power generation and the power supply for the Internet of Things society [1]. The conversion efficiency of the thermoelectric effect (η) is given by $\eta = \eta_c \dfrac{\sqrt{1+ZT}-1}{\sqrt{1+ZT}+T_c/T_h}$, where η_c and ZT indicate the Carnot efficiency and the dimensionless figure of merit, respectively, and T_c and T_h are the temperature of the cold and hot sides. ZT is mathematically given by $ZT = S^2\sigma T/\kappa$, where S, σ, κ, and T are Seebeck coefficient, electric conductivity, thermal conductivity, and temperature, respectively. Since these parameters are material parameters, for the innovation of thermoelectric materials with high efficiency, the precise control of these parameters is required. Unique electronic states such as Dirac electrons, which have been intensively studied in the field of solid-state physics, are of interest [2].

Another method to increase efficiency is to utilize composite systems consisting of several materials. The film-substrate system is a typical example of composite systems. The Seebeck effect in the film-substrate system has been studied in detail [3–6], while the transverse thermoelectric effect, such as the anomalous Nernst effect in the film-substrate system, has been focused on recently, because a huge increase of the anomalous Nernst effect was observed in Fe–Ga alloy on silicon substrate [6].

To analyze the thermoelectric effect theoretically, phenomenological methods such as the Boltzmann equation have been used, while, recently, the microscopic method based on the formalism by Kubo and Luttinger, which can treat the quantum effect exactly, has developed [7–9]. Therefore, first, we introduce the microscopic method based on the Kubo–Luttinger formalism briefly. Then, we apply this method to the thermoelectric effect of type I, II, and III Dirac electrons, which is one of the unique electronic states in solids [10]. Next, we introduce the recent development of film-substrate systems and describe theoretical methods to analyze the thermoelectric effect in film-substrate systems [11].

1.2 Linear Response Theory Based on Kubo–Luttinger and Application to Dirac-electron Systems

1.2.1 Liner Response

When an external electric filed (E) is applied to solids, an electric current is generated. Similarly, applying the temperature gradient (∇T) generates the heat current. These are called Ohm's law and Fourier's law, respectively. On the other hand, the correlated effect between the electric field and the heat current or that between the temperature gradient and the electric current is also known. In the region of the linear response, these phenomena are formally given by

$$J_e = L_{11}E + L_{12}\left(-\frac{\nabla T}{T}\right), \tag{1.1}$$

$$J_Q = L_{21}E + L_{22}\left(-\frac{\nabla T}{T}\right), \tag{1.2}$$

where J_e and J_Q are the densities of the electric current and the heat current, respectively, L_{ij} denote the coefficients of the linear response, and especially L_{11} is called the electric conductivity.

When the temperature gradient is applied to solids, the diffusion of electrons is generated, and, finally, the internal electric field is generated by the diffusion that becomes balanced with the temperature gradient. The ratio between the internal electric field and temperature gradient corresponds to the Seebeck coefficient. Using Eq. (1.1) with the condition of $J_e = 0$, the Seebeck coefficient is given by

$$S = \frac{E}{\nabla T} = \frac{L_{12}}{TL_{11}}. \tag{1.3}$$

Since the Seebeck coefficient is expressed by the coefficients of linear response, the precise estimation of these coefficients is required in theoretical calculations.

1.2.2 Framework of Thermoelectric Theory Based on Kubo–Luttinger Formalism

Electric transport due to the electric field is calculated microscopically by the Kubo formula (i.e., the linear response theory) [7]. On the other hand, the linear response theory of temperature gradient is qualitatively different from that of the electric field. This is because the electric field is a mechanical external field, while the temperature gradient is a statical quantity. Thus, the linear response of the temperature gradient had been difficult to deal with. Luttinger suggested the linear response theory to treat the temperature gradient, using the relationship between the temperature gradient and gravitational potential [8].

Recently, the thermoelectric microscopic theory based on Kubo–Luttinger formalism has been extended [9]. In this section, we introduce the recent progress of the Kubo–Luttinger formalism briefly.

In the Kubo–Luttinger formalism, the thermoelectric effect (precisely speaking L_{12}) is derived based on the heat currents generated depending on the Hamiltonians. Figure 1.1 shows the schematic picture of the Hamiltonians and heat currents generated from each term of the Hamiltonian. Generally, the thermoelectric effects can be divided into three kinds of effects. One is a diffusive thermoelectric effect, as shown in Figure 1.1. The effect is formally given by

$$L_{11} = \int d\epsilon \left(-\frac{\partial f(\epsilon)}{\partial \epsilon}\right) \sigma(\epsilon, T), \tag{1.4}$$

$$L_{12} = \frac{1}{e} \int d\epsilon \left(-\frac{\partial f(\epsilon)}{\partial \epsilon}\right) (\epsilon - \mu)\, \sigma(\epsilon, T), \tag{1.5}$$

where $f(\epsilon)$, $\sigma(\epsilon, T)$, and μ indicate the Fermi distribution function, spectral function, and the chemical potential, respectively. $e\,(< 0)$ is the charge of an electron. Since this equation was first derived by Sommerfeld–Bethe [12], Eq. (1.5) is called the Sommerfeld–Bethe (SB) relation [9]. By using the Sommerfeld expansion in metals at low temperatures, we can derive the Mott formula

$$S = \frac{\pi^2}{3} \frac{k_B^2 T}{e} \frac{\partial ln\sigma(\epsilon_F, 0)}{\partial \epsilon_F}. \tag{1.6}$$

The theory of the thermoelectric effect based on the SB relation has been extended to the analysis of many materials [10, 13–22].

A second thermoelectric effect is a phonon-assisted effect. Applying the temperature gradient to solids, the heat current of phonons is generated. The thermoelectric

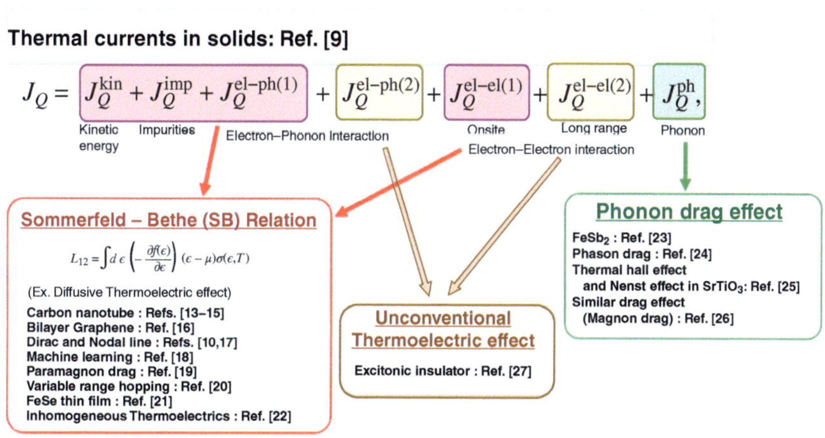

Figure 1.1 Schematic picture of framework and applications of thermoelectric theory based on the Kubo–Luttinger formula.

effect occurs by generating an electric flow through the electron–phonon interaction. This effect is called the phonon drag effect, and the microscopic theory of the drag effect based on Kubo–Luttinger formalism was performed [23–26].

Finally, we describe the thermoelectric effects due to the thermal currents deriving from the many-body effects such as electron–electron interaction and electron–phonon interaction. This thermoelectric effect has not been studied in detail; therefore, this effect is expected to become an unconventional thermoelectric effect [27].

1.2.3 Application to Dirac-electron Systems

Materials that host electrons behaving as relativistic particles which appear in the vicinity of the Fermi surfaces due to a characteristic band dispersion are called Dirac-electron systems. For decades, the Dirac electrons have been known to be realized in bismuth single crystals [28]. Bismuth is a bulk material, and it has a small but finite band gap at the L point. Recently, it was revealed that Dirac electrons also appear in two-dimensional materials such as graphene [29] and quasi-two-dimensional systems (materials have a three-dimensional structure, but the electron conductivity is strong only in the plane) such as an organic conductor α-(BEDT-TTF)$_2$I$_3$ [30]. In these materials, the band gap (ideally) closes; that is, the conduction and the valence bands touch at some points in the momentum space. Such systems are called the massless Dirac electrons, and the band-touching points are called the Dirac points. Particularly in two dimensions, when drawing the dispersion relation as a function of two-dimensional momentum $\left[\boldsymbol{k} = \left(k_x, k_y\right)\right]$, its shape is two cones touching at their apexes. Thus, the spectrum around the Dirac points is called the Dirac cones. Here, we introduce an example of a microscopic study of the thermoelectric effect in two-dimensional Dirac electrons [10].

One of the characteristics of Dirac electrons in solids is that there can be anisotropic dispersion. Dirac cones are classified into three types according to the shape of the equienergy line around the Dirac point, which is determined by the degree of tilting. To see this, let us consider the minimal model describing the tilted massless Dirac cones, whose Hamiltonian is given as a 2×2 matrix as a function of $\boldsymbol{k}$, as

$$H(\boldsymbol{k}) = \lambda \hbar v_{\mathrm{F}} k_x \sigma_0 + \hbar v_{\mathrm{F}} \boldsymbol{k} \cdot \boldsymbol{\sigma}. \tag{1.7}$$

Here, $\boldsymbol{\sigma} = \left(\sigma_x, \sigma_y, \sigma_z\right)$ with $\sigma_x = \begin{pmatrix} 0 & 1 \\ 1 & 0 \end{pmatrix}$, $\sigma_y = \begin{pmatrix} 0 & -i \\ i & 0 \end{pmatrix}$, and $\sigma_z = \begin{pmatrix} 1 & 0 \\ 0 & -1 \end{pmatrix}$ are called the Pauli matrices, and $\sigma_0 = \begin{pmatrix} 1 & 0 \\ 0 & 1 \end{pmatrix}$ is the 2×2 identity matrix. Diagonalizing $H(\boldsymbol{k})$, we obtain the dispersion relation, $\varepsilon_{\boldsymbol{k}}^{\pm} = \hbar v_{\mathrm{F}} \left(\lambda k_x \pm k\right)$ with $k = \sqrt{k_x^2 + k_y^2}$. The superscript $\pm$ is the band index, where $+$ denotes the upper band and $-$ denotes the lower band. Figure 1.2 is the schematic figure of the dispersion relation for several values of λ. For $\lambda < 1$, the tilting is small enough, and the equienergy line around the

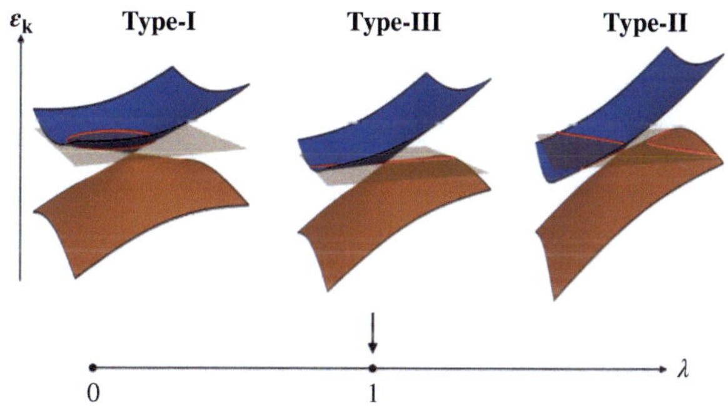

Figure 1.2 Schematic figure of the dispersion relation of the minimal Dirac Hamiltonian with tilted cones in Eq. (1.7).

Dirac point has an elliptical shape. This type of Dirac cone is called the type I Dirac cone, and graphene and α-(BEDT-TTF)$_2$I$_3$ belong to this class. On the other hand, for $\lambda > 1$, the tilting is large enough, and the shape of the equienergy line turns to the hyperbola. This type of Dirac cone is called the type II Dirac cone. The type III Dirac cone is the critical case (i.e., the transition point) between these two types. At the zero energy, the equienergy line is the straight line along a certain direction; that is, the dispersion is completely flat along the direction of tilting (i.e., the k_x direction in this model). Among these three types, when the Fermi level is at zero energy, type I is the gapless semiconductor, and the density of state at the Fermi level vanishes, while the other two types have a finite density of state.

The minimal model of Eq. (1.7) has been used to study the physical properties, such as the transport phenomena and electric/magnetic responses, of the Dirac-electron systems. However, in this model, for types II and III, the Fermi surface (or the line for the two-dimensional systems) extends to infinitely far away from the Dirac point. On the other hand, in the actual solid-state systems, the momentum space is bounded by the Brillouin zone, and the Fermi surface has to have a closed shape. Therefore, in the theoretical studies, tight-binding-type models, where the effect of the Brillouin zone is properly taken into account, are often used. Here, we argue one of such examples [31]. To be specific, we argue the two-orbital model on a square lattice, whose Hamiltonian in the momentum space is given as

$$H(\mathbf{k}) = d_{\mathbf{k}}^0 \sigma_0 + \mathbf{d}_{\mathbf{k}} \cdot \boldsymbol{\sigma}, \tag{1.8}$$

where $d_{\mathbf{k}}^0 = 2\lambda t' \left(\cos k_x a_0 - \cos k_y a_0\right)$, $d_{\mathbf{k}}^x = 2t' \left(\cos k_x a_0 - \cos k_y a_0\right)$, $d_{\mathbf{k}}^y = 0$, and $d_{\mathbf{k}}^z = 2t \left(\cos k_x a_0 + \cos k_y a_0\right)$. Dirac points appear at the momentum where $d_{\mathbf{k}}^x = d_{\mathbf{k}}^z = 0$; that is, $\mathbf{k} = \left(\pm\frac{\pi}{2a_0}, \pm\frac{\pi}{2a_0}\right)$,with all combinations of signs being allowed. a_0 stands for the lattice constant. The parameters t and $t'$are the hopping parameters having the dimension of the energy, while λ in $d_{\mathbf{k}}^0$ is the dimensionless parameters controlling the tilting. Figure 1.3 shows the dispersion relations for the lattice model, where we see that the type I, II, and II Dirac cones are realized at $\varepsilon_{\mathbf{k}} = 0$ for $\lambda < 1$, $\lambda > 1$, and $\lambda = 1$, respectively. In this study, as typical values for electrons in solids, the tight-binding parameters are set as $(t, t' = -1\,\text{eV}, -0.3\,\text{eV})$.

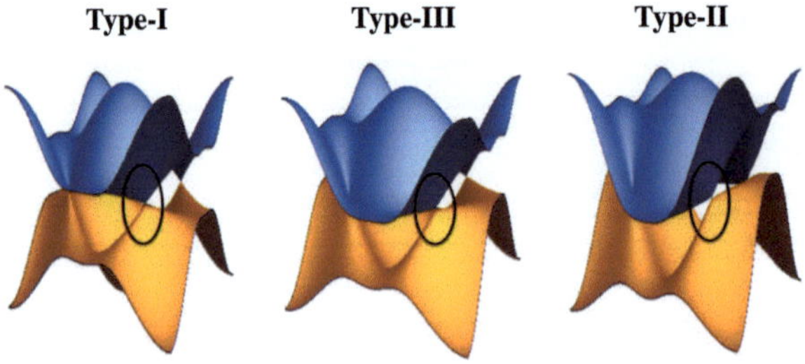

Figure 1.3 Schematic figure of the dispersion relation of the tight-binding model in Eq. (1.8). Black ellipses denote one of four Dirac cones.

In the following, we present the results of the thermoelectric transport based on the Kubo–Luttinger formalism [7, 8]. To be specific, we consider the quasi-two-dimensional system; that is, the layers, whose areas are A and each of which hosts the electrons described by the Hamiltonian of Eq. (1.8), are stacked independently with the interlayer spacing $d_0 = 10$ Å. The spin degrees of freedom are neglected. Under this setup, the spectral conductivity $\sigma(\epsilon)$ is calculated by using the following formula,

$$\sigma(\epsilon) = \frac{\hbar e^2}{2\pi A d_0} \sum_{\boldsymbol{k}} \mathrm{Tr}\left\{G^{(R)}(\boldsymbol{k}, \epsilon)\, v_x(\boldsymbol{k})\, G^{(A)}(\boldsymbol{k}, \epsilon)\, v_x(\boldsymbol{k}) - Re\left[G^{(R)}(\boldsymbol{k}, \epsilon)\, v_x(\boldsymbol{k})\, G^{(R)}(\boldsymbol{k}, \epsilon)\, v_x(\boldsymbol{k})\right]\right\} \tag{1.9}$$

Here $G^{(R)}(\boldsymbol{k}, \epsilon) = [\epsilon + i\Gamma - H(\boldsymbol{k})]^{-1}$ and $G^{(A)}(\boldsymbol{k}, \epsilon) = [\epsilon - i\Gamma - H(\boldsymbol{k})]^{-1}$ are the retarded and advanced Green's functions, respectively, and $v_x(\boldsymbol{k}) = \frac{1}{\hbar}\frac{\partial H(\boldsymbol{k})}{\partial k_x}$ is the velocity operator in the x-direction, which is the direction of the applied electric field and the temperature gradient. Γ denotes the quasi-particle damping rate arising from the impurity scattering, which is approximated as a constant. This approximation is called the relaxation time approximation. For the numerical calculations, Γ is set to be 0.02 eV.

Figure 1.4 shows the spectral conductivity for three values of λ. We see that the conductivity exhibits a dip for the type I and III Dirac cones, whereas it exhibits a peak for the type II cone. This qualitative difference is clearly manifested in the thermoelectric response.

To estimate ZT, assume that the thermal conductivity κ consists only of the electronic contribution. In that case, κ is estimated

$$\kappa = \frac{1}{T}\left(L_{22} - \frac{L_{12}^2}{L_{11}}\right), \tag{1.10}$$

with

$$L_{22} = \frac{1}{e^2}\int d\epsilon \left(-\frac{\partial f(\epsilon)}{\partial \epsilon}\right)(\epsilon - \mu)^2 \sigma(\epsilon, T). \tag{1.11}$$

As our interest is in the thermoelectric responses in the vicinity of the transition point from types I to II, we consider values of λ; namely, $\lambda = 0.95$ (type I), $\lambda = 1$ (type III), and $\lambda = 1.5$ (type II). Figure 1.5 shows the temperature dependence of

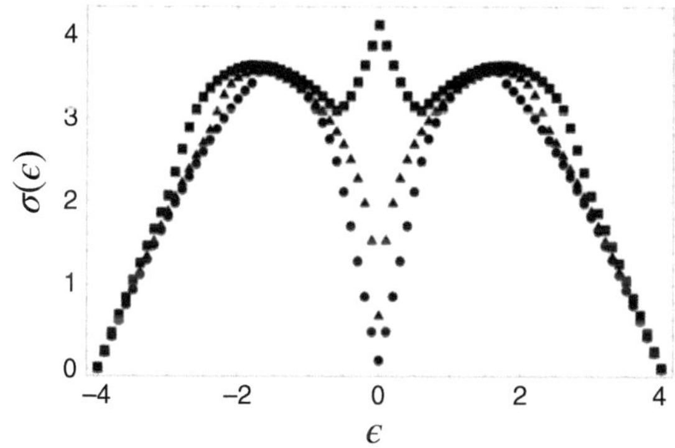

Figure 1.4 Spectral conductivity for the lattice model. The dots, triangles, and squares are for $\lambda = 0.5$ (type I), $\lambda = 1$ (type III), and $\lambda = 1.5$ (type II), respectively. $\sigma(\epsilon)$ is in the unit of $10^6/\Omega$ m, and ϵ is in the unit of eV.

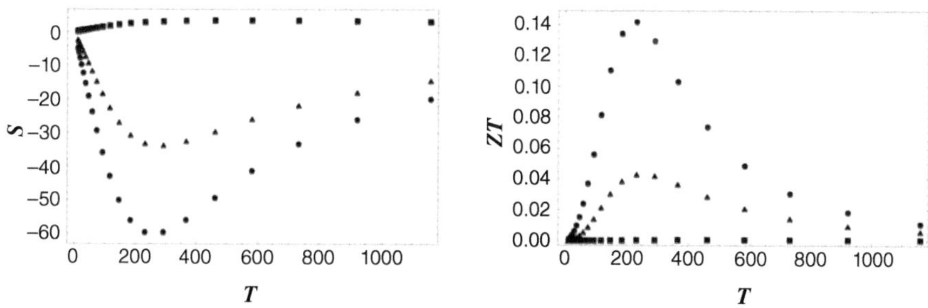

Figure 1.5 Seebeck coefficient (left) and the dimensionless figure of merit (right) as a function of temperature. The dots, triangles, and squares are for $\lambda = 0.95$ (type I), $\lambda = 1$ (type III), and $\lambda = 1.5$ (type II), respectively. S is in the unit of μV/K, and T is in the unit of K.

S and ZT. Here, the chemical potential μ is set to be 0.05 eV. The type I and III cases have noticeably large values of $|S|$ and ZT compared with the type II case, which indicates that the dip of the spectral conductivity around the Dirac points is the key element to obtain the large thermoelectric response. It is also seen that $\lambda = 1$ (type III) does not give the maximal thermoelectric response. In fact, within our numerical investigations, the maximal values of $|S|$ and ZT are obtained for $\mu = 0.05$ eV and $\lambda = 0.89$, and the values are $|S| \sim 70\ \mu$V/K and $ZT \sim 0.18$. These results indicate that the fine-tuning of the electronic structure (i.e., the tilting of the cone in this case) and the chemical potential is needed to optimize the thermoelectric response.

In Figure 1.5, we also see that the overall temperature dependences of S and ZT for the type I and III Dirac cones are very similar. In particular, the peak positions of these two types are almost the same. We discuss the origin of this similarity with respect to the approximated forms of the spectral conductivity. A method of estimating the thermoelectric response by assuming a simple functional form of the spectral conductivity was proposed by Mahan and Sofo [32] and was applied to various systems including the gapped Dirac cones [33]. Here, we apply this idea to the tilted Dirac cones. As we have seen, the spectral conductivity exhibits a dip for the type I and III Dirac cones. So, as the simplest approximation, we employ the following form of the spectral conductivity around the Dirac point,

$$\sigma(\epsilon) = \alpha_0 + \alpha_1 |\epsilon|, \tag{1.12}$$

where α_0 and α_1 are the constants. Substituting Eq. (1.12) into Eqs. (1.4), (1.5), and (1.11), and further considering the limit of $\alpha_0 \to 0$, we find that S and ZT can be given as

$$S = \frac{k_B}{e} \cdot \frac{h_2(-x) - 2xh_1(-x)}{2h_1(-x) - xh_0(-x)}, \tag{1.13}$$

and

$$ZT = \left\{ \frac{[2h_1(-x) - xh_0(-x)]\,[2h_3(-x) - xh_2(-x)]}{[h_2(-x) - 2xh_1(-x)]^2} - 1 \right\}^{-1}, \tag{1.14}$$

where $x = -\frac{\mu}{k_B T}$ and the functions $h_0 - h_3$ are given as

$$h_0(z) = \tanh\frac{z}{2}, \tag{1.15}$$

$$h_1(z) = \ln(1 + e^z) - \frac{ze^z}{e^z + 1}, \tag{1.16}$$

$$h_2(z) = \frac{2z^2 e^z}{e^z + 1} - 4z(1 + e^z) - 4\mathrm{Li}_2(-e^z) - \frac{\pi^2}{3}, \tag{1.17}$$

and

$$h_3(z) = z^2\left[3\ln(1 + e^z) - \frac{ze^z}{e^z + 1}\right] + 6z\mathrm{Li}_2(-e^z) - 6\mathrm{Li}_3(-e^z). \tag{1.18}$$

Here, $\mathrm{Li}_n(z)$ stands for the polylogarithm function of order n. Notably, under this approximation, the Seebeck coefficient and the dimensionless figure of merit are functions of a single parameter, x, and they do not depend on the model-dependent parameter α_1.

Figure 1.6 shows the Seebeck coefficient and the dimensionless figure of merit as a function of $-x$. We see that the minimum of S is at $-x \sim 2$ $\left(\text{i.e., } T \sim \frac{\mu}{2k_B}\right)$, and the maximum of ZT is $-x \sim 2.5$ $\left(\text{i.e., } T \sim \frac{\mu}{2.5\,k_B}\right)$, which are in good agreement with

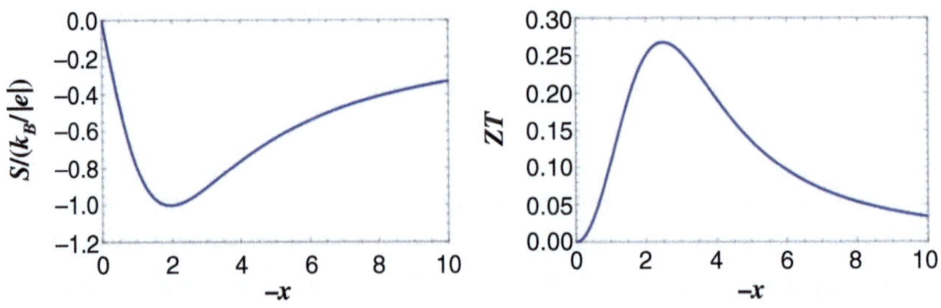

Figure 1.6 Seebeck coefficient (left) and the dimensionless figure of merit (right) as a function of $-x$ of Eqs. (1.13) and (1.14), respectively.

the numerical results for type I and III Dirac cones in Figure 1.5. We also see in Figure 1.6 that the maximal value of $|S|$ is about $k_B/|e| \sim 86\,\mu V/K$ and that of ZT is about 0.27. They are in the same order as the numerical results for the types I and III Dirac cones but do not agree at the quantitative level. This might be because of a non-negligible contribution arising from α_0 , whose value depends on the details of the model parameters and the damping rate.

To summarize this subsection, the microscopic theory based on Kubo–Luttinger formalism provides us with a powerful tool to investigate the thermoelectric transport arising from the characteristic dispersion relation of electrons in solids. In our analysis of the tilted Dirac-electron model, we have found that the optimal response is obtained for the type I Dirac cones with strongly tilted, rather than the type II and III Dirac cones. However, in the actual materials, the fine-tuning of the tilting and the chemical potential is not easy. This is the reason why composite materials such as multilayer systems have been attracting attention, as described in the next section.

1.3 Film-substrate Systems

Composite systems, consisting of multiple components shown in Figure 1.7a, have been of interest in the studies of the thermoelectric effects. The film-substrate systems are a typical example of composite systems as introduced in Figure 1.7b, and their thermoelectric effect has been studied extensively with the development of microfabrication technology.

In the film-substrate system, the Seebeck effect has been studied. More recently, the transverse thermoelectric effect such as the Nernst effect is focused on [6]. The film-substrate system of Fe–Ga alloy film and silicon substrate system shows the huge anomalous Nernst effect of 15 μV/K, which is larger than the bulk value of Fe–Ga alloy and that of Co_2MnGa, which is a record material [6]. The microscopic analysis of the thin-film and substrate system is very difficult, because the consideration of the electronic state and the treatment of the interfacial effect between film and substrate are complicated. Thus, in this section, we introduce the analysis method based on the circuit model [11].

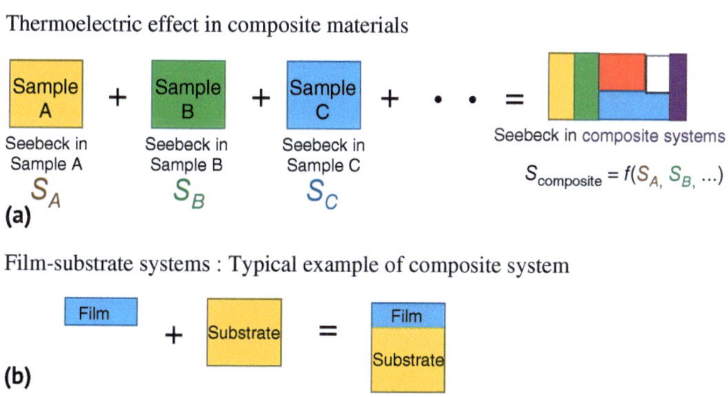

Figure 1.7 Schematic picture of a composite system. (a) Composite systems consisting of many components. (b) A thin-film substrate system as a typical example of a composite system.

1.3.1 Linear Response for Longitudinal and Transverse Thermoelectric Effects

Here, we describe the general linear response including not only the longitudinal but also the transverse thermoelectric effects [11]. When the external electrical and temperature gradients are applied to the x and y directions in system A, the electrical current density and heat current density along the x and y directions are given by

$$\begin{pmatrix} j^x_{e,A} \\ j^y_{e,A} \\ j^x_{Q,A} \\ j^y_{Q,A} \end{pmatrix} = \overline{L}_A \begin{pmatrix} E_{x,A} \\ E_{y,A} \\ -(\nabla T)_{x,A}/T \\ -(\nabla T)_{y,A}/T \end{pmatrix}, \tag{1.19}$$

where $j^{x(y)}_{e,A}$ and $j^{x(y)}_{Q,A}$ indicate the electric current and thermal current along $x(y)$ direction in system A, respectively, $E_{x(y),A}$ and $(\nabla T)_{x(y),A}$ are the external electric field and temperature gradient along the $x(y)$ direction in system A. $\overline{L}_A$ is a 4×4 matrix of linear response defined as

$$\overline{L}_A = \begin{pmatrix} L^{xx}_{11,A} & L^{xy}_{11,A} & L^{xx}_{12,A} & L^{xy}_{12,A} \\ L^{yx}_{11,A} & L^{yy}_{11,A} & L^{yx}_{12,A} & L^{yy}_{12,A} \\ L^{xx}_{21,A} & L^{xy}_{21,A} & L^{xx}_{22,A} & L^{xy}_{22,A} \\ L^{yx}_{21,A} & L^{yy}_{21,A} & L^{yx}_{22,A} & L^{yy}_{22,A} \end{pmatrix}. \tag{1.20}$$

Using the linear response coefficients, the Seebeck coefficient $(S_{xx,A})$, Nernst coefficient $(\nu_{yx,A})$, and anomalous Nernst coefficient $(N_{yx,A})$ are given as

$$S_{xx,A} = \frac{E_x}{(\nabla T)_x} = \frac{L^{xx}_{12}}{TL^{xx}_{11}}, \tag{1.21}$$

$$\nu_{yx,A} = \frac{E_{y,A}}{(\nabla T)_{x,A} B_z} = \frac{L^{xy}_{11,A}L^{xx}_{12,A} - L^{xx}_{11,A}L^{xy}_{12,A}}{TL^{xx}_{11,A}L^{yy}_{11,A}B_z}, \tag{1.22}$$

$$N_{yx,A} = \frac{E_{y,A}}{(\nabla T)_{x,A}} = \frac{L^{xy}_{11,A}L^{xx}_{12,A} - L^{xx}_{11,A}L^{xy}_{12,A}}{T\left(L^{xx}_{11,A}L^{yy}_{11,A} + (L^{xy}_{11,A})^2\right)}. \tag{1.23}$$

1.3.2 Linear Response of Film-substrate System Based on the Circuit Model

Next, we derive the Seebeck coefficient, Nernst coefficient, and anomalous Nernst coefficient in film-substrate systems based on the circuit model [11]. Figure 1.8a shows the schematic picture of film-substrate systems. Here, we set a volume of film (substrate) as $a_x \times a_y \times d_{f(s)}$. As shown in Figure 1.8b, we treat the film-substrate systems as a parallel circuit consisting of film and substrate. Then, from Kirchhoff's law, the electrical potential of the film (substrate) along the $x(y)$ axis $(V_{x(y),f(s)})$ satisfies the relation of

$$V_{x(y),f} = V_{x(y),s}. \tag{1.24}$$

where $V_{x(y),f(s)} = E_{x(y),f(s)}a_{x(y)}$.

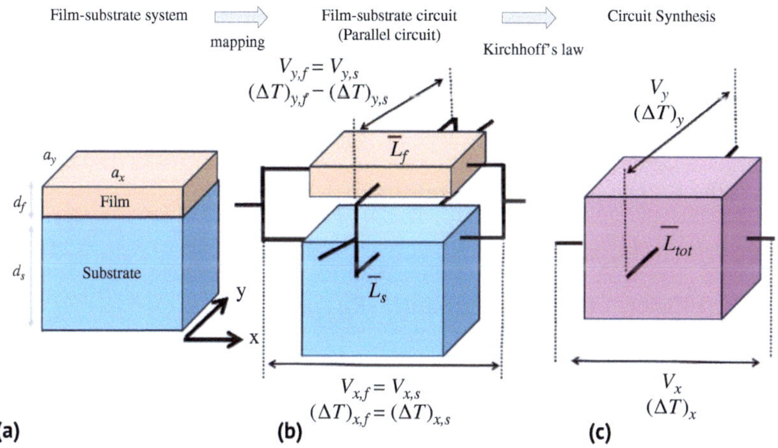

Figure 1.8 Schematic picture of (a) film-substrate system, (b) circuit model, and (c) circuit synthesis.

Next, we study the thermal circuit. Considering the temperature difference in a similar manner to the electric circuit, the temperature differences in film and substrate in the $x(y)$ axis also satisfy the following relation:

$$(\Delta T)_{x(y),f} = (\Delta T)_{x(y),s} \tag{1.25}$$

where $(\Delta T)_{x(y),f(s)} = (\nabla T)_{x(y),f(s)}\, a_{x(y)}$.

In the circuit model, the total of electric and thermal currents is given by the summation of those in the film and the substrate. Thus, the linear response coefficient of the synthesis circuit shown in Figure 1.8c is

$$\bar{L}_{\text{tot}} = \frac{d_f}{d_f + d_s}\bar{L}_f + \frac{d_s}{d_f + d_s}\bar{L}_s. \tag{1.26}$$

Using Eq. (1.21), the Seebeck coefficient in the synthesis circuit is given by

$$S_{\text{tot}} = \frac{S_f + \epsilon_\sigma S_s}{1 + \epsilon_\sigma}, \tag{1.27}$$

where

$$\epsilon_\sigma = \frac{L_{11,s}^{xx} d_s}{L_{11,f}^{xx} d_f} = \frac{L_{11,s}^{yy} d_s}{L_{11,f}^{yy} d_f} = \frac{R_f}{R_s}. \tag{1.28}$$

ϵ_σ corresponds to the ratio of the electric resistance of the film (R_f) and the substrate [5].

In the same manner, the Nernst coefficient and the anomalous Nernst coefficient are

$$\nu_{\text{tot}} = \frac{\nu_f + \epsilon_\sigma \nu_s}{1 + \epsilon_\sigma} + \frac{\epsilon_\sigma}{(1 + \epsilon_\sigma)^2}\nu_s \delta\nu, \tag{1.29}$$

$$N_{\text{tot}} = \frac{N_f + \epsilon_\sigma N_s}{1 + \epsilon_\sigma} + \frac{\epsilon_\sigma}{(1 + \epsilon_\sigma)^2} N_s \delta N, \tag{1.30}$$

where

$$\delta\nu = (S_f - S_s)\,(\theta_{H,s} - \theta_{H,f})/\nu_s B_z, \quad (1.31)$$

$$\delta N = (S_f - S_s)\,(\theta_{H,s} - \theta_{H,f})/N_s, \quad (1.32)$$

and $\theta_{H,f(s)}$ is the Hall angle of the film (substrate) defined as

$$\theta_{H,f(s)} = R_{H,f(s)} L^{xx}_{11,f(s)} B_z, \quad (1.33)$$

and the Hall coefficient of the film (substrate) $(R_{H,f(s)})$ is

$$R_{H,f(s)} = \frac{L^{xy}_{11,f(s)}}{L^{xx}_{11,f(s)} L^{yy}_{11,f(s)} B_z}. \quad (1.34)$$

Noted that a similar expression of anomalous Nernst coefficient was derived in Ref. [6]. Based on the circuit model analysis, the thermoelectric effect of the film and substrate system is analyzed. It should be noted that the interfacial effect between the film and the substrate is not considered in this model.

1.3.3 Model Calculation

Finally, we show ϵ_σ dependence of Seebeck coefficient and Nernst coefficient using Eqs. (1.27) and (1.29). Figure 1.9 shows the ϵ_σ dependence of the Seebeck coefficient (black line) for $\frac{S_f}{S_s} = 100$ and the Nernst coefficient (red line) for $\frac{\nu_f}{\nu_s} = 10$ and $\delta\nu = 100$. The Seebeck coefficient gradually decreases as ϵ_σ increases. On the other hand, the Nernst coefficient exhibits a peak structure at $\epsilon_\sigma \simeq 1.0$. It should be noted that the anomalous Nernst effect behaves the same as the Nernst effect. The part of the Nernst effect in the second term of Eq. (1.29) means that the Nernst effect in the film (substrate) is derived from the Hall angle of the substrate (film). Thus, this is called the cooperative effect, which is a correlated effect between the film and the substrate [11].

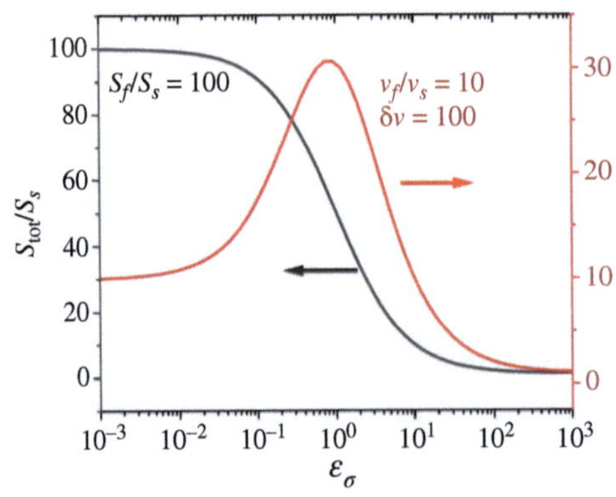

Figure 1.9 ϵ_σ dependence of Seebeck coefficient (black line) and Nernst coefficient (red line).

The peak structure was observed in Fe–Ga alloy film and silicon substrate system in the anomalous Nernst effect [6]. However, there are no experimental results in the Nernst effect, although there is a theoretical prediction in the Bi film and Si substrate system [11].

1.4 Summary

In this section, first, we introduced the microscopic theory of the thermoelectric effect based on the Kubo–Luttinger formalism and its application to the type I, II, and III Dirac systems, and then we show the analysis of film-substrate systems, which is a typical example of composite systems, by the circuit model. We expect that the combination of microscopic and macroscopic methods opens a new way to discover thermoelectric systems with high efficiency.

References

1 Nandihalli, N., Liu C.J. and Mori, T. (2020) Polymer based thermoelectric nanocomposite materials and devices: fabrication and characteristics. *Nano Energy*, 78, 105186.

2 Fu, C., Sun, Y. and Felser, C. (2020) Topological thermoelectrics. *APL Materials*, 8, 040913.

3 Bergman, D.J. and Levy, O. (1991) Thermoelectric properties of a composite medium. *J. Appl. Phys.*, 70, 6821.

4 Bergman, D.J. and Fel L.G. (1999) Enhancement of thermoelectric power factor in composite thermoelectrics. *J. Appl. Phys.*, 85, 8205.

5 Riss, A., Garmroudi, F., Parzer M., Pustogow A., Mori T., and Bauer E. (2024) Thermoelectric power factor of composites. *Phys. Rev. Appl.*, 21, 014002.

6 Zhou, W., Sasaki, T., Uchida, K.-I., and Sakuraba, Y. (2024) Direct-contact Seebeck-driven transverse magneto-thermoelectric generation in magnetic/thermoelectric bilayers. *Adv. Sci.*, 11, 2308543.

7 Kubo, R. (1957) Statistical-mechanical theory of irreversible processes. I. General theory and simple applications to magnetic and conduction problems. *J. Phys. Soc. Jpn.*, 12, 570.

8 Luttinger, J.M. (1964) Theory of thermal transport coefficients. *Phys. Rev.*, 135, A1505.

9 Ogata, M. and Fukuyama, H. (2019) Range of validity of Sommerfeld–Bethe relation associated with Seebeck coefficient and phonon drag contribution. *J. Phys. Soc. Jpn.*, 88, 074703.

10 Mizoguchi, T., Matsuura, H. and Ogata, M. (2022) Thermoelectric transport of type I, II, and III massless Dirac fermions in a two-dimensional lattice model. *Phys. Rev. B*, 105, 205203.

11 Matsuura, H., Riss, A., Garmroudi, F., Parzer, M., and Bauer, E. (2024) Cooperative Nernst effect of multilayer systems: parallel circuit model study. *Phys. Rev. Res.*, 6, 043071.

12 Sommerfeld, A. and Bethe, H. (1933) Elektronentheorie der Metalle, In H. Geiger and K. Scheel (Eds.), Handbuch der Physik, Vol. 24/2, Springer, Berlin/Heidelberg.

13 Yamamoto, T. and Fukuyama, H. (2018) Possible high thermoelectric power in semiconducting carbon nanotubes ~a case study of doped one-dimensional semiconductors~. *J. Phys. Soc. Jpn.*, 87, 024707.

14 Yamamoto, T. and Fukuyama, H. (2018) Bipolar thermoelectric effects in semiconducting carbon nanotubes: description in terms of one-dimensional dirac electrons. *J. Phys. Soc. Jpn.*, 87, 114710.

15 Matsubara, M., Sasaoka, K., Yamamoto, T., and Fukuyama, H. (2021) Optimal thermoelectric power factor of narrow-gap semiconducting carbon nanotubes with randomly substituted impurities. *J. Phys. Soc. Jpn.*, 90, 044702.

16 Horii, H., Matsubara, M., Sasaoka, K., Yamamoto, T., and Fukuyama, H. (2021) Optimization of thermoelectric power factor of bilayer graphene by vertical electric field. *J. Phys. Soc. Jpn.*, 90, 104711.

17 Hosoi, M., Tateishi, I., Matsuura, H., and Ogata, M. (2022) Thin films of topological nodal line semimetals as a candidate for efficient thermoelectric converters. *Phys. Rev. B*, 105, 085406.

18 Hirosawa, T., Schäfer, F., Maebashi, H., Matsuura, H., and Ogata, M. (2022) Data-driven reconstruction of spectral conductivity and chemical potential using thermoelectric transport properties. *J. Phys. Soc. Jpn.*, 91, 114603.

19 Endo, J., Matsuura, H. and Ogata, M. (2022) Effect of paramagnon drag on thermoelectric transport properties: linear response theory. *Phys. Rev. B*, 105, 045101.

20 Yamamoto, T., Ogata, M. and Fukuyama, H. (2022) Thermoelectric effect in Mott variable-range hopping. *J. Phys. Soc. Jpn.*, 91, 044704.

21 Matsubara, M., Yamamoto, T. and Fukuyama, H. (2023) Two-band model with high thermoelectric power factor and its application to FeSe thin film. *J. Phys. Soc. Jpn.*, 92, 104704.

22 Matsubara, M., Yamamoto, T. and Fukuyama, H. (2025) Sommerfeld–Bethe analysis of ZT in inhomogeneous thermoelectrics. *J. Phys. Soc. Jpn.*, 94, 024601.

23 Matsuura, H., Maebashi, H., Ogata, M., and Fukuyama, H. (2019) Effect of Phonon Drag on Seebeck Coefficient Based on Linear Response Theory: Application to $FeSb_2$. *J. Phys. Soc. Jpn.*, 88, 074601.

24 Fukuyama, H. and Ogata, M. (2020) Theory of phason drag effect on thermoelectricity. *Phys. Rev. B*, 102, 205136.

25 Endo, J., Matsuura, H. and Ogata, M. (2024) Phonon drag effect in Nernst and thermal Hall effects: general theory and application to dilute metal $SrTiO_{3-\delta}$. *Phys. Rev. B*, 110, 205121.

26 Matsuura, H., Ogata, M., Mori, T., and Bauer, E. (2021) Theory of huge thermoelectric effect based on a magnon drag mechanism: application to thin-film Heusler alloy. *Phys. Rev. B*, 104, 214421.

27 Takarada, S., Ogata, M. and Matsuura, H. (2021) Theory of thermal conductivity of excitonic insulators. *Phys. Rev. B*, 104, 165122.

28 Fuseya, Y., Ogata, M. and Fukuyama, H. (2015) Transport properties and diamagnetism of Dirac electrons in bismuth. *J. Phys. Soc. Jpn.*, 84, 012001.

29 Castro Neto, A.H., Guinea, F., Peres, N.M.R., Novoselov, K.S., and Geim, A.K. (2009) The electronic properties of graphene. *Rev. Mod. Phys.*, 81, 109.

30 Kajita, K., Nishio, Y., Tajima, N., Suzumura, Y., and Kobayashi, A. (2014) Molecular Dirac fermion systems - theoretical and experimental approaches-. *J. Phys. Soc. Jpn.*, 83, 072002.

31 Mizoguchi, T. and Hatsugai, Y. (2020) Type-III Dirac cones from degenerate directionally flat bands: viewpoint from molecular-orbital representation. *J. Phys. Soc. Jpn.*, 89, 103704.

32 Mahan, G.D. and Sofo, J.O. (1996) The best thermoelectric. *Proc. Natl. Acad. Sci. USA.*, 93, 7436.

33 Hasdeo, E.H., Krisna, L.P.A., Hanna, M.Y., Gunara, B.E., Hung, N.T., and Nugraha, A.R.T. (2019) Optimal band gap for improved thermoelectric performance of two-dimensional Dirac materials. *J. Appl. Phys.*, 126, 035109.

2

Carbon Fiber-reinforced Polymer Piezoelectric Composites

Zhenjin Wang

Graduate School of Environmental Studies, Tohoku University, Sendai, Miyagi, Japan

2.1 Introduction

Piezoelectric materials possess an intrinsic coupling effect between electric fields and mechanical deformation. This inherent property allows them to be polarized by mechanical stress beyond typical dielectric polarization [1]. The phenomenon in which an external force induces polarization and generates an electric charge is known as the piezoelectric effect and is used in sensors [2]. Conversely, when strain is induced by applying a voltage, it is referred to as the inverse piezoelectric effect, which has been used in the development of actuators. Additionally, piezoelectric materials are expected to be used in energy harvesting [3] to power wireless sensor network devices for the Internet of Things, in mass sensing [4] for detecting bacteria and viruses, and in monitoring physical activity and physiological signals [5] for health management.

Piezoelectric ceramics, known for their high energy conversion efficiency, have emerged as promising candidates for wearable energy-harvesting devices. However, their inherent brittleness presents a considerable obstacle. To address this limitation, researchers have developed piezoelectric composite materials [6], which incorporate piezoelectric ceramic particles dispersed in polymers. This innovative approach has spurred active research in the field of wearable sensors and energy harvesting. Figure 2.1 shows a visual representation of the growing interest in piezoelectric composites for energy harvesting. This figure depicts the results of a Scopus search, encompassing article titles, abstracts, and keywords, using the terms "piezoelectric," "composite," and "energy harvesting." Panel (a) illustrates the increasing number of publications, while panel (b) presents the percentage of authors by country. In Japan, Wang and Narita [7] successfully developed a lead-free barium titanate (BTO) piezoelectric particle-dispersed epoxy composite material, demonstrating its potential for unsteady microwind power generation. Research has also elucidated the

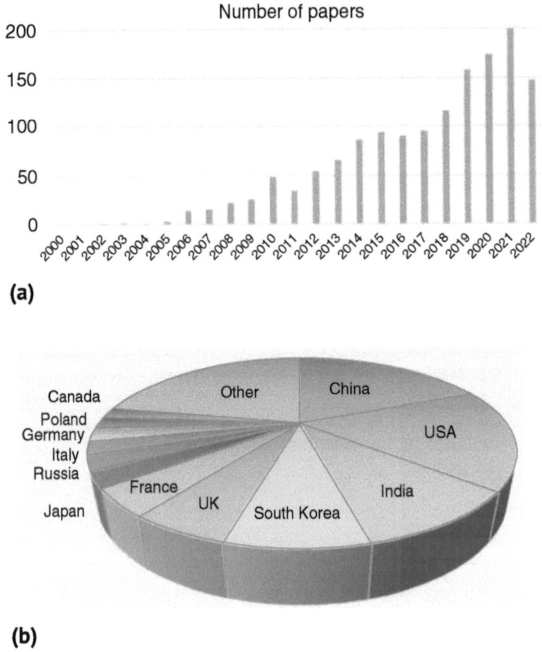

Figure 2.1 Scopus data on piezoelectric, composite, and energy-harvesting research as of September 2022: (a) number of publications and (b) percentage by country.

impact of polarization conditions on the piezoelectric properties of lead-free potassium sodium niobate (KNN) piezoelectric particle-dispersed epoxy composites [8], confirming their long-term stability. Additionally, Wang et al. [9] have initiated the development of gradient-function piezoelectric materials.

Carbon fiber-reinforced polymer (CFRP) composites, known for their high electrical conductivity and exceptional mechanical stability, find widespread application in industries such as aerospace, automotive, and shipbuilding. The emergence of demand for multifunctional composite materials, integrating novel functionalities into CFRP, has spurred interdisciplinary research focused on advancing the design and functionality of these materials [10–12].

This chapter focuses on CFRP composites with piezoelectric properties, covering the fundamentals of piezoelectric materials and providing an overview of current research and development trends. It also presents the latest status of the design, fabrication, and evaluation of CFRP composites incorporating dispersed piezoelectric particles.

2.2 Research and Development Trends in Piezoelectric Laminated CFRP Composites

In the Cartesian coordinate system O-x_1, x_3 depicted in Figure 2.2, we examine a piezoelectric material possessing spontaneous polarization along the x_3 direction. It is assumed that no external electric field is applied. When subjected to perpendicular

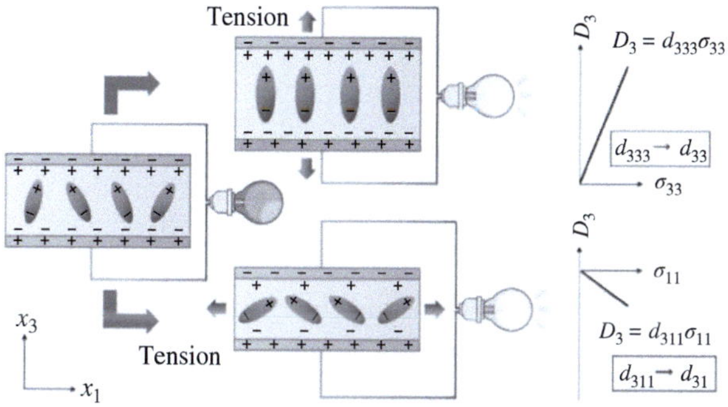

Figure 2.2 Piezoelectric effect and the relationship between electric displacement D_3 and normal stresses σ_{33} and σ_{11}.

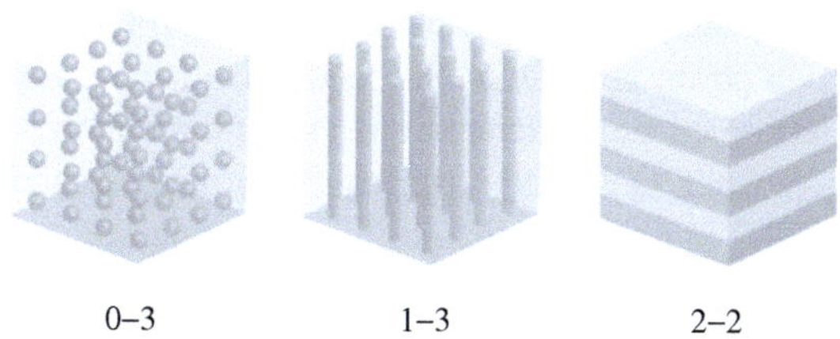

Figure 2.3 Schematic of composite types: 0–3 composite, 1–3 composite, and 2–2 composite configurations.

stresses σ_{33} or σ_{11}, the piezoelectric material experiences a polarization P_3. The linear relationship between the charge density, or flux density D_3, and these stresses is governed by the piezoelectric coefficients d_{333} and d_{311}, respectively. Typically, 33 is written as 3 and 11 as 1 [13], and the piezoelectric coefficients are given by d_{33} and d_{31}.

Newnham et al. [14] performed a comprehensive study of piezoelectric composites comprising two materials. Their analysis categorized the continuity of each material into four dimensions: 0D, 1D, 2D, and 3D. Based on these classifications, they identified 10 distinct types of composites and derived theoretical equations to predict their elastic and piezoelectric coefficients. Figure 2.3 shows three representative examples of piezoelectric composite materials. The composites with piezoelectric particles dispersed throughout a matrix exhibit 0–3 connectivity. When piezoelectric fibers are aligned unidirectionally in the matrix, the connectivity is classified as 1–3. A composite material composed of stacked two-dimensional plates exhibits 2–2 connectivity. In the *i*–*j* notation, *i* represents the dimensionality of the first phase, with 0 denoting a particle (zero-dimensional) and 1 indicating a fiber (one-dimensional). Furthermore, *j* represents the dimensionality of the second phase, with 3 representing a matrix extending in three dimensions. Conversely, the 2–2 connectivity implies that both phases (1 and 2) consist of layers spread across two dimensions arranged in an alternating pattern.

Extensive research has been dedicated to the detection of damage and fracture in CFRP laminates by integrating piezoelectric materials onto or into their surfaces. Andreades et al. [15] performed investigations into damage monitoring under

Figure 2.4 Concept of the piezoelectric CFRP composite. *Source*: [20] with permission of John Wiley & Sons.

tensile–tension fatigue loading by fabricating tensile specimens incorporating lead zirconate titanate (PZT) piezoelectric ceramics embedded in CFRP and covered with glass fibers. Additionally, Philibert et al. [16] achieved successful damage detection by applying a polyvinylidene fluoride piezoelectric polymer coating to CFRP laminates and subjecting them to impact loading.

Beyond damage and fracture detection, considerable research efforts are dedicated to energy harvesting, specifically focusing on vibration and shock power generation. Alsaadi et al. [17] integrated a piezoelectric macrofiber composite (MFC) into a CFRP laminate to evaluate vibration power generation. Subsequently, Jia et al. [18] from the same research group performed finite element analysis and are actively engaged in structural optimization. Chen et al. [19] have also performed research on the topology optimization of CFRP sandwich structures incorporating MFCs to enhance vibration power generation performance.

The 2–2 geometry, where piezoelectric materials are layered on or in the CFRP, often leads to delamination and potential weakening of the CFRP structure. Direct impact on the laminated piezoelectric material can also result in fracture. Conversely, dispersing piezoelectric particles in the matrix or interface of the CFRP, creating a 0–3 configuration, offers advantages such as low cost, lightweight construction, and simplicity. This approach has the potential to yield piezoelectric CFRP composites that maintain mechanical stability and open new avenues for application. Figure 2.4 shows a conceptual diagram of a piezoelectric CFRP composite [20].

2.3 Preparation of the CFRP Composites with Piezoelectric Particle Dispersion

While dispersing piezoelectric particles in CFRP is relatively straightforward, depolarization during the curing process considerably impacts their piezoelectric properties. Therefore, careful consideration must be given to the Curie temperature of the piezoelectric materials (the temperature at which ferroelectricity transitions to paraelectricity) and the curing temperature of the composite materials. Furthermore, polarizing piezoelectric particles dispersed in the matrix or at the interface of the CFRP after curing presents considerable challenges. Owing to the conductive nature of carbon fiber, a substantial electric field does not influence the CFRP, preventing the piezoelectric effect from being achieved. Therefore, a new approach is required to develop CFRP composites exhibiting a piezoelectric effect.

Narita et al. [21] developed piezoelectric CFRP composites by layering CFRP on the top and bottom surfaces of epoxy layers containing dispersed KNN piezoelectric nanoparticles, followed by polarization in silicon oil. A glass fiber-reinforced polymer layer was strategically incorporated at the interface between the epoxy and

CFRP layers to provide insulation and enhance the piezoelectric effect. Wang et al. [22] successfully fabricated piezoelectric nanoparticle-dispersed epoxy layers, using them as electrodes by laminating CFRP on the top and bottom surfaces, and achieved corona discharge polarization.

This chapter provides an example of the fabrication and evaluation of KNN particle–dispersed epoxy/CFRP composites.

2.3.1 Materials and Test Specimens

Figure 2.5 shows the fabrication process for the piezoelectric particle-dispersed epoxy layer. KNN piezoelectric particles (Nippon Chemical Industry Co., Ltd.), with an average diameter of approximately 880 nm, served as the piezoelectric filler. A mixture of 30 vol.% KNN particles and bisphenol F epoxy resin (Daido Corporation) was thoroughly mixed for 30 min and subsequently defoamed using a stirrer for 10 min to ensure uniform dispersion of the KNN particles in the epoxy. A hardener was not introduced at this stage to prevent premature curing of the epoxy during the mixing process. After thorough mixing, the hardener ST-12 (Mitsubishi Chemical Corporation) was added and mixed for 5 min. The resulting mixture was then

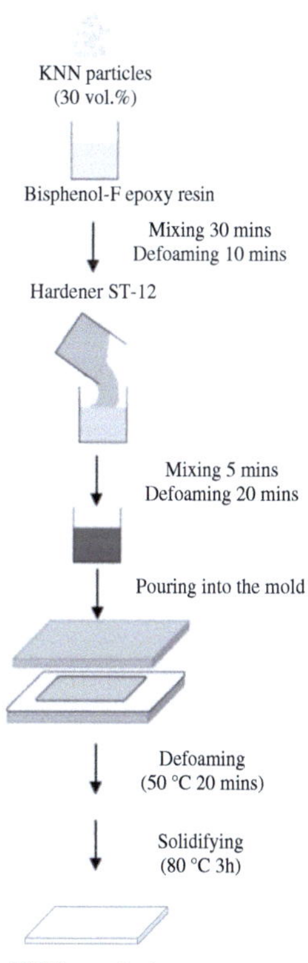

Figure 2.5 Fabrication method of the KNN particle-dispersed epoxy composite sheet.

defoamed for 20 min with the KNN/Bisphenol F solution. The mixture was subsequently poured into a mold, defoamed again at 50 °C for 20 min, and then subjected to a temperature of 80 °C for 3 h to ensure complete curing of the epoxy. After curing, the KNN particle-dispersed epoxy layer was cut and polished to the desired dimensions: 20 or 40 mm in length, 10 mm in width, and 0.5 mm in thickness. The density and apparent porosity of the KNN particle-dispersed epoxy layer were measured following the Japanese standard JIS R 1634: 1998 [23], yielding values of 2.17 g/cm^3 and 2.1%, respectively.

2.3.2 Poling and Lamination

Figure 2.6 shows the poling process for the KNN particle-dispersed epoxy/CFRP composites. Unidirectional carbon fiber prepregs (approximately 0.1 mm thick, 165 g m^{-2}) were symmetrically laminated at a 0°/90°/90°/0° orientation on both the top and bottom surfaces of the KNN particle-dispersed epoxy layer. Following lamination, the KNN particle-dispersed epoxy/CFRP composite was cured at 125 °C for 45 min. The resulting composite material has an overall thickness of approximately 0.8 mm. To induce the piezoelectric effect, the composite material must undergo a polarization process. The corona discharge polarization method, shown in Figure 2.7, was used to polarize the composite material. Two types of specimens were prepared, with the polarization and stacking sequence defined as the poling path, as follows.

1. pe-1: poling after lamination
2. pe-2: laminated after poling

Polarization was applied using a thickness-directed electric field of 16 kV/mm for 30 min at temperatures ranging from 75 °C to 125 °C.

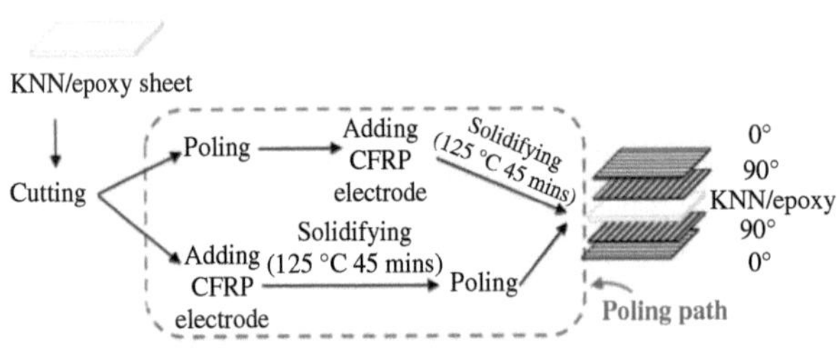

Figure 2.6 Fabrication of KNN/CFRP composite materials.

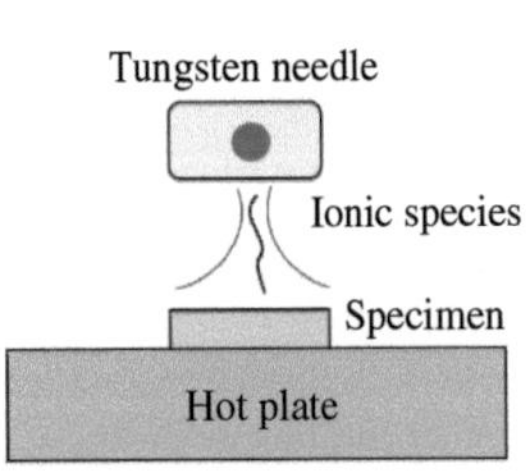

Figure 2.7 Corona poling method.

2.3.3 Characterization

Several methods exist for measuring the piezoelectric coefficients d_{33}, d_{13}, and d_{15} [24]. In this study, d_{33} was measured using a d_{33} meter (YE2730A, Sinocera Piezotronics Inc., China).

To assess the output of KNN particle-dispersed epoxy/CFRP composites under compressive impact and bending vibration, a specialized apparatus was designed and fabricated. Figure 2.8a shows a schematic representation of the specimen, while Figure 2.8b depicts the apparatus used in the impact test. During the impact test, the KNN particle–dispersed epoxy/CFRP composite was secured to a rigid body (brick). A pendulum ball, with a diameter of 13 mm and a weight of 11.0 g, was suspended at a height of 250 mm and then quickly released, impacting the KNN particle-dispersed epoxy/CFRP composite. Experiments were performed with various load resistances (1–9.5 kΩ) connected to the composite material, and the output voltage generated by the impact was recorded using a data logger. The output power was determined using the load resistance and output voltage.

Figure 2.9a and b shows schematic diagrams of the specimen and apparatus used in the vibration tests, respectively. For the vibration test, a KNN particle-dispersed epoxy/CFRP composite, with a free length of 30 mm, a proof mass of 12.0 g, and a load resistance of $R = 5\,\text{k}\Omega$, was mounted on a shaker and subjected to a specific vibration amplitude. The amplitude of the transducer was measured using a laser displacement meter, and the output voltage was recorded with a data logger. The output power can be similarly determined using the load resistance and output voltage.

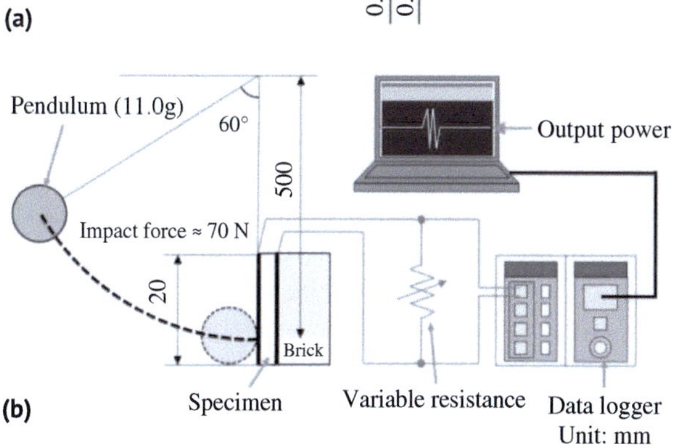

(a)

(b)

Figure 2.8 Schematic of the impact energy-harvesting test: (a) specimen and (b) test setup.

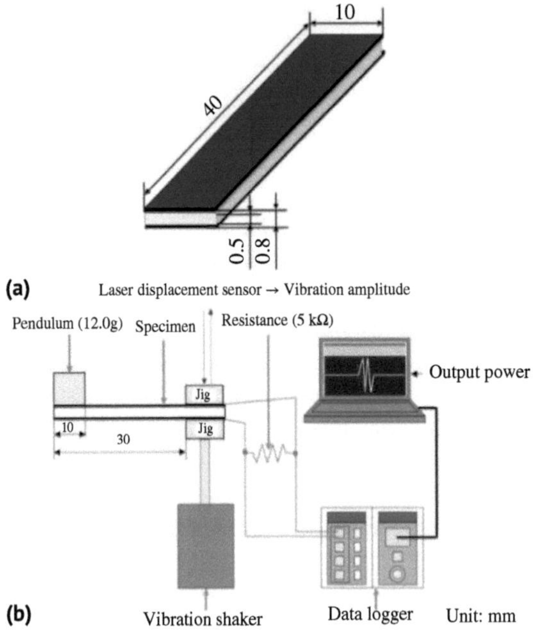

Figure 2.9 Schematic of the vibration energy-harvesting test: (a) specimen and (b) test setup.

2.4 Electromechanical Properties of the Piezoelectric CFRP

This section examines the influence of fabrication conditions on the piezoelectric coefficient, output voltage (sensor performance), and power (energy-harvesting performance) of KNN particle-dispersed epoxy/CFRP composites subjected to compressive and bending forces.

2.4.1 Piezoelectric Properties

Figure 2.10 shows the relationship between the piezoelectric coefficient d_{33} and polarization temperature for three different configurations: a KNN particle-dispersed epoxy/CFRP composite (pe-1) polarized after CFRP stacking [22], a KNN particle-dispersed epoxy/CFRP composite (pe-2) polarized before CFRP stacking, and a KNN particle-dispersed epoxy layer without CFRP lamination for comparison. In all cases, an optimal polarization temperature of approximately 105 °C was observed. Interestingly, the piezoelectric constants vary depending on the order of polarization and stacking. Polarization after CFRP lamination proves to be more advantageous, resulting in a piezoelectric coefficient d_{33} of approximately 5.8 pC N^{-1}.

2.4.2 Shock-induced Voltage and Power

Figure 2.11 shows the impact test results for the pe-2 composite, where CFRP lamination occurred after polarization. Figure 2.11a shows the output voltage of the KNN

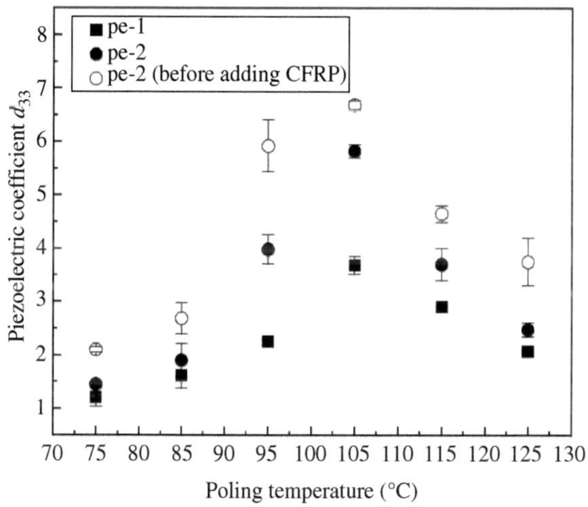

Figure 2.10 Influence of poling temperature and poling sequence on piezoelectric coefficient d_{33}.

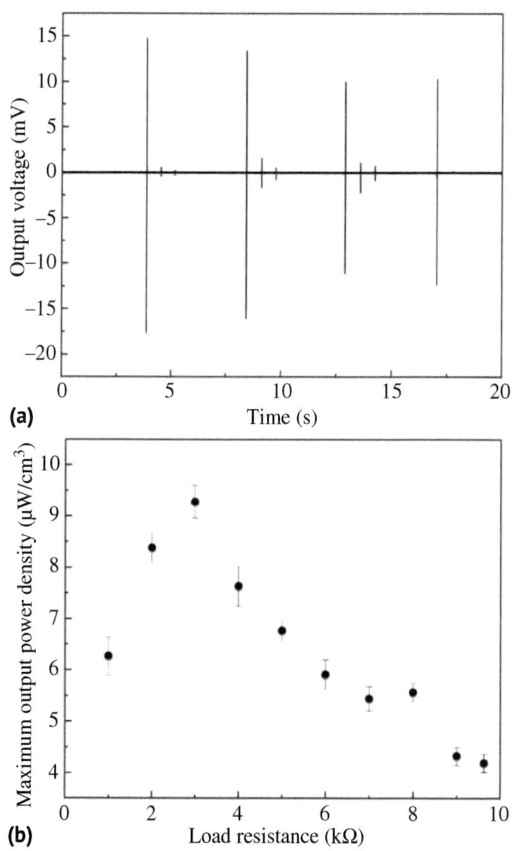

Figure 2.11 Result of the impact energy-harvesting test: (a) real-time output voltage under 1 kΩ and (b) the maximum impact output power density as a function of load resistance.

particle-dispersed epoxy/CFRP composite with a load resistance (R) of 1 kΩ. Upon impact with the pendulum ball, the output voltage increases rapidly. Figure 2.11b shows the relationship between the maximum output power density and load resistance. The output power density increases with increasing load resistance, reaching a maximum value before decreasing. At a load resistance of 3 kΩ, the output power reaches approximately 9.48 cm^{-3}. It is well known that the output power is maximized when the load resistance equals the internal resistance.

The inherent brittleness of piezoelectric ceramics often leads to a reduction in sensor and energy-harvesting performance after impact. However, the KNN particle-dispersed epoxy/CFRP composite demonstrated resilience, exhibiting no damage or delamination after impact testing. The flexural strength of the non-polarized KNN particle-dispersed epoxy/CFRP composite is approximately 900 MPa [25]. From both functional and mechanical perspectives, KNN particle–dispersed epoxy/CFRP composites offer considerable advantages over conventional piezoelectric ceramics. Even KNN particle-dispersed epoxy can be damaged by impact loading [26]; therefore, KNN particle-dispersed epoxy/CFRP composites have the potential to be used safely and sustainably under such conditions.

2.4.3 Vibration-induced Voltage and Power

Figure 2.12 shows the relationship between output voltage, maximum output power density, and frequency for the KNN particle-dispersed epoxy/CFRP composite, with an R of 5 kΩ, for the configuration where CFRP lamination occurred after polarization (pe-2). Because the signal strength transmitted to the shaker is constant, the maximum acceleration is theoretically consistent. The output voltage and power density reached their peak values (approximately 0.6 mV and 0.35 nW cm^{-3}, respectively) at a frequency of approximately 60 Hz. This frequency corresponds to the first resonant frequency of the KNN particle-dispersed epoxy/CFRP composite. While not shown in the figure, the output power density increased with increasing vibration amplitude. For example, at a fixed vibration amplitude of 0.33 mm, the

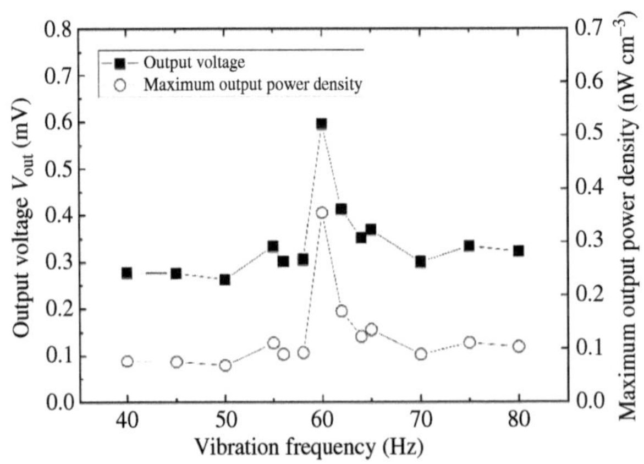

Figure 2.12 Influence of vibration frequency on the output voltage and maximum output power density of KNN particle-dispersed epoxy/CFRP composites.

output power density of pe-2 at 60 Hz was approximately 12 nW cm^{-3}, whereas pe-1 exhibited a power density of only approximately 1 nW cm^{-3}. Therefore, composites with CFRP laminated after polarization demonstrate suitability not only for compressive impacts but also for bending vibrations.

The output power density generated by bending vibration is considerably lower than that produced by impact loading. This is attributed to the placement of the KNN particle-dispersed epoxy layer in the neutral plane of the composite material. A simple modification, such as adjusting the thickness of the CFRP prepreg on one side to shift the KNN particle-dispersed epoxy layer out of the neutral plane, would likely result in a considerable increase in output power density.

2.5 Conclusion

This chapter presents a comprehensive overview of recent research on CFRP composites exhibiting piezoelectric properties. The chapter begins with a review of current trends in piezoelectric laminated CFRP composites. Next, the fabrication process for KNN particle-dispersed epoxy/CFRP composites is detailed, along with the evaluation methods employed. Experimental results on the voltage and power output generated by shock and vibration are presented and discussed. Recent publications in high-impact journals have reported on piezoelectric composites with output voltages reaching approximately 100 V. While these results are noteworthy, it is important to note that the voltage may be amplified through some mechanism, and further investigation of these claims is warranted. Looking ahead, future research endeavors are expected to focus on achieving even higher output from piezoelectric CFRP composites, as well as exploring the integration of magnetostrictive materials [27–29]. Continued development of multifunctional CFRP composite research is highly desirable.

References

1 Uchino, K. (2017). Advanced Piezoelectric Materials, Science and Technology, Woodhead Publishing, India.
2 Narita, F. and Wang, Z. (2023). Piezoelectric Materials, Composites, and Devices: Fundamentals, Mechanics, and Applications, Elsevier, Cambridge.
3 Narita, F. and Fox, M. (2018). A review on piezoelectric, magnetostrictive, and magnetoelectric materials and device technologies for energy harvesting applications. *Adv. Eng. Mater.*, 20, 1700743.
4 Narita, F., Wang, Z., Kurita, H., Li, Z., Shi, Y., Jia, Y., and Soutis, C. (2021). A review of piezoelectric and magnetostrictive biosensor materials for detection of COVID-19 and other viruses. *Adv. Mater.*, 33, 2005448.
5 Wang, Y., Yu, Y., Wei, X. and Narita, F. (2022). Self-powered wearable piezoelectric monitoring of human motion and physiological signals for the postpandemic era: a Review. *Adv. Mater. Tech.*, 7, 2200318. https://doi.org/10.1002/admt.202200318.
6 Yamada, T., Ueda, T. and Kitayama, T.J. (1982). Piezoelectricity of a high-content lead zirconate titanate/polymer composite. *Appl. Phys.*, 53, 4328.

7 Wang, Z. and Narita, F. (2019). Corona poling conditions for barium titanate/epoxy composites and their unsteady wind energy harvesting potential. *Adv. Eng. Mater.*, 21, 1900169.

8 Wang, Z. and Narita, F.J. (2019). Fabrication of potassium sodium niobate nano-particle/polymer composites with piezoelectric stability and their application to unsteady wind energy harvesters. *Appl. Phys.*, 126, 224501.

9 Wang, Z., Maruyama, K. and Narita, F. (2022). A novel manufacturing method and structural design of functionally graded piezoelectric composites for energy-harvesting. *Mater. Des.*, 214, 110371.

10 Li, H., Wang, X., Hu, X., Xiong, J., Han, Q., Wang, X., and Guan, Z. (2021). Vibration and damping study of multifunctional grille composite sandwich plates with an IMAS design approach. *Compos. B*, 223, 109078.

11 Qiu, B., Zhang, X., Xia, S., Sun, T., Ling, Y., Zhou, S., Guang, H., Chen, Y., Xu, Z., Liang, M., and Zou, H. (2022). Magnetic graphene oxide/carbon fiber composites with improved interfacial properties and electromagnetic interference shielding performance. *Compos. A*, 155, 106811.

12 Katsiropoulos, C.V., Pappas, P., Koutroumanis, N., Kokkinos, A. and Galiotis, C. (2022). Enhancement of damping response in polymers and composites by the addition of graphene nanoplatelets. *Compos. Sci. Technol.*, 227, 109562.

13 Nye, J.F. (1960). Physical Properties of Crystals, The Clarendon Press, Oxford.

14 Newnham, R.E., Skinner, D.P. and Cross, L.E. (1978). Connectivity and piezoelectric–pyroelectric composites. *Mater. Res. Bull.*, 13, 525–36.

15 Andreades, C., Meo, M. and Ciampa, F. (2020). Tensile and fatigue testing of impacted smart CFRP composites with embedded PZT transducers for nonlinear ultrasonic monitoring of damage evolution. *Smart Mater. Struct.*, 29, 055034.

16 Philibert, M., Chen, S., Wong, V.-K., Liew, W.H., Yao, K., Soutis, C. and Gresil, M. (2022). Direct-write piezoelectric coating transducers in combination with discrete ceramic transducer and laser pulse excitation for ultrasonic impact damage detection on composite plates. *Struct. Health Monit.*, 21, 1645–60.

17 Alsaadi, A., Shi, Y., Pan, L., Tao, J. and Jia, Y. (2019). Vibration energy harvesting of multifunctional carbon fibre composite laminate structures. *Compos. Sci. Technol.*, 178, 1–10.

18 Jia, Y., Wei, X., Xu, L., Wang, C., Lian, P., Xue, S., Al-Saadi, A. and Shi, Y. (2019). Multiphysics vibration FE model of piezoelectric macro fibre composite on carbon fibre composite structures. *Compos. B*, 161, 376–85.

19 Chen, B., Jia, Y., Narita, F., Wang, C. and Shi, Y. (2022). Multifunctional cellular sandwich structures with optimised core topologies for improved mechanical properties and energy harvesting performance. *Compos. B*, 238, 109899.

20 Takaishi, K., Kubota, Y., Kurita, H., Wang, Z. and Narita, F.J. (2022). Fabrication and electromechanical characterization of mullite ceramic fiber/thermoplastic polymer piezoelectric composites. *J. Am. Ceram. Soc.*, 105, 308–16.

21 Narita, F., Nagaoka, H. and Wang, Z. (2019). Fabrication and impact output voltage characteristics of carbon fiber reinforced polymer composites with lead-free piezoelectric nano-particles. *Mater. Lett.*, 236, 487–90.

22 Wang, Z., Kurita, H., Nagaoka, H. and Narita, F. (2020). Potassium sodium niobate lead-free piezoelectric nanocomposite generators based on carbon-fiber-reinforced

polymer electrodes for energy-harvesting structures. *Compos. Sci. Technol.*, 199, 108331.
23 Japanese Industrial Standard. (1998). Test Methods for Density and Apparent Porosity of Fine Ceramics, Japanese Standard JIS R 1634, Japanese Standards Association.
24 IEEE. (1987). An American National Standard, IEEE Standard on Piezoelectricity, The Institute of Electrical and Electronics Engineers, Inc., New York.
25 Kurita, H., Wang, Z., Nagaoka, H. and Narita, F. (2020). Fabrication and mechanical properties of carbon-fiber-reinforced polymer composites with lead-free piezoelectric nanoparticles. *Sens. Mater.*, 32, 2453–62.
26 Maruyama, K., Kawakami, Y., Mori, K., Kurita, H., Shi, Y., Jia, Y. and Narita, F. (2022). Electromechanical characterization and kinetic energy harvesting of piezoelectric nanocomposites reinforced with glass fibers. *Compos. Sci. Technol.*, 223, 109408.
27 Yang, Z., Wang, Z., Nakajima, K., Neyama, D., and Narita, F. (2021). Structural design and performance evaluation of FeCo/epoxy magnetostrictive composites. *Compos. Sci. Technol.*, 210, 108840.
28 Komagome, R., Katabira, K., Kurita, H. and Narita, F. (2022). Characteristics of carbon fiber reinforced polymers embedded with magnetostrictive Fe–Co wires at room and high temperatures. *Compos. Sci. Technol.*, 228, 109644.
29 Katabira, K., Komagome, R. and Narita, F. (2024). Novel process for monitoring stress in carbon fiber reinforced polymer composites using magnetostrictive wires from cryogenic to high temperatures. *Mech. Adv. Mater. Struct.*, 31(3), 586–594.

3

Energy Harvesting Behavior of Multifunctional Layered Composites

Boyue Chen[1], Constantinos Soutis[2], Yu Jia[3], Yu Shi[1]**

[1] *Leeds Institute of Textiles and Colour, School of Design, University of Leeds, Leeds, UK*
[2] *Department of Materials, The University of Manchester, Manchester, UK*
[3] *School of Engineering and Applied Science, Aston University, Birmingham, UK*

*Corresponding authors: y.jia1@aston.ac.uk; y.shi2@leeds.ac.uk

3.1 Introduction

Advanced composite materials have been widely employed as load-bearing structures in many engineering applications due to the high strength-to-weight ratio and elastic moduli [1–6]. The demand for material systems that perform simultaneously structural and nonstructural functions promoted the study of multifunctional composites [7]. Multifunctional composites are interdisciplinary in nature and functions could include, but are not limited to, electrical and thermal conductivity [8, 9], actuation and sensing [10, 11], electromagnetic interference shielding [12, 13], with the ability for energy harvesting and storage [14–18]. The multifunctional composite can be designed to extract ambient kinematic/thermal/photo energy and convert this into electrical power that can be stored in energy storage elements for continuous or later use [19].

Kinematic energy sources, such as vibration, are abundant in many aerospace, automotive, and machinery engineering applications. The energy conversion for kinematic energy harvesting normally involves the use of electromagnetic [20, 21], electrostatics [22], and piezoelectric transducers [23–26]. Among the aforementioned mechanisms, piezoelectricity stands out since the power generation reduces less significantly than the others; as the dimension of the transducer scales down [27, 28], piezoelectric transducers at micro- or nano-scale are expected to be more efficient. Low-temperature hydrothermal methods have been employed to grow piezoelectric nano-elements, such as ZnO nanowires, on aramid fabrics. The functionalized fabrics are integrated into a structural carbon fiber-reinforced composite (CFRP) laminate [29]. In addition to the fiber engineering scheme, a variety of ferroelectric ceramic particles, such as $BaTiO_3$, lead zirconate titanate (PZT), and $(K,Na)NbO_3$ (KNN) particles, can be dispersed into a polymer matrix [15, 30, 31]. The ceramic/polymer combination either serves as an individual layer embedded between neighboring layers or as the matrix reinforced by glass fabrics [15, 32].

The incorporation of piezoelectricity in a layered composite system is very challenging, and it becomes important to evaluate the energy-harvesting performance in the service environment. So the main focus of this chapter is to review progress and better understand the performance of the selected composite structure and its ability for energy harvesting. Composites with symmetric layup, nonsymmetric bistable configuration, and sandwich structures are reported in this chapter. Subsequently, studies of testing fiber-reinforced polymers (FRPs) with piezoelectric elements in the transportation, wind energy, and prosthetics industries are discussed. This chapter provides a comprehensive understanding of the piezoelectric energy-harvesting potential of structural composites.

3.2 Composite Structural Design for Energy Harvesting

3.2.1 Laminated Composites with Co-cured Piezoelectric Element

Embedment of structural health monitoring sensor nodes within the composites is essential in obtaining condition data from structures such as wind turbine blades and aircraft wings [33]. These data are useful to assess the service conditions and detect invisible microlevel defects within the structure, which indicate the onset and ultimately the prevention of a catastrophic failure. Conventionally, batteries or power cables are required to power such sensors. This suggests interference and extra weight on those engineering structures, increasing the demand for regular maintenance and replacement. Moreover, the difficulty of accessing parts or components in some extreme environments, such as an aircraft during cruising, adds extra cost and danger to the monitoring and maintenance process [34]. To tackle this problem, a vibrational energy-harvesting technique via piezoelectricity has been extensively researched to self-power the sensing system.

Piezoelectric ceramic, such as PZT, has a large piezoelectric coefficient. However, monolithic piezoelectric ceramic is brittle in nature, which raises reliability issues under aerospace-level amplitude acceleration [35, 36]. In addition, the brittleness of piezoelectric ceramic negates its conformation to curved surfaces, such as an airfoil. On the other hand, piezoelectric polymer, such as PVDF and its co-polymer P(VDF-TrFE), possess high mechanical flexibility. Nonetheless, the piezoelectric coefficient is lower than that of their ceramic counterparts by orders of magnitude [30].

Macro-fiber composite (MFC) is a 2-2 composite developed by NASA Langley Research Center [37]. MFC consists of piezoelectric ceramic rod fibers embedded within the epoxy matrix. This assembly enables it to exhibit piezoelectric performance comparable to that of ceramic, without sacrificing mechanical flexibility. The successful combination of the two inherently contradicting properties (hard-soft and flexible) makes MFC an attractive option for energy harvesting on complex-shaped structures.

The integration of MFC onto the host structure is critical because it determines the stress transfer efficiency from the host structure to the piezoelectric transducer. From an energy perspective, a poor integration, such as adhesive outspread, would suggest extra mechanical energy loss that reduces the figure of merit of the whole harvesting system [38]. Up to now, most research on piezoelectric energy harvesting simply used an adhesive to bond the piezoelectric transducers to the substrates [39, 40].

A novel approach was developed in 2017 [38], which was denoted as "co-curing." The host structure in the study was an aerospace-level carbon fiber-reinforced epoxy composite, with a typical quasi-isotropic stacking sequence. After the prepreg was laid up, an MFC element was placed on top of the first ply, with contact pads being protected by tape. The prepreg lay-ups with the MFC element were sealed into a vacuum bag and cured in an autoclave under selected pressure and temperature. The MFC was partially immersed in the epoxy on the surficial lamina, so the use of extra adhesive was avoided. The manufacturing process can be schematically illustrated in Figure 3.1.

As a comparative study, the MFC element has also been directly bonded to the composite structure via adhesive. Two types of bonding schemes were proposed: by mechanical clamping or by the assistance of a vacuum bag. The latter scheme helped to apply uniform external pressure during the curing process, which was to reduce the effect of the adhesive layer on energy harvesting [38].

Cyclic tensile loading of different strain levels and frequencies was applied to the composite structures. It was found that, regardless of the strain amplitude and frequencies, the energy-harvesting system with co-cured MFC element always manifested the highest power output, followed by the scheme of vacuum-bag-assisted adhesion, and the mechanical clamping scheme ranking last. In the case of tensile loading of 340 $\mu\varepsilon$ peak-to-peak magnitude at 340 Hz, the improvement of using co-curing was reported to be as large as 30.7% compared to the case of a direct bonding scheme using mechanical clamping. The result clearly indicated that the co-curing integration scheme effectively helped reduce the mechanical loss in the composite energy-harvesting system [38].

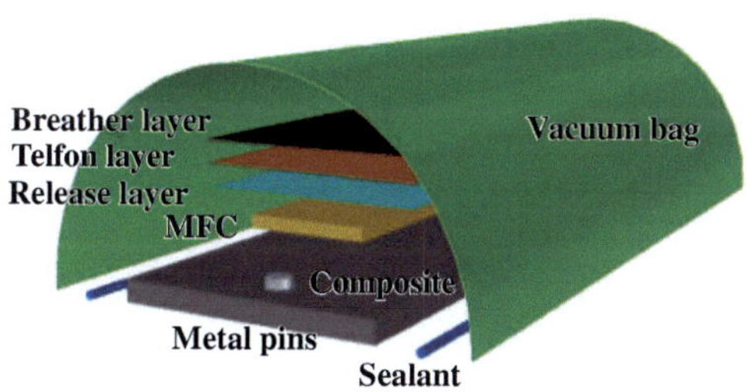

Figure 3.1 MFC element co-curing and composite curing process. *Source:* [41] / IOP publishing / CC BY 3.0.

3.2.2 Composites with Nonsymmetric Layup as Bistable Oscillators

A bistable system is an oscillator with a negative stiffness term for restoring [42]. A generic mathematical representation of a bistable oscillator takes the form of:

$$m\ddot{x} + c\dot{x} - kx + bx^3 = F(t) \tag{3.1}$$

Assuming the oscillator is kept in the elastic regime, the strain energy can be represented as:

$$U(x) = -\frac{1}{2}kx^2 + \frac{1}{4}bx^4 \tag{3.2}$$

Eq. (3.2) suggests that the bistable oscillator has an unstable fixed point at zero displacement and two stable fixed points at nonzero locations. When the vibration is weak, the oscillator is trapped around one of the stable fixed points, operating in the form of intra-well oscillation. As the vibrational energy is sufficiently large for the oscillator to overcome the energy barrier, the oscillator hops between the two potential wells, operating in the form of inter-well oscillation (snap-through state). In combination with a transducer, the electrical energy can be effectively harnessed through either the resonant response of the bistable oscillator or the snap-through oscillation when the vibrational energy is large enough to induce inter-well oscillation.

Within the context of composites, bistable composites are laminated in an asymmetric layup sequence. The bistability arises from the asymmetric thermal residual stress generated during the curing process, which was reported in one of the pioneering works by Hyder in 1981 [43]. The work observed the cured thin composite panels in cylindrical shapes, as opposed to the saddle shapes predicted by the classical lamination theory. The two stable states are schematically illustrated in Figure 3.2.

The bistable composites are often excited by transverse loading, which will lead to complex nonlinear vibrational phenomena. A comprehensive theoretical model

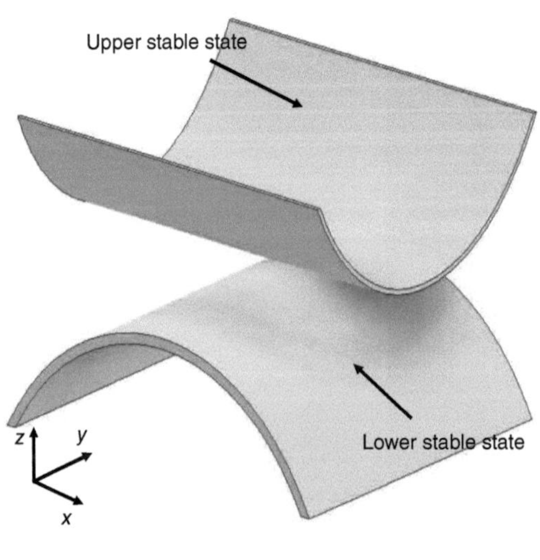

Figure 3.2 Schematic illustration of the two stable states of bistable composite laminates.

for the prediction of the dynamic response of asymmetric composite panels to base excitation was proposed by Zhang et al. [44]. The nonlinear governing dynamic partial differential equations (PDEs) were derived from the Hamilton principle. The PDEs were decoupled into equations of the spatial domain and the temporal domain, respectively, by Galerkin methods. The spatial domain problem, namely the modal problem, was solved by the Rayleigh–Ritz method. The derived mode shapes helped form the governing ordinary differential equations (ODEs) in the temporal domain, and the ODEs were solved numerically. The critical steps in the model were briefly presented in this review to illustrate the outline of solving the dynamic nonlinear characteristics of bistable composite panels. The Hamilton principle was represented as:

$$\int_0^T (\delta K - \delta U + \delta W)\, dt = 0 \tag{3.3}$$

where δK, δU, and δW are the variations of the kinetic energy, strain energy, and the work done by external loading and damping, respectively. To derive the kinetic and strain energy terms in the analytical forms, the stress and strain terms were defined. The strain terms were defined based on von Karman nonlinear strain-displacement relationships.

$$\begin{Bmatrix} \varepsilon_{xx} \\ \varepsilon_{yy} \\ \gamma_{xy} \end{Bmatrix} = \begin{Bmatrix} \varepsilon_{xx}^{(0)} \\ \varepsilon_{yy}^{(0)} \\ \gamma_{xy}^{(0)} \end{Bmatrix} + z \begin{Bmatrix} \varepsilon_{xx}^{(1)} \\ \varepsilon_{yy}^{(1)} \\ \gamma_{xy}^{(1)} \end{Bmatrix} + z^3 \begin{Bmatrix} \varepsilon_{xx}^{(3)} \\ \varepsilon_{yy}^{(3)} \\ \gamma_{xy}^{(3)} \end{Bmatrix}, \begin{Bmatrix} \gamma_{yz} \\ \gamma_{xz} \end{Bmatrix} = \begin{Bmatrix} \gamma_{yz}^{(0)} \\ \gamma_{xz}^{(0)} \end{Bmatrix} + z^2 \begin{Bmatrix} \gamma_{yz}^{(2)} \\ \gamma_{xz}^{(2)} \end{Bmatrix} \tag{3.4}$$

The right-hand sides in Eq. (3.4) can be expanded as:

$$\begin{Bmatrix} \varepsilon_{xx}^{(0)} \\ \varepsilon_{yy}^{(0)} \\ \gamma_{xy}^{(0)} \end{Bmatrix} = \begin{Bmatrix} \dfrac{\partial u_0}{\partial x} + \dfrac{1}{2}\left(\dfrac{\partial w_0}{\partial x}\right)^2 + \kappa_{x,0} w_0 \\ \dfrac{\partial v_0}{\partial y} + \dfrac{1}{2}\left(\dfrac{\partial w_0}{\partial y}\right)^2 + \kappa_{y,0} w_0 \\ \dfrac{\partial u_0}{\partial y} + \dfrac{\partial v_0}{\partial x} + \dfrac{\partial w_0}{\partial x}\dfrac{\partial w_0}{\partial x} \end{Bmatrix}, \begin{Bmatrix} \varepsilon_{xx}^{(1)} \\ \varepsilon_{yy}^{(1)} \\ \gamma_{xy}^{(1)} \end{Bmatrix} = \begin{Bmatrix} \dfrac{\partial \phi_x}{\partial x} \\ \dfrac{\partial \phi_y}{\partial y} \\ \dfrac{\partial \phi_y}{\partial x} + \dfrac{\partial \phi_x}{\partial y} \end{Bmatrix},$$

$$\begin{Bmatrix} \varepsilon_{xx}^{(3)} \\ \varepsilon_{yy}^{(3)} \\ \gamma_{xy}^{(3)} \end{Bmatrix} = -\frac{4}{3h^2} \begin{Bmatrix} \dfrac{\partial \phi_x}{\partial x} + \dfrac{\partial^2 w_0}{\partial x^2} \\ \dfrac{\partial \phi_y}{\partial y} + \dfrac{\partial^2 w_0}{\partial y^2} \\ \dfrac{\partial \phi_y}{\partial x} + \dfrac{\partial \phi_x}{\partial y} + 2\dfrac{\partial^2 w_0}{\partial x \partial y} \end{Bmatrix}, \begin{Bmatrix} \gamma_{yz}^{(2)} \\ \gamma_{xz}^{(2)} \end{Bmatrix} = -\frac{4}{h^2} \begin{Bmatrix} \phi_y + \dfrac{\partial w_0}{\partial y} \\ \phi_x + \dfrac{\partial w_0}{\partial x} \end{Bmatrix} \tag{3.5}$$

$$\begin{Bmatrix} \gamma_{yz}^{(0)} \\ \gamma_{xz}^{(0)} \end{Bmatrix} = \begin{Bmatrix} \phi_y + \dfrac{\partial w_0}{\partial y} - \kappa_{y,0} v_0 \\ \phi_x + \dfrac{\partial w_0}{\partial x} - \kappa_{x,0} u_0 \end{Bmatrix}$$

Eq. 3.5 denotes all the linear and nonlinear strain-displacement relationship terms. u_0, v_0, and w_0 are the in-plane and transverse displacement in the mid-plane. ϕ_x and ϕ_y are the rotations with respect to x and y axis, respectively. And $\kappa_{x,0}$ and $\kappa_{y,0}$ are the curvatures at the two stable states. h is the thickness of the laminate.

To reflect the influence of the thermal residual stress developed during the cooling stage, the thermal expansion terms were included in the constituent relationship, as given in the following.

$$\begin{Bmatrix} \sigma_{xx} \\ \sigma_{yy} \\ \tau_{xy} \\ \tau_{yz} \\ \tau_{xz} \end{Bmatrix}^{(k)} = \boldsymbol{Q}^{(k)} \left\{ \begin{Bmatrix} \varepsilon_{xx} \\ \varepsilon_{yy} \\ \gamma_{xy} \\ \gamma_{yz} \\ \gamma_{xz} \end{Bmatrix} - \begin{Bmatrix} \alpha_{xx} \\ \alpha_{yy} \\ 2\alpha_{xy} \\ 0 \\ 0 \end{Bmatrix} \Delta T \right\} \tag{3.6}$$

where $\boldsymbol{Q}^{(k)}$ corresponds to the global constituent matrix for the k-th laminate layer, which can be derived based on the mechanical properties of the individual lamina and the layup angle via coordinate transfer. α_{xx}, α_{yy}, and α_{xy} are the coefficients of thermal expansion (CTE).

In this work, Reddy's third-order shear deformation theory was used to establish the displacement at arbitrary locations in the composite panels.

$$\begin{aligned} u(x,y,z,t) &= u_0(x,y,t) + z\phi_x(x,y,t) - \frac{4}{3h^2}z^3\left(\phi_x + \frac{\partial w_0}{\partial x}\right) \\ v(x,y,z,t) &= v_0(x,y,t) + z\phi_y(x,y,t) - \frac{4}{3h^2}z^3\left(\phi_y + \frac{\partial w_0}{\partial y}\right) \\ w(x,y,z,t) &= w_0(x,y,t) \end{aligned} \tag{3.7}$$

On the basis of Eqs. (3.5)–(3.7), the kinetic energy and strain energy terms can be represented by:

$$K = \frac{1}{2}\sum \rho_0 \int_A \int_{-\frac{h}{2}}^{\frac{h}{2}} (\dot{u}^2 + \dot{v}^2 + \dot{w}^2)\, dxdydz \tag{3.8}$$

$$U = \frac{1}{2}\sum_{k=1}^{K} \int_A \int_{z_k}^{z_{k+1}} \left(\sigma_{xx}^{(k)}\varepsilon_{xx} + \sigma_{yy}^{(k)}\varepsilon_{yy} + \tau_{xy}^{(k)}\gamma_{xy} + \tau_{yz}^{(k)}\gamma_{yz} + \tau_{xz}^{(k)}\gamma_{xz}\right) dxdydz \tag{3.9}$$

To derive the natural frequencies of the composite laminate and the corresponding mode shapes, the boundary conditions should be clarified. In the work, the bistable composite panel was assumed to be clamped at the center and the four edges were free. This led to the boundary conditions shown here.

$$w_0(x_1, y_1) = 0,\ u_0(x_1, y_1) = 0,\ v_0(x_1, y_1) = 0 \tag{3.10}$$

$$\frac{\partial w_0}{\partial x}(x_1, y_1) = 0;\ \frac{\partial w_0}{\partial y}(x_1, y_1) = 0$$

The influence of the boundary conditions on the strain energy was considered by establishing a Lagrangian equation.

$$U_{\text{mod}} = U + L_1^w w_0(x_1, y_1) + L_1^u u_0(x_1, y_1) + L_1^v v_0(x_1, y_1) + L_1^x \frac{\partial w_0}{\partial x}(x_1, y_1) + L_1^y \frac{\partial w_0}{\partial y}(x_1, y_1) \tag{3.11}$$

where L_1^i $(i = w, u, v, x, and\ y)$ is the Lagrangian multipliers correspond to each boundary conditions.

An eigenmatrix equation could be derived from the following.

$$\frac{\partial\,(U_{\text{mod}} - K)}{\partial q_i} = 0 \tag{3.12}$$

By substituting the modified strain energy term and the kinetic energy term with Eqs. (3.11) and (3.8), respectively, and representing the stress and strain terms as a function of displacement, Eq. (3.12) could be reformatted into the form as follows:

$$\left(\boldsymbol{K} - \omega^2 \boldsymbol{M}\right) \boldsymbol{q} = 0 \tag{3.13}$$

where $\boldsymbol{K}$ **and** $\boldsymbol{M}$ are the stiffness matrix and the mass matrix, respectively. $\boldsymbol{q}$ is the eigenvector. The q_i is the i-th term in the eigenvector. The natural frequencies could be solved by equating the determinant of the eigenmatrix to zero. Based on the obtained natural frequencies, the eigenvector, namely, the mode shapes, could be derived.

After solving the spatial domain problem, Eq. (3.3) could be used in conjunction with the solution of natural frequencies and mode shapes to obtain a series of ODEs, with differentiation with respect to the displacement terms in generalized coordinates. The ODEs could be solved numerically. Hitherto, the nonlinear dynamic response of the bistable composite panels was fully derived.

Based on the nonlinear dynamic oscillation of the bistable composite laminates, piezoelectric elements could be attached to the primary structures for energy harvesting. Experimental works generally observed that as the excitation amplitude increased, the frequency band for effective energy harvesting could be widened [45, 46]. The dynamic characteristic around the critical frequency inclined from single-well oscillation toward double oscillation (either chaotic or periodic) as the excitation increased, which favored energy generation [45, 46].

A later work in 2023 [47] conducted a comprehensive scrutiny of the dynamic regimes of the bistable composite oscillators and the favorable dynamic regimes for energy harvesting. When the bistable composites were subjected to frequency sweeps at a sufficient excitation load, single-frequency intra-well oscillation, limit cycle (a closed loop in the phase portrait to which approaching trajectories will converge), single-period double-well oscillation around the first resonant frequency, multi-period snap-through at the second resonant frequency, chaotic snap-through, and the intermittency between the aforementioned dynamic characteristics could be captured. The limit cycle oscillation was found to be the optimal dynamic regime for energy harvesting. At 11.08 N force amplitude and optimum external resistance, around 4 mW could be obtained at around the first resonance, where the limit cycle oscillation occurred. Interestingly, it was observed that the single-well oscillation around the second mode resonance might outperform the chaotic and multiple-period snap-through oscillation in energy harvesting [47].

Aside from the center-fixed, edge-free configuration, bistable composite harvesters were also reported in the cantilever configuration [48, 49]. The cantilever bistable composite panels had a larger aspect ratio, and therefore, the two stable states are no longer orthosymmetric [49]. With the help of a proof mass, the limit cycle oscillation around the second resonant frequency was found to be most beneficial in terms of energy harvesting. When a 15 mm × 15 mm × 0.2 mm PZT was mounted on a $[0_2/90_2]$ CFRP panel, around 1 mW could be obtained at 59 Hz base excitation at 3.1 g amplitude [49].

3.2.3 Sandwich Structures

Cellular sandwich, as another engineering structure, constitutes a core of different cellular topologies sandwiched between two layers of face sheets [50, 51]. The cellular cores are designed to reduce the overall density of the sandwich structures and improve shear stress transmission efficiency between the two face sheets. On the other hand, the flexural stiffness of the sandwich structures is mainly determined by the face sheets [5]. Conventionally, metal such as aluminum is used for face sheet panels. There has also been extensive research replacing the metal panels with CFRPs in pursuit of a higher stiffness and a further reduction in the weight of the structure [6, 52, 53]. In addition to the static mechanical properties, its superior energy absorption and impact mitigation performance have provoked its application in system shielding and personnel protection [54].

Auxetic cellular structure, characterized by a negative Poisson's ratio, and its application in piezoelectric energy harvesting began to attract research interest in 2017 [55]. Since for piezoelectric material, the in-plane piezoelectric coefficients d_{31} and d_{32} are of the same sign, an auxetic structure as a substrate is advantageous over the non-auxetic counterpart by allowing the piezoelectric contribution from the 1 and 2 in-plane directions to superimpose, instead of cancelling each other out. Based on this principle, substrates of a variety of auxetic topologies have been evaluated in terms of the power output boost compared to non-auxetic plane structures [56–61].

However, in real engineering applications, the auxetic cellular substrate needs to be sandwiched between face sheets for mechanical considerations. A related study investigated the effect of sandwich core topologies on the piezoelectric energy-harvesting performance [62]. Within this study, CFRP panels of $[45/-45]_s$ layup angles were used as face sheets. The composite laminate of the selected stacking sequence possessed a relatively low flexural stiffness and therefore suitable for energy harvesting. An M8528-P2-type MFC element was embedded at the core/top face sheet interface. The core topologies in the study included non-auxetic honeycomb structure, auxetic reentrant honeycomb, and chiral structure, which are illustrated in Figure 3.3.

The designed sandwich prototypes were mounted on the electrodynamic shaker and powered by a power amplifier. Sinusoidal waves were applied to the samples via a functional generator for a preliminary test, at 1 g peak-to-peak magnitude. At

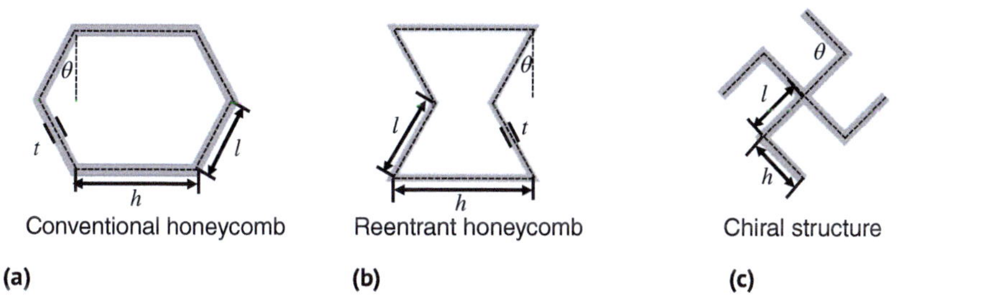

Figure 3.3 Schematic illustration of core topologies of (a) conventional honeycomb, (b) reentrant honeycomb, and (c) chiral structure. *Source*: [62] / with permission of Elsevier.

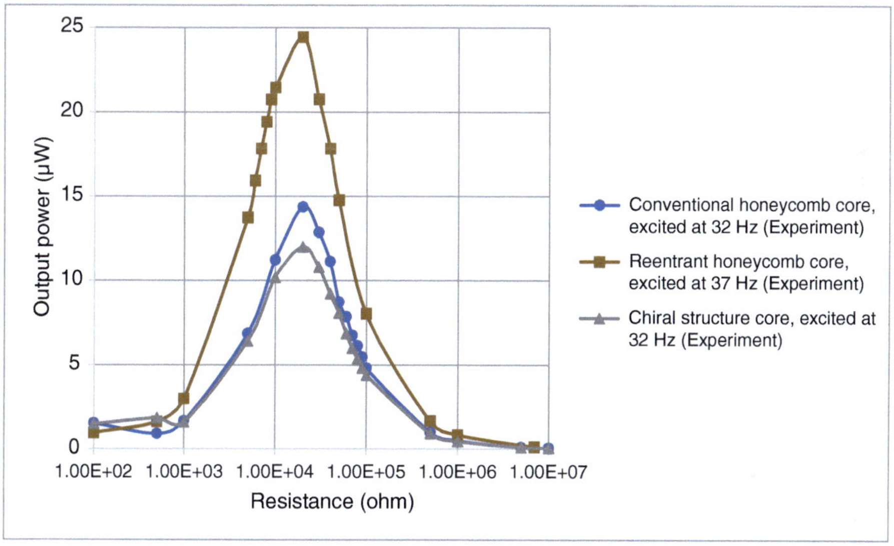

Figure 3.4 Comparison of power output between different sandwiches of different topologies, under 1 g peak-to-peak sinusoidal loading, excited under first-mode resonant frequencies. *Source*: [62] / with permission of Elsevier.

the individual first-mode resonant frequency of each sandwich structure, the resistance sweep result (Figure 3.4) showed that the sandwich with reentrant honeycomb exhibited superior energy-harvesting performance compared to sandwiches with conventional honeycomb and chiral structures, which could be as high as 25 μW [62].

3.3 Case Studies

The traditional composite structures for piezoelectric energy harvesting have been adequately discussed. However, the energy-harvesting performance in the service environment needs to be explored. In this section, works discussing the piezoelectric energy-harvesting characterization of composite structures in various industries are reviewed. The focus is on the transportation, wind energy, and healthcare industries.

3.3.1 Transportation

The high specific strength and specific stiffness make CFRPs the ideal candidate for structures in transportation-related applications. The multifunctional composite with MFC transducer is assessed under a wide variety of vibrational data measured in the transportation sector. These vibrational data are measured on a CGJ aircraft chassis during cruising, a BMW 525d engine beneath the bonnet, a 1600 RPM compressor engine, a section of a rail track, and inside the cabin of a rail carriage. The short-time Fourier transforms of the measured acceleration are listed in Figure 3.5 [63, 64].

The vibration captured within the CGJ aircraft stemmed from both the aircraft body and the engine. The vibration data were broadband without a dominant peak within the spectrum, with the average amplitude as high as 14 g. The measurement within the road car consisted of acceleration toward three axes, arranged as x-axis running along the vehicle, y-axis across the vehicle, and z-axis toward the vehicle upward. The x-axis acceleration exhibited a clear time-invariant peak at around 33 Hz, which corresponded to the engine operational frequency. Some higher frequency harmonics could also be observed at a certain time interval. The mean acceleration level was 1.19 g. Acceleration toward the y and z axes was relatively broadband, both with about 1.57 g mean acceleration. In the case of the compressor motor, acceleration toward all three axes manifested several harmonics and higher vibrational modes, with mean amplitudes of 0.70, 0.48, and 0.77 g, respectively. Acceleration data was finally acquired from the cross girder of the Forth Road suspension bridge (Scotland, UK). The source of the vibration was mainly the traffic passing by during the time span. The acceleration data presented the first resonance peak at around 15 Hz, along with signals exceeding 100 Hz [64].

To evaluate the energy-harvesting performance of the proposed multifunctional composite, an M8525-P2 MFC transducer was co-cured into a carbon-woven fabric-based composite laminate. The prototype was mounted on an electrodynamic shaker. The real-world data mentioned previously were generated in the waveform generator and transmitted to the shaker via a power amplifier. An external resistance of 20 kΩ was connected to the MFC. The testing setup was denoted in Figure 3.6. In some of the applications, the acceleration level exceeded the capacity of the shaker. In such cases, a number of data points were obtained at down-scaled acceleration, and extrapolation was applied to predict the full-scale power output [63].

The measured power output is summarized in Table 3.1. The output from the vibrational acceleration followed the feature of off resonance. Even though it was sufficient to power a number of commercial sensing nodes, such as humidity and temperature sensor HDC2080, Texas Instruments ($\sim 0.89\,\mu$W) [65] and surface-mount pressure sensor MS5607, TE Connectivity ($\sim 1.08\,\mu$W) [66]. Moreover, for realistic applications, the harvesting performance could be easily scaled up by serial or parallel connection of multiple piezoelectric transducers, and the power generated could be stored in supercapacitors or batteries via a rectification circuit.

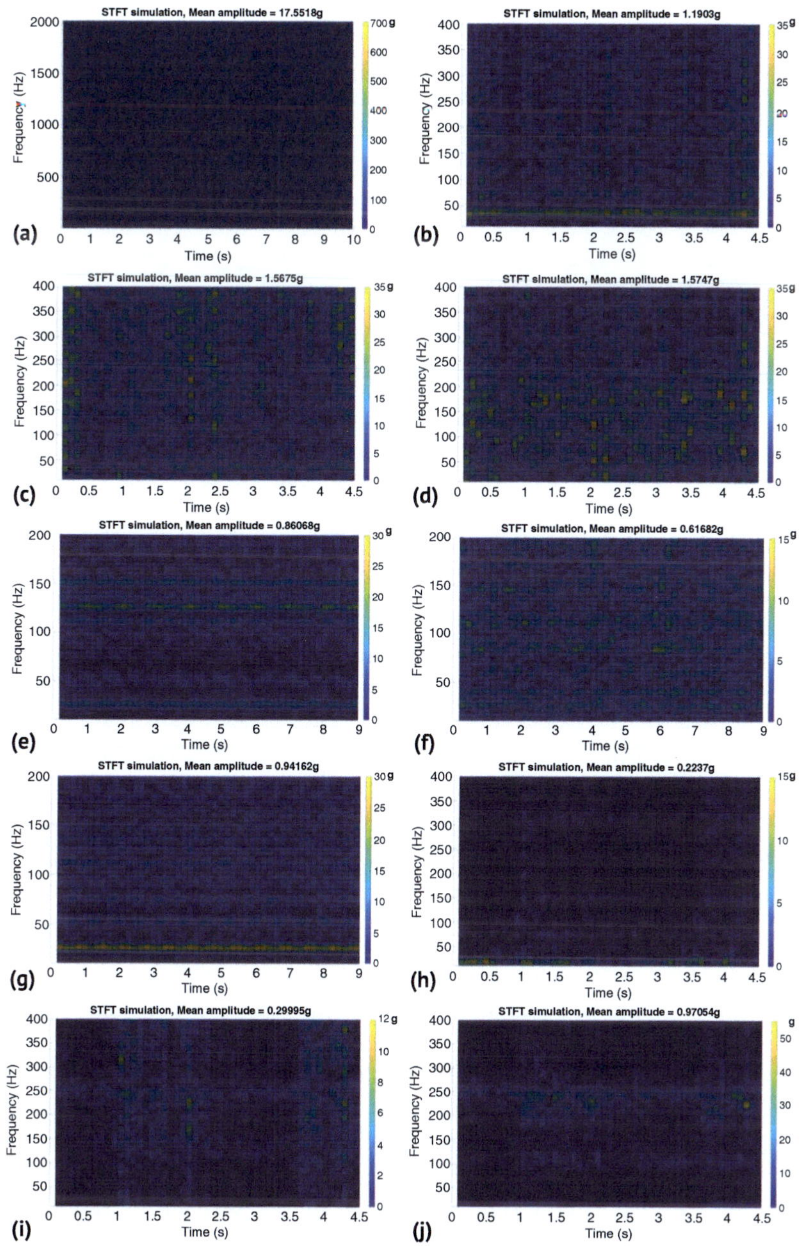

Figure 3.5 Short-time Fourier transform of vibration data under (a) CGJ aircraft, (b) road car, x axis, (c) road car, y axis, (d) road car, z axis, (e) compressor, x axis, (f) compressor, y axis, (g) compressor, z axis, (h) bridge, x axis, (i) bridge, y axis, and (j) bridge, z axis. *Source*: [62] / with permission of Elsevier.

3.3.2 Wind Farm

The large size of turbine blades for the wind turbines suggests a nontrivial material consumption. Therefore, glass fabric-reinforced polymers (GFRPs) are predominantly selected to be the candidate materials for turbine blades. It becomes

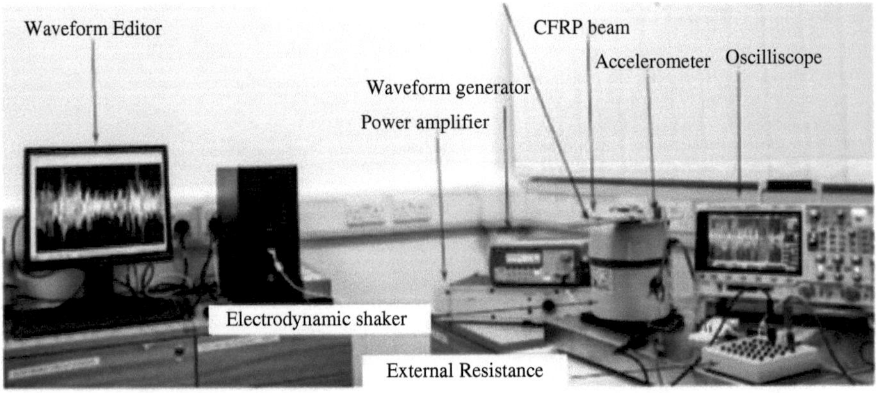

Figure 3.6 Test setup for energy-harvesting performance investigation. *Source*: [63] / with permission of Elsevier.

Table 3.1 Power output of different smart composite designs under a variety of applications.

Power output for different applications				
Aerospace	**Car x-axis**	**Car y-axis**	**Car z-axis**	**Compressor x-axis**
8 mW	0.55 mW	0.55 mW	0.45 mW	0.065 mW
Compressor y-axis	Compressor z-axis	Bridge x-axis	Bridge y-axis	Bridge z-axis
0.075 mW	0.07 mW	0.15 mW	0.52 mW	0.15 mW

Source: [63] / with permission of Elsevier.

outstandingly necessary to study the influence of environmental factors, such as temperature and humidity, on energy-harvesting efficiency. To meet this requirement, a case study was performed based on an MFC/GFRP assembly [67]. A 12-layer GFRP strip, with an M8528-P2 type MFC, was mounted on the shaker and placed in a Weiss VCV3 4060-5 climatic chamber. In this study, the vibrational energy-harvesting test of the GFRP-based system was conducted under a temperature range of −30 to 70 °C (under relative humidity of 50% RH) and humidity ranging from 10% to 90% RH (at 30 °C). At each environmental condition set point, two excitation schemes were considered: a sinusoidal frequency sweep from 5 to 500 Hz at 0.5 g acceleration level, and a vibration profile captured from a wind turbine tower, provided by Titan Wind Power. The experiment setup is illustrated in Figure 3.7.

The RMS voltage response to the sinusoidal frequency sweep at different temperature setpoints is plotted in Figure 3.8, with a trend line via a quadratic fitting. The RMS voltage output reached a maximum of 38.97 mV at the temperature of −10 °C, which was the optimal temperature of energy harvesting for the proposed assembly. The general trend, however, was a lower output at an elevated temperature [67].

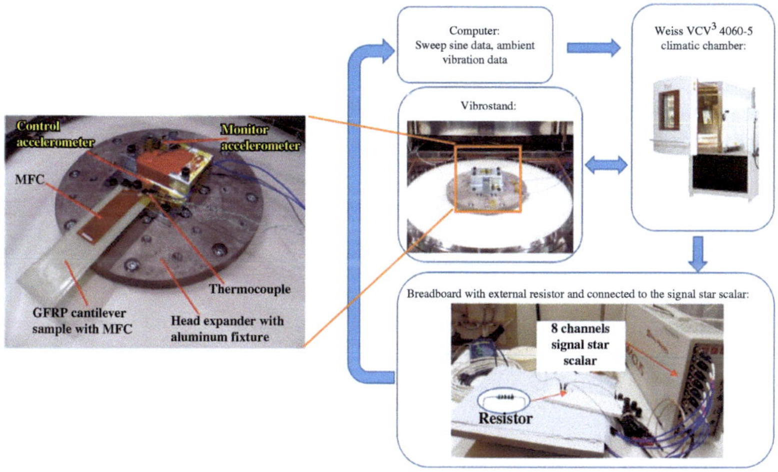

Figure 3.7 Experimental setup for vibrational test on the GFRP/MFC assembly. *Source*: [67] / with permission of Elsevier.

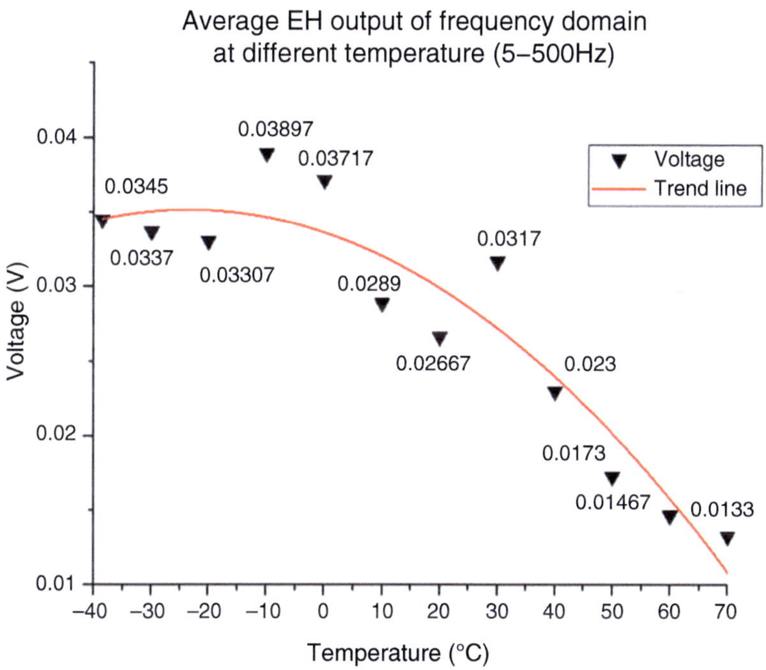

Figure 3.8 The RMS voltage of the frequency sweep test and the trend line for different temperatures. *Source*: [67] / with permission of Elsevier.

Several factors were believed to influence the energy-harvesting performance under varied temperatures. The short-circuit current that can be generated by a piezoelectric patch was denoted by Eq. (3.14) [64, 67]:

$$I_{sc} = \omega(t)\, d_{31}\sigma_{av}A \tag{3.14}$$

where $\omega\ (t)$ is the excitation frequency, d_{31} is the transverse piezoelectric coefficient, σ_{av} is the average stress experienced by the piezoelectric element, and A is the effective area.

With an externally connected impedance Z_{load}, the power output can be calculated by the following [64, 67]:

$$I_{\text{load}}\ (t) = \frac{I_{\text{sc}}^2\ (t)\ Z_{\text{MFC}}}{\sqrt{Z_{\text{MFC}}^2 + Z_{\text{load}}^2}} = \frac{I_{\text{sc}}^2\ (t) \times \frac{1}{2\pi C f}}{\sqrt{\left(\frac{1}{2\pi C f}^2\right) + Z_{\text{load}}^2}} \tag{3.15}$$

$$P(t) = I_{\text{load}}^2\ (t)\ Z_{\text{load}} = \frac{I_{\text{sc}}^4\ (t)\ Z_{\text{load}} Z_i^2}{Z_i^2 + Z_{\text{load}}^2} \tag{3.16}$$

From Eqs. (3.14)–(3.16), it can be found that the parameters that would be affected by the environmental conditions were C, which was the capacitance of the transducer, d_{31}, which reflected the electromechanical coupling, and σ_{av}, which was governed by the mechanical properties of both the GFRP substrate and the MFC element itself. According to early research [68], a positive correlation was found between the d_{31} and the temperature of PZT material, where the specific gradient depended on the chemical composition. Since the functional constituent of MFC was composed of PZT-5A and structural epoxy, it is safe to consider that the rise in temperature would contribute positively to the electromechanical coupling in this work. Aside from the electromechanical coupling coefficient, it was also reported that the internal impedance of the MFC element reduced linearly with temperature, from 20 kΩ at −35 °C to 16 kΩ at 70 °C [69]. An impedance match was required to achieve the maximum output; that is, the external resistance should match the internal impedance of the piezoelectric element. In this regard, a lower impedance, as the temperature rose, would benefit the overall power output [67].

It was deduced from the analysis that the variation of dielectric and electromechanical coupling properties with temperature rise favored a higher power output. However, obviously, Figure 3.8 suggests the opposite, which means the increase in temperature must affect the mechanical performance negatively for energy harvesting. From meso- or micro-mechanics perspective, the CTE mismatch between neighboring laminas or fibers and the polymer matrix would lead to thermal residual stress at the interfaces at both scales. This would lead to stiffness hardening, which might reduce the deformation response under vibration and therefore a lowered electrical output. Moreover, at a higher temperature, an enhanced polymer chain mobility resulted in a decrease in elastic modulus. In this regard, the stress experienced by the MFC element would reduce, which resulted in a drop in the voltage output [67].

Under the vibration obtained from the wind turbine blade, the energy output indicated a similar trend with temperature rise as in the case of sinusoidal excitation, as denoted in Figure 3.9.

Unlike the case of a sinusoidal frequency sweep, where the amplitude was constant, the acceleration captured from the wind farm was relatively broadband and contained several sharp peaks across the frequency spectrum. In a linear dynamic system, the excitation with frequency close to the structure's resonance makes a

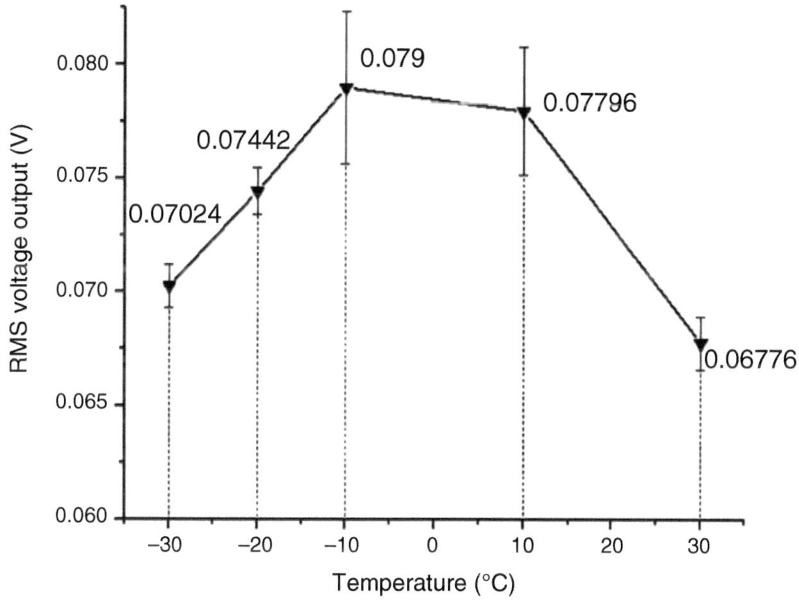

Figure 3.9 The RMS voltage under wind farm vibration and the trend line for different temperatures. *Source*: [67] / with permission of Elsevier.

large contribution to the output, while the off-resonance part of the input is much less effective. The measured resonant frequencies of the first three modes and eigenfrequency simulation are shown in Figure 3.10. It was obvious that the resonant frequencies of the second mode (first torsion) and the third mode (second bending) were more sensitive to temperature variation. The frequency spectrum of the ambient vibration is demonstrated in Figure 3.11. The yellow lines and red lines represented the second- and third-mode resonant frequencies at 10 and 30 °C, respectively. The magnified insets suggested that the resonant frequencies at a lower temperature just matched the ambient frequency spectrum profile, which might explain the fact that the harvester performed better at a relatively lower temperature [67].

The average temperature is approximately 10 °C throughout the year in England and may exceed 30 °C in summer [70]. At these two temperature set points, the GFRP/MFC assembly was assessed at relative humidity ranging from 10 to 90% RH, in terms of its energy-harvesting performance. The RMS voltage output is demonstrated in Figure 3.12. Figure 3.12 clearly suggests that the humidity sensitivity of energy harvesting on relative humidity was much less significant compared to temperature. It is noteworthy that the vibrational test started right after the temperature and humidity reached equilibrium within the chamber. The short-term influence of environmental humidity might not be enough to alter the properties of the composites and therefore affect their energy-harvesting potential. In fact, it has been researched before that the slow diffusion of moisture into the polymer matrix would lead to the degradation of the stiffness of the GFRP composites [71]. In this regard, it is reasonable to expect a more observable voltage change with humidity if the assembly would stay within the environment long enough to reach its own moisture equilibrium [67].

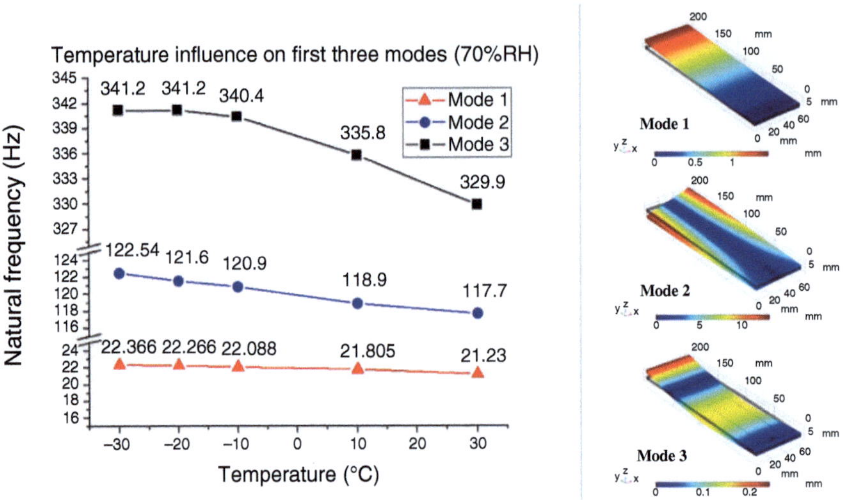

Figure 3.10 Variation of resonant frequencies of the first three modes with increase of temperature, at 70% RH, and mode shapes, indicating that the first mode is a bending mode, the second mode is a torsion mode, and the third mode is a bending mode. *Source*: [67] / with permission of Elsevier.

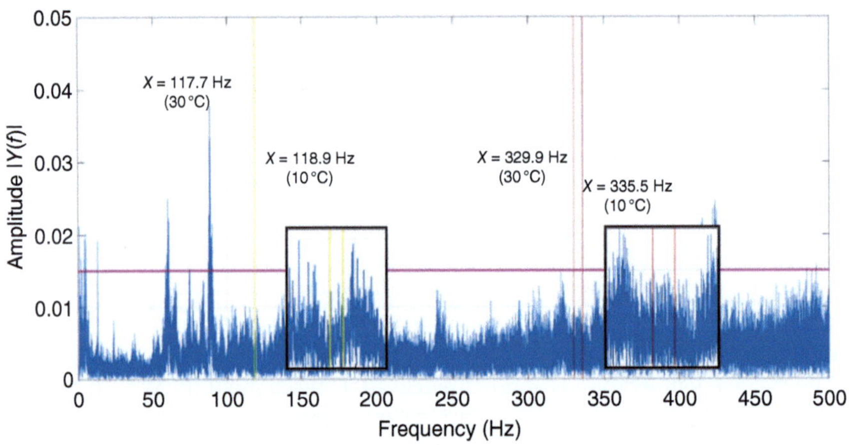

Figure 3.11 Frequency spectrum of the vibration captured from the wind farm, with the purple line denoting the amplitude of the sinusoidal frequency sweep, and the yellow line and red line denoting the second- and third-mode resonant frequency variations, respectively. *Source*: [67] / with permission of Elsevier.

3.3.3 Prosthetics

Lower-limb prosthetic legs are in high demand by amputees to ensure their independent movement and quality of life. One of the earliest contemporary examples was a solid ankle cushion foot reported in the 1950s [72]. The solid design with some soft foam or rubber only allowed the amputees to mimic basic gait function, which was unnatural and would cause repeated stress impulses to injure other parts such

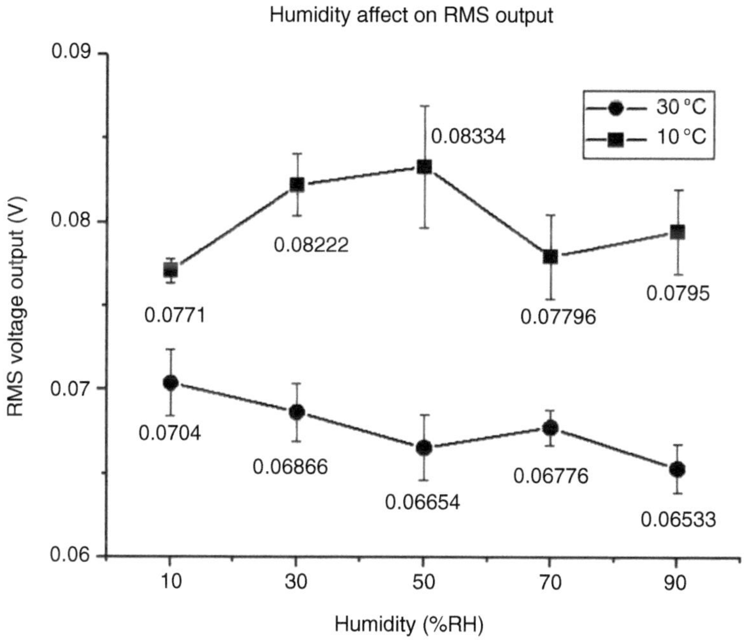

Figure 3.12 The RMS voltage output at different humidity levels, at 10 and 30 °C, respectively. *Source*: [67] / with permission of Elsevier.

as joints or hips [73]. In the following decades, upgraded prosthesis designs with a spring concept began to emerge, which stored mechanical energy and returned to the gait motion [74]. The spring could be either a mechanical spring or a pneumatic spring using air pressure to provide restoring force. The employment of the spring system lowered the stress on the joints of the amputees. CFRPs, in contrast to the conventional metallic material, have been developed as part of the spring structure in low-limb prosthetic legs [75].

Although the spring system was able to provide improved comfort and energy efficiency, there was still room for optimization. Increasing trends have emerged to incorporate motors and actuators integrated with control systems to modulate the motions of the lower limbs of patients [76]. To achieve the actuation of all degrees of freedom of knees and ankles, multiple motors were needed, which required a large amount of power and therefore the walking range of the prosthesis was limited. Moreover, aiming at gait control and condition monitoring, sensors including accelerometers and strain gauges, needed to be integrated and controlled by microcontrollers. All the components mentioned earlier suggested a nontrivial demand on the electrical power from the energy storage system.

Fortunately, during the gait motion, there is a decent amount of kinematic energy being dissipated by the passive components of the prosthesis [77]. The power consumed by the motors and the sensors can be largely supplemented if part of the kinetic energy can be converted into electrical power. A number of transducing mechanisms are feasible, such as piezoelectric [23, 26], electromagnetic [20, 28], and electrostatic [22]. However, it should be noted that careful conversion control is necessary not to burden the patients with parasitic effects [78]. According to a biomechanics study of the prosthetic gait, the restoring cycle of the spring element can be effectively utilized for power conversion, because there would be no need of drawing extra energy from the prosthesis users [79].

To identify the feasibility of utilizing recovered energy from the prosthesis to power typical electronics, a numerical analysis was performed to simulate the amount of power obtained under different gait patterns [80]. A pneumatic air nonlinear spring was included in a lumped-mass model as given in Eq. (3.17) [80].

$$m\ddot{x} + c\dot{x} + kx + P_0\pi r^2\left[\left(\frac{L}{L-x}\right)^{\gamma} - 1\right] = F(t) \tag{3.17}$$

where m is the mass of the body, c is the damping coefficient, and k is the linear stiffness. In terms of the restoring force term by pneumatic air spring, P_0 is the initial pressure, r is the radius of the air cylinder, L is the cylinder length, and γ is the diabatic constant, which equals 1.4 for diabatic gas such as nitrogen. $F(t)$ is the responsive force from the ground at each stride [80].

No specific transducing mechanism was assigned in this study. Instead, a general electrical damping term was assumed to predict the maximum power output at the first-mode resonant frequency as denoted by Eq. (3.18) [80].

$$p(t) = \frac{m_t\zeta_e\omega_0^3 x\,(t)^2}{4\,(\zeta_e + \zeta_m)} \tag{3.18}$$

In this study, real acceleration data measured from a healthy 34-year-old male with a 70-kg body mass were used in the excitation term of the numerical model (Figure 3.13). An accelerometer data logger was placed in the sneaker of the participant's right foot. The one-, two-, and three-axis accelerometers in this work corresponded to the sagittal, longitudinal, and frontal axis of the human body.

The gait patterns being captured were as follows [80]:

- Walking at 3 km/h
- Walking at 6 km/h
- Running at 9 km/h
- Climbing at 30 steps per minute
- Climbing at 60 steps per minute
- Climbing at 90 steps per minute
- Walking at 3 km/h with 20 kg weight in the left hand to simulate limping
- Walking at 3 km/h with 20 kg weight in the right hand to simulate limping

The measured three-axis acceleration data when walking at 3 km/h and walking at 3 km/h with a 20 kg weight in the right hand are presented in Figures 3.14 and 3.15, respectively. It was found that for each gait pattern, there was no acceleration dominant axis and all three axes exhibited a similarly substantial level of acceleration amplitude. The acceleration data when walking at 3 km/h presented several harmonics and the first frequency peak was at around 0.9 Hz. Compared to the acceleration data captured at normal walking, the limping walking pattern exhibited fewer harmonic peaks. Moreover, the time domain plot of the limping pattern suggested a much shorter but more intense impulse (Figures 3.14 and 3.15). The mean acceleration measurements of all cases is summarized in Table 3.2 [80].

The measured acceleration was used as the excitation input of Eq. (3.18), and the power output was summarized in Table 3.3. For most gait patterns, the largest power

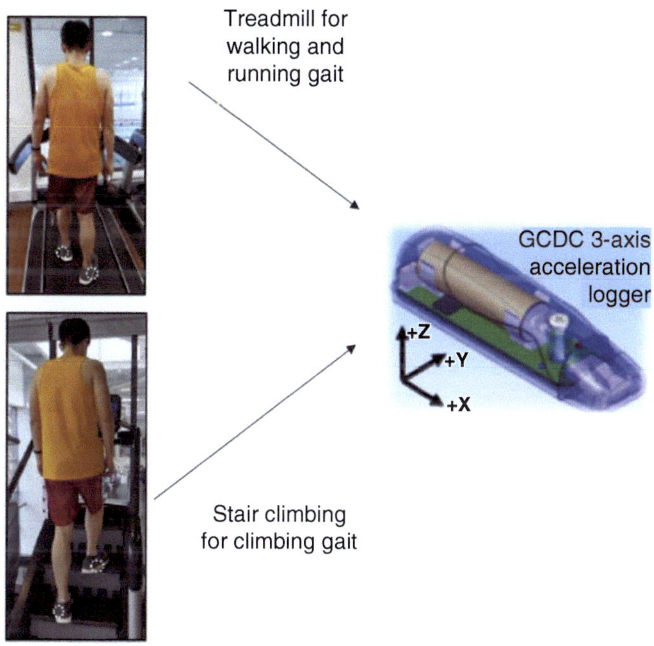

Figure 3.13 Overview of the acceleration measurement scheme. *Source*: [80] / MDPI / CC BY 4.0.

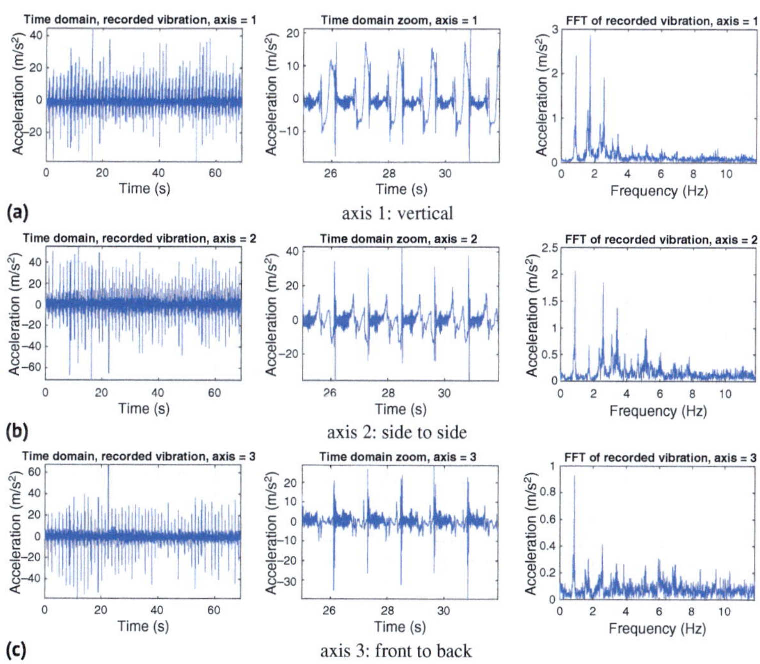

Figure 3.14 Gait acceleration when walking at 3 km/h, where (a), (b), and (c) toward axes 1, 2, and 3 of the accelerometer. *Source*: [80] / MDPI / CC BY 4.0.

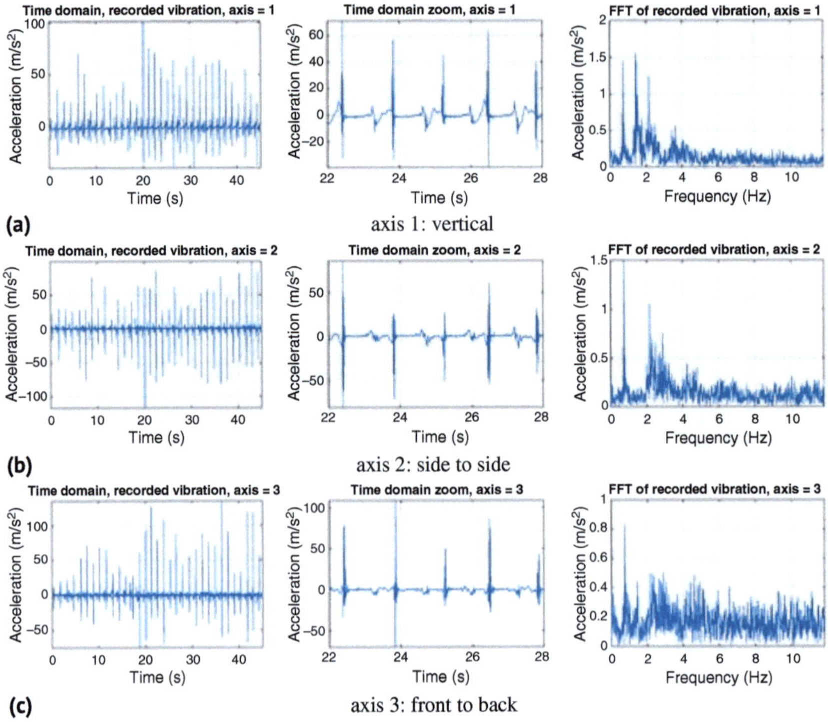

Figure 3.15 Gait acceleration when walking at 3 km/h with 20 kg in the right hand, where (a), (b), and (c) toward axes 1, 2, and 3 of the accelerometer. *Source*: [80] / MDPI / CC BY 4.0.

output was obtained in the longitudinal direction. The power output increased with the speed of walking undoubtedly. The acceleration measured under 9 km/h running generated as high as 392 mW power output on the longitudinal axis. Compared to normal walking, the limped walking pattern produced a higher output with even lower mean acceleration. For example, when walking with a 20 kg weight in the right hand, 102 mW power was generated in the sagittal direction under 2.8 m/s^2 mean acceleration input, which was three times as high as the output from normal walking (4.2 m/s^2). On the other hand, even with comparable mean acceleration, the climbing gait pattern was inferior to most other gait patterns in terms of energy harvesting. This might suggest that the acceleration envelope of the climbing gait was not the most suitable for energy conversion [80].

Following the numerical study, a proof-of-concept prototype was manufactured based on the CFRP composites. The composite prosthetic leg was composed of a foot and thin segments. Each segment was fabricated out of prepregs laid on specifically designed molds and cured in a programmable oven. An M8528-P2 MFC piezoelectric element was mounted on the foot segment. Normally, the MFC element should be located where the stress is maximum under gait motion to achieve the highest output. In this work, however, the MFC element was applied to another stress-rich region, considering its flexural endurance [81].

The measured acceleration toward the longitudinal direction of different gait patterns (Table 3.2) was used to vibrate the CFRP prosthetic leg. The voltage output

Table 3.2 Three-axis acceleration measurement of different gait patterns.

Gait condition	Axis	Mean acceleration (m/s^2)
Walking at 3 km/h	Sagittal	4.2
	Longitudinal	4.2
	Frontal	2.3
Walking at 6 km/h	Sagittal	7.9
	Longitudinal	8.1
	Frontal	6.1
Running at 9 km/h	Sagittal	11.7
	Longitudinal	11.3
	Frontal	9.4
Climbing at 60 steps per minute	Sagittal	2.3
	Longitudinal	3.7
	Frontal	1.9
Walking with a 20 kg weight in the left hand	Sagittal	2.3
	Longitudinal	2.4
	Frontal	2.5
Walking with a 20 kg weight in the right hand	Sagittal	2.8
	Longitudinal	2.7
	Frontal	2.6

Source: Adapted from [80].

across a 20 kΩ external resistance was monitored by the oscilloscope, and the power output was calculated by Ohm's law [81].

The measured voltage across the external resistance and the calculated power were listed in Table 3.3. Likewise, the gait pattern of running at 9 km/h produced the highest amount of power output, which was mainly due to its highest intensity in acceleration input. Moreover, the limping gait pattern when a 20 kg weight was on the right hand recovered larger power than normal walking, which was consistent with the numerical finding. Unlike the numerical outcome, there was a huge discrepancy between power output from the two limped gait patterns, which was mainly attributed to some human measurement factors [81].

It can be easily noted that orders of magnitude are different between the measured power output and the simulation outcome. Due to the limitation of the stroke of the shaker, full-scale acceleration might not be reachable for several gait patterns. This would unquestionably lead to a lower measured output. In addition, the options of locations where the MFC could be mounted were constrained by the flexural endurance of the piezoelectric transducer. The non-optimized mounting place significantly reduced the energy conversion efficiency [81]. Even though the real measured output was non-optimized, it was still sufficient to power controlling units such as 32-bit ARM Cortex MPU (320 μW in active mode) [82] and MEMS sensors such as a 9 DOF motion sensor from InvenSense (50 μW) [83].

Table 3.3 Numerical and experimental power outcomes from different gait motions.

Gait condition	Axis	Numerical power output (mW)	Experimental measurement (μW)
Walking at 3 km/h	Sagittal	31	—
	Longitudinal	45	14.85
	Frontal	34	—
Walking at 6 km/h	Sagittal	96	—
	Longitudinal	75	206.04
	Frontal	57	—
Running at 9 km/h	Sagittal	341	—
	Longitudinal	392	432.18
	Frontal	241	—
Climbing at 60 steps per minute	Sagittal	29	—
	Longitudinal	41	151.38
	Frontal	27	—
Walking with a 20 kg weight in the left hand	Sagittal	87	—
	Longitudinal	52	32.88
	Frontal	62	—
Walking with a 20 kg weight in the right hand	Sagittal	102	—
	Longitudinal	69	204.02
	Frontal	103	—

Source: Adapted from [81].

3.4 Summary

This chapter presented and discussed some examples of combining piezoelectric energy-harvesting techniques and smart fiber-reinforced composites. It started with the design of such structures and the effort to enhance energy-harvesting performance. Bistable composites exhibited a broadened effective frequency band for energy harvesting. Carbon-woven FRP laminates and sandwich configurations with several topological cores, namely, conventional honeycomb, reentrant honeycomb, and chiral structure, have been studied, with an MFC element used as the piezoelectric transducer. Subsequently, various case studies were reviewed regarding the piezoelectric energy harvesting of fiber-reinforced composites in engineering and healthcare sector applications, including transportation, wind farms, and prosthetics. The MFC/CFRPs assembly was tested under acceleration, captured in a variety of industrial scenarios in terms of energy harvesting. Uniquely, the influence of temperature and humidity on the energy-harvesting performance of MFC/GFRP assembly was assessed when the assembly was tested under acceleration measured on a wind turbine blade when in service. Finally, another exemplary application of smart composites was reported where CFRPs were used in the construction of a prosthetic leg with integrated MFC. The energy-harvesting performance was evaluated under certain gait patterns, including normal and limped walking at various speeds and climbing. Much progress has been made, but as new materials and sensors are continuously developed, further improvements in structural performance and functionality are expected, especially as artificial intelligence and deep machine learning [84] can speed up the optimization of material selection and the overall design process.

References

1 Yang, M., Hu, Y., Zhang, J., Ding, G., and Song, C. (2018). Analytical model for flexural damping responses of CFRP cantilever beams in the low-frequency vibration. *J. Low Freq. Noise Vib. Act. Control*, 37, 669–81. https://doi.org/10.1177/1461348418756024.

2 Shi, Y., Swait, T., and Soutis, C. (2012). Modelling damage evolution in composite laminates subjected to low velocity impact. *Compos. Struct.*, 94, 2902–13. https://doi.org/10.1016/j.compstruct.2012.03.039.

3 Hwang, J.H., Jin, C.K., Seo, H.Y., and Kang, C.G. (2019). Effect of the number of CFRP prepregs and roughness at the bonding area on the spring-back and flexural strength of hybrid composites of CFRP combined with CR980. *Metals (Basel)*, 9. https://doi.org/10.3390/met9101054.

4 Srivastava, V.K., Gries, T., Veit, D., Quadflieg, T., Mohr, B., and Kolloch, M. (2017). Effect of nanomaterial on mode I and mode II interlaminar fracture toughness of woven carbon fabric reinforced polymer composites. *Eng. Fract. Mech.*, 180, 73–86. https://doi.org/10.1016/j.engfracmech.2017.05.030.

5 Xiao, Y., Hu, Y., Zhang, J., Song, C., Huang, X., Yu, J., and Liu, Z. (2018). The bending responses of sandwich panels with aluminium honeycomb core and CFRP skins used in electric vehicle body. *Adv. Mater. Sci. Eng.* https://doi.org/10.1155/2018/5750607.

6 Farooq, U., Ahmad, M.S., Rakha, S.A., Ali, N., Khurram, A.A., and Subhani, T. (2017). Interfacial mechanical performance of composite honeycomb sandwich panels for aerospace applications. *Arab. J. Sci. Eng.*, 42, 1775–82. https://doi.org/10.1007/s13369-016-2307-z.

7 Gibson, R.F. (2010). A review of recent research on mechanics of multifunctional composite materials and structures. *Compos. Struct.*, 92, 2793–810. https://doi.org/10.1016/j.compstruct.2010.05.003.

8 Galooyak, S.S., Palassi, M., Farahani, H.Z., and Goli, A. (2015). Effect of carbon nanotube on the rheological properties of bitumen. *Pet. Coal.*, 57, 556–64. https://doi.org/10.1063/5.0108538.

9 Thostenson, E.T., Ziaee, S., and Chou, T.W. (2009). Processing and electrical properties of carbon nanotube/vinyl ester nanocomposites. *Compos. Sci. Technol.*, 69, 801–4. https://doi.org/10.1016/j.compscitech.2008.06.023.

10 Lin, Y. and Sodano, H.A. (2008). Concept and model of a piezoelectric structural fiber for multifunctional composites. *Compos. Sci. Technol.*, 68, 1911–8. https://doi.org/10.1016/j.compscitech.2007.12.017.

11 Lin, Y. and Sodano, H.A. (2009). Electromechanical characterization of a active structural fiber lamina for multifunctional composites. *Compos. Sci. Technol.*, 69, 1825–30. https://doi.org/10.1016/j.compscitech.2009.03.022.

12 Huang, Y., Li, N., Ma, Y., Du, F., Li, F., He, X., Lin, X., Gao, H., and Chen, Y. (2007). The influence of single-walled carbon nanotube structure on the electromagnetic interference shielding efficiency of its epoxy composites. *Carbon*, 45, 1614–21. https://doi.org/10.1016/j.carbon.2007.04.016.

13 Yang, Z., Peng, H., Wang, W., and Liu, T. (2010). Crystallization behavior of poly(ε-caprolactone)/layered double hydroxide nanocomposites. *J. Appl. Polym. Sci.*, 116, 2658–67. https://doi.org/10.1002/app.

14 Bowland, C., Zhou, Z., and Sodano, H.A. (2014). Multifunctional barium titanate coated carbon fibers. *Adv. Funct. Mater.*, 24, 6303–8. https://doi.org/10.1002/adfm.201401417.

15 Narita, F., Nagaoka, H., and Wang, Z. (2019). Fabrication and impact output voltage characteristics of carbon fiber reinforced polymer composites with lead-free piezoelectric nano-particles. *Mater. Lett.*, 236, 487–90. https://doi.org/10.1016/j.matlet.2018.10.174.

16 Karalis, G., Tzounis, L., Lambrou, E., Gergidis, L.N., and Paipetis, A.S. (2019). A carbon fiber thermoelectric generator integrated as a lamina within an 8-ply laminate epoxy composite: efficient thermal energy harvesting by advanced structural materials. *Appl. Energy,* 253. https://doi.org/10.1016/j.apenergy.2019.113512.

17 Han, S. and Chung, D.D.L. (2013). Through-thickness thermoelectric power of a carbon fiber/epoxy composite and decoupled contributions from a lamina and an interlaminar interface. *Carbon*, 52, 30–9. https://doi.org/10.1016/j.carbon.2012.08.071.

18 Sung, D.H., Kang, G.H., Kong, K., Kim, M., Park, H.W., and Park, Y. Bin. (2016). Characterization of thermoelectric properties of multifunctional multiscale composites and fiber-reinforced composites for thermal energy harvesting. *Compos. Part B Eng.*, 92, 202–9. https://doi.org/10.1016/j.compositesb.2016.02.050.

19 Dagdeviren, C., Yang, B.D., Su, Y., Tran, P.L., Joe, P., Anderson, E., Xia, J., Doraiswamy, V., Dehdashti, B., Feng, X., Lu, B., Poston, R., Khalpey, Z., Ghaffari, R., Huang, Y., Slepian, M.J., and Rogers, J.A. (2014). Conformal piezoelectric energy harvesting and storage from motions of the heart, lung, and diaphragm. *Proc. Natl. Acad. Sci. U S A*, 111, 1927–32. https://doi.org/10.1073/pnas.1317233111.

20 Bolat, F.C., Basaran, S., and Sivrioglu, S. (2019). Piezoelectric and electromagnetic hybrid energy harvesting with low-frequency vibrations of an aerodynamic profile under the air effect. *Mech. Syst. Signal Process.,* 133. https://doi.org/10.1016/j.ymssp.2019.106246.

21 Rajarathinam, M. and Ali, S.F. (2018). Investigation of a hybrid piezo-electromagnetic energy harvester [Untersuchung eines hybriden piezo-elektromagnetischen Energy-Harvesters]. *Tech. Mess.*, 85, 541–52. https://doi.org/10.1515/teme-2017-0086.

22 Jia, Y., Yan, J., Soga, K., and Seshia, A.A. (2013). Parametrically excited MEMS vibration energy harvesters with design approaches to overcome the initiation threshold amplitude. *J. Micromech. Microeng.,* 23. https://doi.org/10.1088/0960-1317/23/11/114007.

23 Gljušćić, P. and Zelenika, S. (2021). Experimental characterization of optimized piezoelectric energy harvesters for wearable sensor networks. *Sensors,* 21. https://doi.org/10.3390/s21217042.

24 Ponnamma, D., Parangusan, H., Tanvir, A., and AlMa'adeed, M.A.A. (2019). Smart and robust electrospun fabrics of piezoelectric polymer nanocomposite for self-powering electronic textiles. *Mater. Des.*, 184, 108176. https://doi.org/10.1016/j.matdes.2019.108176.

25 Zhang, J. and Shi, F. (2022). Preparation of organic ferroelectric material P (Vdf-Trfe) nanostructure and its application in process simplification. *Integr. Ferroelectr.*, 229, 138–57. https://doi.org/10.1080/10584587.2022.2074229.

26 Zhao, P., Wang, S., and Kadlec, A. (2016). Piezoelectric and dielectric properties of nanoporous polyvinylidence fluoride (PVDF) films. *Behav. Mech. Multifunct. Mater. Compos.*, 9800, 98000P. https://doi.org/10.1117/12.2218896.

27 Dal Bo, L., Gardonio, P., and Turco, E. (2020). Analysis and scaling study of vibration energy harvesting with reactive electromagnetic and piezoelectric transducers. *J. Sound Vib.*, 484, 115510. https://doi.org/10.1016/j.jsv.2020.115510.

28 Dal Bo, L. and Gardonio, P. (2018). Energy harvesting with electromagnetic and piezoelectric seismic transducers: unified theory and experimental validation. *J. Sound Vib.*, 433, 385–424. https://doi.org/10.1016/j.jsv.2018.06.034.

29 Malakooti, M.H., Patterson, B.A., Hwang, H.S., and Sodano, H.A. (2016). ZnO nanowire interfaces for high strength multifunctional composites with embedded energy harvesting. *Energy Environ. Sci.*, 9, 634–43. https://doi.org/10.1039/c5ee03181h.

30 Shepelin, N.A., Glushenkov, A.M., Lussini, V.C., Fox, P.J., Dicinoski, G.W., Shapter, J.G., and Ellis, A.V. (2019). New developments in composites, copolymer technologies and processing techniques for flexible fluoropolymer piezoelectric generators for efficient energy harvesting. *Energy Environ. Sci.*, 12, 1143–76. https://doi.org/10.1039/c8ee03006e.

31 Kurita, H., Wang, Z., Nagaoka, H., and Narita, F. (2020). Fabrication and mechanical properties of carbon-fiber-reinforced polymer composites with lead-free piezoelectric nanoparticles. *Sensors Mater.*, 32, 2453–62. https://doi.org/10.18494/SAM.2020.2820.

32 Maruyama, K., Kawakami, Y., Mori, K., Kurita, H., Shi, Y., Jia, Y., and Narita, F. (2022). Electromechanical characterization and kinetic energy harvesting of piezoelectric nanocomposites reinforced with glass fibers. *Compos. Sci. Technol.*, 223, 109408. https://doi.org/10.1016/j.compscitech.2022.109408.

33 Narayana, K.J. and Gupta Burela, R. (2018). A review of recent research on multifunctional composite materials and structures with their applications. *Mater. Today Proc.*, 5, 5580–90. https://doi.org/10.1016/j.matpr.2017.12.149.

34 Anton, S.R. and Sodano, H.A. (2007). A review of power harvesting using piezoelectric materials (2003-2006). *Smart Mater. Struct.*, 16. https://doi.org/10.1088/0964-1726/16/3/R01.

35 Rua Taborda, M.I., Elissalde, C., Chung, U.C., Maglione, M., Fernandes, E., Salehian, A., Santawitee, O., and Debéda, H. (2020). Key features in the development of unimorph stainless steel cantilever with screen-printed PZT dedicated to energy harvesting applications. *Int. J. Appl. Ceram. Technol.*, 17, 2533–44. https://doi.org/10.1111/ijac.13588.

36 Lapeine, D., Ferrandis, J.Y., Gatsa, O., Pascal, F., and Combette, P. (2016). Study and optimization of screen-printed PZT transducers for acoustic applications. In *Proceedings of theProc 2016 IEEE International Conference on Dielectrics Int Conf Dielectr ICD,* 1, 253–6. https://doi.org/10.1109/ICD.2016.7547592.

37 Smart Material. (2017). Macro Fiber Composite (MFC), Datasheet 2017, 8.

38 Shi, Y., Hallett, S.R., and Zhu, M. (2017). Energy harvesting behaviour for aircraft composites structures using macro-fibre composite: part I – integration and experiment. *Compos. Struct.*, 160, 1279–86. https://doi.org/10.1016/j.compstruct.2016.11.037.

39 Yang, Y., Tang, L., and Li, H. (2009). Vibration energy harvesting using macro-fiber composites. *Smart Mater. Struct.*, 18. https://doi.org/10.1088/0964-1726/18/11/115025.

40 Arms, S.W., Townsend, C.P., Churchill, D.L., Moon, S.M., and Phan, N. (2006). Energy harvesting wireless sensors for helicopter damage tracking. *Annu. Forum Proc. - AHS Int.*, II, 1335–40.

41 Shi, Y., Piao, C., El Fadlaoui, D., Al-Saadi, A., and Jia, Y. (2019). Integration and characterisation of piezoelectric macro-fibre composite on carbon fibre composite for vibration energy harvesting. *J. Phys. Conf. Ser.*, 1407. https://doi.org/10.1088/1742-6596/1407/1/012080.

42 Jia, Y. (2020). Review of nonlinear vibration energy harvesting: duffing, bistability, parametric, stochastic and others. *J. Intell. Mater. Syst. Struct.*, 31, 921–44. https://doi.org/10.1177/1045389X20905989.

43 Hyer, M.W. (1981). Some observations on the cured shape of thin unsymmetric laminates. *J. Compos. Mater.*, 15, 175–94. https://doi.org/10.1177/002199838101500207.

44 Zhang, W., Liu, Y.Z., and Wu, M.Q. (2019). Theory and experiment of nonlinear vibrations and dynamic snap-through phenomena for bi-stable asymmetric laminated composite square panels under foundation excitation. *Compos. Struct.*, 225. https://doi.org/10.1016/j.compstruct.2019.111140.

45 Betts, D.N., Bowen, C.R., Alicia Kim, H.A., Guyer, R.A., Le Bas, P.Y., and Inman, D.J. (2014). Modelling the dynamic response of bistable composite plates for piezoelectric energy harvesting. In *55th AIAA/ASME/ASCE/AHS/ASC Structures, Structural Dynamics, and Materials Conference*, 1–13. https://doi.org/10.2514/6.2014-0154.

46 Syta, A., Bowen, C.R., Kim, H.A., Rysak, A., and Litak, G. (2015). Experimental analysis of the dynamical response of energy harvesting devices based on bistable laminated plates. *Meccanica,* 50, 1961–70. https://doi.org/10.1007/s11012-015-0140-1.

47 Dong, T., Cao, D., and Dong, M. (2023). The dynamic regimes and the energy harvesting of the energy harvester based on the bistable piezoelectric composite laminate. *Thin-Walled Struct.*, 188. https://doi.org/10.1016/j.tws.2023.110777.

48 Pan, D., Dai, F., and Li, H. (2015). Piezoelectric energy harvester based on bi-stable hybrid symmetric laminate. *Compos. Sci. Technol.*, 119, 34–45. https://doi.org/10.1016/j.compscitech.2015.09.018.

49 Li, H., Dai, F., and Du, S. (2015). Broadband energy harvesting by exploiting nonlinear oscillations around the second vibration mode of a rectangular piezoelectric bistable laminate. *Smart Mater. Struct.*, 24, 45024. https://doi.org/10.1088/0964-1726/24/4/045024.

50 Thomas, T. and Tiwari, G. (2019). Crushing behavior of honeycomb structure: a review. *Int. J. Crashworthiness,* 24, 555–79. https://doi.org/10.1080/13588265.2018.1480471.

51 Lakes, R.S. (1989). Cellular solids. *J. Biomech.*, 22, 397. https://doi.org/10.1016/0021-9290(89)90056-0.

52 Luo, C.J., Zhang, H.L., Yu, W.Z., He, K., and Du, R.X. (2014). Study on preparation method and energy-absorbing characteristics of carbon fiber composite energy absorber. *Appl. Mech. Mater.*, 442, 98–103. https://doi.org/10.4028/www.scientific.net/AMM.442.98.

53 Zaharia, S.M., Pop, M.A., Semenescu, A., Florea, B., and Chivu, O.R. (2017). Mechanical properties and fatigue performances on sandwich structures with CFRP skin and Nomex honeycomb core. *Mater. Plast.*, 54, 67–72. https://doi.org/10.37358/mp.17.1.4788.

54 Rebelo, H.B., Lecompte, D., Cismasiu, C., Jonet, A., Belkassem, B., and Maazoun, A. (2019). Experimental and numerical investigation on 3D printed PLA sacrificial honeycomb cladding. *Int. J. Impact Eng.*, 131, 162–73. https://doi.org/10.1016/j.ijimpeng.2019.05.013.

55 Li, Q., Kuang, Y., and Zhu, M. (2017). Auxetic piezoelectric energy harvesters for increased electric power output. *AIP Adv.*, 7. https://doi.org/10.1063/1.4974310.

56 Krishnaswamy, J.A., Buroni, F.C., Melnik, R., Rodriguez-Tembleque, L., and Saez, A. (2020). Design of polymeric auxetic matrices for improved mechanical coupling in lead-free piezocomposites. *Smart Mater. Struct.*, 29, 54002. https://doi.org/10.1088/1361-665X/ab7e35.

57 Umino, Y., Tsukamoto, T., Shiomi, S., Yamada, K., and Suzuki, T. (2018). Development of vibration energy harvester with 2D mechanical metamaterial structure. *J. Phys. Conf. Ser.*, 1052. https://doi.org/10.1088/1742-6596/1052/1/012103.

58 Chandrasekharan, N. and Thompson, L.L. (2016). Increased power to weight ratio of piezoelectric energy harvesters through integration of cellular honeycomb structures. *Smart Mater. Struct.*, 25. https://doi.org/10.1088/0964-1726/25/4/045019.

59 Eghbali, P., Younesian, D., and Farhangdoust, S. (2020). Enhancement of piezoelectric vibration energy harvesting with auxetic boosters. *Int. J. Energy Res.*, 44, 1179–90. https://doi.org/10.1002/er.5010.

60 Eghbali, P., Younesian, D., and Farhangdoust, S. (2020). Enhancement of the low-frequency acoustic energy harvesting with auxetic resonators. *Appl. Energy*, 270, 115217. https://doi.org/10.1016/j.apenergy.2020.115217.

61 Eghbali, P., Younesian, D., Moayedizadeh, A., and Ranjbar, M. (2020). Study in circular auxetic structures for efficiency enhancement in piezoelectric vibration energy harvesting. *Sci. Rep.*, 10, 1–11. https://doi.org/10.1038/s41598-020-73425-1.

62 Chen, B., Jia, Y., Narita, F., Wang, C., and Shi, Y. (2022). Multifunctional cellular sandwich structures with optimised core topologies for improved mechanical properties and energy harvesting performance. *Compos. Part B Eng.*, 238, 109899. https://doi.org/10.1016/j.compositesb.2022.109899.

63 Alsaadi, A., Shi, Y., Pan, L., Tao, J., and Jia, Y. (2019). Vibration energy harvesting of multifunctional carbon fibre composite laminate structures. *Compos. Sci. Technol.*, 178, 1–10. https://doi.org/10.1016/j.compscitech.2019.04.020.

64 Jia, Y., Wei, X., Xu, L., Wang, C., Lian, P., Xue, S., Al-Saadi, A., and Shi, Y. (2019). Multiphysics vibration FE model of piezoelectric macro fibre composite on carbon fibre composite structures. *Compos. Part B Eng.*, 161, 376–85. https://doi.org/10.1016/j.compositesb.2018.12.081.

65 Texas Instruments. (2019). HDC2010 low-power humidity and temperature digital sensors.

66 Torri, A., Foken, T., and Bange, J. (2021). Pressure sensors. In: Foken, T. (ed.) Springer Handbook of Atmospheric Measurements. Springer Handbooks, 273–96. https://doi.org/10.1007/978-3-030-52171-4_10.

67 Wen, T., Ratner, A., Jia, Y., and Shi, Y. (2021). Parametric study of environmental conditions on the energy harvesting efficiency for the multifunctional composite structures. *Compos. Struct.*, 255, 112979. https://doi.org/10.1016/j.compstruct.2020.112979.

68 Wolf, R.A. and Trolier-McKinstry, S. (2004). Temperature dependence of the piezoelectric response in lead zirconate titanate films. *J. Appl. Phys.*, 95, 1397–406. https://doi.org/10.1063/1.1636530.

69 Płaczek, M. and Kokot, G. (2019). Modelling and laboratory tests of the temperature influence on the efficiency of the energy harvesting system based on MFC piezoelectric transducers. *Sensors (Switzerland),* 19. https://doi.org/10.3390/s19071558.

70 Kendon, M., McCarthy, M., Jevrejeva, S., Matthews, A., Sparks, T., and Garforth, J. (2020). State of the UK Climate 2019. *Int. J. Climatol.*, 40, 1–69. https://doi.org/10.1002/joc.6726.

71 Lipscomb, I.P., Weaver, P.M., Swingler, J., and McBride, J.W. (2009). The effect of relative humidity, temperature and electrical field on leakage currents in piezo-ceramic actuators under dc bias. *Sens.ors Actuators, A Phys.*, 151, 179–86. https://doi.org/10.1016/j.sna.2009.01.017.

72 Staros, A. (1957). The SACH (solid-ankle cushion-heel) foot. *Orthop. Prosthet. Appl. J.*, 11, 23–31.

73 Pailler, D., Sautreuil, P., Piera, J.B., Genty, M., and Goujon, H. (2004). Evolution in prostheses for sprinters with lower-limb amputation. *Ann. Readapt. Med. Phys.*, 47, 374–81. https://doi.org/10.1016/j.annrmp.2004.05.007.

74 Chiriac, O.A. and Bucur, D. (2020). From conventional prosthetic feet to bionic feet. A review. *Lect. Notes Networks Syst.*, 143, 130–8. https://doi.org/10.1007/978-3-030-53973-3_14.

75 Scholz, M.S., Blanchfield, J.P., Bloom, L.D., Coburn, B.H., Elkington, M., Fuller, J.D., Gilbert, M.E., Muflahi, S.A., Pernice, M.F., Rae, S.I., Trevarthen, J.A., White, S.C., Weaver, P.M., and Bond, I.P. (2011). The use of composite materials in modern orthopaedic medicine and prosthetic devices: a review. *Compos. Sci. Technol.*, 71, 1791–803. https://doi.org/10.1016/j.compscitech.2011.08.017.

76 Liu, M., Datseris, P., and Huang, H. (2012). Prototype for smart prosthetic legs-analysis and mechanical design. *Adv. Mater. Res.*, 403–408, 1999–2006. https://doi.org/10.4028/www.scientific.net/AMR.403-408.1999.

77 Tucker, M.R. and Fite, K.B. (2010). Mechanical damping with electrical regeneration for a powered transfemoral prosthesis. In *IEEE/ASME International Conference on Advanced Intelligent Mechatronics,* 13–8. https://doi.org/10.1109/AIM.2010.5695828.

78 González, J.L., Rubio, A., and Moll, F. (2002)., Human powered piezoelectric batteries to supply power to wearable electronic. *Int. J. Soc. Mater. Eng. Res.*, 10, 34–40.

79 Selles, R.W., Bussmann, J.B.J., Wagenaar, R.C., and Stam, H.J. (1999). Effects of prosthetic mass and mass distribution on kinematics and energetics of prosthetic gait: a systematic review. *Arch. Phys. Med. Rehabil.*, 80, 1593–9. https://doi.org/10.1016/S0003-9993(99)90336-2.

80 Jia, Y., Wei, X., Pu, J., Xie, P., Wen, T., Wang, C., Lian, P., Xue, S., and Shi, Y. (2019). A numerical feasibility study of kinetic energy harvesting from lower limb prosthetics. *Energies*, 12, 1–17. https://doi.org/10.3390/en12203824.

81 Pu, J., Shi, Y., and Jia, Y. (2020). Energy harvesting from kinetics of prosthetic leg 2020. In *19th International Conference on Micro and Nanotechnology for Power Generation and Energy Conversion Applications (PowerMEMS)*, 1–6. https://doi.org/10.1109/powermems49317.2019.20515807357.

82 Jose, S. (2014). MPU-9250 product specification, 1.

83 STMicroelectronics. (2017). STM32L4R5xx Ultra-low-power Arm® Cortex®-M4 32-bit MCU+FPU, 150DMIPS,up to 2MB Flash, 640KB SRAM, LCD-TFT & MIPI DSI, ext. SMPS 2017.

84 Wang, Y., Soutis, C., Ando, D., Sutou, Y., & Narita, F. (2022). Application of deep neural network learning in composites design. European Journal of Materials, 2(1), 117–170. https://doi.org/10.1080/26889277.2022.2053302.

4

High-performance, Flexible Energy-harvesting Nanogenerators and Sensors for Revolutionizing Healthcare

Ponnusamy Dhivya[1], *Keerthnasre Dhandapani*[2, 3], *Mani Ganesh Kumar*[4, 5] *Kazuyoshi Tsuchiya*[4] *and Ando Yutaka*[6]

[1] *Education Center for Emerging Energy Technology on Climate Change, Chungnam National University, Daejeon, Republic of Korea*
[2] *Department of IT–Energy Convergence (BK21 PLUS), Korea National University of Transportation, Chungju, Republic of Korea*
[3] *Department of Material Science and Engineering, Korea National University of Transportation, Chungju, Republic of Korea*
[4] *Micro/Nano Technology Center, Tokai University, Shonan Campus, Hiratsuka, Kanagawa, Japan*
[5] *Department of Physics, Korea Advanced Institute of Science and Technology (KAIST), Daejeon, Republic of Korea*
[6] *Department of Mechanical Engineering, Tokai University, Shonan Campus, Hiratsuka, Kanagawa, Japan*

4.1 Overview of Flexible Energy-harvesting Devices

The healthcare sector is experiencing a significant transformation, propelled by the emergence of flexible energy-harvesting devices that provide real-time health tracking, diagnoses, and therapeutic treatments [1, 2]. These devices are crucial for delivering tailored medical care, enhancing patient outcomes, and managing diverse persistent medical conditions. Nonetheless, a notable constraint of contemporary devices is dependence on traditional power sources, including primary or rechargeable batteries. Although batteries have historically been used to power these gadgets, there are several issues with their limited longevity, size restrictions, and requirement for frequent substitution or recharging [3, 4]. These difficulties include growing discomfort among patients, the possibility of problems following replacement surgery, and restrictions on the usefulness and structure of the device. To get over these obstacles, the creation of flexible energy-harvesting systems has become a viable solution [5, 6].

These devices can absorb and transform thermal, mechanical, or electromagnetic energy from the environment into electrical power, which could lead to the development of self-sustaining systems. These implanted energy-harvesting systems combine microelectronic parts for interaction, data collection and storage, sensors, actuators, and energy sources. They make it possible to perform a variety of tasks, such as nerve stimulation, electromechanical actuation, and constant tracking, which is causing a revolution in the way healthcare is provided globally.

The inclusion of these systems has resulted in substantial advantages, such as early identification of illnesses and diagnosis, focused therapeutic interventions for efficient management of chronic illnesses [7], lifesaving procedures [8], tailored rehabilitation [9, 10], and interconnected health [11], among others. These implanted energy-harvesting technologies not only overcome the constraints of conventional power sources but also provide the potential to extract energy directly from the human body, facilitating a renewable method for supplying medical devices.

4.1.1 Energy Harvesting from Human Body and Power Demands for Healthcare Devices

The extraction of energy from the human body is a viable solution to meet the power requirements of contemporary healthcare devices, facilitating uninterrupted functioning independent of conventional power sources. The human system produces multiple energy types, including mechanical (from actions like walking or cardiac contractions), thermal (from body temperature), and bioelectrical energy [12–14].

Modern harvesting systems can be transformed into various consumable electric energy using sophisticated harvesting methods, removing the necessity for regular battery replacements or other power supplies. Piezoelectric materials may produce electricity from mechanical strain, whereas triboelectric nanogenerators harness friction between surfaces to generate energy [15, 16]. Moreover, thermoelectric materials transform temperature differentials into electrical energy, and pyroelectric materials generate electricity from dynamic temperature changes [17–19]. These technologies, incorporated into wearable and implantable devices, provide an economical and patient-focused solution by obviating the necessity for regular battery changes or external recharging, hence ensuring continuous performance. Healthcare supplies, including wearable biosensors, fitness trackers, and implanted medical devices such as pacemakers or neural stimulators, exhibit diverse power requirements based on their level of complication and performance [20]. Wearable devices often require minimal power for the ongoing assessment of metrics such as heart rate, oxygen saturation, or body temperature [21, 22]. Conversely, implanted devices frequently necessitate reliable and sustained power sources to execute essential functions such as nerve stimulation or distributing drugs. Energy extracted from the human body can adequately fulfill these requirements, particularly as improvements in the efficiency of energy conversion and material technology progress. The integration of these devices not only mitigates the operational constraints of conventional power sources but also facilitates the increasing inclination toward individualized and remote healthcare treatments.

4.1.2 Implantable Medical Devices and Their Power Demands

Implantable medical devices (IMDs) are essential to contemporary healthcare, offering remedies for individuals with chronic or life-threatening ailments [23, 24]. These devices are surgically inserted to execute essential activities, such as monitoring vital signs, administering medications, activating nerves, or aiding organ functioning. IMDs must be compact, dependable, and able to function autonomously with minimal intervention, posing distinct issues concerning their power needs. The power requirements of IMDs fluctuate based on their complexity, functionality, and usage duration. Fundamental devices, like pacemakers, necessitate minimal power for ongoing heart rhythm regulation, whereas more sophisticated devices, such as deep brain stimulators or cochlear implants, demand greater power for complex functions like signal transmission or nerve stimulation at precise frequencies and intensities [25].

These devices must function consistently for prolonged durations, frequently spanning several years, and have a durable power source. Previously, batteries have supplied power IMDs; however, their dimensions, short lifespan, and necessity for replacement operations pose considerable obstacles [26]. As batteries diminish, patients may necessitate more procedures for replacement or recharging, which entails both medical hazards and financial expenditures. Researchers are concentrating on the development of energy-harvesting systems that directly absorb energy from the human body to overcome these restrictions. This method provides a renewable energy source, obviating the necessity for regular battery changes or external charging, while accommodating the growing demand for personalized and remote medical treatments.

In this chapter, we explore the developments in flexible energy-harvesting technologies with an emphasis on how they are used in implanted and wearable devices for healthcare. Among them, piezoelectric polymers and flexible pyroelectric materials are particularly noteworthy as breakthrough technologies. Body-generated heat can be efficiently converted into electric current by pyroelectric materials, like barium titanate (BTO)/poly-L-lactic acid (PLLA) composites, providing wearable biosensors with cost-effective and lightweight alternatives. Like this, piezoelectric polymers such as P(VDF-TrFE) show remarkable effectiveness in transforming mechanical energy from bodily motions into electrical energy, which can be used. By addressing significant issues with conventional batteries, including their weight and short lifespan, these technologies open the door for self-powered electronics. By incorporating these energy-harvesting systems, the coming generation of medical devices promises uninterrupted functioning, improved comfort for patients, and major advancements in proactive and customized treatment.

4.2 Flexible Pyroelectric Nanogenerator for Healthcare

4.2.1 Motivation

The growth of cutting-edge technology targeted at enhancing medical outcomes and lowering medical expenses has been encouraged by the growing emphasis on preventive healthcare. Wearable sensors have become a key achievement among these developments, allowing for continuous tracking of vital physiological indicators, including body temperature, physical activity levels, and cardiovascular activity. These sensors have tremendous potential to improve early detection, encourage proactive medical procedures, and enhance patient care.

Despite their many advantages, the use of traditional batteries to power wearable sensors is still a major drawback. Because of their essential size limitations, traditional batteries are heavy, need to be changed frequently, and prevent wearable technology from becoming smaller. This restriction reduces wearable sensors' usefulness and accessibility, resulting in them being impractical for prolonged, continuous usage. Therefore, there is an urgent need for self-sustaining, flexible, and productive alternatives that can produce energy precisely on the surface of the body [27].

Flexible pyroelectric materials provide an innovative remedy to this problem by transforming thermal energy produced by our bodies into electrical current. These materials have the mechanical adaptability needed for smooth integration with wearable technology, in addition to being biodegradable and ultralight. An important advancement in the development of self-powered, adhesive-free devices for wearable applications is the creation of freestanding nanosheets, such as BTO/PLLA composites [28]. This invention represents a significant breakthrough in wearable medical technology and opens the door to more effective, dependable, and environmentally friendly energy sources.

4.2.2 Investigation of the Pyroelectric Coefficient in Fabricated Flexible BTO/PLLA Nanosheets

4.2.2.1 Fabrication of Freestanding BTO/PLLA Nanosheet

A freestanding nanosheet was developed by first applying a sacrificial layer of polyvinyl alcohol (PVA) onto a silica substrate employing spin coating at a speed of 3000 rpm for 30 seconds. A BTO/PLLA mixture was produced with a molar ratio of 1:1 for both BTO and PLLA [29]. Prior to adding BTO into PLLA, a 1 molar ratio of BTO was subjected to ball milling for 60 minutes at a speed of 600 rpm, resulting in fine BTO powder. The average particle size of BTO powder was determined to be 100 nm utilizing dynamic light scattering analysis. The 100 nm BTO powder was mixed with a PLLA mixture in a 1:1 molar ratio and spin-coated at a speed of 3000 rpm for 60 seconds onto the pre-coated PVA sacrificial layer.

The sacrificial layer (PVA) was removed by immersing it in water. The water-soluble PVA layer has been removed, resulting in a freestanding BTO/PLLA nanosheet, and the complete procedure is schematically shown in Figure 4.1. It has been suggested that these BTO/PLLA nanosheets, known for their exceptional flexibility and biocompatibility, may adhere to the target surface only via physical adsorption, therefore eliminating the need for adhesives or reactive functional groups. The resultant freestanding BTO/PLLA sheet was removed from the water using a glass slide and placed in ambient conditions for drying.

4.2.2.2 Results and Discussion

The surface of freestanding PLLA and BTO/PLLA nanosheet was observed under a laser microscope and is shown in Figure 4.2.

The results reveal that the PLLA nanosheet exhibited a highly smooth surface. The incorporation of BTO into the PLLA nanosheet resulted in improved structural

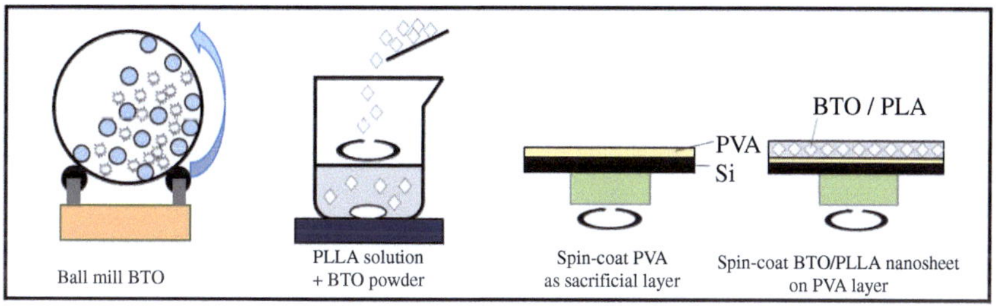

Figure 4.1 Schematic representation of BTO/PLLA nanosheet fabrication.

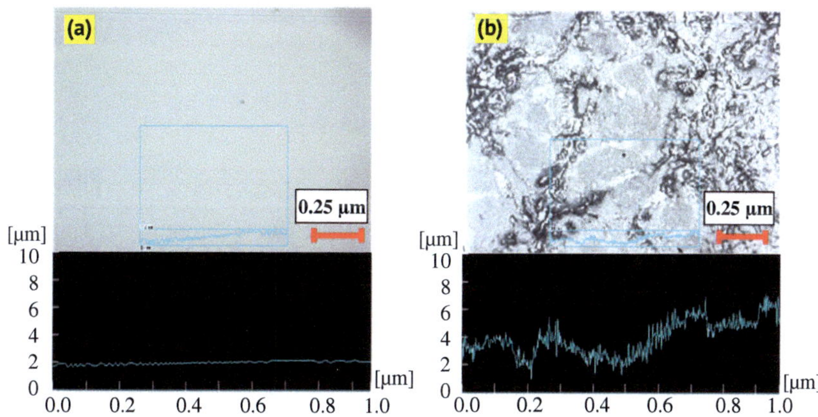

Figure 4.2 Laser microscope image: (a) PLLA nanosheet; (b) BTO/PLLA nanosheet.

features of the grains, with variations in surface texture confirming the uniform distribution of BTO within the sheet. The elemental composition was determined using an energy dispersive X-ray spectrometer, which confirmed the presence of barium, titanium, oxygen, and carbon in the following mass percentages: 23.5%, 5.5%, 6.41%, and 64.59%, respectively.

The structural characteristics of BTO/PLLA nanosheets were examined by X-ray diffraction (XRD), shown in Figure 4.3. The PLLA peaks were seen at lower Bragg's angles with less intensity, indicating the amorphous characteristics of PLLA. The BTO exhibited distinct peaks, indicating its polycrystalline nature. The positions and intensities indicate crystallographic planes, such as (100) and (110), associated to the perovskite structure of BTO [30].

The thickness of the fabricated BTO/PLLA nanosheet was measured using a tabletop stylus profiling device at three distinct locations (upper region, central region, and lower region), with the average of these measurements calculated to get a more accurate thickness value. The PLLA nanosheets exhibit an average thickness of 4.567 μm. The BTO/PLLA nanosheet had an average thickness of 5.115 μm. The incorporation of BTO resulted in an average thickness increase of approximately 0.548 μm relative to the PLLA nanosheet.

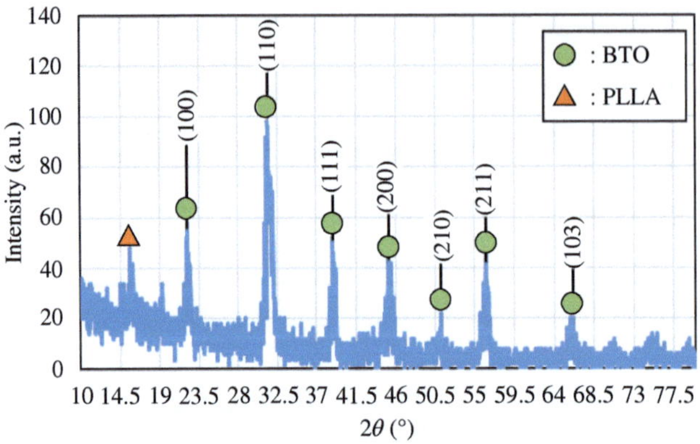

Figure 4.3 XRD pattern of BTO/PLLA nanosheet.

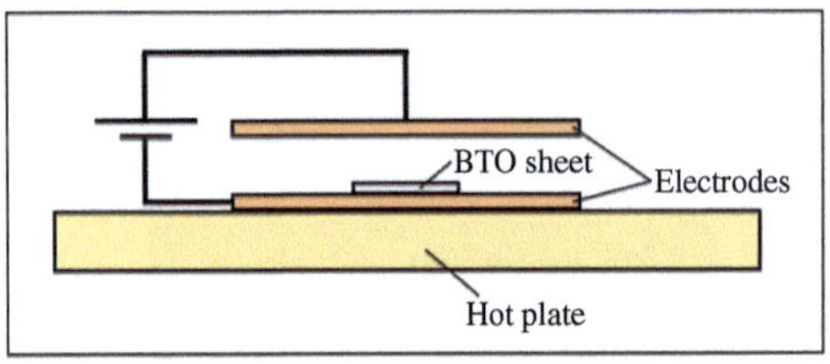

Temperature [°C]	**130**
Electrodes distance [cm]	**2**
Voltage [kV]	**12**
Time [min]	**30**

Figure 4.4 Experimental setup for polarization treatment and its condition.

4.2.2.2.1 Evaluation of Pyroelectric Properties of BTO/PLLA Nanosheet

Before conducting pyroelectric tests, the fabricated BTO/PLLA nanosheet underwent polarization. Figure 4.4 illustrates the arrangement and conditions of the polarization treatment.

BTO/PLLA nanosheets were positioned on the hot plate to increase their temperature to the target of 130 °C. The interelectrode distance was maintained at 2 cm, and a voltage of 12 kV was supplied for polarization treatment. Subsequently, the BTO/PLLA nanosheets were dried at ambient temperature to investigate their pyroelectric properties.

Figure 4.5 shows a schematic representation of the BTO/PLLA sheet for pyroelectric current measurement. The glass substrate is initially covered with a layer of kapton tape, known for its superior insulation and heat resistance. Subsequently, a conductive copper (Cu) tape is applied as the lower electrode, measuring 1 mm square, onto which the BTO/PLLA sheet, is affixed. Then, conductive Cu tape is affixed as an electrode to the upper section of the sheet, and the upper and lower Cu tapes are linked to the conductors. Currently, kapton tape is employed to secure the sheet and electrodes. The electrodes were connected to digital multimeter to assess the voltage variation on the sheet with respect to temperature change. Four different temperatures (40 °C, 50 °C, 60 °C, and 70 °C) were taken to study the pyroelectric current on the developed BTO/PLLA nanosheet. The measurement time is set at 120 seconds, comprising the following sequence: wait for 30 seconds, commence

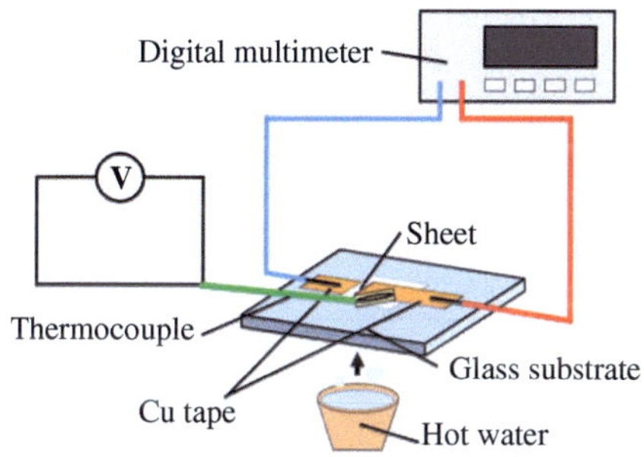

Figure 4.5 Experimental setup for measurement of pyroelectric current.

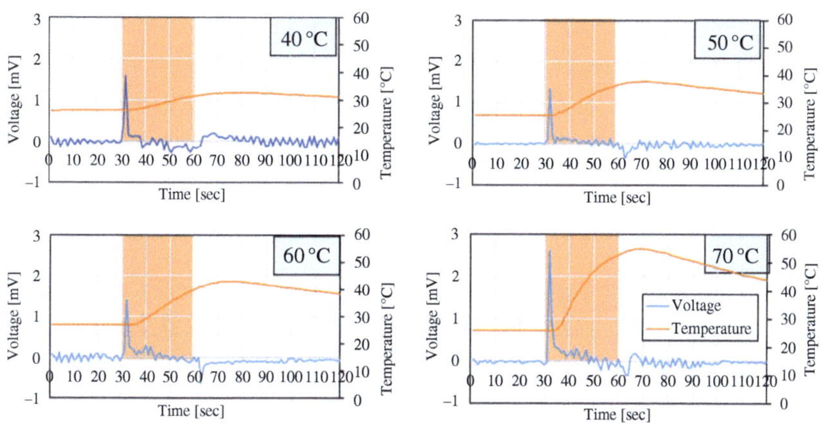

Figure 4.6 Results of pyroelectric measurements under different temperatures.

measurement, heat for 30 seconds, keep 60 seconds standby, and conclude the measurement. This study utilized four distinct temperatures of hot water (40 °C, 50 °C, 60 °C, and 70 °C) to assess the variation in voltage response attributable to the temperature gradient and the resultant pyroelectric coefficient.

From the results (Figure 4.6) the increase in the voltage was confirmed when the BTO/PLLA nanosheet was heated. From these results the pyroelectric coefficient [31] was calculated using Eq. (4.1). The conductance and sheet resistance of BTO/PLLA nanosheets were found to be 0.02 nS and 5G Ω, respectively, which were used for the calculation.

$$i_p = Ap\left(\frac{dT}{dt}\right)[A] \rightarrow p = \frac{i_p A}{(dT/dt)} = \frac{V_p A}{(dT/dt)\,R}\,[C/(m^2 \cdot K)]$$

where i_p is the pyroelectric current (A), A is the electrode area (m^2), p is the pyroelectric coefficient ($C/(m^2 \cdot K)$), T is the temperature (K), t is the time (sec), V_p is the pyroelectric voltage (V), and R is the resistance (Ω).

The pyroelectric coefficient values of the BTO/PLLA sheet with respect to different temperatures, 40 °C, 50 °C, 60 °C, and 70 °C, are shown in Table 4.1.

Table 4.1 The pyroelectric coefficient values of the BTO/PLLA sheet with respect to different temperatures: 40 °C, 50 °C, 60 °C, and 70 °C.

Temperature (°C)	40	50	60	70	Average
Pyroelectric coefficient [nC/(cm^2·K)]	0.00811	0.00696	0.00784	0.00608	0.00725

The pyroelectric coefficient of the developed sheets is 0.00734 nC/(cm^2·K), indicating a significant decrease from the literature for pure BTO thin films [32]. This results from the increased resistivity of the PLLA nanosheet, approximately 6.0×10^{12} Ω·cm. The pyroelectric coefficient is known to decrease with an increase in resistance. The quantity of PLLA in the sheet after the spin coating is considerably higher than that of BTO. In addition, it is probable that the sheet resistance has significantly increased, providing an additional reason for a decrease in the pyroelectric coefficient. The forthcoming research will focus on investigating the impact of altering the ratio of BTO to PLLA in the composite. A thorough investigation might determine an ideal composition that integrates pyroelectric performance and mechanical flexibility while maintaining a moderate resistance.

4.3 Flexible Piezoelectric Nanogenerator for Healthcare

4.3.1 Motivation

In recent years, there has been an important rise in interest in preventive medicine, motivated by increased understanding of its effect on improving health outcomes and reducing medical costs. A significant advancement in this field is the development of wearable sensors, which offer continuous monitoring of multiple physiological parameters, such as heart rate, body temperature, and activity levels.

However, wearable sensors experience a considerable challenge: the need for a reliable electrical source. Conventional batteries, typically utilized as external power sources, require regular replacement, making them inconvenient and unsuitable for continuous usage. The design of these batteries limits their reduction in size beyond a certain limit, hence limiting the downsizing of wearable devices.

Consequently, there is an immediate necessity for a device that is both lightweight and capable of generating energy on the body's surface. This device could eliminate dependence on traditional batteries, hence improving the functionality and ease of wearable sensors. This significant development is essential for advancing wearable technology and enhancing the affordability and efficiency of preventive healthcare.

To address the challenges of powering wearable sensors, poly(vinylidene fluoride-trifluoroethylene) P(VDF-TrFE) is used in wearable sensors due to its high piezoelectricity, which allows it to efficiently convert mechanical energy from body movements into electrical energy [33]. This property makes it ideal for self-powered devices, eliminating the need for traditional batteries and enabling continuous operation. Additionally, P(VDF-TrFE) (PVDF) is flexible and lightweight, making it suitable for integration into wearable technology.

4.3.2 Development of Wearable Nano Power Generators Using Nonwoven PVDF Fibers

4.3.2.1 Fabrication of PVDF Nanofibers: Electrospinning

PVDF nanofiber polymer material can be processed into nonwoven fabric using the electrospinning method [34]. The solution is prepared by dissolving 15 wt% of PVDF in acetone and stirring for one hour at a speed of 600 rpm at room temperature. The solution is transferred into a 10 ml syringe. The injection volume of the syringe pump is 100 μl / min. The distance from the needle tip to the collector was fixed at 100 mm, and the applied voltage was established at 12 kV. The electrospinning parameters were tuned to achieve continuous, defect-free, and uniform nonwoven PVDF nanofibers.

4.3.2.2 Results and Discussion

The surface morphology of electrospun PVDF nonwoven fibers is shown in Figure 4.7. The SEM micrograph confirms the formation of uniform, continuous electrospun nanofibers characterized by smooth surfaces and consistent diameters. The absence of bead defects results in confirming the enhanced quality of electrospun nonwoven nanofiber. The fibers are arranged randomly, creating a dense, interconnected network.

4.3.2.2.1 Piezoelectric Measurement on PVDF Nonwoven Fibers

A flexible PVDF nonwoven fiber piezoelectric sensor was constructed onto an ITO/PET substrate. A copper electrode was affixed to the top of the PET substrate, acting as an electrode. Nonwoven fibers made of PVDF were positioned over the Cu electrode and coated with aluminum foil, which serves as an additional electrode. The entire apparatus was encased in high resistance kapton tape to enhance its reliability. It is shown in Figure 4.8. Experiments are carried out to validate the piezoelectric properties of PVDF nonwoven fabrics. The piezoelectric measurement was conducted by applying displacement in the in-plane direction using fingers. The appropriate voltage was measured using a nanovoltmeter (Fluke 8846A). The 10 000 plots were collected with 0.02 PLC and measured for a duration of 10 seconds.

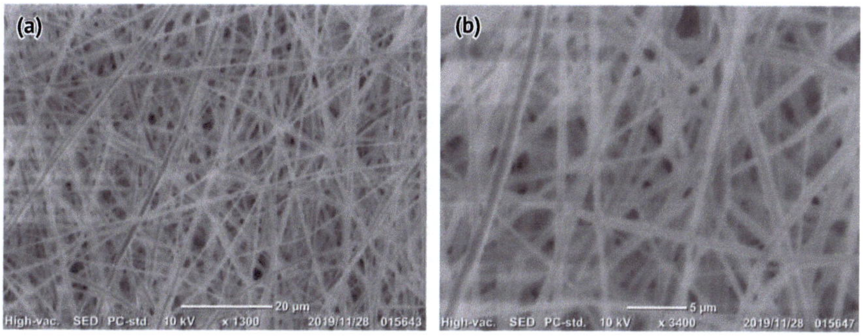

Figure 4.7 Surface morphology of (a) low and (b) high magnification of electrospun PVDF nonwoven fibers.

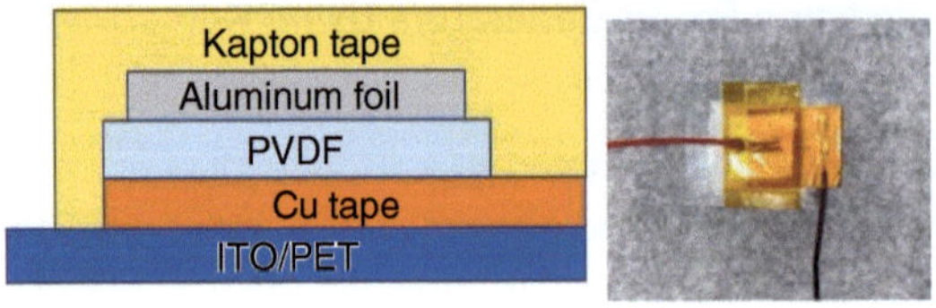

Figure 4.8 Schematic and photographic representations of a flexible piezoelectric sensor.

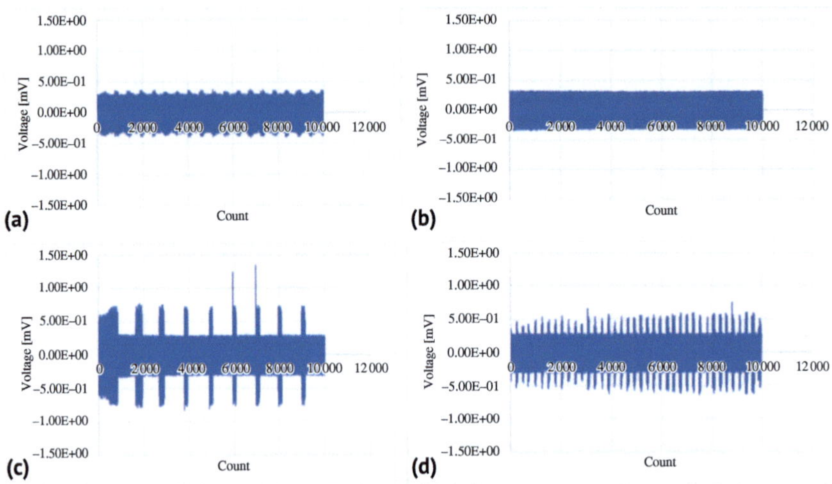

Figure 4.9 (a) Noise, (b) steady state, (c) tapping one second, and (d) tapping five times per second.

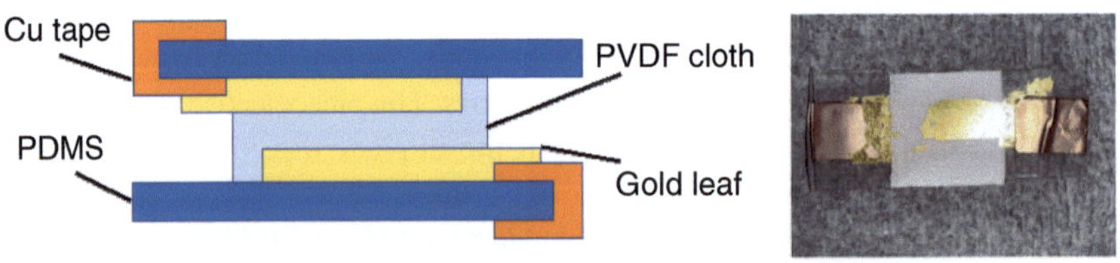

Figure 4.10 Schematic and photographic representations of a piezoelectric sensor modified with gold leaf.

The sample was tapped with a finger at two different speeds: about once per second and five times per second, while the corresponding voltage was recorded.

Initially, noise in the measurement was noticed, and once the noise was stabilized, the piezoelectric studies were carried out as shown in Figure 4.9a–d.

By studying piezoelectric current, the nonwoven fabric must have adequate flexibility and must not hinder its deformation. Furthermore, given that the output of PVDF nonwoven fabric is as minimal as millivolts, low resistance is advantageous. We evaluated the use of silver ink and gold leaf materials for printing.

4.3.2.2.2 Piezoelectric Measurement on PVDF Nonwoven Fibers with Gold Leaf

The previous set up was modified with gold leaf to increase the conductivity of the piezoelectric sensor to get enhanced performance (Figure 4.10). Gold leaf acts

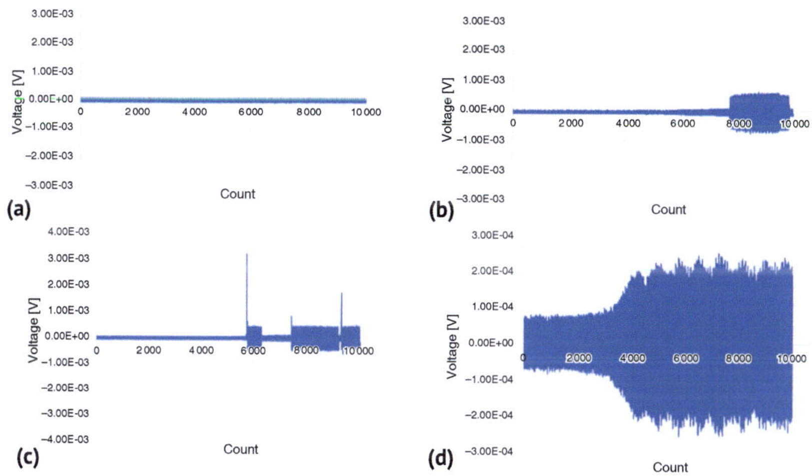

Figure 4.11 Measurement of voltage in (a) steady state and (b) one, (c) two, and (d) eight times of stretching the gold leaf-modified structure.

as a flexible and low-resistance electrode in the measurement of the piezoelectric effect on in-plane deformation. The resistance and average thickness were found to be 2.05 Ωm and 0.1–0.2 μm, respectively. A polydimethylsiloxane (PDMS) serves as the substrate, due to its lack of ability to directly attach gold leaf to nonwoven fabric; samples were created utilizing PDMS with an approximate thickness of 0.5 mm as a substrate with Cu tape affixed at the terminus to secure the electrode. Gold leaf is cut into sections measuring 10 mm in width and 25 mm in length to ensure conductivity with the copper tape and then affixed to the PDMS. Upon validating the conductivity of the gold leaf and Cu tape, they are assembled by interleaving 20 mm squares of PVDF nonwoven fabric. The voltage is measured when the displacement is applied stretching one, two, and eight times within a duration of 10 seconds, as seen in Figure 4.11a–d. The results confirmed that the voltage created from the PVDF nonwoven fabric can be measured using a gold foil electrode, as the voltage stabilizes after generation. It was confirmed that the measured gold leaf was not conductive.

The voltage measurement confirms that gold foil electrodes are effective for piezoelectric measurement of in-plane deformation. It is necessary to process the gold leaf into a shape that allows it to stretch easily against displacement in the in-plane direction. It is important to investigate how the gold leaf incision influences its efficacy in facilitating easy stretching against displacement in the in-plane direction.

4.3.2.2.3 Piezoelectric Measurement Using Incision Gold Leaf Electrodes

This section emphasizes the fabrication of gold foil electrodes, in which the gold leaf is easily deformed due to in-plane displacement. The PVDF nonwoven fiber is positioned between the two gold leaf electrodes. PDMS is applied to the upper and lower surfaces of the leaf electrodes to provide adequate sealing as shown in Figure 4.12. Piezoelectric measurements are conducted on in-plane deformation with processed gold foil electrodes to validate the efficacy of the incision shape.

The PVDF piezoelectric device, modified with an incised gold leaf electrode, was connected to a nanovoltmeter (Fluke 8846A) for measurement. The sample was

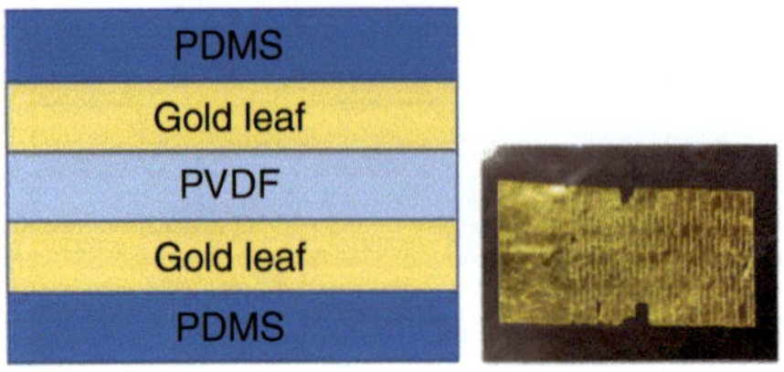

Figure 4.12 Incision gold leaf electrode and device structure.

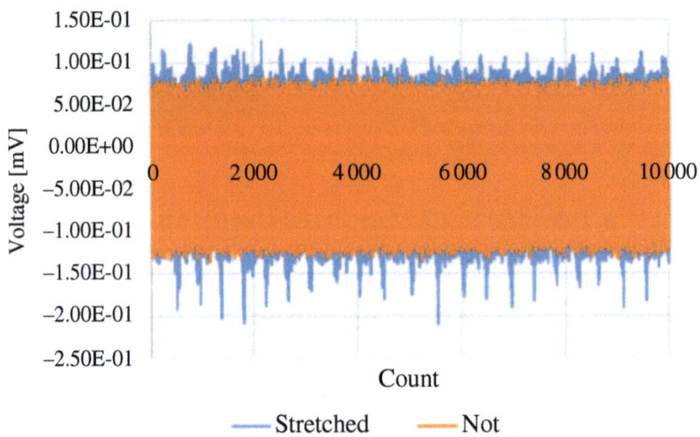

Figure 4.13 Piezoelectric measurement on a device modified with an incised gold leaf electrode.

elongated approximately 5 mm using a finger at one-second intervals and thereafter measured. The prepared sample underwent piezoelectric measurement. The measurement was conducted at a constant value by stretching with a finger approximately once per second, and the resulting voltage was recorded. The measurement results are presented in Figure 4.13. A voltage above the steady value is produced at a specific interval. Nevertheless, as the electromotive force produced by finger pressure is also measured, it is essential to assess the electromotive force in the pure in-plane direction. Furthermore, it was confirmed that the gold leaf electrode exhibited conductivity throughout a broad region, in contrast to its condition prior to preparation. It is necessary to measure the electromotive force due to expansion and contraction only.

4.3.2.2.4 Measurement Using a Telescopic Device

The telescopic device is designed to perform piezoelectric measurements of constant periodic in-plane displacement using only expansion and contraction. The telescopic device was developed using 3D printing. The piezoelectric measurement was conducted on the nonwoven fabric by moving it in the in-plane direction over a specified duration. The displacement was established at 5 mm with an interval of 1 Hz, and a servo motor (SG-90) was used, using a scotch yoke mechanism. The objective of this study is to create a device that creates electricity on the body's surface, with

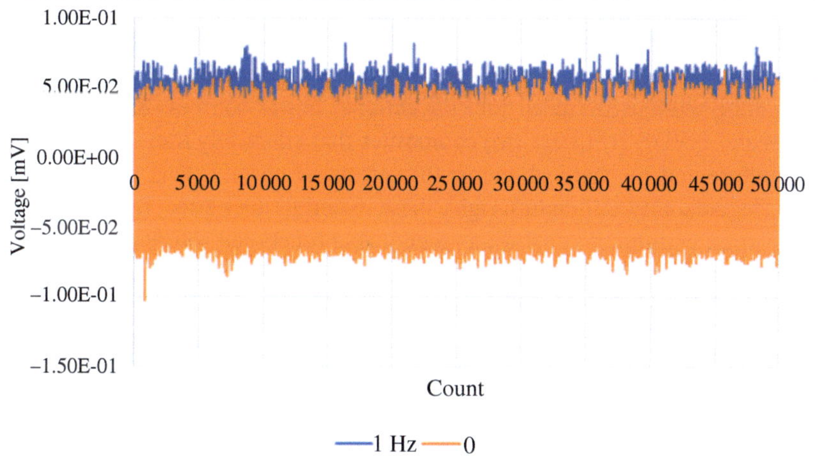

Figure 4.14 Piezoelectric measurement on a telescopic device.

the target value being human walking, and the average duration is 0.853 ± 0.14 seconds, around 1 second [35].

The measurement results indicate that the piezoelectric action is confirmed, although sufficient peaks have not been confirmed, as shown in Figure 4.14. This is believed to result from the press of the gold foil electrode and the PVDF nonwoven fabric with fingers carried out in previous results, shown in Figures 4.9 and 4.11, leading to insufficient contact between the PVDF nonwoven fabric and the electrode during displacement. The nonwoven fabric is believed to have been unable to return to its previous shape after shrinking, resulting in a reduction in the negative output. Upon the application of a stretching device just to the PVDF nonwoven fabric, it was observed that the cloth elongated by 1 mm. This can be attributed to the PVDF nonwoven fabric being bonded with threads, resulting in numerous holes inside the fabric. Therefore, it is crucial to identify a technique for attaching gold leaf to the PVDF nonwoven fabric and establish a measurement mechanism.

The resulting findings were derived from the production of nonwoven fibers utilized in wearable power generators by the electrospinning technique and piezoelectric evaluations. The findings of piezoelectric tests with incised gold leaf demonstrated its efficacy for in-plane directional displacement. It was confirmed that when the nonwoven fabric is elongated and contracted, it stretches, resulting in a positive bias in the output electromotive force. It is essential to identify an approach to establish adhesion between the nonwoven fabric and PDMS by measurements obtained with a telescopic device.

4.4 Summary and Outlook

The incorporation of flexible energy-harvesting devices has resulted in notable progress in the development of sustainable technologies for medicine, especially in healthcare applications. Batteries and other conventional power sources are limited in size, longevity, and frequency of replacement. This has prompted the creation of energy-harvesting systems, which use materials like piezoelectric and pyroelectric technologies to capture energy from human beings and enable uninterrupted device functioning without the need for conventional power sources. BTO/PLLA composites and P(VDF-TrFE) are two composites that effectively transform the body's temperature and mechanical energy into electrical current. Due to their flexibility, biodegradability, and light weight, these materials are perfect for incorporation into medical equipment, providing an enduring functional solution that improves patient comfort and replaces traditional batteries. Independent, adaptable energy-harvesting devices promise to revolutionize healthcare in the coming years by facilitating preventive health control, real-time ongoing illness monitoring, and individualized therapy. These devices get rid of the necessity of manual recharging and intrusive battery changes. Likewise, as material science and energy conversion technologies continue to progress, these devices will transform healthcare by enhancing patient care, saving expenses, and offering effective, environmentally friendly medical care. To make these technologies commercially feasible, research toward improving material effectiveness, resilience, and adaptability will be essential.

References

1 Shuvo, M.M.H., Titirsha, T., Amin, N. and Islam, S.K. (2022) Energy harvesting in implantable and wearable medical devices for enduring precision healthcare. *Energies*, 15(20), 7495.

2 Gao, Z., Zhou, Y., Zhang, J., Foroughi, J., Peng, S., Baughman, R.H., Wang, Z.L. and Wang, C.H. (2024) Advanced energy harvesters and energy storage for powering wearable and implantable medical devices. *Adv. Mater.*, 36(42), 2404492.

3 Goodenough, J.B. (2013) Evolution of strategies for modern rechargeable batteries. *Acc. Chem. Res.*, 46(5), 1053–61.

4 Liang, Y., Zhao, C.Z., Yuan, H., Chen, Y., Zhang, W., Huang, J.Q., Yu, D., Liu, Y., Titirici, M.M., Chueh, Y.L., Yu, H. and Zhang, Q. (2019) A review of rechargeable batteries for portable electronic devices. *InfoMat.*, 1(1), 6–32.

5 Fan, F.R., Tang, W. and Wang, Z.L. (2016) Flexible nanogenerators for energy harvesting and self-powered electronics. *Adv. Mater.*, 28(22), 4283–305.

6 Khan, A.S. and Khan, F.U. (2022) A survey of wearable energy harvesting systems. *Int. J. Energy Res.*, 46(3), 2277–329.

7 Pons-Faudoa, F.P., Ballerini, A., Sakamoto, J. and Grattoni, A. (2019) Advanced implantable drug delivery technologies: transforming the clinical landscape of therapeutics for chronic diseases. *Biomed. Microdevices*, 21(2), 1–22.

8 Wang, S., Cui, Q., Abiri, P., Roustaei, M., Zhu, E., Li, Y.R., Wang, K., Duarte, S., Yang, L., Ebrahimi, R., Bersohn, M., Chen, J. and Hsiai, T.K. (2023) A self-assembled implantable microtubular pacemaker for wireless cardiac electrotherapy. *Sci. Adv.*, 9(42), eadj0540. https://doi.org/10.1126/sciadv.adj0540.

9 Li, P., Lee, G.H., Kim, S.Y., Kwon, S.Y., Kim, H.R. and Park, S. (2021) From diagnosis to treatment: recent advances in patient-friendly biosensors and implantable devices. *ACS Nano*, 15(2), 1960–2004.

10 Zhang, J., Das, R., Zhao, J., Mirzai, N., Mercer, J. and Heidari, H. (2022) Battery-free and wireless technologies for cardiovascular implantable medical devices. *Adv. Mater. Technol.*, 7(6), 2101086.

11 Loncar-Turukalo, T., Zdravevski, E., da Silva, J.M., Chouvarda, I. and Trajkovik, V. (2019) Literature on wearable technology for connected health: Scoping review of research trends, advances, and barriers. *J. Med. Internet Res.*, 21(9), e14017.

12 Dieffenderfer, J., Goodell, H., Mills, S., McKnight, M., Yao, S., Lin, F., Beppler, E., Bent, B., Lee, B., Misra, V., Zhu, Y., Oralkan, O., Strohmaier, J., Muth, J., Peden, D. and Bozkurt, A. (2016) Low-power wearable systems for continuous monitoring of environment and health for chronic respiratory disease. *IEEE J. Biomed. Health Inform.*, 20(5), 1251–1264.

13 Kwon, S.H. and Dong, L. (2022) Flexible sensors and machine learning for heart monitoring. *Nano Energy*, 102, 107632.

14 Al-Nabulsi, J.I., Al-Doori, H.A. and Salawy, N.T. (2017) Human motion to recharge implantable devices. *2017 10th Jordan International Electrical and Electronics Engineering Conference, JIEEEC 2017.*

15 Zheng, Q., Shi, B., Li, Z. and Wang, Z.L. (2017) Recent progress on piezoelectric and triboelectric energy harvesters in biomedical systems. *Adv. Sci.*, 4(7), 1700029.

16 Han, M., Wang, H., Yang, Y., Liang, C., Bai, W., Yan, Z., Li, H., Xue, Y., Wang, X., Akar, B., Zhao, H., Luan, H., Lim, J., Kandela, I., Ameer, G.A., Zhang, Y., Huang, Y. and Rogers, J.A. (2019) Three-dimensional piezoelectric polymer microsystems for vibrational energy harvesting, robotic interfaces and biomedical implants. *Nat. Electron.*, 2(1), 26–35.

17 Zhang, D., Wu, H., Bowen, C.R. and Yang, Y. (2021) Recent advances in pyroelectric materials and applications. *Small*, 17(51), 2103960.

18 Nozariasbmarz, A., Collins, H., Dsouza, K., Polash, M.H., Hosseini, M., Hyland, M., Liu, J., Malhotra, A., Ortiz, F.M., Mohaddes, F., Ramesh, V.P., Sargolzaeiaval, Y., Snouwaert, N., Özturk, M.C. and Vashaee, D. (2020) Review of wearable thermoelectric energy harvesting: From body temperature to electronic systems. *Appl. Energy*, 258, 114069.

19 Korkmaz, S. and Kariper, A. (2021) Pyroelectric nanogenerators (PyNGs) in converting thermal energy into electrical energy: Fundamentals and current status. *Nano Energy*, 84, 105888.

20 Khan, Y., Ostfeld, A.E., Lochner, C.M., Pierre, A. and Arias, A.C. (2016) Monitoring of vital signs with flexible and wearable medical devices. *Adv. Mater.*, 28(22), 4373–95.

21 Iqbal, S.M.A., Mahgoub, I., Du, E., Leavitt, M.A. and Asghar, W. (2021) Advances in healthcare wearable devices. *npj Flex. Electron.*, 5(1), 1–14.

22 Hong, Y.J., Jeong, H., Cho, K.W., Lu, N. and Kim, D.H. (2019) Wearable and implantable devices for cardiovascular healthcare: from monitoring to therapy based on flexible and stretchable electronics. *Adv. Funct. Mater.*, 29(19), 1808247.

23 Joung, Y.H. (2013) Development of implantable medical devices: from an engineering perspective. *Int. Neurourol. J.*, 17(3), 98.

24 Cao, H., Li, A.L., Nguyen, C.M., Peng, Y.B. and Chiao, J.C. (2012) An integrated flexible implantable micro-probe for sensing neurotransmitters. *IEEE Sens. J.*, 12(5), 1618–24.

25 Wood, M.A. and Ellenbogen, K.A. (2002) Cardiac pacemakers from the patient's perspective. *Circulation*, 105(18), 2136–8.

26 Wei, X. and Liu, J. (2008) Power sources and electrical recharging strategies for implantable medical devices. *Front. Energy Power Eng. China*, 2(1), 1–13.

27 Mani, G.K., Nimura, Y. and Tsuchiya, K. (2020) Advanced artificial electronic skin based pH sensing system for heatstroke detection. *ACS Sens.*, 5(4), 911–6.

28 Schönlein, R., Larrañaga, X., Panfilo, A., Li, Y., Larrañaga, A., Liu, G., Müller, A.J., Aguirresarobe, R. and Ugartemendia, J.M. (2025) Enhanced piezoelectric properties of poly(l-lactide) nanocomposite microfiber scaffolds due to polydopamine-coating of barium titanate nanoparticles. *Adv. Mater. Interfaces,* 12(1), 2400546. https://doi.org/10.1002/admi.202400546.

29 Okamura, Y., Kabata, K., Kinoshita, M., Saitoh, D. and Takeoka, S. (2009) Free-standing biodegradable poly(lactic acid) nanosheet for sealing operations in surgery. *Adv. Mater.*, 21(43), 4388–92.

30 Biglar, M., Gromada, M., Stachowicz, F. and Trzepieciński, T. (2017) Synthesis of barium titanate piezoelectric ceramics for multilayer actuators (MLAs). *Acta Mech. Autom.*, 11(4), 275–9.

31 Jachalke, S., Mehner, E., Stöcker, H., Hanzig, J., Sonntag, M., Weigel, T., Leisegang, T. and Meyer, D.C. (2017) How to measure the pyroelectric coefficient? *Appl. Phys. Rev.*, 4(2), 021303. https://doi.org/10.1063/1.4983118.

32 Puli, V.S., Li, P., Adireddy, S. and Chrisey, D.B. (2015) Crystal structure, dielectric, ferroelectric and energy storage properties of La-doped $BaTiO_3$ semiconducting ceramics. *J. Adv. Dielectr.*, 05(03), 1550027.

33 Kumakiri, Y., Mani, G.K. and Tsuchiy, K. (2018) Development of Pasted Sheet Sensor Using P(VDF-TrFE). *2018 International Symposium on Micro-NanoMechatronics and Human Science (MHS)*, 1–4.

34 Noyori, M., Neo, Y. and Mimura, H. (2015) Single-crystalline poly(vinylidene fluoride–trifluoroethylene) nanofiber webs fabricated by electrospinning. *Jpn J. Appl. Phys.*, 54(2), 021601.

35 Wanezaki, Y., Suzuki, A., Takakubo, Y., Nakajima, T., Toyono, S., Toyoshima, S., Fukushima, S., Yamamoto, T., Ito, T. and Takagi, M. (2023) Lower limb alignment in healthy Japanese adults. *J. Orthop. Sci.*, 28(1), 200–3.

5

Digital Self-powered Energy Harvester Based on Vibration Estimation and Control

Kanjuro Makihara

Department of Aerospace Engineering, Tohoku University, Sendai, Miyagi, Japan

5.1 Introduction

Extensive research has been conducted on various energy sources in the context of sustainable development. These clean power sources can contribute significantly to the mitigation of global warming by decreasing the dependence on oil for energy requirements. Energy harvesting presents considerable potential for sustainable energy production as it facilitates energy capture from various sources. These batteries can replace batteries and reduce the requirement for battery exchange in low-power units. This chapter analyzes various methods for harvesting energy from mechanical vibrations using electromagnetic materials [1], lead zirconate titanate (PZT) materials, macro-fiber composites, poly-vinylidene di-fluoride [2], and magnetostrictive materials [3]. Additionally, piezoelectric actuators have been widely used in energy-harvesting devices. Conventional energy-harvesting methods employ a diode bridge as the primary component, presenting a straightforward approach for converting alternating current into direct current (DC) [4].

The synchronized switch harvesting on an inductor (SSHI) was proposed to enhance the output power when compared with typical energy-harvesting circuits [5]. This approach involves integrating a nonlinear electric element into a conventional harvesting circuit. It enhances the output power when compared with conventional energy-harvesting circuits using a circuit comprising on–off switches. In particular, the SSHI utilizes the on–off switches to bring two conditions: one in which electric current flows and the other in which it does not. Consequently, several studies have been carried out to enhance the performance [6–13]. For instance, reference [6] reported a four-time enhancement in the energy harvested using a conventional circuit through the implementation of a synchronized circuit. A selector switch was used to enhance the compactness and sophistication of the synchronized circuit [7], thereby improving the harvesting performance by limiting the superfluous energy consumption. The energy-harvesting performance is represented by the effective electromechanical coupling coefficient, which is dependent on the geometric parameters of the harvester, as demonstrated in [8]. The mechanical performance parameters can be considered as variables for various boundary conditions.

A novel harvesting method that involves using micro-harvesters, which can be referred to as the enhanced harvesting approach, was proposed in reference [9]. This approach significantly increased the harvested power by approximately three times when compared with the conventional harvesting method. A systematic analysis of the power-extraction circuits was proposed in [10], which enabled an increase in the damping under both ideal and realistic constraint conditions. A comparative analysis was conducted between the various harvesting methods. It is demonstrated that the effectiveness of the methods analyzed for enhancing the power output is beyond that of a conventional bridge rectifier. This was achieved through the inversion and charging of a piezoelectric capacitor via an inductor, which involves using advanced techniques.

The underlying effects of the synchronous charge extraction methods helped establish mathematical models used to evaluate the efficiency of these harvesting methods. They demonstrated various mechanisms used by switching circuits to meet the switch-harvesting requirements. Their theoretical output power predictions were compared with the experimental results for a small harvesting device, thereby demonstrating its high efficiency. The effectiveness of the control delay on the performance of the predicted practical output power was also analyzed [13]. The performance varied between the positive and negative control delays; small delays yielded negligible performance. It was reported that the serial SSHI shows symmetric behavior, while the parallel one shows asymmetric behavior.

In this section, the advantages of the proposed digital harvester over conventional systems were discussed based on experiments. Furthermore, the underlying concept behind a self-powered harvester is presented, and a detailed description of the fabrication of a commercial prototype is also presented. First, the concept of a self-powered harvester was explained, and its features were explained and elucidated. Subsequently, the fabricated prototype was presented. The harvesting performance was evaluated in comparison with that of the conventional passive harvesters. Lastly, the limitations of vibrational energy harvesters were addressed.

5.2 Concept of Self-powered Harvester

Figure 5.1 depicts the working principle of a conventional passive harvester, which forms the basis for the proposed self-powered harvester. The circuit comprises a piezoelectric actuator, a diode bridge, and a storage device. The actuator integrated within the structure generates alternating voltages. The passive harvester typically exhibits suboptimal performance owing to its inability to meet substantial harvesting demands. Therefore, more effective harvesting methods must be developed. Consequently, the harvesting technique was developed [5], which enhances the harvesting performance. The method is implemented with the circuit comprising an inductor

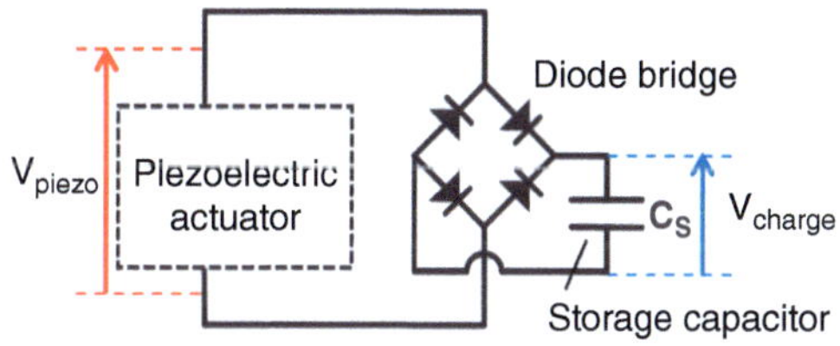

Figure 5.1 Diagram of a conventional harvester.

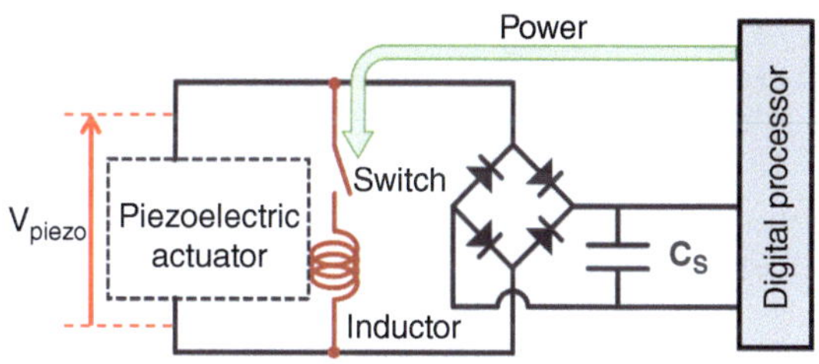

Figure 5.2 Self-powered harvester.

and switches; it inverts the piezoelectric charge synchronously with the mechanical vibrations. Subsequently, the voltage generated by the actuator was amplified.

A digital processor is a potential solution for addressing this requirement. The harvester integrates a passive harvester as a component of a self-powered system, as shown in Figure 5.2. The processor serves as a control unit. A portion of the electrical power obtained by the harvester was utilized to operate the processor. The processor estimated the modal properties based on the sensor information. Subsequently, the signal outputs were transmitted according to the switching principle, as explained in greater detail. A self-powered harvester can realize active harvesting, requiring a power supply or external control mechanism. The detailed configuration is described in [14–16].

5.3 Fabrication of Harvester Prototype

Figure 5.3 depicts the digital processor used in this controller. Figure 5.4 depicts a prototype of the proposed digital self-powered harvester. It was designed and fabricated by a team of academic and business groups using a digital processor, a DC–DC converter, and harvesting switches. The circuit was densely packed to form a compact commercial harvester. The processor was strategically placed at the core of the second layer to overcome potential interference caused by external devices, and digital devices were integrated into the circuit board. The configuration of the harvesting settings can be realized in a sophisticated manner by using setting jumpers. Furthermore, the software for self-powered harvesters was developed for a digital processor that controls the harvesting devices. The implementation of a digital control device facilitates the integration of sophisticated control logic, which enhances the functionality of the system.

Figure 5.3 Processor for self-powered energy harvester.

Figure 5.4 Prototype of self-powered harvesters with digital devices.

5.3.1 Switching Control Based on LQR Theory

The method effectively suppressed the multimodal vibrations by converting the vibrational energy into electrical energy. Vibration suppression rapidly reduces vibrational energy. The method used for controlling the vibration suppression was changed to collect different types of vibrations. The basic equations for piezoelectric materials are:

$$f_p = k_p u - b_p Q, \quad V_p = -b_p u - C_p^{-1} Q \tag{5.1}$$

where k_p, b_p, and C_p denote the stiffness, piezoelectric coefficient, and capacitance of the actuator, respectively. The equation of motion for the vibrating structure can be given by:

$$\mathbf{M}\ddot{\vec{x}} + \mathbf{D}\dot{\vec{x}} + \mathbf{K}\vec{x} = \mathbf{B}_p Q + \vec{w} \tag{5.2}$$

In the modal coordinates, the equation of motion is expressed as follows:

$$\ddot{\vec{\xi}} + \Xi\dot{\vec{\xi}} + \mathbf{\Omega}\vec{\xi} = \mathbf{\Phi}^T \mathbf{B}_p Q + \mathbf{\Phi}^T \vec{w}, \tag{5.3}$$

where $\vec{x} \equiv \mathbf{\Phi}\vec{\xi}$. Subsequently, Eq. (5.3) can be written as:

$$\dot{\vec{z}} = \mathbf{A}\vec{z} + \mathbf{B}Q + \mathbf{E}\vec{w}, \tag{5.4}$$

If Q in Eq. (5.4) is considered as an active control input with index:

$$J \equiv \int_0^\infty \left(\vec{z}^{\mathrm{T}}\mathbf{W}_1\vec{z} + QW_2Q\right) \mathrm{d}t \tag{5.5}$$

is minimized by:

$$Q = Q_T \equiv -\mathbf{F}\vec{z}. \tag{5.6}$$

The values of the weighting factors varied significantly from those in the conventional control guidelines. Semi-active control of mechanical vibration was realized by selecting the connection point rather than actively supplying a control charge or voltage to the actuators. In the method, Q_T was considered as the switching standard. The significance of the weighting matrices for the switching control is detailed in [17]. For accurate calculation, the modal estimation of a multi-degree-of-freedom structure was implemented as a modal truncation function. For multimodal estimation, a Kalman filter was derived, and harvesting was realized. A sensor equation is expressed as follows:

$$V_{\mathrm{sen}} = \mathbf{C}\vec{z}. \tag{5.7}$$

The Kalman filter to estimate $\vec{z}$ is expressed as follows:

$$\dot{\hat{\vec{z}}} = \mathbf{A}\hat{\vec{z}} + \mathbf{B}Q + \Gamma\left(V_{\mathrm{sen}} - \mathbf{C}\hat{\vec{z}}\right). \tag{5.8}$$

The Kalman filter estimates $\vec{z}_{\mathrm{ex}}$. Subsequently, $\hat{\vec{z}}_{\mathrm{ex}}$ is used for calculating the switching standard as follows:

$$Q_T \equiv -\mathbf{F}\hat{\vec{z}}_{\mathrm{ex}}. \tag{5.9}$$

The optimal input cannot be achieved as the proposed energy harvester does not use an energy source as an active input. The switches were regulated to ensure that Q was as close to Q_T as we can.

5.4 Performance of Harvester Prototype

5.4.1 Experimental Setup

The harvesting experiment was carried out via vibration excitation, as shown in Figure 5.5. The setup comprised masses, springs, and an actuator. The pantograph structure is used as a spring. The actuator was used to transfer mechanical energy into electrical energy. Subsequently, it was placed in the pantograph structure. The pantograph structure lengthens the actuator along the horizontal direction and widens it along the vertical direction. This is significant as the stretching range of the actuators is limited. The actuator lengthens in proportion to the displacement.

The machine vibrated using a vibration shaker. The excited force was generated by combining two trigonometric functions. The natural frequencies at a constant electric charge were 20 and 36 Hz. This excited force causes the structure to vibrate naturally in two ways. The vibration strengths at the frequencies were considerably high, and they contained almost all the vibration energy. This study primarily focused on

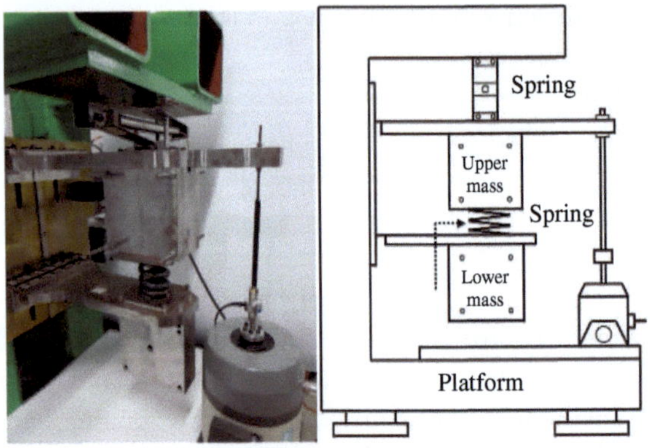

Figure 5.5 Two degrees of freedom experimental apparatus.

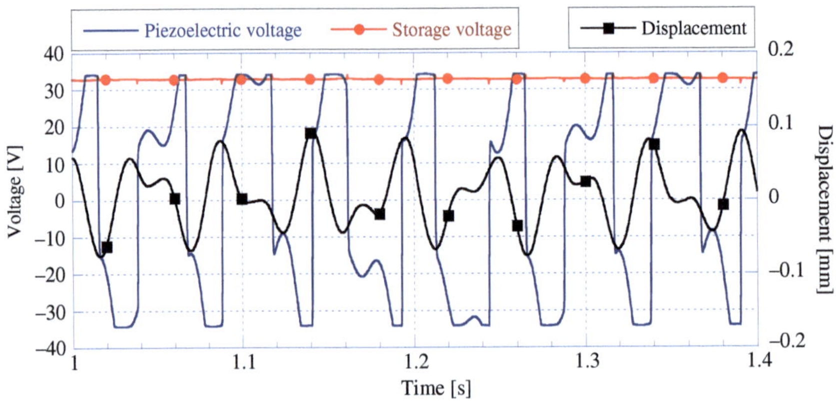

Figure 5.6 Energy-harvesting operations for multimodal vibration.

obtaining energy from these two types of vibrations and exciting the structure using a special force.

5.4.2 Experimental Results

Figure 5.6 depicts time histories of the vibrations for the self-powered harvester, piezoelectric and harvested voltages, and displacement of the upper mass. The time required for voltage inversion varied based on the vibrational state. This is essential for advanced harvesting as it enables adaptation to vibrations across several frequencies, which cannot be achieved with simple switching that occurs at equally spaced intervals. Additionally, the piezoelectric voltage remains high as the switch timing was designed to prevent energy wastage. The piezoelectric voltage remained relatively stable, as indicated by the plateaus observed at the summit of the voltage waveforms. The space among these vibration levels and the capacitor voltages are attributed to the forward-voltage of the rectifier. The upper-mass displacement waveform was observed to be complex, as shown in Figure 5.6. These results clearly

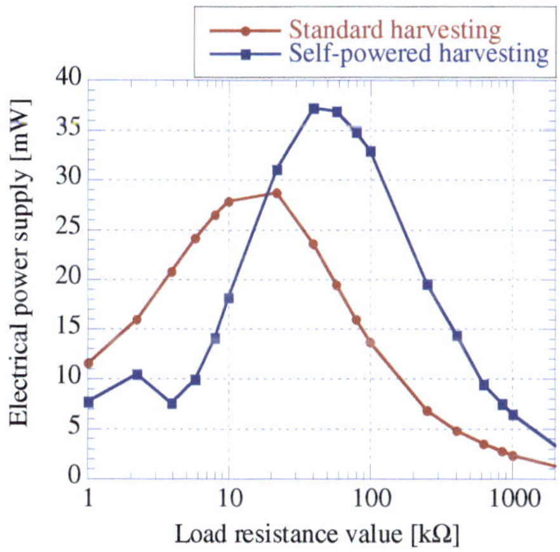

Figure 5.7 Comparison of harvested power for multimodal vibration.

demonstrate the effectiveness of the digital energy harvester in capturing the electrical energy from complex vibrations, such as random or multimodal vibrations.

Figure 5.7 depicts the power consumption of the various load values. The improvement of the power supply is dependent on some reasons, for example, the level of vibration and the mode ratio.

5.5 Modified Synchronized Switch Harvesting on an Inductor

5.5.1 Theory of the Strategy

A switching strategy for the modified synchronized switch harvesting was proposed. This method was originally proposed in reference [18] for analog controllers and displacement measurements. Figure 5.8a and b depicts the modified control flow and waveform of the piezoelectric voltage, respectively, which comprises three stages: (I) voltage inversion, (II) voltage proportional to the displacement, and (III) energy supply. The end of the energy supply stage corresponds to the instant of peak displacement.

5.5.2 Harvesting Circuit for Measurement

Figure 5.9 depicts the measurement circuit. The harvester comprised three sections. The first is a switch that uses the actuator to supply energy to the storage capacitor and load. The second section includes a controller containing a digital device that outputs switch-control signals corresponding to the switch-control strategy. It requires the harvested voltage, V_{load}, to be converted into the microcontroller voltage, V_{drive}, using a DC–DC converter. The third section includes the proposed measurement circuit. Figure 5.12 depicts the waveforms of the circuit. The piezoelectric voltage represents the difference between two measured signals. The measurement circuit comprised six resistances: $R_1 - R_6$. Electrical insulation between the actuator

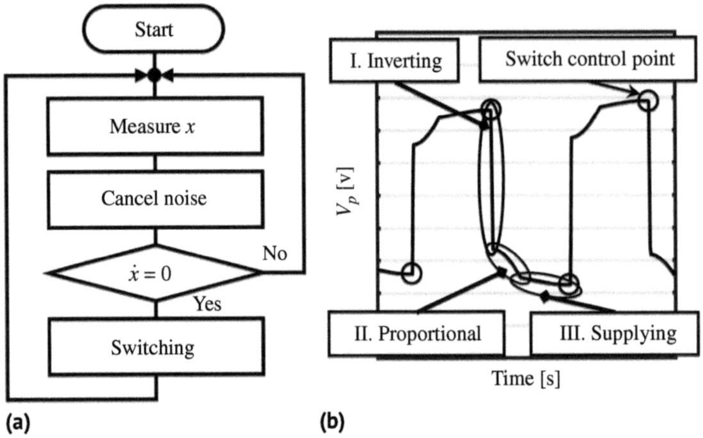

Figure 5.8 (a) Control flow of the modified strategy and (b) voltage history.

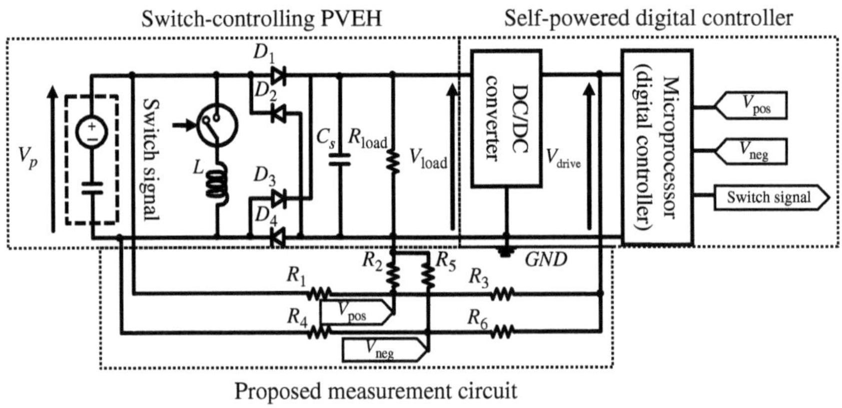

Figure 5.9 Measurement circuit and digital control.

and the digital controller was realized using diodes, D_2 and D_4. Additionally, the circuit could modify the amplitudes of the measured signals. The range of the analog-to-digital converter in the microcontroller was assumed to be $0 - V_{drive}$. The level of the voltage was significantly higher than the upper end value in this range. The proposed circuit divides the voltage by exploiting resistance. Resistances, $R_1 - R_3$, divided the positive voltage, whereas $R_4 - R_6$ divided the negative voltage to provide a proper voltage range to the converter. Furthermore, the measurement circuit applied a bias voltage to the measured signals, as both the measured signals exhibited negative values based on the forward-voltages of the diodes. Consequently, the harvester microcontroller exhibited intensities of 0 V when the measured signals exhibited intensities below 0 V, which causes information loss. Figure 5.10a and b depicts the time histories of the measured signals.

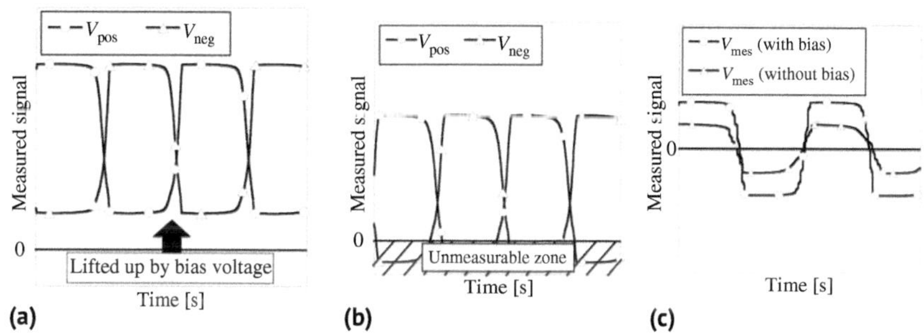

Figure 5.10 Signals (a) with-bias-case, (b) without-bias-case, and (c) differential voltages.

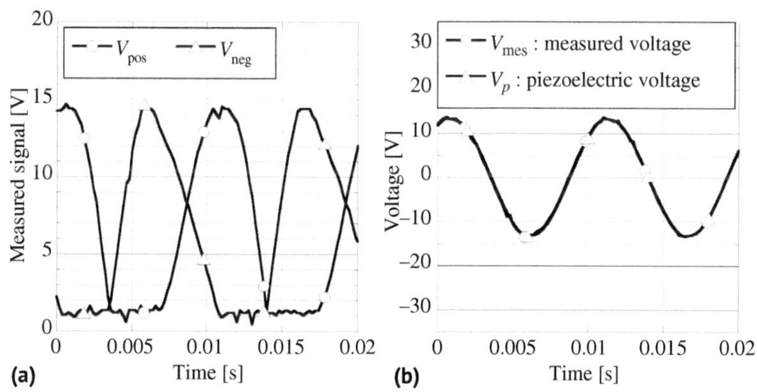

Figure 5.11 (a) Signals measured by the proposed circuit. (b) Piezoelectric voltages.

5.5.3 Measurements for Validation

The circuit operation was first measured. Figure 5.11a depicts the measured signals without switch control. The signals were measured based on the sign of the piezoelectric voltage and exhibited a bias increase that enabled accurate measurement in the microcontroller. Figure 5.11b depicts the voltage measured using the proposed circuit, which was equal to the piezoelectric voltage.

5.5.4 Energy-harvesting Evaluation

Figure 5.12 depicts the comparison in power consumption. The consumption of power was significantly larger than that without switch control for a sufficiently high load resistance. The power for the modified method was approximately equal to that supplied by the conventional method. Additionally, the detection of the peak piezoelectric voltage is difficult for low-load resistances. The rate of energy supply was approximately zero at low resistance values. In this case, the digital controller starts switching before the optimal timing and is unable to supply more energy than that

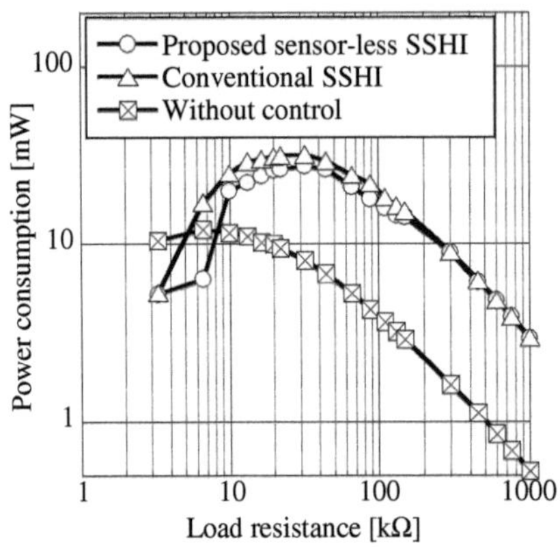

Figure 5.12 Energy consumption with various load values.

supplied using the conventional method. The power supply of the proposed one was the lowest at an approximate load of 6.6 kΩ, as the load consumption was dominant, and the digital controller was not operated in this range.

5.6 Future Development of Self-powered Harvester

Harvesting experiments were conducted under identical excitation conditions. This helps the self-powered devices suppress vibrations while harvesting energy [19]. Therefore, the vibration suppression effect is useful from an engineering perspective. However, from an energy-harvesting perspective, a decrease in the vibration magnitude is detrimental as it leads to a decrease in the harvested energy. A new control method was designed to overcome this limitation [20–23]. This method occasionally pauses the harvesting action and increases the final amount of energy harvested. The novel control method aims to restore the vibration amplitude to its original level during a break. Future works must aim to integrate predictive control for predicting vibrations with self-powered vibration control [24–26]. The prediction of vibrations can help in conserving a significant amount of control energy.

Conventionally, the target control input for harvesting control was formulated under the assumption that the excitation angle frequency was known. However, in practice, situations may arise where the excitation angle frequency is unknown or where the known excitation angle frequency undergoes changes. Consequently, conventional control methodologies predicated on target control inputs are rendered ineffective in such circumstances. Furthermore, the issue of low disturbance estimation accuracy persists. The author proposes four methods to solve these problems for self-powered harvesting.

1. The first solution involves the incorporation of an excitation angle frequency identification method into the harvesting control. The integration of an identification method based on an objective function enables the extension of the harvesting control method to construct target control inputs even in scenarios

where the excitation angle frequency is unknown. The incorporation of the identified excitation angle frequency into the target control input is expected to facilitate the construction of target inputs tailored to the characteristics of the disturbance. The construction of the target control input is predicated on the capacity to account for future state variables, thereby enabling effective harvesting control. The utilization of the model predictive control concept facilitates the incorporation of future state variables into the target control input.

2. The second solution endeavors to enhance the precision of state estimation by integrating pseudo-observations into the observer. The integration of pseudo-observations into the observed quantities has been demonstrated to enhance the accuracy of state estimation. Conventional harvesting control systems have been plagued by the inaccuracy of disturbance estimation in disturbance estimators, which hinders the precise acquisition of disturbance state quantities. The incorporation of state quantities as pseudo-observations into the observed quantities is expected to enhance estimation accuracy for disturbance and other state quantities.
3. The third solution involves incorporating a Gaussian model that predicts disturbances input to the controlled structure. Disturbances have an impact on changes in the state of the controlled system. To achieve effective harvesting control, the implementation of control strategies that compensate for these disturbance effects is recommended. A model predicting the future trajectory of disturbances is constructed, with the intention of combining it with model predictive control.
4. The fourth solution involves using a learning switching standard to decrease the computational time of prediction and optimization processes using a tree structure. The harvesting performance of switching at any given timing is predicted and evaluated. A method is introduced that accumulates data on this evaluation metric and interprets it as a probability density function. The learning switching criterion is expected to yield high harvesting effectiveness.

5.7 Conclusions

The prototype of a self-powered harvester was designed and developed. The proposed harvester incorporated an integrated digital processor powered by the harvested energy, thereby enabling self-powered digital control. The self-powered harvester showed a higher level of harvesting when compared with passive harvesters. This represents a significant amount of energy for vibration-based harvesters. The energy consumption of control devices was assessed, including digital processors and DC–DC converters. The proposed circuit can precisely measure the piezoelectric voltage. This voltage was then utilized in a modified synchronized switch harvesting to improve harvesting performance. Additionally, the power output of the system was analogous to that of the conventional method that employed a displacement sensor. Future work must aim to integrate predictive control for predicting vibrations with self-powered vibration control.

Acknowledgment

This work was supported by a Grant-in-Aid for Scientific Research (KAKENHI) (grant numbers 23K22945, 23K04240, and 23K26061) from the Japan Society for the Promotion of Science and the JSPS Core-to-Core Program, A. Advanced Research Networks (JPJSCCA20200005).

References

1 Glynne-Jones, P., Tudor, M.J., Beeby, S.P. and White, N.M. (2004) An electromagnetic vibration-powered generator for intelligent sensor systems. *Sens. Actuator A-Phys.*, 110, 344–9.
2 Zhao, Q., Yang, L., Chen, L., Ma, Y., Peng, Q., Ji, H. and Qiu, J. (2020) Flexible textured MnO_2 nanorods/PVDF hybrid films with superior piezoelectric performance for energy harvesting application. *Compos. Sci. Technol.*, 199(20), 108330.
3 Staley, M.E. and Flatau, A.B. (2005) Characterization of energy harvesting potential of Terfenol-D and Galfenol. *Proc. SPIE*, 5764, 630–40.
4 Kazmierski, T.J. and Beeby, S. (2010) Energy Harvesting Systems: Principles Modeling and Applications, London, Springer.
5 Guyomar, D., Badel, A., Lefeuvre, E. and Richard, C. (2005) Toward energy harvesting using active materials and conversion improvement by nonlinear processing. *IEEE Trans. Ultrason. Ferroelectr. Freq. Control.*, 52, 584–95.
6 Badel, A., Guyomar, D., Lefeuvre, E. and Richard, C. (2006) Piezoelectric energy harvesting using a synchronized switch technique. *J. Intell. Mater. Syst. Struct.*, 17, 831–9.
7 Makihara, K., Miyakawa, T. and Onoda, J. (2006) Low energy dissipation electric circuit for energy harvester. *Smart Mater. Struct.*, 15(5), 1493–8.
8 Kan, J., Qiu, J., Tang, K., Zhu, K. and Shao, C. (2009) Modeling and simulation of piezoelectric composite diaphragms for energy harvesting. *Int. J. Appl. Electromagn. Mech.*, 30, 95–106.
9 Shen, H., Qiu, J., Ji, H., Zhu, K. and Balsi, M. (2010) Enhanced synchronized switch harvesting, a new energy harvesting scheme for efficient energy extraction. *Smart Mater. Struct.*, 19(11), 115017.
10 Dicken, J., Mitcheson, P.D., Stoianov, I. and Yeatman, E.M. (2012) Power-extraction circuits for piezoelectric energy harvesters in miniature and low-power applications. *IEEE Trans. Power Electron.*, 27, 4514–29.
11 Zhu, L., Chen, R. and Liu, X. (2012) Theoretical analyses of the electronic breaker switching method for nonlinear energy harvesting interfaces. *J. Intell. Mater. Syst. Struct.*, 23, 441–51.
12 Guyomar, D., Sebald, G. and Kuwano, H. (2011) Energy harvester of 1.5 cm^3 giving output power of 2.6 mw with only 1 G acceleration. *J. Intell. Mater. Syst. Struct.*, 22, 415–20.
13 Lallart, M., Wu, Y.-C. and Guyomar, D. (2012) Switching delay effects on nonlinear piezoelectric energy harvesting techniques. *IEEE Trans. Ind. Electron.*, 59, 464–72.

14 Yamamoto, Y., Yoshimizu, K. and Makihara, K. (2015) Synthetic assessment of self-powered energy-harvesting including robustness evaluation. *Mech. Eng. J.*, 2(5), 14-00549.

15 Hara, Y., Saito, K. and Makihara, K. (2019) Compact, digital and self-powered piezoelectric vibration energy harvester with generation control using voltage measurement circuit. *Sens. Actuator A-Phys.*, 299, 111609.

16 Hara, Y., Yamamoto, Y. and Makihara, K. (2020) Self-sensing state estimation of SSHI energy harvesters. *J. Intell. Mater. Syst. Struct.*, 31(20), 2326–41.

17 Makihara, K., Onoda, J. and Minesugi, K. (2007) Comprehensive assessment of semi-active vibration suppression including energy analysis. *Trans. ASME, J. Vib. Acoust.*, 129(1), 84–93.

18 Makihara, K., Takeuchi, S., Shimose, S., and Onoda, J. (2012) Innovative digital self-powered autonomous system for multimodal vibration suppression. *AIAA J.*, 50(9), 2004–11.

19 Lesieutre, G.A., Ottman, G.K. and Hofmann, H.F. (2004) Damping as a result of piezoelectric energy harvesting. *J. Sound Vib.*, 269, 991–1001.

20 Makihara, K., Yamamoto, Y., Yoshimizu, K., Horiguchi, C., Sakaguchi, H. and Fujimoto, K. (2015) A novel controller to increase harvested energy from negating vibration-suppression effect. *Smart Mater. Struct.*, 24, 037005.

21 Yoshimizu, K., Yamamoto, Y., Asahina, K. and Makihara, K. (2017) Strategy for enhancing the active harvesting of piezoelectric energy. *J. Intell. Mater. Syst. Struct.*, 28(8), 1059–70.

22 Hara, Y., Zhou, M., Li, A., Otsuka, K. and Makihara, K. (2021) Piezoelectric energy enhancement strategy for active fuzzy harvester with time-varying and intermittent switching. *Smart Mater. Struct.*, 30(1), 015038.

23 Hara, Y., Otsuka, K. and Makihara, K. (2021) Adaptive and robust operation with active fuzzy harvester under nonstationary and random disturbance conditions. *Sensors*, 21(11), 3913.

24 Takamoto, I., Abe, M., Hara, Y., Otsuka, K. and Makihara, K. (2022) Comprehensive predictive control for vibration suppression based on piecewise constant input formulation. *J. Intell. Mater. Syst. Struct.*, 33(7), 901–17.

25 Abe, M., Hara, Y., Otsuka, K. and Makihara, K. (2023) Semi-active switching vibration control with tree-based prediction and optimization strategy. *J. Intell. Mater. Syst. Struct.*, 34(4), 440–60.

26 Abe, M., Mishima, K., Hara, Y., Otsuka, K. and Makihara, K. (2023) Statistically oriented predictive optimization and disturbance prediction strategy for semi-active vibration control with a piezoelectric actuator. *IEEE Trans. Control Syst. Technol.*, 33(1), 165–80.

6

Nonlinear Finite Element Analysis and Wind Tunnel Experiment of Flutter Energy Harvesting

Keisuke Otsuka

Department of Aerospace Engineering, Tohoku University, Aoba-Ward, Sendai, Japan

6.1 Introduction

Flutter is a self-excited vibration of a structure placed in a fluid, caused by a combination of aerodynamic force, elastic force, and inertia force of the structure. Flutter vibration leads to serious incidents of structures. Therefore, many studies have been conducted to predict the flutter speed and improve the flutter resistance in the design of structures such as bridges and wings. Meanwhile, from the viewpoint of energy harvesting, flutter vibration can be considered a very attractive energy source [1]. This is because structural or aerodynamic nonlinearity causes limit cycle oscillation (LCO) that maintains high flutter amplitude while avoiding serious destruction.

A rotating wind turbine can harvest relatively large amounts of electricity. However, it can be installed only on vast lands where the annual wind speed is around 6 m/s, and it is also costly to transport the wind turbine components. In addition, the wind turbine components must be inspected once a year or every half year. Therefore, it cannot be installed anywhere. On the other hand, a flutter energy harvester is useful for generating small amounts of power. Since there is no restriction on the installation position, power generation is possible anywhere if there is wind. Flutter vibration power generation can be carried out by installing it in high places, such as bridges and towers, where the flow velocity is large. The flutter-harvested energy can be used to operate small-sized sensors and microprocessors used for structural health monitoring, for example. Further, it can be used indoors by installing it in a pipe through which air flows. There are many methods to convert vibration energy into electrical energy, such as the use of electromagnetic induction, electrostatic induction, and magnetostrictive and piezoelectric materials. Vibration energy harvesting using these conversion methods has been actively studied over the past few decades. The piezoelectric harvester is advantageous because the power output per volume of the power generation device is large compared with other power generation methods. A thin plate with a piezoelectric patch is widely used as a harvester.

A thin piezoelectric film can be designed in an arbitrary shape depending on the application and cause flutter at a relatively low flow speed. Therefore, it is desirable that the piezoelectric film can be used as a harvester component. Many conventional studies have focused on flag-type flutter harvesters [2, 3] in which the longitudinal direction is parallel to the flow direction. A wing-type flutter harvester whose longitudinal direction is perpendicular to the flow direction has not been fully studied yet, especially when it undergoes large deformation.

To design both the flag-type and wing-type flutter harvesters, an efficient analysis method considering large deformation is necessary. Structural nonlinearity needs to be considered in the analysis method to express LCO. Absolute nodal coordinate formulation (ANCF) [4, 5] is one of the efficient ways to consider structural nonlinearity. ANCF is a non-incremental finite element method that was developed in the multibody dynamics community in the late 1990s. In most nonlinear finite element methods, position vector and rotation parameters (e.g., Euler angles or Euler parameters) are used as generalized coordinates [6]. On the other hand, ANCF uses position vector gradients as generalized coordinates instead of rotation parameters. The use of the position vector gradients retains the mass matrix constant and centrifugal/Coriolis forces equivalently zero. In addition, aerodynamic mesh on the structural surface modeled by ANCF can be easily generated by a simple multiplication of a shape function and a generalized coordinate vector. Because of these advantages, ANCF has been widely used for various engineering applications and fluid-structure interaction analysis [7]. For example, Otsuka et al. [8, 9] coupled ANCF beam elements with an aerodynamic model based on the medium-fidelity unsteady vortex lattice method (UVLM) [10]. Because UVLM is based on the potential flow assumption, it cannot describe viscous and compressible phenomena, such as separation and shock. However, UVLM calculates the unsteady aerodynamic force acting on a structure undergoing nonlinear large deformation. The coupling analysis results of ANCF and UVLM were in good agreement with wind tunnel experiment results [11, 12]. ANCF beam elements can be coupled with much higher fidelity aerodynamic models, such as the smoothed particle hydrodynamics method and lattice Boltzmann method, if necessary. There are some studies that couple ANCF and piezoelectricity at the strain level [13]. Mukogawa et al. [14] coupled ANCF and piezoelectricity at the strain level, and then they conducted flutter energy harvesting analysis by using UVLM. The strain-level coupling between ANCF and piezoelectricity is relatively complex because the strain is written as a highly nonlinear function of generalized coordinates in ANCF. Therefore, a more simplified method to couple ANCF and piezoelectricity is desired, especially for the conceptual design of the flutter harvester.

This chapter has two objectives. The first objective is to investigate the characteristics of a wing-type flutter energy harvester by conducting wind tunnel experiments. The second objective is to present a more simplified method that couples ANCF and piezoelectricity.

6.2 Flutter Energy Harvesting Experiment

A wind tunnel at the Institute of Fluid Science of Tohoku University, in Figure 6.1, was used to investigate the characteristics of the wing-type flutter harvester. This wind tunnel is a single-channel flow type, and air flows from a blower inside the wind tunnel. Air flow from the blower is rectified by wire mesh or honeycomb, and then it is sent to the test section by the contraction nozzle. The flow velocity distribution in the axial direction of the channel becomes uniform in the test section. This low-turbulence heat transfer wind tunnel is characterized by extremely low-turbulence intensity of the air flow in the test section, and the maximum turbulence intensity is 0.05%. The flow velocity in the wind tunnel test section is determined by the rotation speed of the blower.

By setting the rotational speed of the blower using the electronic control panel, the flow speed of the wind tunnel test section can be selected from 5 to 80 m/s. Since the flow velocity varies with the air temperature, humidity, and atmospheric pressure, it is adjusted to an arbitrary flow velocity by using a Pitot tube. Figure 6.2a shows the flutter harvester used for the wind tunnel experiment. A piezoelectric film is attached to the black vinyl chloride plate fixed by a jig. Table 6.1 shows the specification of the black plate. Four plates with different lengths were used. Figure 6.2b shows the piezoelectric film used for the experiment. The piezoelectric film TP-110 (Elmech Electronics Industry) is packaged with electrodes. The piezoelectric element is polarized to generate an electric field when the thickness changes. in which the electric field generated in the piezoelectric element causes electric charges to be biased toward the electrode portion, thereby generating a voltage between the terminals.

Figure 6.3 shows how to place the flutter harvester in the open test section of the wind tunnel. The jig is mounted to a rotation table to change the harvester direction against the flow direction, namely angle of attack. Figure 6.4 shows the measurement devices used for the experiment. A laser displacement sensor

Figure 6.1 Low-turbulence heat transfer wind tunnel.

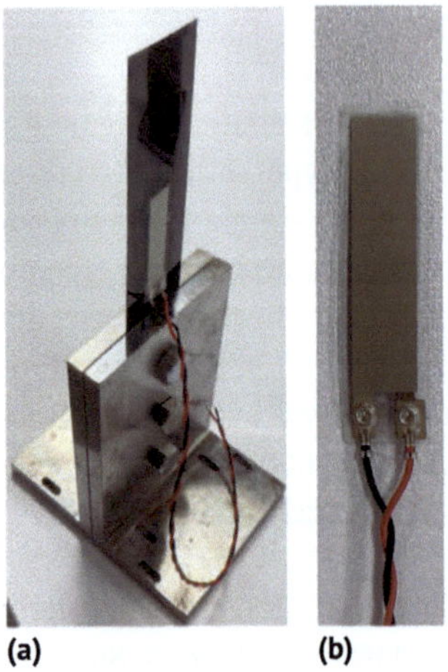

Figure 6.2 (a) Flutter harvester and (b) piezoelectric film.

Table 6.1 Specifications of black plate part of harvester.

Parameters	Value
Length h (mm)	130, 170, 210, 290
Width b (mm)	40
Thickness t (mm)	0.5
Density (kg/m^3)	1400

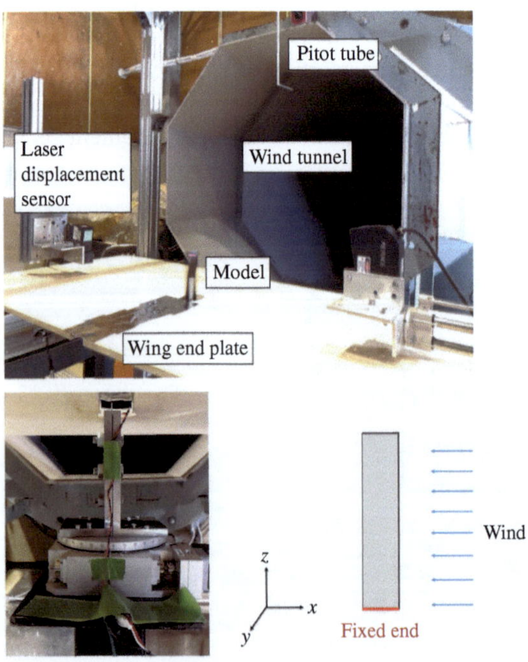

Figure 6.3 Experimental equipment for wing flutter power generation.

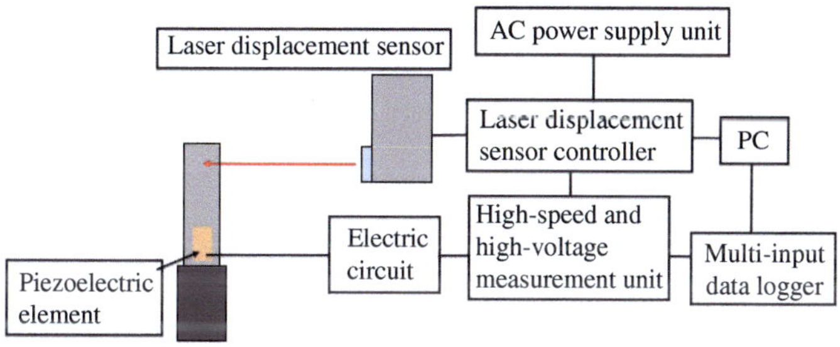

Figure 6.4 Connection diagram of the measurement device.

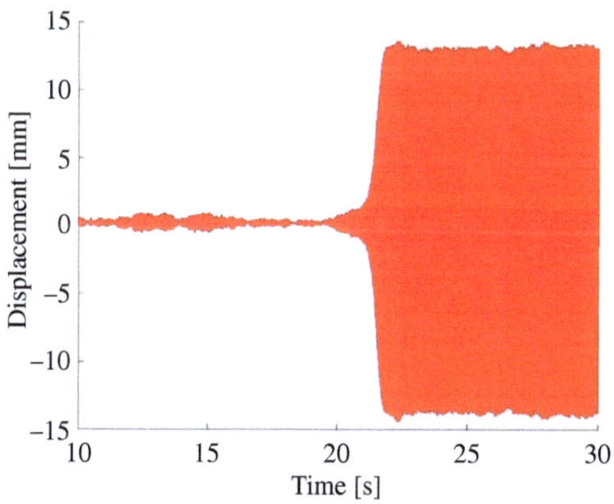

Figure 6.5 An example of laser displacement sensor data when flutter occurs.

(LK-G400, KEYENCE) was used to measure the black plate displacement y. The measurement position was the position of half the length of the model. A high-voltage measurement device (NR-HV04, KEYENCE) was used to measure the voltage V_{load} of the resistive element connected to the piezoelectric element. When the flow velocity is gradually increased, flutter occurs suddenly at a certain flow velocity and the amplitude increases drastically. Figure 6.5 shows the time history of the plate displacement when flutter occurs. Flow velocity U at this time is defined as the flutter flow velocity U_f in this study. Flow velocity was fixed at an arbitrary flow velocity, and the harvested voltage was measured. The effects of the plate length h, angle of attack θ, and flow velocity U on the root mean squares (RMSs) y_{RMS} and V_{RMS} for the displacement y and voltage V_{load}, respectively, were investigated.

6.2.1 Plate Length *h*

The wing-type flutter harvester characteristic with respect to the plate length h was investigated. The electric loading resistance $R_{\text{load}} = 5$ MΩ was used. The angle of attack was set to 0°. Figure 6.6 shows the relationship between plate length h, flutter flow velocity U_f, and voltage V_{RMS}. As the plate length h becomes long, the flutter velocity U_f decreases mainly because stiffness becomes low. Although the voltage decreases up to $h = 210$ mm, it increases from $h = 210$ to $h = 290$ mm. This is

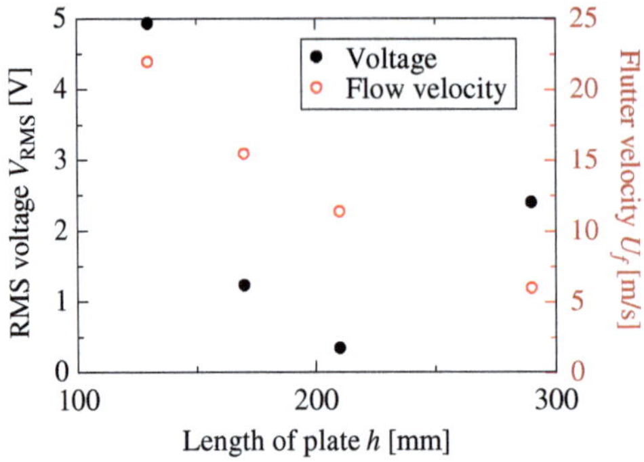

Figure 6.6 Voltage V_{RMS} and flutter velocity U_f for different plate lengths h.

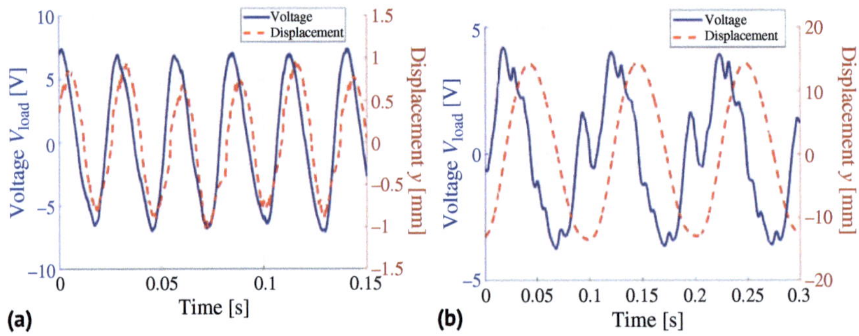

Figure 6.7 Time histories of voltage V_{load} and displacement y of flutter for different plate lengths. (a) $h = 130$ mm, $U_f = 22.0$ m/s, $R_{load} = 5$ MΩ and (b) $h = 290$ mm, $U_f = 6.0$ m/s, $R_{load} = 5$ MΩ.

probably because bending and torsional coupling modes in flutter change when the plate length increases from 210 to 290 mm. Figure 6.7 shows the time histories of the voltage and displacement in the cases of $h = 130$ and 290 mm when flutter occurs. The amplitudes of the voltage and displacement are different between $h = 130$ and $h = 290$. Figure 6.8 shows the fast Fourier transformation results of the voltages in Figure 6.7. The peak frequency difference between $h = 130$ and $h = 290$ indicates that different bending and torsional modes are coupled in flutter. According to these results, there may be an optimal plate length that has a low flutter speed and high voltage.

6.2.2 Angle of Attack θ

The wing-type flutter harvester characteristic with respect to the angle of attack θ was investigated. Figure 6.9 shows the relationship between the angle of attack θ, flutter speed U_f, and voltage V_{RMS}. As the angle of attack θ increases, the flutter speed U_f decreases. This is because nonzero angle of attack causes static deflection resulting in torsional frequency variation and different bending-torsional coupling in flutter. On the other hand, the piezoelectric voltage shows an inconsistent trend against the angle of attack. Figure 6.10 shows the time histories of the voltage and

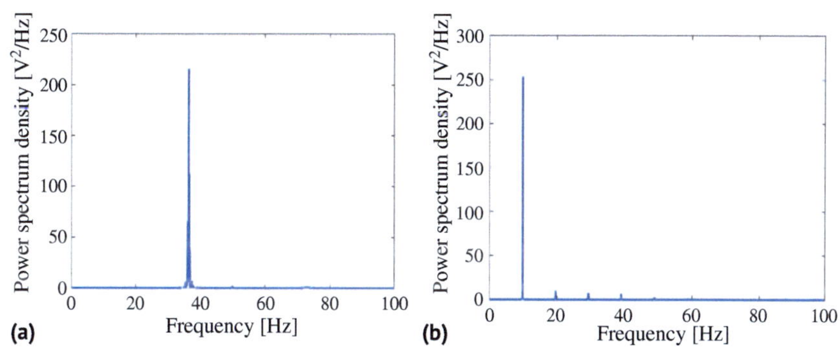

Figure 6.8 Fast Fourier transformation results of voltages V_{load}. (a) $h = 130$ mm, $U_f = 22.0$ m/s, $R_{\text{load}} = 5$ MΩ and (b) $h = 290$ mm, $U_f = 6.0$ m/s, $R_{\text{load}} = 5$ MΩ.

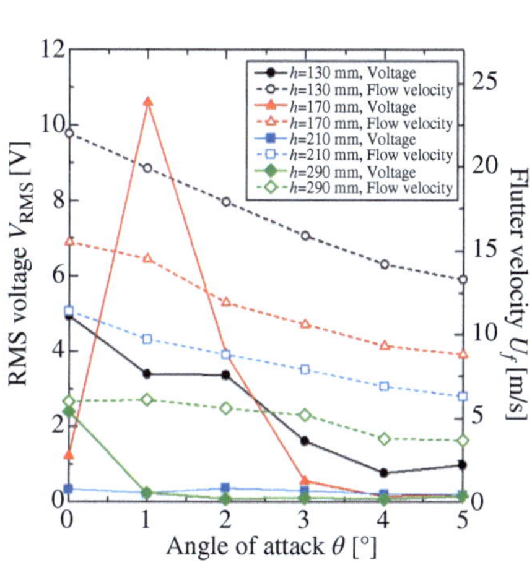

Figure 6.9 Voltage V_{RMS} and flutter velocity U_f for different angles of attack.

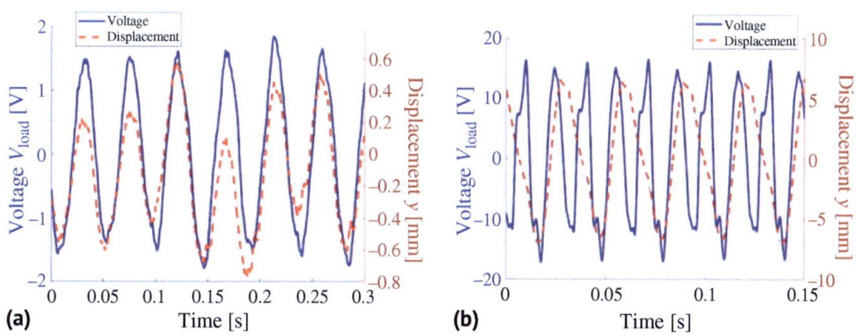

Figure 6.10 Time histories of voltage V_{RMS} and displacement y of flutter at different angles of attack. (a) $h = 170$ mm, $\theta = 0°$, $U_f = 15.5$ m/s, $R_{\text{load}} = 5$ MΩ and (b) $h = 170$ mm, $\theta = 1°$, $U_f = 14.5$ m/s, $R_{\text{load}} = 5$ MΩ.

displacement of a harvester with the same plate length $h = 170$ mm and different angles of attack $\theta = 0$ and $1°$ when flutter occurs. Such a small angle of attack variation results in large voltage variation. According to these results, although flutter

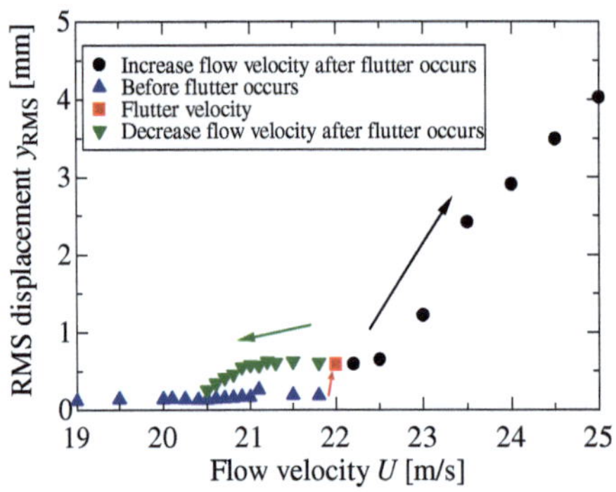

Figure 6.11 Displacement y_{RMS} against flow velocity U.

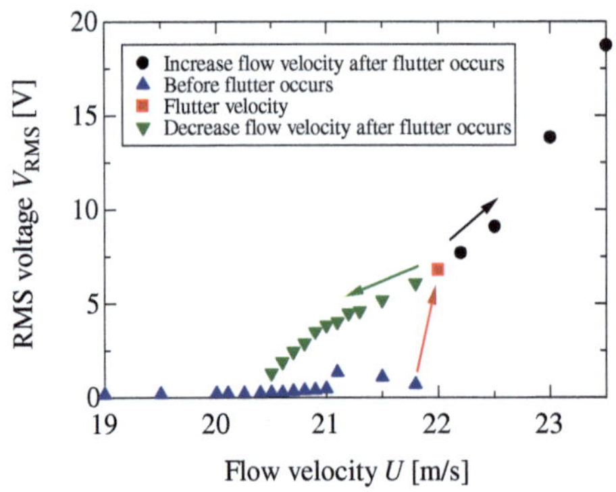

Figure 6.12 Voltage V_{RMS} against flow velocity U.

harvesting can be achieved at a lower flow speed by increasing the angle of attack θ, the output voltage shows inconsistent trend, which shows the difficulty in designing a wing-type flutter harvester based on an experimental try and error method.

6.2.3 Flow Speed

The wing-type flutter harvester characteristic with respect to the flow speed U was investigated. Even if the flow speed U is beyond the flutter speed U_f, flutter harvesting is possible while avoiding structural destruction because of nonlinearity. Therefore, it is valuable to investigate the flow speed effect on flutter harvesting. Here, the plate length h and angle of attack θ are fixed to 130 mm and 0°, respectively. Figure 6.11 shows the relationship between the flow velocity U and displacement y. Figure 6.12 shows the relationship between the flow velocity U and the voltage V_{load}. The blue plots show that flutter did not occur. The red plot shows flutter occurrence. The black plots show the cases where the wind tunnel flow speed increased after flutter occurred. Even if the flow speed exceeds the flutter speed, structural destruction did not occur, and much higher voltage was generated. On the other hand, the green plots show the cases where the wind tunnel flow speed decreased after flutter occurred. This hysteresis indicates that flutter harvesting can be achieved even if flow speed becomes less than the flutter speed once flutter occurs. Mukogawa et al. [14]

pointed out that predicting this hysteresis trend by using an analytical or numerical method is difficult. Although such prediction is beyond the scope of this chapter, a theoretical or numerical prediction method for the hysteresis would be useful for designing a more efficient flutter harvester.

6.3 Flutter Energy Harvesting Simulation

This section presents a flutter harvesting analysis method based on ANCF coupled with piezoelectricity in a simpler way than the method developed by Mukogawa et al. [14]. The simpler method is suitable for the conceptual design of the flutter harvester.

6.3.1 ANCF

To describe various vibration modes, an ANCF thin plate model is employed in this study. Figure 6.13 shows the thin plate element. This element has four nodes A, B, C, and D, respectively. The position vector on the element mid surface is defined as

$$\mathbf{r} = \mathbf{S}(x, y)\mathbf{e}. \tag{6.1}$$

$\mathbf{S}$ is a shape function, and $\mathbf{e}$ is a generalized nodal coordinate defined as

$$\mathbf{e} \equiv \begin{bmatrix} \mathbf{e}^{\mathrm{A}} \\ \mathbf{e}^{\mathrm{B}} \\ \mathbf{e}^{\mathrm{C}} \\ \mathbf{e}^{\mathrm{D}} \end{bmatrix}, \mathbf{e}^{i} \equiv \begin{bmatrix} \mathbf{r}^{i} \\ \mathbf{r}_{x}{}^{i} \\ \mathbf{r}_{y}{}^{i} \end{bmatrix}. \tag{6.2}$$

$\mathbf{r}_x$ and $\mathbf{r}_y$ are the position vector gradients that describe an arbitrary large element rotation. Because these position vectors and gradient vectors are defined in the global coordinate system, rotational coordinate transformation is not necessary for ANCF. Therefore, the equation of motion of the element is written as

$$\mathbf{M}\ddot{\mathbf{e}} + \mathbf{F}_{\text{elastic}} = \mathbf{F}_{\text{external}}, \tag{6.3}$$

with a constant mass matrix $\mathbf{M}$. $\mathbf{F}_{\text{elastic}}$ is a nonlinear elastic force vector. The detailed derivation of the elastic force can be found in [15]. $\mathbf{F}_{\text{external}}$ is an external force vector including the damping force and aerodynamic force calculated by UVLM.

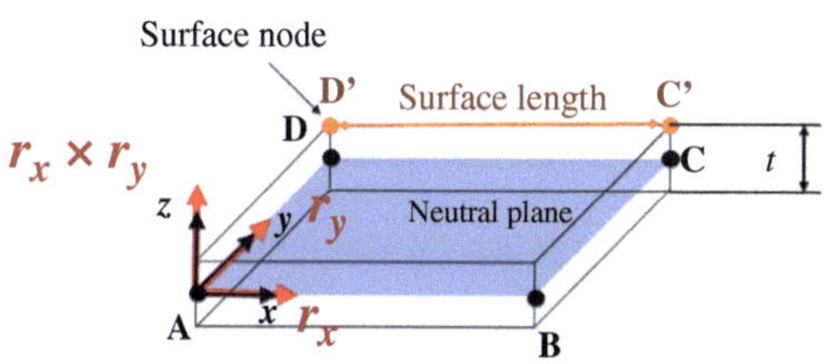

Figure 6.13 ANCF plate element.

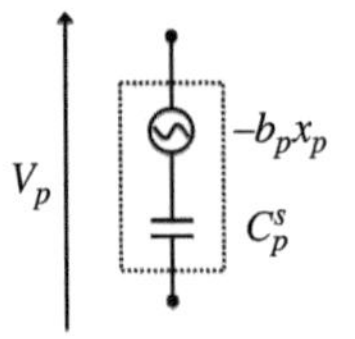

Figure 6.14 Electrical model of a piezoelectric element.

6.3.2 Piezoelectricity

The piezoelectric film is polarized to produce an electric field (i.e., voltage) when the thickness is changed. This is expressed as

$$V_p = -b_p x_p + \frac{Q_p}{C_p^s}. \tag{6.4}$$

V_p is the voltage generated in the piezoelectric element, b_p is the piezoelectric coefficient, x_p is the expansion and contraction of the piezoelectric element in the polarized direction, Q_p is the accumulated charges of the piezoelectric element, and C_p^s is the constant-strain capacitance. Since the piezoelectric element is bonded to the plate in this study, it is considered that the deformation of the piezoelectric element completely follows the deformation of the plate. The piezoelectric element can be expressed as an electric element in which a voltage source and a capacitor are connected in series, as shown in Figure 6.14. This property of the piezoelectric element serves to convert vibration energy into electrical energy.

6.3.3 Electric Circuit

To evaluate how much electric energy could be extracted from the vibration of the structure, it is necessary to evaluate whether electricity is stored in an external capacitor or used in an external resistance. In this study, an electric circuit is used for the evaluation in Figure 6.15. Here, R_{load} is the resistive element used to evaluate work, and V_{load} is the electric potential applied to the resistive element. This resistive element is directly connected to a piezoelectric element. The Kirchhoff's second law for this circuit is written as

$$V_p + R_{\text{load}} \frac{dQ_p}{dt} = 0. \tag{6.5}$$

Substituting Eq. (6.4) into Eq. (6.5) yields

$$\frac{dQ_p}{dt} + \frac{1}{R_{\text{load}} C_p^s} Q_p = \frac{b_p}{R_{\text{load}}} x_p. \tag{6.6}$$

6.3.4 Electricity Generation Mechanism of Piezoelectric Harvester

Figure 6.16 shows the expansion and contraction of the piezoelectric device when the plate vibrates. Figure 6.16a displays that the upper surface of the plate is extended. Figure 6.16b shows that the upper surface of the plate is compressed. Since the piezoelectric element is adhered to the surface of the plate, the expansion and contraction of the piezoelectric element causes the Poisson effect and the thickness

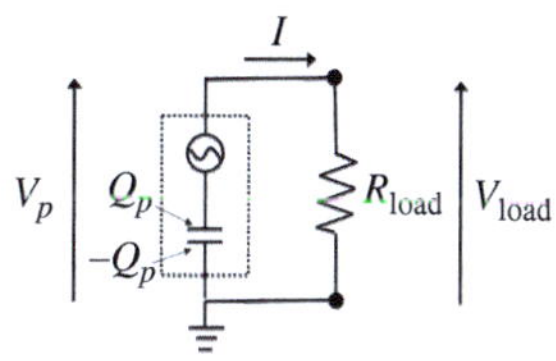

Figure 6.15 Electrical circuit to evaluate flutter harvesting performance.

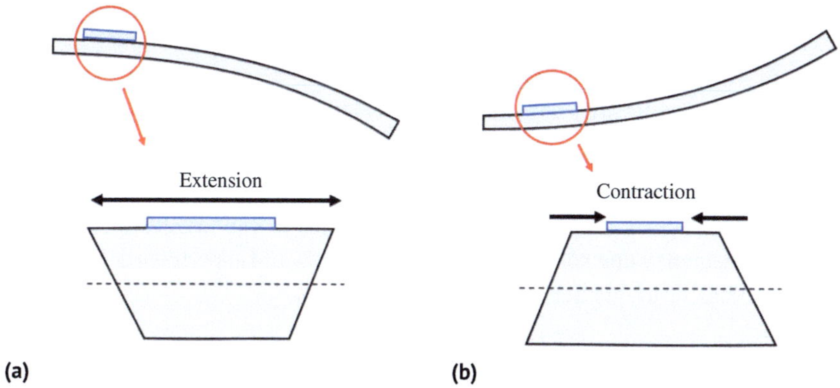

Figure 6.16 (a) Extension and (b) contraction of the plate surface with a piezoelectric patch.

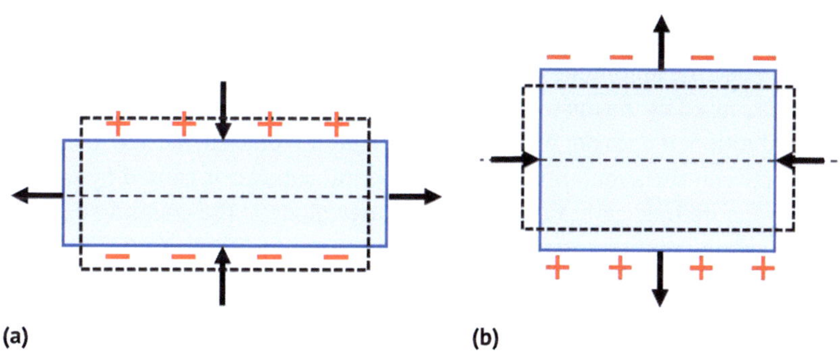

Figure 6.17 (a) Extension and (b) contraction of the piezoelectric patch.

changes, as shown in Figure 6.17. Since the thickness in the polarization direction changes, an electric field is generated. Based on this mechanism, it becomes possible to generate electricity from the vibration of the plate.

6.3.5 Flutter Harvesting Analysis Method

The flutter harvesting analysis method is based on two assumptions. The first assumption is that the piezoelectric force on the plate is negligible. Although this assumption is not valid in some harvester configurations, it allows us to solve the equation of motion in Eq. (6.3) and the electric circuit equation in Eq. (6.6) separately. The second assumption is that the piezoelectric patch displacement x_p is proportional to the plate displacement y at a user-defined point. This proportional relationship can be written as $b_p x_p = b_y y$. Thus, the electric equation in Eq. (6.6) can be transformed into

$$\frac{dQ_p}{dt} + \frac{1}{R_{\text{load}} C_p^s} Q_p = \frac{b_y}{R_{\text{load}}} y. \tag{6.7}$$

Based on the aforementioned assumptions, the harvested voltage V_p is calculated in two steps. The first step is to solve the equation of motion in Eq. (6.3) by using a numerical integration method to generate the time history of the plate deflection y. Substituting the time history of y in Eq. (6.7) yields the time history of the piezoelectric electric charge Q_p that results in the voltage V_p.

6.3.6 Simulation Results of Free-vibration Harvesting

Before conducting the flutter harvesting simulation, a free-vibration harvesting simulation was conducted. Table 6.2 shows the simulation parameters of the harvester plate. b_y is 3000 V/m. C_p^s is 58 pF. R_{load} is 5 MΩ. Figure 6.18 shows the time histories of the displacement y and voltage V_{load}. The simulation and experimental results are in good agreement in this free vibration.

6.3.7 Simulation Results of Flutter Harvesting

Figure 6.19 shows the time histories of the displacement and voltage when flutter occurs at 22.2 m/s. Although the simulation describes LCO and calculates the voltage, the simulation and experimental voltages are not in good agreement. There are two possible reasons why there is a difference between the simulation and the experiment. First, the amplitude difference between the simulation and experimental voltages is caused by the mistuning of b_p. Because $b_y = 3000$ V/m was correlated in the free vibration, it does not work well in the flutter vibration. Second, the phase difference between the simulation and experimental voltages is caused by the second assumption defined as $b_p x_p = b_y y$. The flutter mode is the coupling between the second bending and first torsional modes, as shown in Figure 6.20. Therefore, the extensional displacement x_p of the piezoelectric film attached to the plate root and the displacement y of the plate center do not have the same phase. Third, the

Table 6.2 Simulation parameters of the harvester plate.

Parameters	Value
Length h (mm)	1.30×10^2
Width b (mm)	4.00×10^1
Thickness t (mm)	5.00×10^{-1}
Density (kg/m^3)	1.40×10^3
Young's modulus (GPa)	4.73×10^0
Poisson's ratio (–)	3.80×10^{-1}
Damping α (–)	3.50×10^0
Damping β (–)	1.52×10^{-4}
Flow velocity U (m/s)	2.22×10^1

phase difference is caused by the internal resistance that is not considered in the simulation.

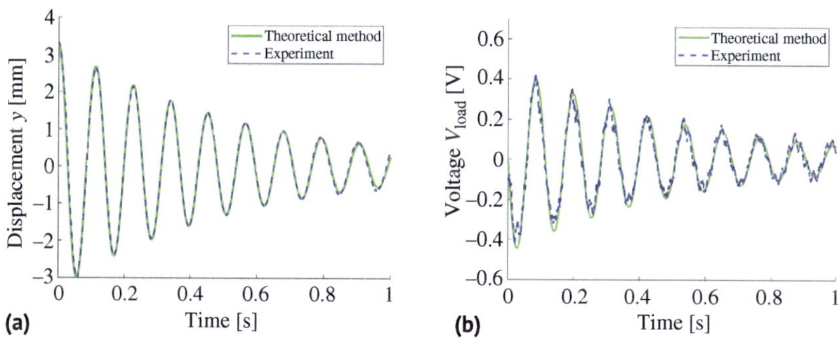

Figure 6.18 Time histories of (a) displacement and (b) voltage in free vibration.

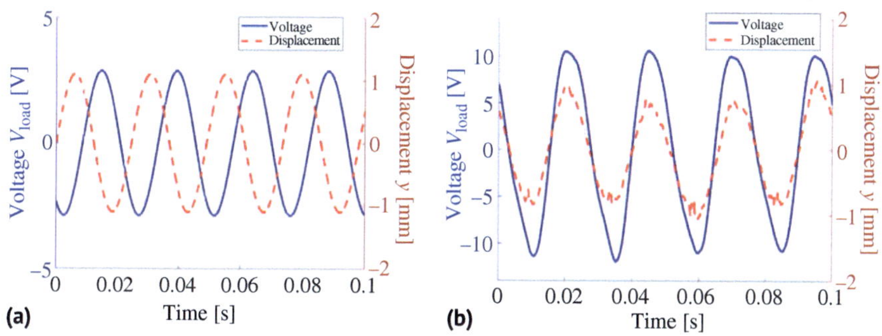

Figure 6.19 Time histories of displacement and voltage in flutter. (a) Simulation and (b) Experiment.

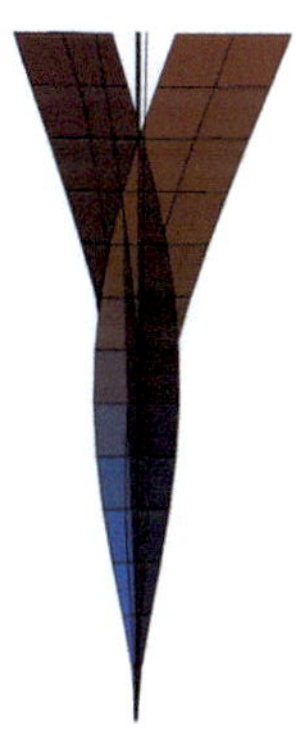

Figure 6.20 Simulated flutter mode shape.

6.4 Conclusions

This study focused on the wing-type flutter harvester composed of a plate with a piezoelectric film.

The characteristics of the harvester were investigated by conducting wind tunnel experiments. The findings of the experiment are as follows:

a. As the plate length increases, the flutter speed decreases. On the other hand, the generated voltage is not a simple linear function of the plate length.
b. As the angle of attack increases, the flutter speed decreases. On the other hand, the generated voltage is not a linear, simple function of the angle of attack.
c. Even if the flow speed is beyond flutter speed, flutter harvesting can be achieved while avoiding structural destruction.
d. Even if the flow speed becomes lower than flutter speed after flutter occurs, flutter harvesting can be achieved by leveraging the hysteresis phenomenon.

According to the findings (a) and (b), it is difficult to design a flutter harvester that has a low flutter speed and a high generated voltage by an experimental trial-and-error method. Therefore, a simplified flutter harvesting analysis method was developed for the conceptual design of the flutter harvester. The method is based on two assumptions. The first assumption is that the piezoelectric force does not affect the flutter vibration, which enables us to solve the equation of motion and the electric equation independently. The second assumption is that the plate bending displacement at a user-defined point is proportional to the piezoelectric film displacement, which enables easy correlation between the simulation and the experiment. In free-vibration harvesting, the simulation result was in good agreement with the experimental result. In flutter harvesting, the simulation method could calculate the voltage generated from LCO. However, the simulation result was not in good agreement with the experimental result. This is because the second assumption did not work well for the flutter vibration with the second bending mode. The removal of the second assumption would improve the simulation accuracy.

References

1 Abdelkefi, A. (2016) Aeroelastic energy harvesting: a review. *Int. J. Eng. Sci.*, 100, 112–35. https://doi.org/10.1016/j.ijengsci.2015.10.006.

2 De Marqui, C., Tan, D. and Erturk, A. (2018) On the electrode segmentation for piezoelectric energy harvesting from nonlinear limit cycle oscillations in axial flow. *J. Fluids Struct.*, 82, 492–504. https://doi.org/10.1016/j.jfluidstructs.2018.07.020.

3 Tang, D.M., Levin, D. and Dowell, E.H. (2019) Experimental and theoretical correlations for energy harvesting from a large flapping flag response. *J. Fluids Struct.*, 86, 290–315. https://doi.org/10.1016/j.jfluidstructs.2019.02.018.

4 Shabana, A.A., Hussien, H.A. and Escalona, J.L. (1998) Application of the absolute nodal coordinate formulation to large rotation and large deformation problems. *J. Mech. Des.*, 120, 188–95. https://doi.org/10.1115/1.2826958.

5 Gerstmayr, J., Sugiyama, H. and Mikkola, A. (2013) Review on the absolute nodal coordinate formulation for large deformation analysis of multibody systems. *J. Comput. Nonlinear Dyn.*, 8, 031016. https://doi.org/10.1115/1.4023487.

6 Simo, J.C. and Vu-Quoc, L. (1988) On the dynamics in space of rods undergoing large motions – a geometrically exact approach. *Comput. Methods Appl. Mech. Eng.*, 66, 125–61. https://doi.org/10.1016/0045-7825(88)90073-4.

7 Otsuka, K., Makihara, K. and Sugiyama, H. (2022) Recent advances in the absolute nodal coordinate formulation: literature review from 2012 to 2020. *J. Comput. Nonlinear Dyn.*, 17, 080803. https://doi.org/10.1115/1.4054113.

8 Otsuka, K., Wang, Y. and Makihara, K. (2021) Three-dimensional aeroelastic model for successive analyses of high-aspect-ratio wings. *J. Vib. Acoust.*, 143, 061006. https://doi.org/10.1115/1.4050276.

9 Otsuka, K., del Carre, A. and Palacios, R. (2022) Nonlinear aeroelastic analysis of high-aspect-ratio wings with a low-order propeller model. *J. Aircr.*, 59, 293–306. https://doi.org/10.2514/1.C036285.

10 Murua, J., Palacios, R. and Graham, J.M.R. (2012) Applications of the unsteady vortex-lattice method in aircraft aeroelasticity and flight dynamics. *Prog. Aerosp. Sci.*, 55, 46–72. https://doi.org/10.1016/j.paerosci.2012.06.001.

11 Otsuka, K., Wang, Y., Fujita, K., Nagai, H. and Makihara, K. (2019) Multifidelity modeling of deployable wings: multibody dynamic simulation and wind tunnel experiment. *AIAA J.*, 57, 4300–11. https://doi.org/10.2514/1.J058676.

12 Dong, S., Otsuka, K. and Makihara, K. (2023) Hamiltonian formulation with reduced variables for flexible multibody systems under linear constraints: theory and experiment. *J. Sound Vib.*, 547, 117535. https://doi.org/10.1016/j.jsv.2022.117535.

13 Nada, A.A. and El-Assal, A.M. (2012) Absolute nodal coordinate formulation of large-deformation piezoelectric laminated plates. *Nonlinear Dyn.*, 67, 2441–54. https://doi.org/10.1007/s11071-011-0158-4.

14 Mukogawa, T., Shimura, K., Dong, S., Fujita, K., Nagai, H., Kameyama, M., Shi, Y., Jia, Y., Soutis, C., Kurita, H., Narita, F., Hara, Y., Makihara, K. and Otsuka, K. (2024) Piezoelectric flutter energy harvesting: absolute nodal coordinate formulation model and wind tunnel experiment. *Mech. Res. Commun.*, 143, 104351. https://doi.org/10.1016/j.mechrescom.2024.104351.

15 Otsuka, K. and Makihara, K. (2018) Deployment simulation using absolute nodal coordinate plate element for next-generation aerospace structures. *AIAA J.*, 56, 1266–76. https://doi.org/10.2514/1.J056477.

7

Thermoelectric Materials for Flexible Devices

Souichiro Koga[1], *Daichi Shimokawa*[1], *Takanobu Watanabe*[2], *and Yukio Furukawa*[1]

[1] *Department of Chemistry and Biochemistry, Graduate School of Advanced Science and Engineering, Waseda University, Shinjuku-ku, Tokyo, Japan*
[2] *Department of Electronic and Physical Systems, Graduate School of Fundamental Science and Engineering, Waseda University, Shinjuku-ku, Tokyo, Japan*

7.1 Introduction

In modern society, humans generate a large amount of energy through their activities. A significant portion of the generated energy dissipates in the form of heat into the surroundings. Thermoelectric energy conversion has attracted the attention of researchers worldwide as a promising energy harvesting technology in various fields such as the "Internet of Things" (IoT). A thermoelectric generator (TEG) can convert waste heat directly into useful electrical energy [1–4]. Although inorganic materials have dominated the field, organic polymer materials are candidates for low-temperature (<200 °C) applications because of their solution processability, great mechanical flexibility, low cost, light weight, and nature-friendly characteristics [5]. To evaluate the performance of thermoelectric materials, a dimensionless figure of merit ZT has been introduced and is determined by the electrical conductivity σ, the Seebeck coefficient S, the thermal conductivity κ, and the absolute temperature T [5]:

$$ZT = \frac{\sigma S^2 T}{\kappa} \tag{7.1}$$

where σ is related to the carrier density n, the charge of a carrier q, and the carrier mobility μ, expressed as:

$$\sigma = nq\mu \tag{7.2}$$

Notably, the κ of polymers is relatively low (<0.5 W/m K), and their measurement is difficult. Thus, the performance of thermoelectric materials can also be evaluated by the power factor PF:

$$PF = S^2\sigma \tag{7.3}$$

Promising thermoelectric materials should possess high electrical conductivity and a high Seebeck coefficient, in addition to low thermal conductivity. However, there is a trade-off relationship between σ and S because σ increases with n, whereas S changes inversely [2]. Thus, there is a maximum in ZT versus n plots or PF versus n plots [2].

Polymers with π-conjugated electrons exhibit high electrical conductivities when they are doped with electron acceptors or electron donors. Acceptor-doped polymers exhibit p-type behavior, whereas donor-doped polymers exhibit n-type behavior. Undoped neutral conjugated polymers are used as active materials in flexible light-emitting diodes, transistors, and solar cells. Doped conducting polymers are expected to be used as active materials for flexible TEGs. A conducting polymer can hold two kinds of carriers: a polaron with charge $q = +e$ or $-e$ and spin of 1/2 and a bipolaron with $q = +2e$ or $-2e$ and no spin. The bipolaron is an anomalous spinless carrier. The type of carrier can be clarified using Raman spectroscopy or vis–NIR absorption spectroscopy [6–8]. The mobility of positive polarons is the same as that of positive bipolarons because the density of bipolarons is half that of polarons, whereas the charge of a bipolaron is twice that of a polaron. Thus, we consider the "charge number density" instead of the carrier density to avoid the polaron/bipolaron problem.

Among many highly conducting polymers, the current best p-type polymer thermoelectric material is doped poly(ethylenedioxythiophene):poly (4-styrenesulfonate) (PEDOT:PSS) [9–12]. The chemical structures of PEDOT and PSS are shown in Figure 7.1a and b. PEDOT:PSS films are stable in air. PEDOT chains are oxidized (doped), and the positive charges are compensated with the pendant SO_3^- groups of PSS. The positive charges on PEDOT chains can be compensated with anions such as ClO_4^- and tosylate anions. The typical values of the electrical conductivity, the Seebeck coefficient, and the thermal conductivity are 1000 S/cm, 30 μV/K, and 0.3 W/mK, respectively. Thus, PF and ZT at 300 K are calculated to be 90 $\mu W/mK^2$ and 0.09, respectively. For Bi–Te materials, the typical values of the electrical conductivity, Seebeck coefficient, and thermal conductivity are 800 S/cm, 200 μV/K, and 1.5 W/mK, respectively. The typical PF and ZT at 300 K are 3200 $\mu W/mK^2$ and 0.64, respectively.

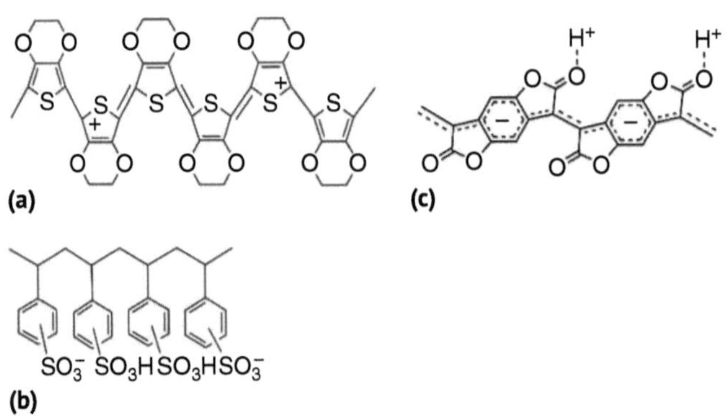

Figure 7.1 Chemical structures: (a) doped PEDOT; (b) PSS; and (c) doped PBFDO.

Both p-type and n-type material legs are required to build an efficient π-type TEG [5]. Few n-type thermoelectric polymers have been reported because almost all donor-doped polymers are unstable in air. The most important issue in polymer TEGs is the development of high-performance n-type thermoelectric polymers. Recently, Tang et al. [13] synthesized n-type doped poly(benzodifurandione) (PBFDO). Doped PBFDO showed a fairly good *PF* of 90 $\mu W/mK^2$ [12]. In this chapter, we present a review of the structure and properties of n-type doped PBFDO and the performance of a TEG consisting of p-type doped PEDOT and n-type doped PBFDO.

7.2 Doped PBFDO

Negatively doped PBFDO was synthesized via the oxidative polymerization of 3,7-dihydrobenzo[1,2-b:4,5-b′]difuran-2,6-dione in the presence of tetramethyl-1,4-benzoquinone [13]. During the oxidative polymerization process, tetramethyl-1,4-benzoquinone is reduced to tetramethyldihydroquinone, as shown in Figure 7.2. This compound reduces the resulting polymer, as shown in Figure 7.3. Thus, n-type doped PBFDO is formed. The negative charges generated on the polymer chains are compensated with protons attached to the carbonyl groups. A film of n-type doped PBFDO can be prepared from a DMSO solution using the drop-cast method or the spin-cast method. A similar polymerization method based on $Cu(OAc)_2$ has also been reported [14]. PBFDO is stable in air. The electrical conductivity of an unencapsulated PBFDO film remained at almost 95% of the initial value after 35 days of exposure to air (25 °C, 60% relative humidity).

Figure 7.4 shows the vis–NIR absorption spectrum of the as-prepared doped PBFDO film prepared on a glass plate. Absorption extending to the infrared region

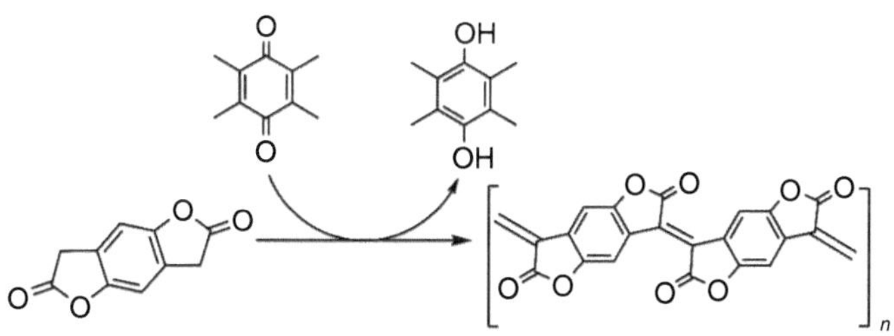

Figure 7.2 Oxidative polymerization of 3,7-dihydrobenzo[1,2-b:4,5-b′]difuran-2,6-dione.

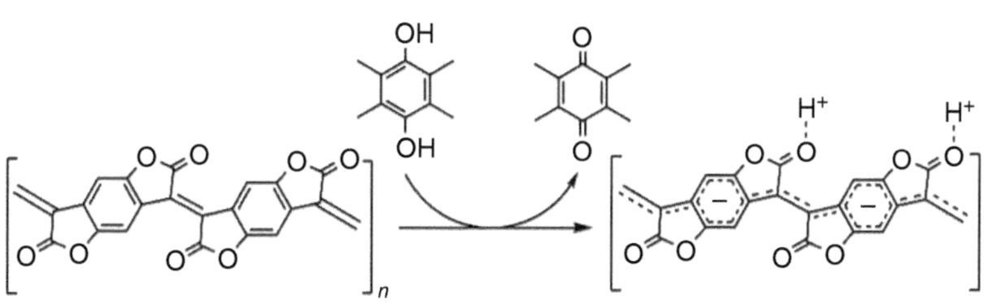

Figure 7.3 Reductive n-doping by tetramethyldihydroquinone.

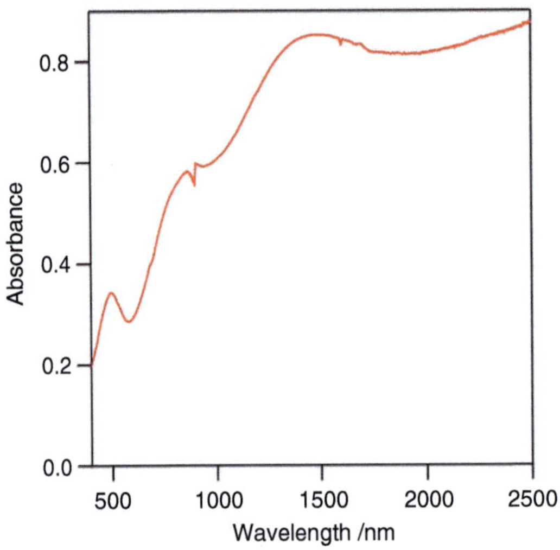

Figure 7.4 Vis–NIR absorption spectrum of the as-prepared PBFDO film prepared on a glass plate.

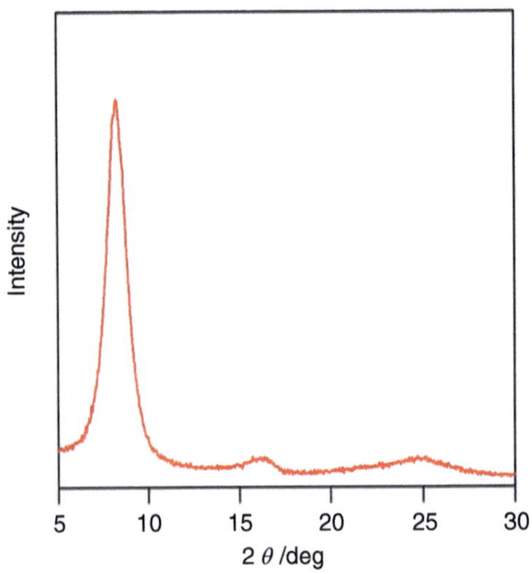

Figure 7.5 GIXRD pattern of the as-prepared doped PBFDO film.

is observed, which is attributable to the negative carriers, polarons/bipolarons. The undoped neutral PBFDO exhibited absorption bands at 610 and 900 nm [13]; the 900-nm absorption band was attributed to the HOMO–LUMO transition [15].

Figure 7.5 shows the GIXRD pattern of the as-prepared doped PBFDO film. A strong line is observed at $2\theta = 8.31°$, indicating that the PBFDO film is crystalline. This pattern is quite similar to that of doped PEDOT, which is explained by the orthorhombic crystal system [16]. Assuming that doped PBFDO also takes the orthorhombic crystal system, the peak at $2\theta = 8.31°$ is attributable to the (100) plane, corresponding to the lamellar stacking distance of 1.06 nm. Tang et al. [13] reported that the lamellar stacking distance is 1.118 nm and that the π–π stacking distance is 0.330 nm. A PBFDO chain has an edge-on orientation.

The resonance Raman spectra of doped PBFDO measured with excitation at 830, 633, and 532 nm are shown in Figures 7.6–7.8, respectively. The vis–NIR absorption of doped PBFDO is associated with carriers (negative polaron/bipolaron) generated by doping. Thus, the observed Raman spectra are probably associated with the carriers. Further studies are necessary to explain the observed bands attributed to polarons/bipolarons. The Raman spectra depend on the excitation laser wavelength, indicating that some types of carriers are generated. The strongest bands are observed at 1506, 1509, and 1505/cm for 830, 633, and 532 nm excitation, respectively. These bands are attributable to the displacement of carbon atoms in the five-membered ring [15].

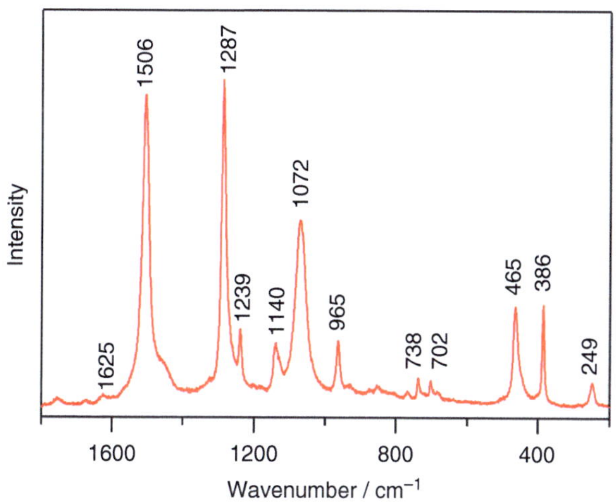

Figure 7.6 Raman spectrum of a negatively doped PBFDO film excited at 830 nm.

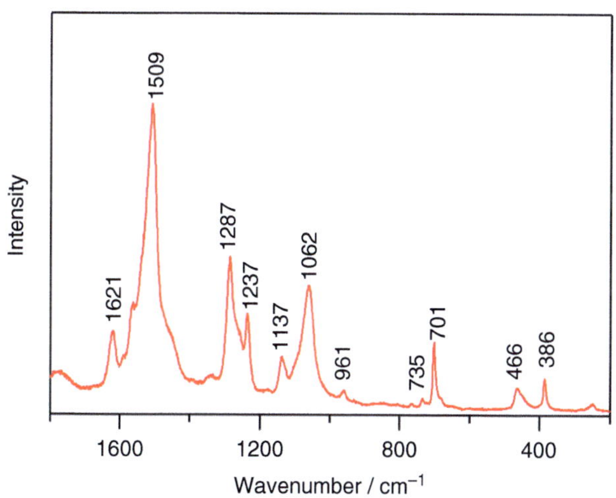

Figure 7.7 Raman spectrum of a negatively doped PBFDO film excited at 633 nm.

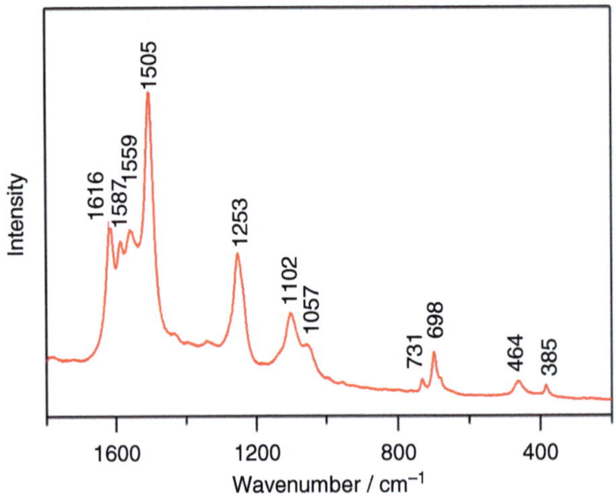

Figure 7.8 Raman spectrum of a negatively doped PBFDO film excited at 532 nm.

7.3 Thermoelectric Polymers

PEDOT:PSS and doped PBFDO exhibit good thermoelectric properties. Other conducting polymers have been studied; a summary of high-performance thermoelectric polymers is listed in Table 7.1. Since a trade-off relationship between σ and S holds, Bubnova et al. [9] optimized the doping level (charge number density) of doped PEDOT, and the ZT at room temperature was 0.25. Park et al. [17] optimized the doping level of doped PEDOT to a record PF of 1270 $\mu W/mK^2$. Fan et al. [18] reported that ion accumulation of an ionic liquid on the PEDOT:PSS surface increases the Seebeck coefficient and has a high PF of 754 $\mu W/mK^2$, corresponding to a ZT value of 0.75. This value is comparable to those of inorganic thermoelectric materials.

7.4 Thermoelectric Generators

A thermoelectric generator (TEG) structure is shown in Figure 7.9. The device consists of p-type PEDOT:PSS and n-type PBFDO films on a glass plate. A 5 vol% ethylene glycol was added to an aqueous suspension of PEDOT:PSS (Clevios PH1000, 1.0–1.3 wt% aqueous suspension). The addition of ethylene glycol increases the electrical conductivity and ZT [10]. A PEDOT:PSS film was prepared from this suspension via the drop-cast method. The sizes of these two films were 1.2 cm × 1.0 cm; the thicknesses of these films were 4–6 μm. The electrical conductivity and Seebeck coefficient of the PEDOT:PSS film at room temperature were 755 S/cm and 25 $\mu V/K$, respectively, and the PF was calculated to be 47 $\mu W/mK^2$. The PBFDO film was prepared via the drop-cast method from a DMSO solution of PBFDO. The electrical conductivity and Seebeck coefficient of the PBFDO film at room temperature were 1290 S/cm and -29 $\mu V/K$, respectively, and the PF was calculated to be 108 $\mu W/mK^2$. Tang et al. [13] reported

Table 7.1 Summary of high-performance thermoelectric polymers and their properties.

Material[a]	σ (S/cm)	S (μV/K)	κ (W/mK)	PF (μW/mK^2)	ZT (300 K)	References
1. p-type						
IL/PEDOT:PSS	1720	65	0.3	754	0.75	[18]
PEDOT:PSS	923	117	—	1270	—	[17]
FTS/PBTTT	1000	33	—	110	—	[19]
F_6TCNQQ-doped PBTTT	2430	47	—	530	—	[20]
$FeCl_3$-doped TBDOPV-T	400	—	—	100	—	[21]
Poly [Cu_x (Cu-ett)]	0.76	80	—	6.5	0.005	[22]
2. n-type						
PBFDO	2000	−21	—	90	—	[13]
N-DMBI-doped TBDOPV-T	100	—	—	200	—	[21]
Poly [K_x (Ni-ett)]	1.12	−121.6	—	26	0.1	[22]

a. Abbreviations: IL = ionic liquid; PBTTT = poly(2,5-bis(3-alkylthiophen-2-yl)thieno[3,2-*b*]thiophene); FTS = (tridecafluoro-1,1,2,2-tetrahydrooctyl)trichlorosilane; F_6TCNQQ = 1,3,4,5,7,8-hexafluoro-tetracyanonaphthoquinodimethane; TBDOPV-T = thiophene-fused benzodifurandione oligo(p-phenylenevinylene); *N*-DMBI = 4-(1,3-dimethyl-2,3-dihydro-1*H*-benzimidazol-2-yl)phenyl)dimethylamine; ett = 1,1,2,2-ethenetetrathiolate.

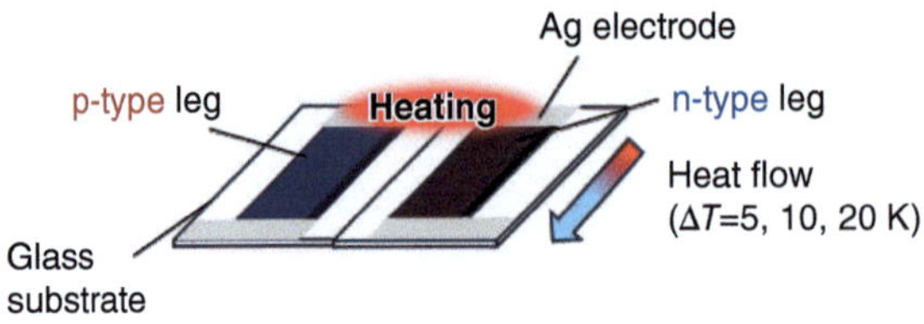

Figure 7.9 Schematic structure of a TEG.

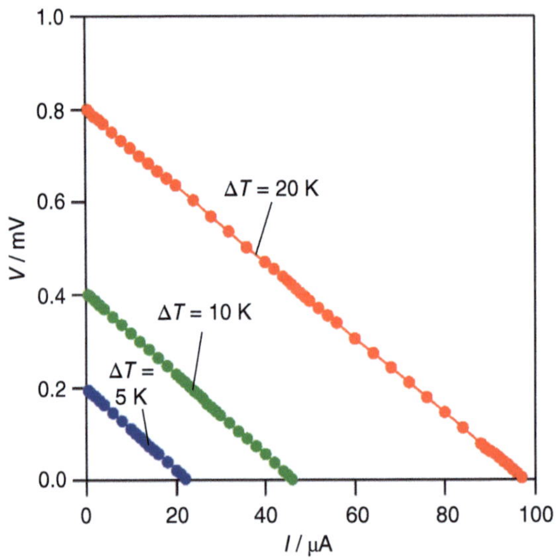

Figure 7.10 Current versus voltage relationship.

that the electrical conductivity is ca. 2000 S/cm, the Seebeck coefficient is −21 μV/K, and the *PF* was calculated to be 88 $\mu W/mK^2$. The charge number density obtained via the Hall effect is $1.3 \times 10^{22}/cm^3$ [13]. Thermal gradients, $\Delta T = 5$, 10, and 20 K, were made by heating the top side of the device, as shown in Figure 7.9.

The relationship between the voltage and the current is shown in Figure 7.10. The open-circuit voltage of the TEG was 0.80 mV at a temperature gradient ΔT of 20 K. The inner resistance of the TEG at room temperature was 9.2 Ω, which is low among organic TEGs. Parabolas are observed at 5, 10, and 20 K in the *P–V* plots (Figure 7.11). The maximum output power P_m at $\Delta T = 20$ K was 19 nW. Sarabia-Riquelme et al. [23] reported that a two-pair TEG consisting of PEDOT:PSS fibers and PBFDO fibers was 2.40 nW at $\Delta T = 22$ °C. Tang et al. [13] reported that the performance of a six-pair TEG consisting of PEDOT:PSS and PBFDO legs was 1.94 μW at a hot plate temperature of 160 °C.

Yu et al. [21] fabricated a corrugated flexible TEG consisting of three pairs with p-type and n-type doped TBDOPV-T legs on a polyimide film. Only the single polymer TBDOPV-T was used. The temperature of the heat source was 150 °C; $\Delta T = 45.1$ °C. This temperature difference originates from the corrugated structure. The open-circuit voltage and the short-circuit current were 28.1 mV and 57.2 μA,

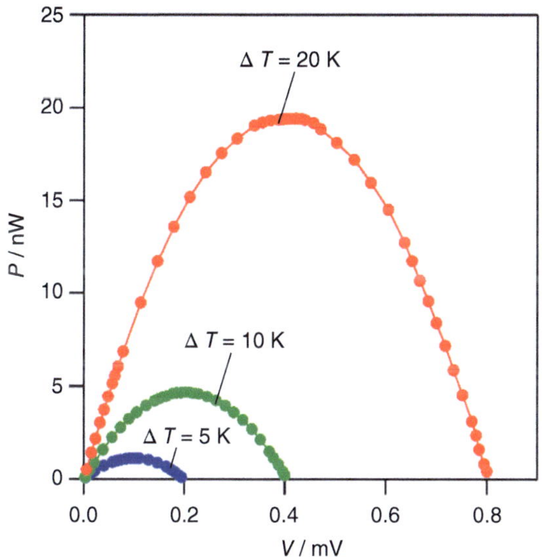

Figure 7.11 Power versus voltage relationship.

respectively. The maximum power output of this TEG was 404 nW. Sun et al. [22] fabricated a TEG consisting of 35 p–n pairs based on poly [Na_x (Ni-ett)] and poly [Cu_x (Cu-ett)] on an aluminum nitride wafer. When T_{hot} = 150 °C and ΔT = 82 °C, the open-circuit voltage and short-circuit current were 0.26 V and 10.1 mA, respectively. The maximum output power of the TEG was 750 μW.

7.5 Concluding Remarks

The thermoelectric properties of conducting polymers have notably improved over the past decade. The *ZT* of doped p-type PEDOT is the same as that of inorganic thermoelectric materials. Doped n-type PBFDO, which was recently reported, is stable in air, whereas most n-type conducting polymers are very unstable in air. Further investigation is necessary to determine a better *ZT*. A TEG consisting of a PEDOT:PSS leg and a PBFDO leg was fabricated, and it exhibited good performance. Polymer TEGs are promising for applications in wearable sensors and IoT devices.

References

1 Bell, L.E. (2008) Cooling, heating, generating power, and recovering waste heat with thermoelectric systems. *Science*, 321, 1457–61.

2 Haras, M. and Skotnicki, T. (2018) Thermoelectricity for IoT – a review. *Nano Energy*, 54, 461–76.

3 Petsagkourakis, I., Tybrandt, K., Crispin, X., Ohkubo, I., Satoh, N. and Mori, T. (2018) Thermoelectric materials and applications for energy harvesting power generation. *Sci. Tech. Adv. Mater.*, 19, 836–62.

4 Zhao, W., Ding, J., Zou, Y., Di, C.-A. and Zhu, D. (2020) Chemical doping of organic semiconductors for thermoelectric applications. *Chem. Soc. Rev.*, 49, 7210–28.

5 Hasan, Md. N., Nafea, M., Nayan, N. and Ali, M.S.M. (2022) Thermoelectric generator: materials and applications in wearable health monitoring sensors and internet of things devices. *Adv. Mater. Technol.*, 7, 2101203.

6 Furukawa, Y. (1996) Electronic absorption and vibrational spectroscopies of conjugated conducting polymers. *J. Phys. Chem.*, 100, 15644–53.

7 Yamamoto, J. and Furukawa, Y. (2015) Electric and vibrational spectra of positive polarons and bipolarons in regioregular poly(3-hexylthiophene) doped with ferric chloride. *J. Phys. Chem. B*, 119, 4788–94.

8 Furukawa, Y. and Shimokawa, D. (2023) Polarons, bipolarons, and electrical properties of crystalline conducting polymers. *Bull. Chem. Soc. Jpn.*, 96, 1243–51.

9 Bubnova, O., Khan, Z. U., Malti, A., Braun, S., Fahlman, M., Berggren, M. and Crispin, X. (2011) Optimization of the thermoelectric figure of merit in the conducting polymer poly(3,4-ethylenedioxythiophene). *Nat. Mater.*, 10, 429–33.

10 Kim, G.-H. Shao, L., Zhang, K. and Pipe, K.P. (2013) Engineered doping of organic semiconductors for enhanced thermoelectric efficiency. *Nat. Mater.*, 12, 719–23.

11 Bubnova, O., Khan, Z.U., Wang, H., Braun, S., Evans, D.R., Fabretto, M., Hojati-Talemi, P., Dagnelund, D., Arlin, J.-B., Geerts, Y.H., Desbief, S., Breiby, D.W., Andreasen, J.W., Lazzaroni, R., Chen, W.M., Zozoulenko, I., Fahlman, M., Murphy, P.J., Berggren, M. and Crispin, X. (2014) Semi-metallic polymers. *Nat. Mater.*, 131, 190–4.

12 Fan, Z., Li, P., Du, D. and Ouyang, J. (2017) Significantly enhanced thermoelectric properties of PEDOT:PSS films through sequential post-treatments with common acids and bases. *Adv. Energy Mater.*, 7, 1602116.

13 Tang, H., Liang, Y., Liu, C., Hu, Z., Deng, Y., Guo, H., Yu, Z., Song, A., Zhao, H., Zhao, D., Zhang, Y., Guo, X., Pei, J., Ma, Y., Cao, Y. and Huang, F. (2022) A solution-processed n-type conducting polymer with ultrahigh conductivity, *Nature*, 611, 271–7.

14 Ke, Z., Abtahi, A., Hwang, J., Chen, K., Chaudhary, J., Song, I., Perera, K., You, L., Baustert, K. N., Graham, K. R. and Mei, J. (2023) Highly conductive and solution-processable n-doped transparent organic conductor. *J. Am. Chem. Soc.*, 145, 3706–15.

15 Li, N., Sheng, H., Sun, Y. and Wang, J. (2023) Spectroscopic study on size-dependent optoelectronics of N-type ultra-high conductive polymer PBFDO. *Spectrochim. Acta Part A: Mol. Biomol. Spectrosc.*, 298, 122744.

16 Aasmundtveit, K.E., Samuelson, E.J., Pettersson, L.A.A., Inganäs, O., Johansson, T. and Freidenhans, R. (1999) Structure of thin films of poly(3,4-ethylenedioxy-thiophene). *Synth. Met.*, 101, 561–4.

17 Park, T., Park, C., Kim, B., Shin, H. and Kim, E. (2013) Flexible PEDOT electrodes with large thermoelectric power factors to generate electricity by the touch of fingertips. *Energy Environ. Sci.*, 6, 788–92.

18 Fan, Z., Du, D., Guan, X. and Ouyang, J. (2018) Polymer films with ultrahigh thermoelectric properties arising from significant Seebeck coefficient enhancement by ion accumulation on surface. *Nano Energy*, 51, 481–8.

19 Patel, S.N., Glaudell, A.M., Kiefer, D. and Chabinyc, M.L. (2016) Increasing the thermoelectric power factor of a semiconducting polymer by doping from the vapor phase. *ACS Macro Lett.*, 5, 268–72.

20 Vijayakumar, V., Durand, P., Zeng, H., Untilova, V., Herrmann, L., Algayer, P., Leclerc, N. and Brinkmann, M. (2020) Influence of dopant size and doping method on the structure and thermoelectric properties of PBTTT films doped with F_6TCNNQ and F_4TCNQ. *J. Mater. Chem. C*, 8, 16470–82.

21 Yu, Z.-D., Lu, Y., Wang, Z. Y., Un, H.-I., Zelewski, S. J., Cui, Y., You, H. Y., Liu, Y., Xie, K. F., Yao, Z. F., He, Y. C., Wang, J. Y., Hu, W. B., Sirringhaus, H. and Pei, J. (2023) High n-type and p-type conductivities and power factors achieved in a single conjugated polymer. *Sci. Adv.*, 9, eadf3495.

22 Sun, Y., Sheng, P., Di, C., Jiao, F., Xu, W., Qiu, D. and Zhu, D. (2012) Organic thermoelectric materials and devices based on p- and n-type poly(metal 1,1,2,2-ethenetetrathiolate)s. *Adv. Mater.*, 24, 932–7.

23 Sarabia-Riquelme, R., Noble, L.E., Espejo, P.A., Ke, Z., Graham, K.R., Mei, J., Paterson, A.F. and Weisenberger, M.C. (2024) Highly conductive n-type polymer fibers from the wet-spinning of n-doped PBDF and their application in thermoelectric textiles. *Adv. Funct. Mater.*, 34, 2311379.

8

Origami Thermoelectric Generator

Shingo Terashima and Eiji Iwase

School of Fundamental Science and Engineering, Waseda University, Tokyo, Japan

8.1 Introduction

To realize good power generation performance in a thermoelectric power generator (TEG), it goes without saying that the thermoelectric (TE) properties of the TE material are important, but the device structure is also important. While research on TE materials is actively being conducted, little research focuses on device structure. Many conventional TEGs do not pay particular attention to device structure and are constructed using rigid, flat substrates. In this case, TEGs can be used only for flat heat sources and cannot be used, for example, when the heat source has a curved surface. In this section, we will introduce a TEG that uses a device structure called an origami structure to realize good thermal contact even with curved heat sources such as the heat piping or the human body, which can be bent and stretched. Conventional general TEGs often consist of a TE element and a highly thermally conductive insulating substrate, such as a ceramic substrate. Ceramic substrates have high thermal conductivity, but because of their high rigidity, they cannot be bent or stretched [1, 2]. Recently, TEGs using rubber-based stretchable substrates such as polydimethylsiloxane have been studied to attach to curved heat sources such as heat piping or the human body [3–10]. However, a stretchable substrate made of rubber has a low thermal conductivity, so there is a problem in that thermal loss in the substrate increases. Therefore, instead of using a stretchable substrate to make the entire device deformable, we devised a method to achieve stretchability by making the substrate an origami structure, as shown in Figure 8.1. The origami structure of the substrate can provide elasticity to the entire device by bending only at the folding hinges. This origami design means that the substrate material itself does not need to be stretchable, enabling the use of materials with excellent thermal performance. In addition, since there are flat plate parts in the origami substrate, the TE material does not need stretchability, and TE materials that are hard and have good TE performance can be used.

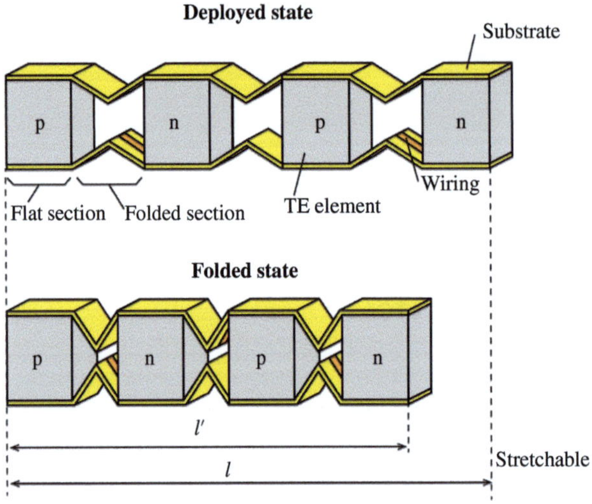

Figure 8.1 Stretchable TEG using origami structure. *Source*: Adapted from [11]/CC BY 4.0.

8.2 Design of Origami Structure TEG

To realize a flexible TEG, we focused on the structure of the substrate. Note that the word "flexibility" is used here to refer to the property with both bendability and stretchability. Using materials with high thermal conductivity, such as metals, leads to significantly different difficulty levels in achieving a bendable TEG and a stretchable TEG. The bendable TEG can be realized by applying a thin substrate. However, the stretchable TEG cannot be realized even if a thin substrate is applied. Therefore, we adopted an origami structure as the substrate structure of the TEG. Hereafter, we refer to TEGs that apply origami structures as "origami TEG." The structure of the origami TEG initially proposed in this research is shown in Figure 8.1. Folding occurs through localized bending, allowing the entire origami TEG to have stretchability. Additionally, since there are nondeforming areas in the origami TEG, rigid TE elements with high performance can be mounted in nondeforming areas. A copper–polyimide laminate was selected as the structural substrate because it offers flexibility for bending while maintaining dimensional rigidity and stability across a broad temperature range. The stretchability of the origami TEG is realized by folding the crease parts. In this study, TE elements of the p-type and n-type were placed in an alternating sequence and connected in series to create a π-shaped module.

The manufacturing steps are outlined in Figure 8.2. As illustrated in Figure 8.2a, the polyimide–copper substrate, consisting of a 12.5 μm-thick polyimide layer and a 9 μm-thick copper layer, was patterned using a cutting plotter (GRAPHTEC, CE6000-40). The upper and lower substrates are both 16 × 16 mm squares with nine 2 mm square holes in each substrate. The cut polyimide-Cu substrate was fixed to a glass substrate, and Cu was patterned into a wiring shape by photolithography. Then, both substrates were folded into a shape consisting of mountain and valley folding. Finally, the TE elements (Toshima Seisakusho, p-type: $Bi_{0.3}Sb_{1.7}Te_3$,

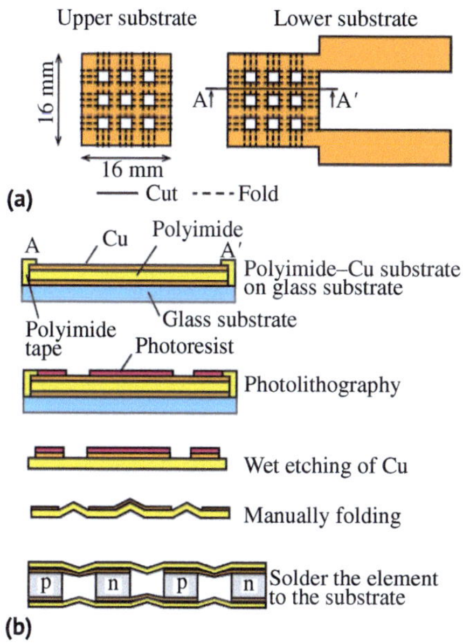

Figure 8.2 Overview of the design concept and manufacturing sequence for the origami TEG. (a) Layout illustrating the patterned cutting paths and designated fold lines. (b) Outline of the microfabrication steps carried out using photolithographic techniques. *Source*: Adapted from [11]/CC BY 4.0.

n-type: Bi_2Te_3) were fixed to the flat parts of the origami substrate using cream solder (Sun Hayato Co., Ltd., SMX-H05). The size of each TE element is 2 × 2 × 1 mm.

8.3 Evaluation Methods of Origami TEG

We evaluated the output characteristics of the origami TEG. The output power of the origami TEG with respect to the heat source temperature before and after folding was measured. The open-circuit voltage V_{TEG} (V) of the origami TEG was clarified from the temperature difference applied to the TEG and the Seebeck coefficient of the TE element. The Seebeck coefficient of the TE element can be determined directly from the open-circuit voltage V_{TEG}. The origami TEG also possesses an internal resistance, R_{TEGc}. When a load is connected, a current flows through the circuit, producing a voltage drop across this internal resistance and altering the delivered power. As a result, the relationship between output voltage, output current, and the internal resistance can be explicitly characterized. By placing a source meter in the load resistance part, we measured the output power of the origami TEG. First, a temperature difference was applied between the top substrate and the bottom substrate of the origami TEG. As shown in Figure 8.3, this was achieved by placing the origami TEG on the ceramic heater and placing an aluminum block of 20 × 21 × 30 mm on the top surface of the origami TEG. The origami TEG was connected to a source meter (Keithley, 2614B) and voltage was applied, and the current was measured.

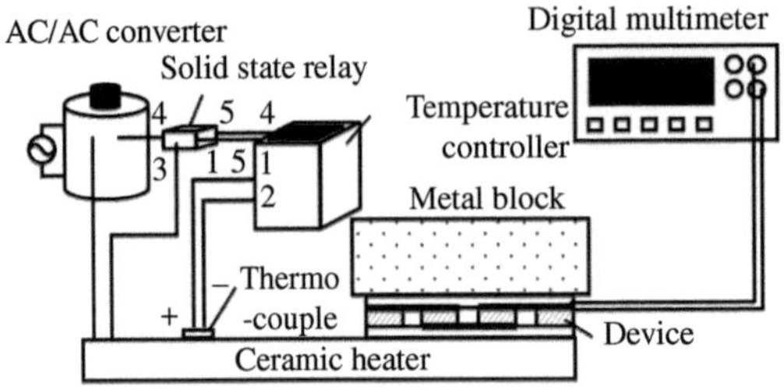

Figure 8.3 Schematics of measuring the current–voltage characteristics of the origami TEG. *Source*: Adapted from [11]/CC BY 4.0.

The ceramic heater was operated at temperatures ranging from 313 to 393 K, increasing in steps of 20 K. The ambient temperature during the experiments was 295 K.

The output voltage of the origami TEG with the eight-pair with respect to the heat source temperature before and after folding was measured. A ceramic heater (Sakaguchi Heat Transfer, MS-2) was driven by an AC/AC converter (Yamato Electric Co., Bolt Slider) to generate heat. The heater temperature was monitored using a thermocouple and regulated through feedback control with a temperature controller (Omron, E5CC-RW0AUM-000). The fabricated origami TEG was then placed on the ceramic heater, and a metal block (65 mm in diameter, 28 mm in height) was positioned on top to promote cooling of the upper surface. The output voltage of the origami TEG was measured using a digital multimeter (Keithley Instruments, Model 2000). The heater temperature was varied from 313 to 393 K in increments of 20 K, and the corresponding output voltage was recorded. The room temperature was 298 K.

8.4 Output Characteristics of Origami TEG

First, the stretchability of the fabricated origami TEG was evaluated. Figure 8.4 shows the completed origami TEG. As illustrated in Figure 8.4c, the device is shown in its unfolded state after the eight p–n element pairs have been installed. The overall active length of the structure, not including the padding, is 15 mm. The TE elements were separated 1.8 mm long in the flat state. In the folded device, the length between the TE elements was 1.0 mm. On the other hand, Figure 8.4d shows the folded state, and the total length of the device was 12 mm. This result indicates that an origami TEG capable of 20% stretchable deformation was achieved.

Next, the basic characteristics of the origami TEG with a single pair of TE elements were measured. Figure 8.5a and b shows current–voltage characteristics of the flat and folded states, respectively. Both the flat and folded configurations exhibit clear linear I–V behavior. The voltage–current relation for the TEG can be described as follows:

$$V_L = -R_{TEG}I + V_{TEG} \tag{8.1}$$

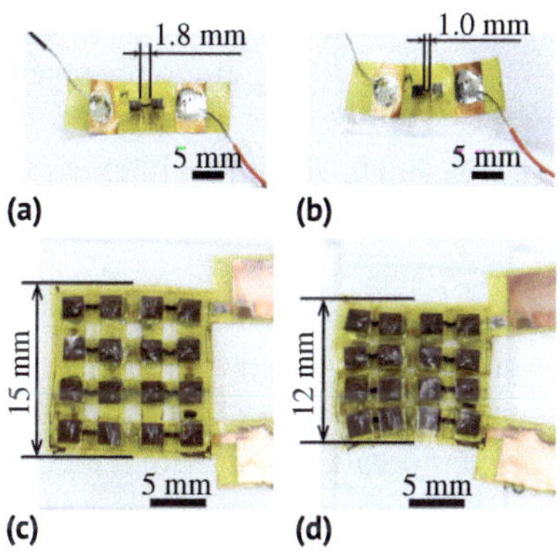

Figure 8.4 Photographs of the fabricated origami TEG. A single pair of the origami TEG in (a) the flat state and (b) the folded state. Eight pairs of the origami TEG in (c) the flat state and (d) the folded state. *Source*: Adapted from [11]/CC BY 4.0.

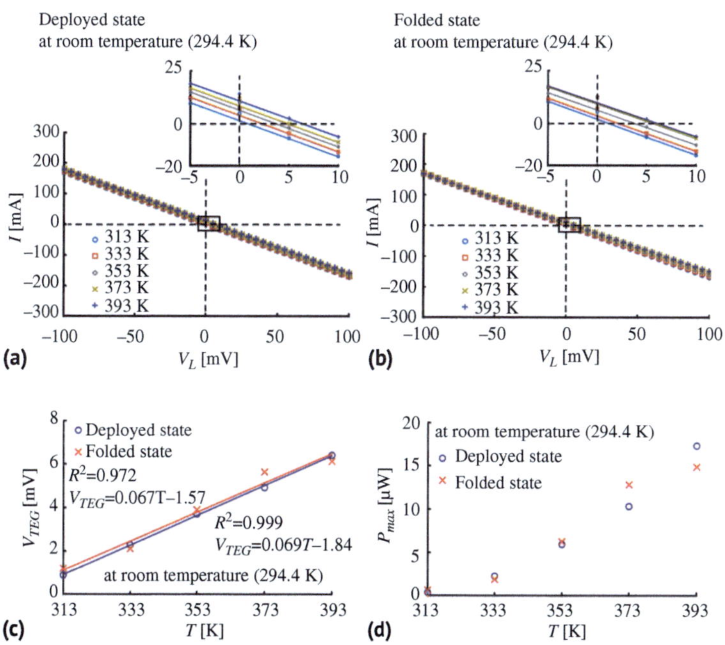

Figure 8.5 Measurement results of the origami TEG in the case of a single pair of p-type and n-type TE elements. Relationship between the output current and the input voltage for (a) the deployed state and (b) the folded state. (c) Relationship between open-circuit voltage and temperature. (d) Relationship between maximum output power with respect to the heat source temperature. *Source*: Adapted from [11]/CC BY 4.0.

In this expression, V_{TEG} represents the open-circuit voltage generated by the origami TEG, V_L is the voltage applied by the source meter, R_{TEG} denotes the device's internal resistance, and I is the measured current. Consequently, the internal resistance of the TEG consisting of a single p–n pair was obtained from the slopes of the linear fits shown in Figure 8.5a and b. From Eq. (8.1), the open-circuit voltage with respect to the temperature can be obtained. Figure 8.5c shows the open-circuit voltage of the origami TEG of a single pair of TE elements with respect to the heat source temperature obtained from Figure 8.5a and b. The fitting lines of the deployed and folded states have $R^2 = 0.999$ and $R^2 = 0.972$, respectively. When the heat source temperature was 393 K, the deployed origami TEG showed 6.41 mV, and the folded device showed 6.12 mV open-circuit voltage. When the heat source temperature increased, a temperature difference was generated between the top and the bottom of the origami TEG, so that the open-circuit voltage increased. The relationship between the output power and the heat source temperature is shown in Figure 8.5d. As the temperature difference increases, the output power also increases. When the heat source temperature was 393 K, the output power of the unfolded state and the folded state were 17.3 and 14.9 μW, respectively. Although the output power is proportional to the square of the temperature difference, since the heat-radiation surface is not actually exposed to room temperature and the temperature difference is not sufficient, the output power is considered to be decreasing. These results indicate that the origami TEG that can be stretched can be fabricated by using folded substrates (wirings).

To evaluate the device under a more practical configuration, the open-circuit voltage of a TEG incorporating eight p–n pairs was measured, since the effective installation area differs between a single-pair and an eight-pair module. The temperature-dependent output voltage is presented in Figure 8.6. Both the flat and folded state of the origami TEG exhibited highly linear fitting curves, and the output increased proportionally with the temperature of the heat source, confirming that the device functions effectively as a TEG. The folded device, however, produced a lower voltage than the flat configuration. This reduction is attributed to the presence of small air gaps formed between the folded substrate and the heater or metal block, which impede heat transfer. Notably, when the open-circuit voltage of a single deployed p–n pair is multiplied by eight, the resulting value closely matches the voltage measured from the eight-pair module. This indicates that a sufficiently large temperature drop was established across the entire folded structure. Overall, these results demonstrate that the TEG can be successfully implemented on a substrate shaped through folding.

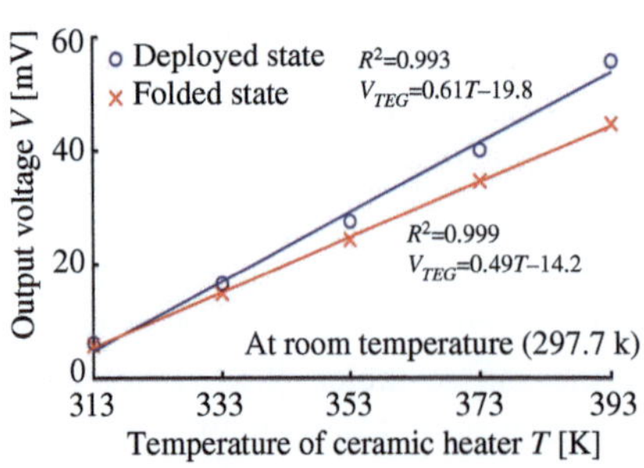

Figure 8.6 Measured output voltage with respect to the heat source temperature. *Source*: Adapted from [11]/CC BY 4.0.

8.5 Design of Origami Fin for Origami TEG

In the origami TEG mentioned previously, the folded structure was a valley fold. However, we thought that if the folded structure was a mountain fold, the folded structure could function as a heat-radiation fin (origami fin) [12]. Figure 8.7 shows the origami TEG with the origami fin, which we proposed. Hereafter, we refer to the origami TEG with origami fins as the origami fin TEG. The origami-fin TEG has stretchability similar to the origami TEG described earlier. Therefore, a non-stretchable copper–polyimide film can be used as the substrate. Furthermore, since the origami-fin TEG has flat regions, rigid TE elements can be mounted on these flat areas. In this study, BiTe-based elements were selected.

Figure 8.8 shows the fabrication process of the origami-fin TEG. Fabrication began with preparing the top and bottom substrates from copper–polyimide laminates, consisting of a 25 μm polyimide layer and an 8 μm copper layer. The wiring pattern was then defined by selectively ablating the copper layer with a UV laser system (OPI Corporation, OLMUV-335-5A-K). As explained in the previous section, a cutting plotter, photolithography, and etching were used to cut the substrate for the origami TEG. However, this time, a UV laser cutting machine was used to cut the substrate, taking into consideration both processing time and processing accuracy. With a UV laser cutting machine, processing can be completed in a single process, significantly shortening the fabrication time. High-precision processing is also possible, as the laser line width is 20 μm. After the substrate is cut with the UV laser, it is folded in a mountain fold. Then, TE elements (TOSHIMA Manufacturing, p-type: $Bi_{0.3}Sb_{1.7}Te_3$, n-type: Bi_2Te_3) were soldered onto the electrodes. TE elements with dimensions of 1.5 mm × 1.5 mm × 1.0 mm were implemented as a π-type TEG.

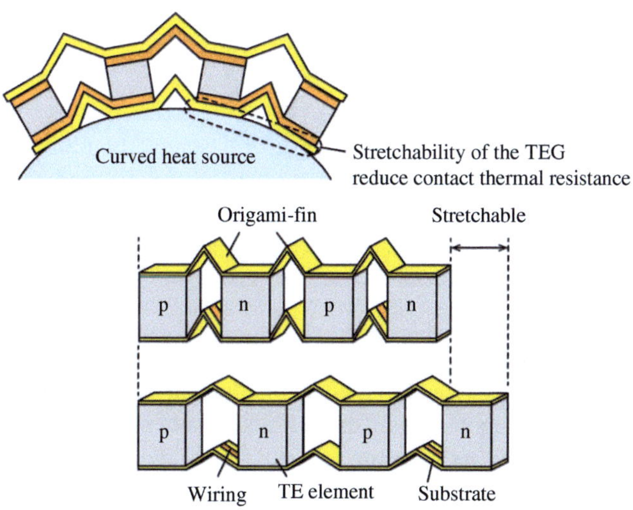

Figure 8.7 Schematics of the origami fin for origami TEG. The origami fin between the TE elements has high stretchability and heat-radiation performance. *Source*: Adapted from [12]/CC BY 4.0.

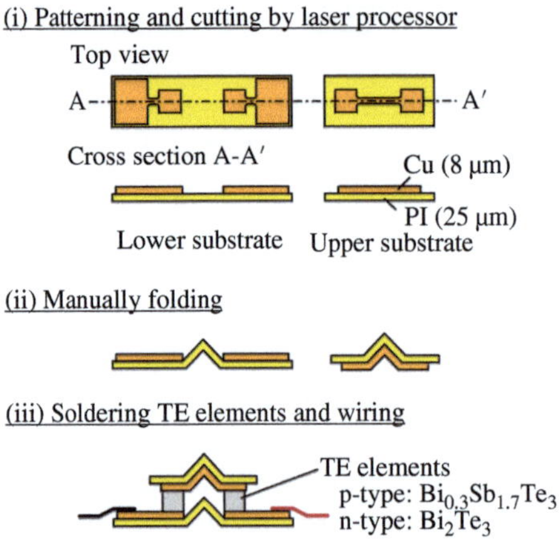

Figure 8.8 Fabrication process of the origami TEG with origami fin. *Source*: Adapted from [12]/CC BY 4.0.

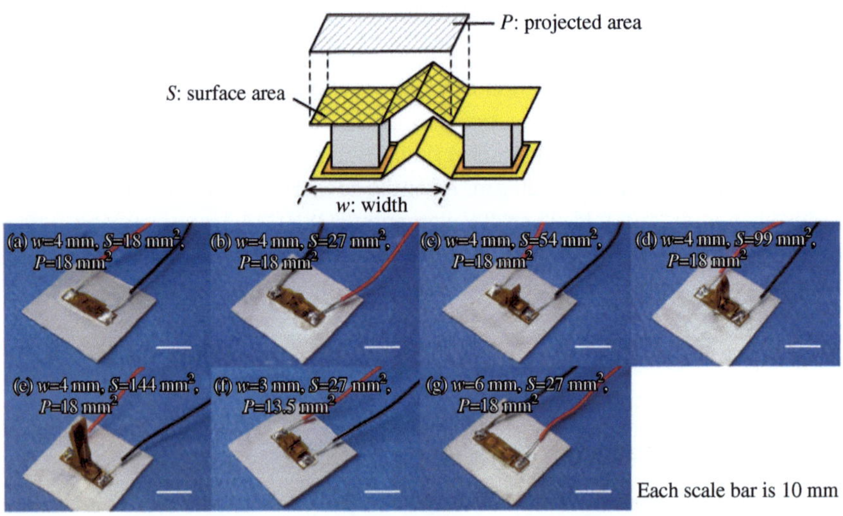

Figure 8.9 Overview of the evaluation unit and examples of origami-fin TEGs with different geometries. (a)–(e) Devices with a constant width of 4 mm and surface areas of 18, 27, 54, 99, and 144 mm². (f, g) Devices with widths of 3 and 6 mm, each having a surface area of 27 mm². *Source*: Adapted from [12]/CC BY 4.0.

The fabricated origami-fin TEGs are shown in Figure 8.9. As shown at the top of Figure 8.9, an evaluation unit was defined to include one flat part and one folded part. In this unit, the width is denoted as w (mm), the substrate surface area as S (mm^2), and the projected area of the substrate as $w \times P$ (mm^2). Each origami-fin TEG was attached to a 0.5 mm-thick aluminum plate using a high thermal conductivity adhesive (Cemedine SX1008) so that its shape could be maintained. To evaluate the heat-radiation performance, five origami-fin TEGs with different surface areas ($S = 18$, 27, 54, 99, and 144 mm^2) were fabricated (Figure 8.9a and b). All TEGs had the same width of 4 mm, so their projected areas were identical. The fin with

$S = 18\ \text{mm}^2$ was kept completely flat, meaning that its surface area and projected area were equal (Figure 8.9a). The other four fins increased their surface areas, ranging from 1.5 to 8 times the projected area, by folding (Figure 8.9b–e). To evaluate output stability during stretching and contraction, three additional fins were fabricated using the same surface area $(S = 27\ \text{mm}^2)$ but with different widths of 3, 4, and 6 mm (Figure 8.9b, f, and g). The fin with $w = 6$ mm was fully stretched to a flat state, while the other two were folded to reduce their widths.

8.6 Output Characteristics of Origami TEG with Origami Fin

Evaluation of the origami-fin TEGs involved determining their open-circuit voltage V_{TEG}, which served as an indicator of both the fins' thermal dissipation effectiveness and the consistency of the electrical output. Using the setup shown in Figure 8.10a, the current–voltage $(I–V)$ characteristics of the origami-fin TEGs were measured, and the open-circuit voltage V_{TEG} of the origami-fin TEGs was derived from these measurements. Good thermal contact between the heater and the origami-fin TEG was ensured by applying a high thermal conductivity grease (Shin-Etsu Silicones, G-747). Using a Peltier temperature controller (VICS, VTH1.8K-70S), the heater temperature was swept from 313 to 353 K in 10 K increments. The ambient temperature during testing was approximately 298 K. The current–voltage characteristics were obtained by sweeping the applied voltage with a source measurement unit (Keithley Instruments, 2614B). As an example, Figure 8.10b presents the $I–V$ curve measured for the origami-fin TEG with a width of 4 mm and a surface area of 27 mm². The open-circuit voltage V_{TEG} was determined from the x-intercept of the fitted regression line. Although the measurements were performed at several heater temperatures to evaluate temperature dependence, the value at 323 K was used for comparison among devices.

To verify the reliability of the measurement results, the open-circuit voltage V_{TEG} was also estimated using finite element analysis. As illustrated in Figure 8.11, a two-dimensional model was constructed with the same dimensions and materials as the

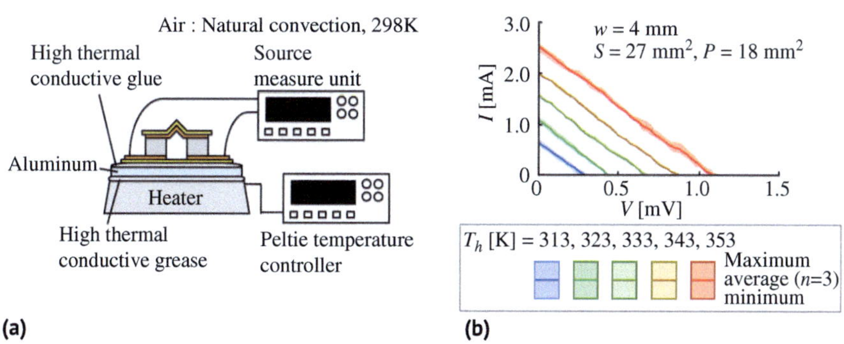

Figure 8.10 (a) Experimental setup for measuring V_{TEG} of TEGs. (b) The $I–V$ characteristics of the origami TEG with origami fin with w of 4 mm and S of 18 mm². *Source*: Adapted from [12]/CC BY 4.0.

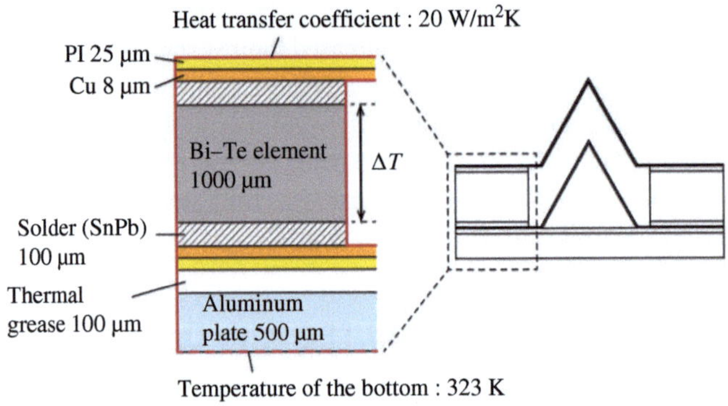

Figure 8.11 Configuration of 2D simulation models. *Source*: Adapted from [12]/CC BY 4.0.

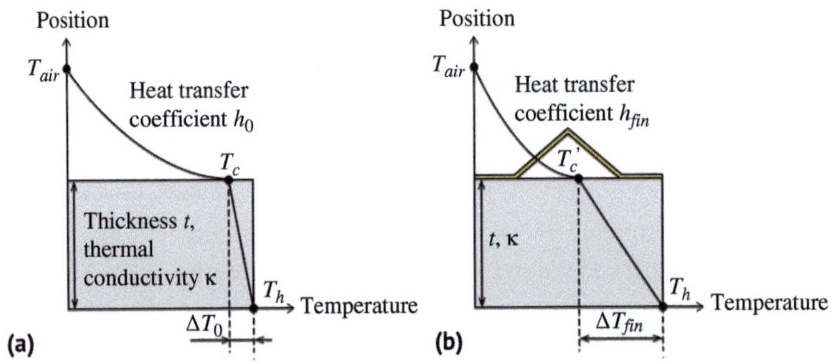

Figure 8.12 Conceptual representations used to derive h_{fin}. (a) TE element modeled without an attached *fin*. (b) TE element modeled with an added fin structure. *Source*: Adapted from [12]/CC BY 4.0.

origami-fin TEGs. In the simulation, the bottom surface of the TEG was fixed at 323 K, and a heat transfer coefficient of 20 W/m^2K was applied to the outer boundaries. This value falls within the typical range for natural convection from a solid surface to air. Thermal-contact resistances between the individual materials were not considered. A steady-state temperature distribution was then computed, and V_{TEG} was obtained by multiplying the temperature difference across the BiTe-based TE element (ΔT) by its Seebeck coefficient. According to the material data sheet, the Seebeck coefficient difference at 323 K is $Sp - Sn = 317.2 \times 10^{-6}$ V/K, where Sp and Sn denote the Seebeck coefficients of the p-type and n-type, respectively.

A heat transfer coefficient (h_{fin}) between the origami fin and the surrounding air can be obtained using the previously obtained V_{TEG}. This heat transfer coefficient (h_{fin}) evaluates the heat-radiation performance of the origami fin. A simple model, as shown in Figure 8.12a and b, was used to derive h_{fin}. The bottom temperature of the TE element was denoted as T_h (K), the top temperature as T_c (K), and the ambient air temperature as T_{air} (K). The thermal conductivity of the TE element was κ (W/mK), and the thickness of the TE element was t (m). With a constant heat flux and a heat transfer coefficient h_0 (W/m^2K) between the TE element and air, the resulting temperature difference ΔT_0 $(= T_h - T_c)$ can be obtained as follows:

$$\Delta T_0 = T_h - T_c = \frac{t/\kappa}{t/\kappa + 1/h_0}\left(T_h - T_{air}\right) \tag{8.2}$$

With an origami fin attached to the top of the TE element, the heat transfer coefficient to air is h_{fin}, and the applied temperature difference is ΔT_{fin}. This value is obtained in the same way as ΔT_0:

$$\Delta T_{fin} = T_h - T_c' = \frac{t/\kappa}{t/\kappa + 1/h_{fin}}\left(T_h - T_{air}\right) \tag{8.3}$$

where T_c' is the top temperature of the TE element in the presence of the fin. The relation between ΔT_{fin} and ΔT_0 is therefore expressed as:

$$\frac{\Delta T_{fin}}{\Delta T_0} = \frac{h_{fin}\left(\kappa/t + h_0\right)}{h_0\left(\kappa/t + h_{fin}\right)} \tag{8.4}$$

The BiTe-based TE element has a thermal conductivity κ of roughly 1 W/m·K. With an element thickness of about 1 mm, κ/t is on the order of 10^3 W/m^2·K. Meanwhile, natural-convection heat transfer between solids and air is only 10^0–10^1 W/m^2·K, which is negligible compared to κ/t. Thus, Eq. (8.4) simplifies to:

$$\frac{\Delta T_{fin}}{\Delta T_0} \approx \frac{h_{fin}}{h_0} \tag{8.5}$$

If V_0 represents the V_{TEG} of the flat TEG with the minimum surface area $\left(S = 18\ \text{mm}^2\right)$ and V_{fin} represents that of the origami-fin TEGs, the ratio V_{fin}/V_0 can be written as follows:

$$\frac{V_{fin}}{V_0} = \frac{\Delta T_{fin}}{\Delta T_0} \tag{8.6}$$

Because $V_{TEG} = \Delta T\left(S_p - S_n\right)$, Eqs. (8.5) and (8.6) allow h_{fin} to be expressed using the measured V_0, V_{fin}, and an arbitrarily selected h_0 as follows:

$$h_{fin} = \frac{V_{fin}}{V_0} h_0 \tag{8.7}$$

For the analysis, h_0 was fixed at 20 W/m^2·K, and h_{fin} was subsequently computed. Figure 8.13 shows the results derived from the measured and simulated V_{TEG}.

Figure 8.13a shows that the origami *fin* increased both the measured and simulated V_{TEG}, demonstrating its role as a heat-radiation *fin*. In this simulation, V_{TEG} was improved with increasing surface area S. Measurements revealed that all fins with $S > 54$ mm^2 generated a higher V_{TEG} than the flat TEG with $S = 18$ mm^2. The fin with $S = 144$ mm^2 yielded a V_{TEG} 1.9 times higher than that of the flat, corresponding to an effective heat transfer coefficient h_{fin} 1.9 times larger than that of the unfolded substrate. The V_{TEG} value depends on the specific TE element, whereas h_{fin} offers a constant index of the heat-radiation capability of the origami fin and changes only if h_0 is reset. As shown in Figure 8.13b, both the measured and simulated values of V_{TEG} and h_{fin} remained nearly unchanged with respect to the width w. Experimentally, the V_{TEG} of the origami TEG with $w = 3$ mm was 92% of that of the flat TEG with $w = 6$ mm. This indicates that even a 50% reduction in fin

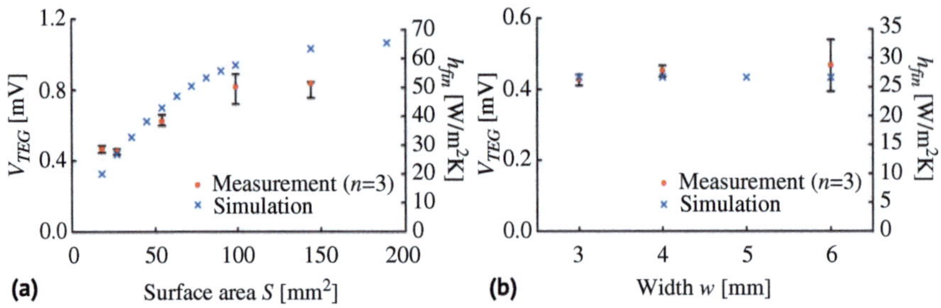

Figure 8.13 Measurement and simulation results. The temperature of the heater was 323 K, and that of the air was 298 K. (a) V_{TEG} and h_{fin} with respect to the surface area S. (b) V_{TEG} and h_{fin} with respect to the width w. *Source*: Adapted from [12]/CC BY 4.0.

width results in only 8% decrease of h_{fin}. These findings demonstrate that the stability of the origami fin is maintained under stretching deformation. Consequently, an origami fin can be effectively incorporated into a stretchable TEG while preserving its heat-radiation functionality.

8.7 Design of Self-folded Origami TEG

In the following sections, we fabricated an array of origami-fin TEG and evaluated its performance [13]. The folding parts in the origami-fin TEG described earlier had been fabricated by hand. However, folding all of the folding parts of an array of origami-fin TEG by hand would be extremely time-consuming. Therefore, we consider folding them automatically using self-folding by applying a shrinkable layer.

Figure 8.14 presents a schematic of the self-folded origami TEG. As shown in Figure 8.14a, the TE element was positioned between the top and bottom substrates, and a slit was introduced in the folding region to reduce bending rigidity. To avoid unintended folding triggered by the heat-shrinkable film used for self-folding, the shrink layer should be attached only after all fabrication steps involving thermal stimuli—such as TE element mounting—have been completed. The design should allow the shrink layer to be attached through a linkage mechanism so that all hinges of the self-folding origami TEG can reliably fold into the intended geometry. When the entire device is heated, the shrink layer produces a driving force that actuates the hinge and transforms the structure into the origami configuration shown in Figure 8.14b. The folded region can still be deformed to various angles even after the self-folding process. Consequently, as shown in Figure 8.14c, the self-folded origami TEG exhibits high stretchability and bendability despite the rigidity of the mounted TE element, enabling it to conform to curved high-temperature surfaces of arbitrary shape while maintaining low contact thermal resistance.

Figure 8.15 outlines the modeling approach used to represent the linkage mechanism of the self-folding origami TEG and the method used to design its hinges. When a heat-shrink film is applied to one side of a substrate, that side naturally forms a valley fold because the shrinking layer pulls the opposing material inward during

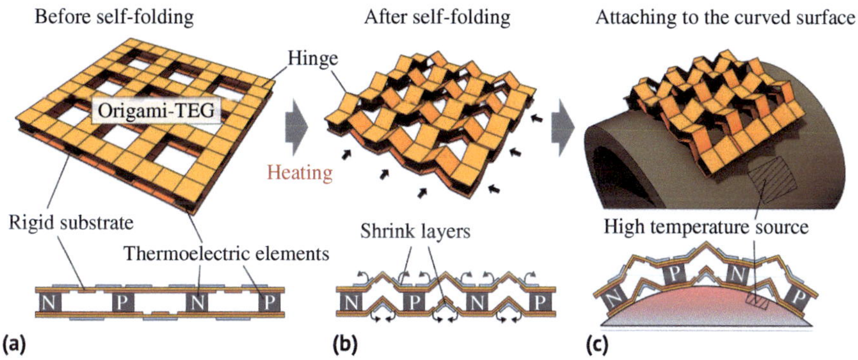

Figure 8.14 Conceptual illustrations of the self-folding origami TEG. (a) Configuration of the device prior to activation. (b) Structure formed after thermal actuation, where the shrink layer contracts and induces autonomous folding. (c) Self-folded device conforming to a curved thermal surface; the grid-based bellows geometry provides substantial flexibility. *Source*: Adapted from [13]/CC BY 4.0.

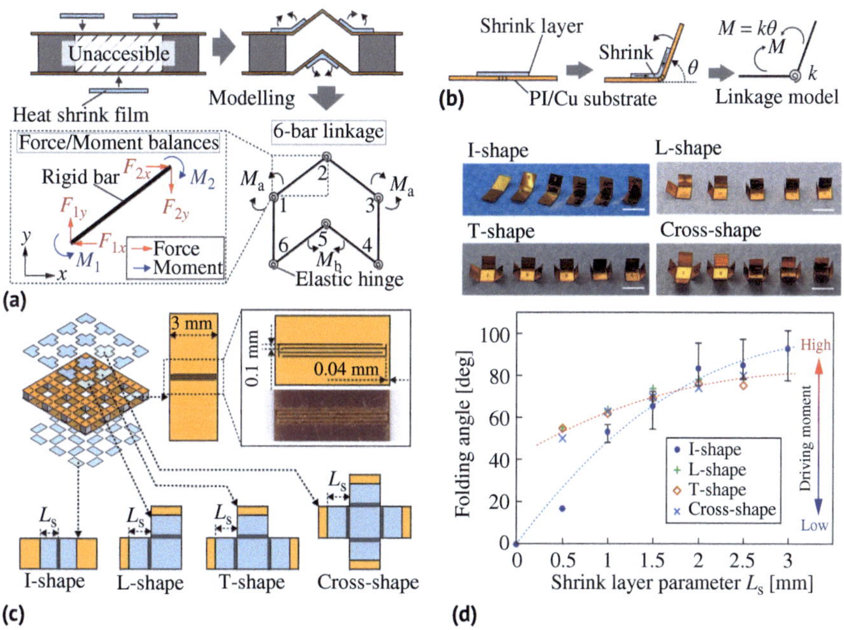

Figure 8.15 Overview of the self-folding based on a linkage-driven approach. (a) Mechanical representation of the paired origami TEG system, including force and moment balancing in the linkage configuration. (b) Conceptual model of the hinge elements. (c) Design of hinge prototypes and definition of the shrink-layer parameter L_s. (d) Variation of the folding angle θ as a function of L_s for each hinge. Scale bar: 5 mm. *Source*: Adapted from [13]/CC BY 4.0.

heating. For this reason, placing the shrink layer on only one face prevents the structure from forming both mountain and valley folds. Once the TE element is mounted, the inner region of the folded structure, illustrated in Figure 8.15a, can no longer be accessed. To analyze how the folding motion propagates through the structure, the pair of origami TEGs was simplified into a six-bar linkage composed of rigid segments and elastic hinges. In this model, a driving force applied to one hinge causes the corresponding hinge on the opposite side to rotate. Force and moment balance analysis showed that all six hinges reach their intended folding angles when only three hinges—numbers 1, 3, and 5—are actively driven. The forces required at each hinge were obtained from static equilibrium equations, which account for the loads on each link and the moment balance at every joint. Because the structure contains no built-in actuators and ultimately settles into a mechanically stable configuration, these equilibrium relationships fully determine the required driving forces.

As shown in Figure 8.15b, when the hinge is modeled as an elastic joint, the moment M generated at the hinge can be expressed by its elastic modulus k and the folding angle θ. This implies that, for a constant elastic modulus, the driving force at the hinge is controlled by regulating the folding angle. Accordingly, as illustrated in Figure 8.15c, the folding angle is controlled by adjusting the attachment configuration of the heat-shrink film. Four types of shrink-layer geometries—I-, L-, T-, and cross-shaped—were employed. For each type, the parameter L_s, which specifies the effective attachment area, was defined, and the driving force generated at the hinge was tuned by varying L_s. The test specimens and the self-folding origami TEG were fabricated using three primary materials. A 33-μm polyimide–copper laminate (Toray Advanced Materials Korea, Metaloyal®) served as the substrate, and a 12-μm polyolefin heat-shrink film (Taiyo Electric Industry, HS-2520) acted as the active layer. A 25-μm double-sided adhesive film (NEION Film Coating, MHM-SI25) was placed between the substrate and the shrink film to secure their attachment. The crease regions were patterned using a UV laser cutting machine to locally reduce bending stiffness while maintaining electrical connectivity. Figure 8.15d presents the folding results after heating the structures at 82 °C for 3 min (Yamato Scientific Co., DK300) with different values of L_s. The photographs show that the folding angle θ increased with L_s for both hinges. Notably, the relationship between L_s and θ for the I-shaped shrink layer differed from that of the other three geometries. Therefore, separate approximate curves were fitted to the experimental results of the I-shaped and non-I-shaped specimens, and the required shrink-layer parameter L_s for each hinge was subsequently determined from these fitted relationships.

8.8 Output Characteristics of Self-folded Origami TEG

The fabrication process of the self-folded origami TEG is outlined in Figure 8.15. First, the polyimide–copper substrate was patterned using a UV laser cutting machine to form hinge cuts and wiring traces by selectively removing the copper layer (Figure 8.16a–i). TE elements—p-type $Bi_{0.3}Sb_{1.7}Te_3$ and n-type Bi_2Te_3, each measuring 1.5 × 1.5 × 1 mm—were precoated with a special solder (Cerasolza, Kuroda Techno Co., 168936) using an ultrasonic soldering machine (Kuroda Techno Co., USM-560) to enhance adhesion. These elements were then placed onto the cream-solder-coated substrate and reflowed on a hot plate at 300 °C (Figure 8.16a-ii). The shrink layer and double-sided tape were first laminated onto a weak-adhesive sheet. A laser was then used to cut only the shrink layer and tape, leaving the underlying sheet intact. After removing the excess material, the remaining pattern was

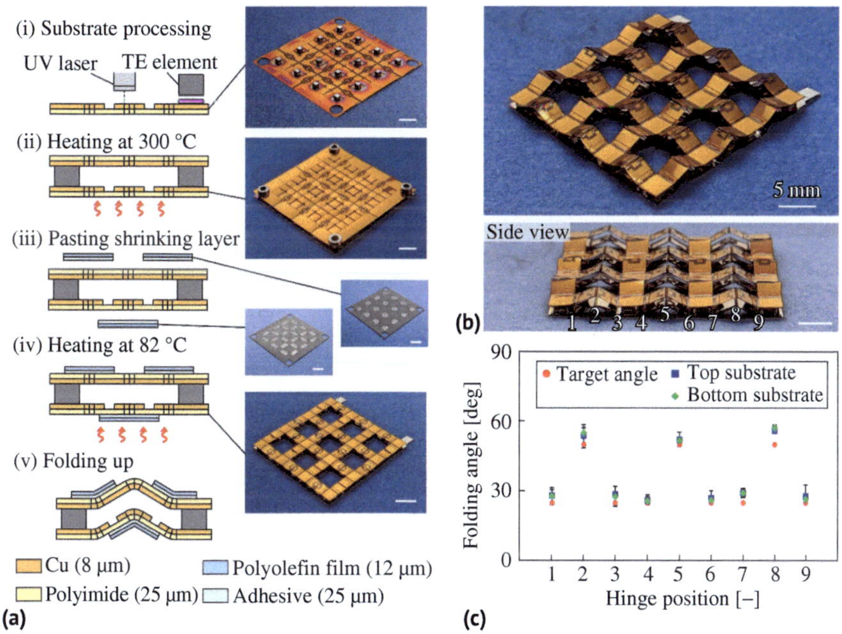

Figure 8.16 Overview of the fabrication sequence for the self-folding origami TEG. (a) Step-by-step process used to construct the self-folding structure. (b) Photograph of the completed origami TEG after self-folding. (c) Evaluation of hinge-folding precision for each joint. Scale bar: 5 mm. *Source*: Adapted from [13]/CC BY 4.0.

transferred onto the self-folded origami TEG (Figure 8.16a-iii). The device was subsequently heated in an oven at 82 °C for three minutes (Figure 8.16a-iv), during which the polyolefin film contracted and generated the driving force required for folding, transforming the structure into its designed configuration (Figure 8.16a-v). A photograph of the completed self-folded origami TEG is shown in Figure 8.16b. The device measured 30 mm × 30 mm before self-folding and 27 mm × 27 mm after folding. The origami TEG was self-folded nearly as intended. Figure 8.16c shows the folding angles measured for both the upper and lower substrates at each hinge. Although hinges 1, 3, 4, and 6 were intended to rotate upward by 25° from the flat configuration, the upper substrate differed by an average of only 3.2°, and the lower substrate by about 3°. Considering that hinge specimens generally displayed an angular spread of roughly 5.7°, the linkage-based design approach can be regarded as highly precise in realizing the targeted folding geometry. In addition, the bellows-like arrangement of the substrate provides the TEG with substantial mechanical flexibility, allowing it to follow curved surfaces with nonzero Gaussian curvature while keeping mechanical stress low.

Performance testing of the self-folded origami TEG was carried out by examining its electrical behavior under controlled thermal conditions. Figure 8.17a outlines the three thermal-contact arrangements considered: a folded device pressed against a planar heater, an unfolded device placed on the same planar surface, and a folded device interfaced with a curved heater. A schematic of the measurement system is provided in Figure 8.17b. The temperature of the heating unit was stepped from 40 to 80 °C in 10 °C intervals. Electrical characterization was conducted with a

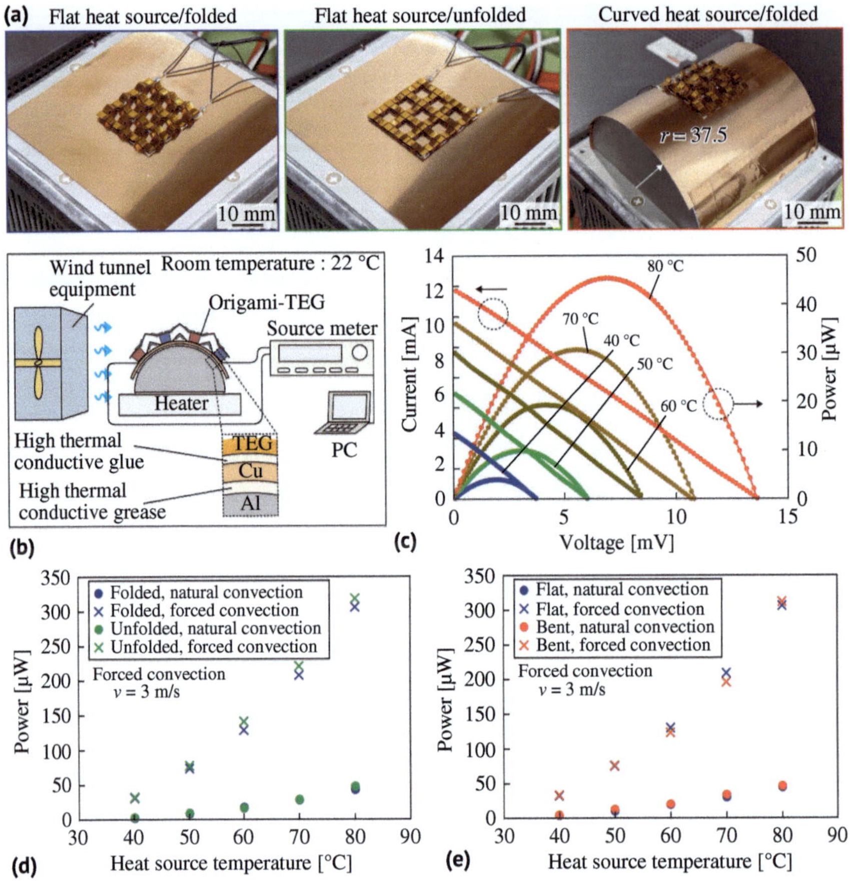

Figure 8.17 Performance assessment of the origami TEG. (a) Images of the device in three different test configurations. (b) Instrumentation used for evaluating the fundamental electrical characteristics. (c) I–V and power–voltage responses of the folded device on a planar heat source. (d) Output-power robustness during stretching deformation. (e) Output-power robustness under bending deformation. *Source*: Adapted from [13]/CC BY 4.0.

source meter while the device was exposed either to air or to a 3 m/s airflow generated by a wind tunnel (Kanomax, S0522195). For the curved-heater condition, a semi-cylindrical aluminum block with a radius of 37.5 mm served as the thermal interface. Figure 8.17c displays the response obtained in natural-convection conditions. The I–V curves remained linear throughout the temperature sweep, and the power output rose steadily with heater temperature, reaching a peak of 44.7 μW at 80 °C.

Figure 8.17d compares the peak power outputs of the origami TEG in both its folded and unfolded forms while being placed on a flat heater. Under natural convection, the difference in peak power between the two states was 8.5%, and under forced convection it was 5.3%, showing that the folding state has little influence on overall performance. This result suggests that the origami architecture imparts stretchability without compromising device output. Figure 8.17e compares the performance of the self-folded origami TEG when placed on flat and curved heat sources. The average relative differences in maximum output were 10.9% under natural convection and 3.2% under forced convection, demonstrating that attachment to a curved surface has only a negligible impact on performance. These findings confirm that the self-folded origami TEG functions effectively as a flexible device capable of maintaining stable output under both stretching and bending deformations.

Finally, the stability of the internal resistance during deformation of the self-folded origami TEG was evaluated, as summarized in Figure 8.18. Figure 8.18a shows the rate of change in internal resistance ($\Delta R/R_0$) for a single unit as the folding angle was varied. The maximum change was only 0.2%, demonstrating that the device maintains highly stable electrical characteristics under stretching deformation. For bending angles between 0° and 80°, the structure exhibits a stretching ratio of 83%. Given that most conventional flexible TEGs achieve less than 50% stretchability, the self-folded origami TEG demonstrates markedly superior resistance to stretching deformation [6–8].

Figure 8.18b shows that the resistance stays almost unchanged even after 1000 stretching cycles at 20% strain. Figure 8.18c presents the change in resistance ($\Delta R/R_0$) when the self-folded origami TEG was bent in both convex and concave directions at various bending radii r. Even at $r = 18.6$ mm, $\Delta R/R_0$ was limited to only 0.6% for both bending modes. Figure 8.18d indicates that the internal resistance shows almost no variation, even after 100 bending cycles performed across

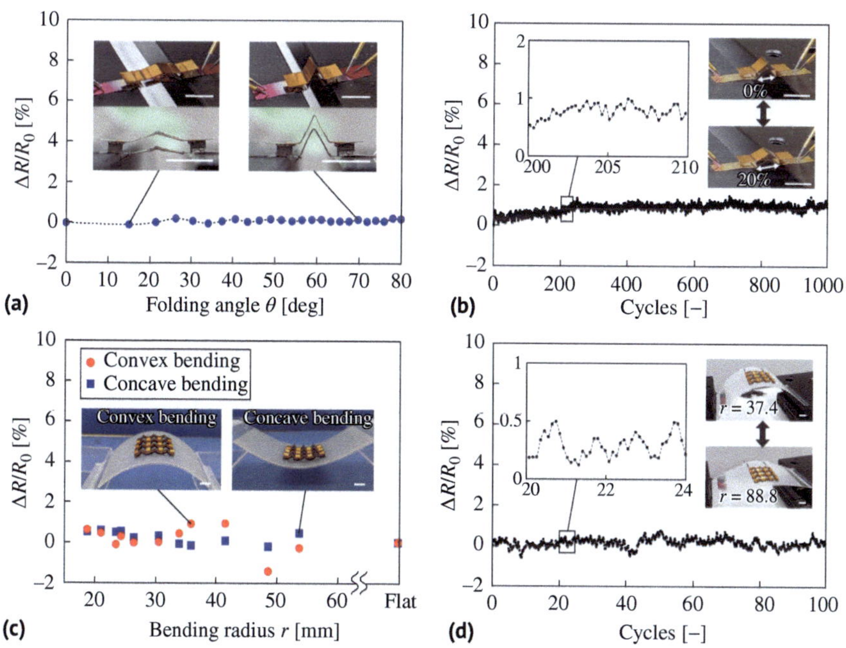

Figure 8.18 Assessment of electrical stability under mechanical deformation. (a) Variation in resistance as a function of folding angle. (b) Results of cyclic stretching at 20% strain. (c) Changes in resistance during concave and convex bending. (d) Resistance behavior under repeated bending cycles. Scale bar: 5 mm. *Source*: Adapted from [13]/CC BY 4.0.

radii ranging from 37.4 to 88.8 mm. Since the self-folded origami TEG is typically deformed only once during manual attachment to a curved high-temperature surface—and only a few additional times if repositioned—the results of these two cyclic tests indicate that the device possesses more than sufficient durability for practical use. Overall, the findings confirm that the origami TEG maintains excellent internal-resistance stability under both stretching and bending deformation.

References

1 Rowe, D.M. and Min, G. (1996) Design theory of thermoelectric modules for electrical power generation. *IEE Proc. Sci. Meas. Technol.*, 143, 351–6.

2 Snyder, G.J. and Toberer, E.S. (2008) Complex thermoelectric materials. *Nat. Mater.*, 7, 105–14.

3 Suemori, K., Hoshino, S. and Kamata, T. (2013) Flexible and lightweight thermoelectric generators composed of carbon nanotube-polystyrene composites printed on film substrate. *Appl. Phys. Lett.*, 103, 153902.

4 Kim, S.J., We, J.H. and Cho, B.J. (2014) A wearable thermoelectric generator fabricated on a glass fabric. *Energy Environ. Sci.*, 7, 1959–65.

5 Jeong, S.H., Cruz, F.J., Chen, S., Gravier, L., Liu, J., Wu, Z., Hjort, K., Zhang, S.L. and Zhang, Z.B. (2017) Stretchable thermoelectric generators metallized with liquid alloy. *ACS Appl. Mater. Interfaces*, 9, 15791–7.

6 Glatz, W., Schwyter, E., Durrer, L. and Hierold, C. (2009) Bi_2Te_3-based flexible micro thermoelectric generator with optimized design. *J. Microelectromechanical Syst.*, 18, 763–72.

7 Kim, M.K., Kim, M.S., Jo, S.E. and Kim, Y.J. (2012) Flexible thermoelectric generator for human body heat energy harvesting. *Electron. Lett.*, 48, 1015–7.

8 Francioso, L., De Pascali, C., Sglavo, V., Grazioli, A., Masieri, M., and Siciliano, P. (2017) Modelling, fabrication and experimental testing of a heat sink free wearable thermoelectric generator. *Energy Convers. Manag.*, 145, 204–13.

9 Yang, C., Souchay, D., Kneiß, M., Bogner, M., Wei, H.M., Lorenz, M., Oeckler, O., Benstetter, G., Fu, Y.Q. and Grundmann, M. (2017) Transparent flexible thermoelectric material based on non-toxic earth-abundant p-type copper iodide thin film. *Nat. Commun.*, 8, 4–10.

10 Shi, Y., Wang, Y., Mei, D., Feng, B. and Chen, Z. (2018) Design and fabrication of wearable thermoelectric generator device for heat harvesting. *IEEE Robot. Autom. Lett.*, 3, 373–8.

11 Fukuie, K., Iwata, Y. and Iwase, E. (2018) Design of substrate stretchability using origami-like folding deformation for flexible thermoelectric generator. *Micromachines*, 9(7), 315.

12 Akuto, M. and Iwase, E. (2020) An origami heat radiation fin for use in a stretchable thermoelectric generator. *Micromachines*, 11(3), 263.

13 Sato, Y., Terashima, S. and Iwase, E. (2023) Origami-type flexible thermoelectric generator fabricated by self-folding. *Micromachines*, 14(1), 218.

9

Flexible Thermoelectric Composites for Energy Harvesting

Wei Wang

School of Mechanical Engineering, Beijing Institute of Technology, Beijing, China

9.1 Introduction

The 21st century has witnessed an unprecedented surge in global energy demand, driven by rapid industrialization, urbanization, and the proliferation of electronic devices. However, this growing demand has also highlighted the limitations of traditional energy sources, such as fossil fuels, which are not only finite but also contribute significantly to environmental degradation and climate change. The need for sustainable energy solutions has never been more urgent, prompting researchers and engineers to explore alternative energy-harvesting technologies that can provide clean, renewable, and efficient power sources.

One of the most promising avenues in this quest is the development of thermoelectric materials, which can directly convert waste heat into usable electricity. This capability is particularly valuable in a world where a significant amount of energy is lost as heat in industrial processes, automotive systems, and even everyday electronic devices. Thermoelectric energy harvesting offers a way to recover this otherwise wasted energy, improving overall energy efficiency and reducing reliance on nonrenewable resources.

The role of thermoelectric materials in energy harvesting is rooted in the thermoelectric effect, specifically the Seebeck effect, which describes the generation of an electric voltage in response to a temperature gradient across a material. This phenomenon enables the creation of devices that can generate power from heat sources without moving parts, making them highly reliable and suitable for a wide range of applications, from large-scale industrial systems to small-scale wearable electronics.

Despite their potential, traditional thermoelectric materials, such as bismuth telluride (Bi_2Te_3) and lead telluride (PbTe), are often rigid, brittle, and expensive, limiting their applicability in flexible and lightweight systems. This has spurred interest in the development of flexible thermoelectric composites, which combine the desirable properties of thermoelectric materials with the mechanical flexibility and versatility needed for modern applications.

Flexible thermoelectric composites are particularly well-suited for emerging technologies such as wearable electronics, the Internet of Things (IoT), and portable medical devices, where the ability to conform to curved surfaces and withstand mechanical deformation is crucial. By enabling energy harvesting from low-grade heat sources, such as body heat or ambient environmental heat, these materials have the potential to revolutionize the way we power small electronic devices, reducing the need for batteries and contributing to a more sustainable energy ecosystem.

9.2 Overview of Thermoelectric Energy Harvesting

Thermoelectric energy harvesting is a process that leverages the thermoelectric effect to convert thermal energy (heat) directly into electrical energy. This phenomenon is based on the fundamental principles of thermoelectricity, which include the Seebeck effect, Peltier effect, and Thomson effect. The Seebeck effect is the most critical for energy-harvesting applications, as it enables the generation of electricity from a temperature gradient.

9.2.1 Principle of the Thermoelectric Effect (Seebeck Effect)

The Seebeck effect, discovered by Thomas Johann Seebeck in 1821, describes the generation of an electric voltage (thermoelectric voltage) when a temperature difference is applied across a material. This occurs because charge carriers (electrons or holes) in the material diffuse from the hot side to the cold side, creating an electric potential. The magnitude of this voltage is proportional to the temperature gradient and is expressed by the Seebeck coefficient (S), a material-specific property. The governing equation for the Seebeck effect is:

$$V = S \cdot \Delta T \tag{9.1}$$

where:

- V is the generated thermoelectric voltage (in volts),
- S is the Seebeck coefficient (in volts per Kelvin, V/K),
- ΔT is the temperature difference across the material (in Kelvin, K).

The Seebeck coefficient measures how effectively a material can convert a temperature gradient into an electric voltage. Materials with a high Seebeck coefficient are desirable for thermoelectric applications.

9.2.2 Key Parameters in Thermoelectric Energy Harvesting

The performance of thermoelectric materials is evaluated based on several key parameters, which determine their efficiency in converting heat into electricity. These parameters are interrelated and often involve trade-offs, making the design of high-performance thermoelectric materials a complex challenge.

9.2.2.1 Seebeck Coefficient (*S*)

As described earlier, the Seebeck coefficient quantifies the voltage generated per unit temperature gradient. It is influenced by the material's electronic structure, carrier concentration, and bandgap. Higher ***S*** values lead to greater voltage generation for a given temperature difference.

9.2.2.2 Electrical Conductivity (σ)

Electrical conductivity measures how easily electric charge can flow through a material. High electrical conductivity is essential to minimize energy losses due to electrical resistance. It is expressed as:

$$\sigma = \frac{1}{\rho} \tag{9.2}$$

where ρ is the electrical resistivity (in ohm-meters, Ω m).

9.2.2.3 Thermal Conductivity (*k*)

Thermal conductivity measures a material's ability to conduct heat. It is defined as:

$$k = \frac{Q \cdot L}{A \cdot \Delta T} \tag{9.3}$$

where

- Q is the heat flow (in watts, W),
- L is the length of the material (in meters, m),
- A is the cross-sectional area (in square meters, m^2)
- ΔT is the temperature difference (in Kelvin, K).

Low thermal conductivity is desirable to maintain a large temperature gradient across the material, which enhances energy conversion efficiency.

9.2.2.4 Figure of Merit (*ZT*)

The performance of a thermoelectric material is often summarized by its dimensionless figure of merit, ZT, which combines the Seebeck coefficient, electrical conductivity, and thermal conductivity:

$$ZT = \frac{S^2\sigma \cdot T}{k} \tag{9.4}$$

where

- S is the Seebeck coefficient,
- σ is the electrical conductivity,
- k is the thermal conductivity,
- T is the absolute temperature (in Kelvin, K).

A higher ZT value indicates better thermoelectric performance. Currently, materials with $ZT > 1$ are considered suitable for practical applications, but ongoing research aims to achieve $ZT > 2$ or higher for improved efficiency.

9.2.2.5 Interplay Between Key Parameters

The optimization of thermoelectric materials involves balancing the three key parameters (S, σ, k) to maximize ZT. However, these parameters are often interdependent:

- Increasing the Seebeck coefficient (S) typically requires reducing carrier concentration, which can lower electrical conductivity (σ).
- Reducing thermal conductivity (k) is desirable, but it must be achieved without significantly compromising electrical conductivity.

9.2.2.6 Need for Flexible Thermoelectric Composites

The development of flexible thermoelectric composites has emerged as a critical area of research in response to the growing demand for energy-harvesting solutions that can integrate seamlessly into modern technologies. Unlike traditional rigid thermoelectric materials, flexible composites offer unique advantages in terms of adaptability, lightweight design, and compatibility with unconventional surfaces. However, their development also presents significant challenges that must be addressed to unlock their full potential.

9.3 Advantages of Flexibility in Wearable Electronics, IoT Devices, and Unconventional Surfaces

a. **Wearable Electronics**

   Wearable devices, such as smartwatches, fitness trackers, and medical sensors, require energy sources that can conform to the human body and withstand mechanical deformation (e.g., bending, stretching, and twisting). Flexible thermoelectric composites can harvest energy from body heat, providing a continuous and sustainable power source for these devices. The governing equation for power generation from body heat is based on the Seebeck effect:

$$P = \frac{S^2\sigma \cdot (\Delta T)^2}{4} \tag{9.5}$$

   where
   - P is the power generated,
   - S is the Seebeck coefficient,
   - σ is the electrical conductivity, and
   - ΔT is the temperature difference between the body and the environment.

   By maintaining flexibility, these materials can ensure optimal contact with the skin, maximizing heat transfer and energy conversion efficiency.

b. **IoT Devices**

   IoT devices, such as environmental sensors and wireless communication modules, are often deployed in remote or hard-to-reach locations where battery replacement is impractical. Flexible thermoelectric composites can harvest energy from ambient heat sources, such as pipelines, machinery, or even sunlight-warmed surfaces, to power these devices. The ability to conform to curved or irregular surfaces ensures efficient thermal contact and energy harvesting.

c. **Unconventional Surfaces**
Many real-world applications involve surfaces that are not flat or rigid, such as vehicle interiors, industrial equipment, or building structures. Flexible thermoelectric composites can be integrated into these surfaces without compromising their functionality or aesthetics, enabling energy harvesting from waste heat in diverse environments.

9.4 Challenges in Developing Flexible Thermoelectric Materials

Despite their advantages, the development of flexible thermoelectric composites faces several challenges that must be overcome to achieve practical and scalable applications:

a. **Trade-offs Between Flexibility and Thermoelectric Performance**
Traditional thermoelectric materials, such as bismuth telluride (Bi_2Te_3), exhibit high thermoelectric performance but are inherently rigid and brittle. Incorporating flexibility often requires the use of polymers or other soft materials, which typically have lower electrical conductivity and Seebeck coefficients, reducing the overall figure of merit (ZT). The challenge lies in optimizing the material's composition to achieve both flexibility and high thermoelectric efficiency.

b. **Mechanical Durability**
Flexible thermoelectric composites must withstand repeated mechanical stress, such as bending, stretching, and compression, without degrading their performance. The mechanical properties of the composite, such as tensile strength and elasticity, are governed by the material's microstructure and the interactions between its components.

For example, the stress–strain relationship in a flexible composite can be described by:

$$\sigma_{mech} = E \cdot \epsilon \tag{9.6}$$

where
- σ_{mech} is the mechanical stress,
- E is the Young's modulus (a measure of stiffness), and
- ϵ is the strain (deformation).

Achieving a balance between flexibility and durability is critical for long-term performance.

c. **Thermal and Electrical Contact Resistance:**
In flexible thermoelectric devices, maintaining low thermal and electrical contact resistance at the interfaces between materials is essential for efficient energy conversion. Poor contact can lead to significant energy losses, reducing

the overall efficiency of the device. The thermal contact resistance (R_{th}) and electrical contact resistance (R_{el}) can be expressed as:

$$R_{th} = \frac{\Delta T}{Q}, R_{el} = \frac{V}{I} \tag{9.7}$$

where

- ΔT is the temperature drop at the interface,
- Q is the heat flow,
- V is the voltage drop, and
- I is the current.

d. **Scalability and Cost**
   The fabrication of flexible thermoelectric composites often involves complex processes, such as nanostructuring, doping, and composite synthesis, which can be expensive and difficult to scale. Developing cost-effective manufacturing techniques is essential for widespread adoption.
e. **Environmental Stability**
   Flexible thermoelectric materials must maintain their performance under varying environmental conditions, such as humidity, temperature fluctuations, and exposure to UV radiation. Degradation of material properties over time can limit the lifespan of the device.

9.5 Comparison of Rigid Versus Flexible Thermoelectric Materials

Thermoelectric materials can be broadly categorized into rigid and flexible types, each with distinct properties, advantages, and limitations. Traditional rigid thermoelectric materials, such as bismuth telluride (Bi_2Te_3), have been widely used in thermoelectric devices due to their high performance. However, their rigidity and brittleness limit their applicability in modern technologies, such as wearable electronics and IoT devices.

In contrast, flexible thermoelectric composites offer unique advantages, including adaptability, lightweight design, and compatibility with unconventional surfaces. This section compares rigid and flexible thermoelectric materials, highlighting the limitations of rigid materials and the advantages of flexible composites.

9.5.1 Limitations of Rigid Thermoelectric Materials

a. **Brittleness and Lack of Flexibility**
   Rigid thermoelectric materials, such as bismuth telluride (Bi_2Te_3) and lead telluride (PbTe), are inherently brittle and cannot withstand mechanical deformation (e.g., bending, stretching, or twisting). This limits their use in applications requiring flexibility, such as wearable electronics or curved surfaces.
b. **High Thermal Conductivity**
   Many rigid thermoelectric materials have relatively high thermal conductivity (k), which reduces the temperature gradient across the material and

lowers energy conversion efficiency. For example, bismuth telluride has a thermal conductivity of approximately 1.5–2.0 W/m·K, which is higher than desired for optimal thermoelectric performance.

c. **High Cost and Scalability**
Rigid thermoelectric materials often require complex and expensive fabrication processes, such as crystal growth and vacuum deposition. These processes are difficult to scale for mass production, limiting their widespread adoption.

d. **Environmental and Safety Concerns**
Some rigid thermoelectric materials, such as lead-based compounds, pose environmental and health risks due to their toxicity. This has led to increased regulatory scrutiny and a push for safer alternatives.

9.5.2 Advantages of Flexible Thermoelectric Composites

9.5.2.1 Mechanical Flexibility and Durability

Flexible thermoelectric composites can withstand mechanical deformation, such as bending, stretching, and compression, without losing functionality. This makes them ideal for applications requiring conformability, such as wearable electronics and IoT devices.

9.5.2.2 Low Thermal Conductivity

Flexible composites often incorporate polymers or nanostructured materials, which have inherently low thermal conductivity. This helps maintain a large temperature gradient across the material, enhancing energy conversion efficiency. The thermal conductivity of a composite material can be approximated using the rule of mixtures:

$$k_{composite} = \sum f_i \cdot k_i \tag{9.8}$$

where

- f_i is the volume fraction of the i-th component, and
- k_i is the thermal conductivity of the i-th component.

9.5.2.3 Lightweight and Scalable Fabrication

Flexible composites are typically lightweight, making them suitable for portable and wearable applications. They can be fabricated using cost-effective and scalable techniques, such as solution processing, printing, and roll-to-roll manufacturing. Flexible thermoelectric composites can be integrated into a wide range of surfaces, including curved, irregular, or dynamic surfaces. This enables energy harvesting from diverse heat sources, such as body heat, ambient environmental heat, or industrial waste heat.

Many flexible composites use nontoxic and environmentally friendly materials, such as conductive polymers (e.g., PEDOT:PSS) or carbon-based nanomaterials (e.g., graphene, carbon nanotubes (CNTs)). This reduces environmental and health risks compared to traditional rigid materials.

9.6 Materials for Flexible Thermoelectric Composites

9.6.1 Conductive Polymers

Conductive polymers are a class of organic materials that combine the electrical properties of metals with the mechanical flexibility and processability of plastics. They are widely used in flexible thermoelectric composites due to their unique combination of properties, including tunable electrical conductivity, low thermal conductivity, and ease of fabrication. Among the most prominent conductive polymers for thermoelectric applications are poly(3,4-ethylenedioxythiophene) polystyrene sulfonate (PEDOT:PSS), polyaniline (PANI), and polypyrrole (PPy).

9.6.1.1 Poly(3,4-Ethylenedioxythiophene) PEDOT:PSS

PEDOT:PSS is one of the most widely used conductive polymers in flexible thermoelectric composites. It consists of a conjugated polymer (PEDOT) doped with a polyelectrolyte (PSS), which enhances its electrical conductivity and processability. The electrical conductivity (σ) of PEDOT:PSS can vary significantly, ranging from 10^{-3} to 10^{3} S/cm, depending on the doping level and processing conditions. Its Seebeck coefficient (S) typically falls within the range of 10 to 50 μV/K, while its thermal conductivity (κ) remains very low due to the amorphous nature of the polymer.

The advantages of PEDOT:PSS are notable. It exhibits high flexibility, allowing films to be bent, stretched, and twisted without significant loss of electrical conductivity. This makes it highly suitable for applications in flexible electronics. Additionally, its solution processability enables easy deposition using techniques such as spin coating, inkjet printing, or spray coating, facilitating scalable and cost-effective fabrication. The properties of PEDOT:PSS are also tunable; its electrical conductivity and Seebeck coefficient can be optimized through chemical doping, solvent treatment, or blending with other materials.

However, there are limitations to its use. The figure of merit (ZT) of PEDOT:PSS is typically low ($ZT < 0.1$), primarily due to its moderate Seebeck coefficient and electrical conductivity. This limits its overall thermoelectric performance. Furthermore, PEDOT:PSS is sensitive to environmental factors such as humidity and temperature, which can degrade its performance over time. Despite these limitations, the unique combination of properties offered by PEDOT:PSS makes it a promising material for various applications in the field of flexible thermoelectric composites.

9.6.1.2 PANI and PPy

PANI is another widely studied conductive polymer for thermoelectric applications, known for its ease of synthesis, environmental stability, and tunable electrical properties. Its electrical conductivity (σ) ranges from 10^{-2} to 10^{2} S/cm, depending on the doping level and processing conditions, while its Seebeck coefficient (S) typically falls between 10 and 30 μV/K. Like other conductive polymers, PANI exhibits very low thermal conductivity (κ), which is advantageous for thermoelectric performance.

The advantages of PANI include its superior environmental stability compared to PEDOT:PSS, making it suitable for long-term applications. Additionally, PANI can be synthesized using simple and inexpensive chemical methods, contributing to its low cost. Its electrical conductivity and Seebeck coefficient can also be optimized through doping or blending with other materials. However, PANI has limitations, including its moderate thermoelectric performance, characterized by a low figure of merit ($ZT < 0.1$). Furthermore, PANI is less soluble than PEDOT:PSS, making it more challenging to process into thin films or coatings.

PPy is another conductive polymer known for its high electrical conductivity and ease of synthesis, often used in combination with other materials to enhance thermoelectric performance. PPy's electrical conductivity (σ) ranges from 10^{-1} to 10^{2} S/cm, depending on the doping level and processing conditions, while its Seebeck coefficient (S) typically falls between 10 and 40 μV/K. Like PANI and PEDOT:PSS, PPy exhibits very low thermal conductivity (κ).

PPy offers several advantages, including its high electrical conductivity, which surpasses that of PANI and PEDOT:PSS, making it suitable for applications requiring high power output. It can be synthesized using simple chemical or electrochemical methods and is highly flexible, allowing it to be integrated into wearable devices or curved surfaces. However, PPy also has limitations. Its thermoelectric performance is moderate, with a low figure of merit ($ZT < 0.1$). Additionally, PPy is sensitive to oxidation and degradation under ambient conditions, which can limit its long-term stability.

9.6.2 Advantages and Limitations of Polymer-based Thermoelectric

9.6.2.1 Advantages of Polymer-based Thermoelectric

Polymer-based thermoelectrics offer several unique advantages that make them attractive for a wide range of applications. Their inherent flexibility and lightweight nature are particularly noteworthy, as these properties enable seamless integration into wearable electronics, IoT devices, and other systems that require mechanical adaptability. This makes them ideal for applications where traditional rigid materials would be impractical.

Another significant advantage is their low thermal conductivity, which stems from the polymers' amorphous structure. This property helps maintain a large temperature gradient across the material, thereby enhancing energy conversion efficiency, which is a crucial factor for effective thermoelectric performance.

In addition to their mechanical and thermal properties, conductive polymers benefit from excellent solution processability. They can be processed using a variety of low-cost and scalable techniques, such as spin coating, inkjet printing, and roll-to-roll manufacturing. This not only simplifies fabrication but also makes large-scale production more feasible and cost-effective.

Furthermore, the properties of conductive polymers are highly tunable. Their electrical conductivity, Seebeck coefficient, and mechanical properties can be optimized through chemical doping, blending with other materials, or nanostructuring. This tunability allows researchers and engineers to tailor the materials to specific application requirements, maximizing their performance and versatility.

9.6.2.2 Limitations of Polymer-based Thermoelectric

Despite their numerous advantages, polymer-based thermoelectrics also face several limitations that currently restrict their widespread adoption. One of the most significant challenges is their relatively low figure of merit (ZT), which is a key metric for evaluating thermoelectric performance. Conductive polymers typically exhibit ZT values below 0.1, significantly lower than those of inorganic materials. This lower ZT value limits their overall efficiency and makes them less competitive in high-performance applications.

Another limitation is their environmental sensitivity. Many conductive polymers are susceptible to degradation due to humidity, temperature fluctuations, and oxidation. This sensitivity can lead to a decline in performance over time, particularly in outdoor or uncontrolled environments, thereby reducing their reliability and longevity.

Optimizing the properties of polymer-based thermoelectrics often involves trade-offs. For example, enhancing electrical conductivity might come at the expense of reducing the Seebeck coefficient or compromising mechanical flexibility. These trade-offs complicate material design and require careful balancing to achieve optimal performance across multiple properties. While polymer-based thermoelectrics offer significant benefits in terms of flexibility, low thermal conductivity, processability, and tunability, their low figure of merit, environmental sensitivity, and property trade-offs present ongoing challenges that need to be addressed to fully realize their potential in practical applications.

9.7 Inorganic Nanomaterials

Inorganic nanomaterials play a crucial role in the development of flexible thermoelectric composites due to their exceptional electrical and thermal properties. These materials, which include bismuth telluride (Bi_2Te_3), antimony telluride (Sb_2Te_3), and carbon-based materials such as graphene and CNTs, offer high thermoelectric performance and can be integrated into flexible matrices to enhance the overall properties of the composite.

9.7.1 Bismuth Telluride (Bi_2Te_3), Antimony Telluride (Sb_2Te_3), and Their Nanostructures

Bismuth telluride (Bi_2Te_3) and antimony telluride (Sb_2Te_3) are among the most widely used inorganic thermoelectric materials, primarily due to their high figure of merit (ZT) near room temperature. These materials are predominantly employed in their bulk form for rigid thermoelectric devices. However, their nanostructured forms are increasingly being explored for integration into flexible composites, expanding their potential applications.

Bi_2Te_3 and Sb_2Te_3 exhibit high Seebeck coefficients (S), typically in the range of $200-300\ \mu V/K$. This high Seebeck coefficient is crucial for efficient thermoelectric energy conversion. Additionally, these materials possess high electrical conductivity (σ), around 10^3–10^4 S/cm, stemming from their narrow bandgaps and high carrier mobility. Their thermal conductivity (k) is relatively low and can be further reduced through nanostructuring, which enhances their thermoelectric performance by maintaining a larger temperature gradient across the material.

The development of nanostructures, such as nanoparticles and nanowires, has significantly advanced the application potential of Bi_2Te_3 and Sb_2Te_3. Reducing these materials to the nanoscale enhances phonon scattering, thereby reducing lattice thermal conductivity (k_l) without significantly affecting electrical conductivity. This phonon engineering approach is key to improving the overall thermoelectric efficiency. Moreover, nanostructured thin films of Bi_2Te_3 and Sb_2Te_3 can be deposited on flexible substrates using advanced techniques such as sputtering or chemical vapor deposition, enabling the fabrication of flexible thermoelectric devices.

The primary advantage of Bi_2Te_3 and Sb_2Te_3 is their high thermoelectric performance, characterized by ZT values of around 1.0–1.5 near room temperature. This makes them highly suitable for energy-harvesting applications, particularly in environments with small temperature gradients. Additionally, the thermoelectric properties of these materials can be further optimized through doping or alloying, such as creating Bi_2Te_3-Sb_2Te_3 alloys. This tunability allows for precise tailoring of the materials' performance to meet specific application requirements.

Despite their excellent thermoelectric properties, Bi_2Te_3 and Sb_2Te_3 face several limitations. In their bulk form, these materials are brittle and cannot withstand mechanical deformation, which restricts their use in flexible devices. Furthermore, the fabrication of nanostructured Bi_2Te_3 and Sb_2Te_3 can be expensive and challenging to scale up for large-scale production. These challenges need to be addressed to fully leverage the potential of these materials in emerging flexible thermoelectric applications.

Bi_2Te_3 and Sb_2Te_3 offer high thermoelectric performance and tunable properties, making them attractive for energy-harvesting applications. However, their brittleness and the cost and scalability challenges associated with nanostructuring limit their widespread adoption in flexible devices.

9.7.2 Carbon-based Materials: Graphene and CNTs

Carbon-based materials, such as graphene and CNTs, have emerged as promising candidates for flexible thermoelectric composites due to their exceptional electrical, thermal, and mechanical properties. These materials offer unique advantages that make them well-suited for applications in flexible electronics and energy-harvesting devices.

9.7.2.1 Graphene

Graphene, a single layer of carbon atoms arranged in a hexagonal lattice, exhibits remarkable properties that make it an attractive material for thermoelectric applications. Its electrical conductivity (σ) is extremely high, reaching up to 10^6 S/cm, primarily due to its high carrier mobility. However, its Seebeck coefficient (S) is typically low, ranging from 10 to 30 μV/K, although it can be enhanced through chemical doping or functionalization. Graphene also has very high thermal conductivity, which, while beneficial for some applications, is a disadvantage for thermoelectric materials, as it reduces the temperature gradient across the material and lowers energy conversion efficiency.

Graphene's high flexibility allows it to be integrated into flexible substrates, making it suitable for wearable electronics and other flexible devices. Additionally, graphene's thermoelectric properties can be optimized through various strategies, such as chemical doping, strain engineering, or hybridization with other materials. However, its intrinsic low Seebeck coefficient and high thermal conductivity pose significant challenges to achieving high thermoelectric performance.

9.7.2.2 Carbon Nanotubes

CNTs, which can be either metallic or semiconducting depending on their chirality, also offer unique properties for thermoelectric applications. Their electrical conductivity (σ) is high, ranging from 10^3 to 10^4 S/cm, depending on the type and alignment of the CNTs. The Seebeck coefficient (S) of CNTs typically ranges from 20 to 60 μV/K, with values influenced by the doping level and chirality. CNTs exhibit high thermal conductivity along their axis, but this can be reduced in composite materials due to interfacial phonon scattering.

CNTs share some advantages with graphene, such as high flexibility, allowing them to be integrated into flexible substrates or even woven into fabrics. Their thermoelectric properties can also be tuned through doping, functionalization, or blending with other materials. However, like graphene, the high thermal conductivity of CNTs is a disadvantage for thermoelectric applications. Additionally, achieving uniform dispersion and alignment of CNTs within a composite matrix remains a significant challenge, as it directly affects the overall performance of the material.

For nanostructured materials, the thermal conductivity can be further broken down into electronic (k_e) and lattice (k_l) contributions:

$$K = k_e + k_l \tag{9.9}$$

Nanostructuring reduces lattice thermal conductivity (k_l) by increasing phonon scattering at interfaces and boundaries, while maintaining or enhancing electrical conductivity (σ).

9.8 Hybrid Composites

Hybrid composites represent a promising approach to developing high-performance flexible thermoelectric materials by combining the advantages of organic polymers and inorganic nanomaterials. These composites leverage the flexibility, processability, and low thermal conductivity of polymers with the high electrical conductivity and thermoelectric performance of inorganic materials.

9.8.1 Polymer-inorganic Nanocomposites

Polymer-inorganic nanocomposites represent a class of hybrid materials that integrate inorganic nanomaterials, such as nanoparticles, nanowires, or nanosheets, into a polymer matrix. This unique combination leverages the mechanical flexibility and processability of polymers while benefiting from the enhanced thermoelectric properties of the inorganic fillers, making them highly attractive for applications in flexible thermoelectric devices.

The nature of the interactions between the polymer matrix and the inorganic fillers significantly influences the performance of polymer-inorganic nanocomposites. These interactions can be broadly categorized into physical and chemical interactions, each contributing to the composite's overall stability and functionality.

Physical interactions between the polymer matrix and inorganic fillers include van der Waals forces, hydrogen bonding, or electrostatic interactions. These noncovalent interactions are relatively weak but can still provide sufficient binding strength to enhance the dispersion of inorganic fillers within the polymer matrix. For example, electrostatic interactions can be particularly effective in stabilizing the composite by preventing the agglomeration of nanoparticles.

Chemical interactions, such as covalent bonding or coordination bonding, offer a stronger and more stable connection between the polymer and inorganic phases. These interactions not only improve the dispersion of the inorganic fillers but also enhance the overall compatibility and mechanical integrity of the composite. For instance, in a PEDOT:PSS-Bi_2Te_3 nanocomposite, the sulfonate groups in PSS can form electrostatic interactions with the surface of Bi_2Te_3 nanoparticles. This interaction not only improves the dispersion of Bi_2Te_3 nanoparticles within the polymer matrix but also enhances interfacial adhesion, leading to better overall thermoelectric performance.

The design of polymer-inorganic nanocomposites involves careful consideration of both the type of polymer matrix and the inorganic filler and the method of incorporating the filler into the matrix. Techniques such as physical mixing, in situ synthesis, or chemical bonding are commonly employed to achieve optimal dispersion and interaction. The choice of method depends on the specific materials used and the desired properties of the composite.

For example, in situ polymerization can grow inorganic nanoparticles directly within the polymer matrix, ensuring uniform distribution and strong interfacial interactions. Alternatively, chemical bonding can functionalize the surface of inorganic fillers, enhancing their compatibility with the polymer matrix and improving the overall stability of the composite.

In hybrid composites, the effective electrical conductivity (σ_{eff}) and thermal conductivity (k_{eff}) can be estimated using the rule of mixtures:

$$\sigma_{eff} = \sum f_i \cdot \sigma_i,\ k_{eff} = \sum f_i \cdot k_i, \tag{9.10}$$

where

- f_i is the volume fraction of the i-th component, and
- σ_i and k_i are the electrical and thermal conductivities of the i-th component.

9.8.2 Advantages of Polymer-inorganic Nanocomposites

Polymer-inorganic nanocomposites offer significant advantages that make them highly attractive for applications in flexible thermoelectric devices. These hybrid materials integrate the inherent flexibility and processability of polymers with the enhanced thermoelectric performance provided by inorganic fillers. The inorganic components, such as nanoparticles, nanowires, or nanosheets, create pathways for charge transport within the composite, significantly improving electrical conductivity compared to the polymer alone. This enhancement is crucial for achieving higher efficiency in thermoelectric applications.

The polymer matrix and the interfaces between the polymer and inorganic phases play a crucial role in reducing thermal conductivity. These interfaces scatter phonons, which are the primary carriers of heat in solids, thereby maintaining a larger temperature gradient across the material. This reduction in thermal conductivity is particularly advantageous for applications requiring high energy conversion efficiency.

The polymer matrix also imparts mechanical flexibility to the composite, allowing it to be integrated into devices with curved or dynamic surfaces. This flexibility is essential for applications in wearable electronics, flexible sensors, and other devices that require mechanical adaptability. The ability to bend, stretch, and twist without significant performance loss makes polymer-inorganic nanocomposites highly suitable for next-generation flexible technologies.

One of the most significant advantages of polymer-inorganic nanocomposites is the tunability of their thermoelectric properties. By adjusting the type, size, and concentration of the inorganic fillers, researchers and engineers can optimize the composite's electrical conductivity, Seebeck coefficient, and thermal conductivity. This tunability allows for the customization of the composite to meet specific application requirements, making it a versatile material for a wide range of thermoelectric applications.

As research in this field continues to advance, polymer-inorganic nanocomposites are poised to play a crucial role in the development of next-generation wearable and flexible energy-harvesting devices. These devices combine the best of both worlds: the flexibility and processability of polymers with the enhanced thermoelectric properties of inorganic materials.

9.8.3 Strategies to Enhance ZT in Hybrid Materials

a. **Optimizing Filler Dispersion**
Uniform dispersion of inorganic fillers in the polymer matrix is critical for achieving high electrical conductivity and effective phonon scattering. Techniques such as sonication, surface functionalization, and in situ synthesis are used to improve dispersion and interfacial adhesion.

b. **Enhancing Electrical Conductivity**
Enhancing the electrical conductivity of polymer-inorganic nanocomposites is critical to optimizing their performance for thermoelectric applications. One effective approach is the incorporation of highly conductive inorganic fillers such as Bi_2Te_3, Sb_2Te_3, or CNTs. These materials provide additional pathways for charge transport within the composite, significantly improving its overall electrical conductivity.

Chemical doping is another strategy for enhancing electrical conductivity. By doping either the polymer or the inorganic phase, the carrier concentration and mobility can be increased. This process introduces additional charge carriers into the material, facilitating more efficient charge transport and thereby improving the composite's electrical properties.

Achieving a percolation network of conductive fillers is also essential for maximizing electrical conductivity. When the concentration of conductive fillers reaches a critical threshold, known as the percolation threshold, a continuous pathway for charge transport is formed throughout the composite. This interconnected network ensures efficient charge movement and significantly boosts the composite's electrical performance.

The percolation threshold (f_c) can be estimated using:

$$\sigma_{eff} = \sigma_0 \left(f - f_c\right)^t \tag{9.11}$$

where

- σ_0 is the conductivity of the filler,
- f is the volume fraction of the filler, and
- t is a critical exponent.

c. **Reducing Thermal Conductivity**

Nanostructuring plays a pivotal role in optimizing the thermoelectric performance of polymer-inorganic nanocomposites by significantly reducing lattice thermal conductivity (κ_l). By incorporating nanoscale fillers such as nanoparticles or nanowires, phonon scattering at the interfaces between the polymer matrix and the inorganic fillers is increased. This enhanced scattering disrupts the coherent propagation of phonons, effectively reducing thermal conductivity while minimally affecting electrical conductivity. This selective reduction in thermal conductivity is crucial for improving the thermoelectric figure of merit (ZT) of the composite.

Interfacial engineering further enhances this effect by creating interfaces with mismatched acoustic properties. For instance, when spherical nanoparticles are embedded within a polymer matrix, the difference in acoustic impedance between the filler and the matrix amplifies phonon scattering at the interfaces. This phenomenon can be quantitatively modeled using the Maxwell–Garnett equation, which describes the effective thermal conductivity of a composite based on the volume fraction and properties of the fillers.

$$K_{eff} = k_m \left(\frac{k_f + 2k_m + 2f\left(k_f - k_m\right)}{k_f + 2k_m - f\left(k_f - k_m\right)} \right) \tag{9.12}$$

where

- K_m is the thermal conductivity of the matrix,
- K_f is the thermal conductivity of the filler, and
- f is the volume fraction of the filler.

By carefully designing these interfaces and optimizing the size and concentration of the nanoscale fillers, the thermal conductivity of the composite can be tailored to achieve a desirable balance between electrical and thermal transport properties. This approach is particularly valuable for developing high-performance thermoelectric materials that can efficiently convert waste heat into electrical energy while maintaining mechanical flexibility and processability.

d. **Enhancing the Seebeck Coefficient and Improving Mechanical Properties**
Enhancing the Seebeck coefficient and improving the mechanical properties are two critical aspects of optimizing polymer-inorganic nanocomposites for thermoelectric applications. Energy filtering can be employed to enhance the Seebeck coefficient by introducing energy barriers at the interfaces between the polymer and inorganic phases. These barriers selectively scatter low-energy carriers while allowing high-energy carriers to pass through, effectively increasing the Seebeck coefficient. Additionally, band engineering through the modification of the electronic structure of either the inorganic fillers or the polymer matrix can also enhance the Seebeck coefficient. For example, incorporating quantum dots or other nanostructures with discrete energy levels can increase the density of states near the Fermi level, thereby improving the thermoelectric performance of the composite.
In parallel, improving the composite's mechanical properties is essential for ensuring its durability and flexibility. Chemical crosslinking of the polymer matrix can significantly enhance the composite's mechanical strength and durability, making it more robust for practical applications. Furthermore, using flexible inorganic fillers such as graphene or CNTs ensures that the composite maintains its flexibility even under mechanical stress. This combination of enhanced Seebeck coefficient and improved mechanical properties makes polymer-inorganic nanocomposites highly attractive for flexible and wearable thermoelectric applications.

9.8.4 Examples of Polymer-inorganic Nanocomposites

9.8.4.1 PEDOT:PSS-Bi_2Te_3 Composites

PEDOT:PSS provides flexibility and low thermal conductivity, while Bi_2Te_3 nanoparticles enhance electrical conductivity and thermoelectric performance. The composite achieves a ZT of up to 0.3, making it suitable for flexible thermoelectric devices.

9.8.4.2 PANI–CNT Composites

PANI offers tunable electrical properties, while CNTs provide high electrical conductivity and mechanical strength. The composite exhibits a ZT of up to 0.1, with potential for further improvement through doping or nanostructuring.

9.8.4.3 Graphene-polymer Composites

Graphene nanosheets enhance electrical conductivity, while the polymer matrix ensures flexibility and low thermal conductivity. The composite achieves a ZT of up to 0.2, with potential for further enhancement through energy filtering or band engineering.

9.9 Emerging Materials

The field of flexible thermoelectric composites is rapidly evolving, with emerging materials such as MXenes, metal-organic frameworks (MOFs), and 2D materials offering exciting new possibilities. These materials exhibit unique properties, including high electrical conductivity, tunable electronic structures, and mechanical flexibility, making them ideal for next-generation thermoelectric devices. Additionally, self-healing and stretchable thermoelectric materials are being developed to enhance the durability and adaptability of flexible thermoelectric devices.

9.9.1 MXenes

MXenes are a family of 2D transition metal carbides, nitrides, and carbonitrides with the general formula $M_{n+1}X_nT_x$, where M is a transition metal, X is carbon or nitrogen, and T_x represents surface functional groups (e.g., –OH, –F, –O).

MXenes, a unique class of two-dimensional (2D) materials, have garnered significant attention for their potential applications in flexible thermoelectric composites. These materials are composed of alternating layers of transition metals and carbon or nitrogen atoms, forming a structure held together by strong covalent bonds within the layers and weaker van der Waals interactions between them. The presence of surface functional groups (denoted as T_x) not only modifies the electronic properties of MXenes but also enhances their stability, making them suitable for various applications.

In terms of electrical properties, MXenes exhibit high electrical conductivity (σ), reaching up to 10^4 S/cm. This remarkable conductivity is attributed to their inherent metallic nature and high carrier mobility. Additionally, the Seebeck coefficient (S) of MXenes typically ranges from 10 to 50 μV/K, depending on their specific composition and surface functionalization. This tunability of the Seebeck coefficient through surface modification is a key advantage for optimizing thermoelectric performance.

Thermal conductivity (κ) in MXenes is relatively low, primarily due to phonon scattering at the interfaces and the presence of surface functional groups. This low thermal conductivity is beneficial for thermoelectric applications, as it helps maintain a larger temperature gradient across the material, thereby enhancing energy conversion efficiency.

Given these properties, MXenes are particularly well-suited for use in flexible thermoelectric composites. Their high electrical conductivity, tunable surface chemistry, and mechanical flexibility make them ideal candidates for integrating into devices that require both efficient energy conversion and mechanical adaptability. This combination of attributes positions MXenes as a promising material for next-generation flexible thermoelectric applications, where they can contribute to the development of wearable electronics, energy-harvesting systems, and other innovative technologies. The electrical conductivity of MXenes can be modeled using the Drude model:

$$\sigma = ne\mu \tag{9.13}$$

where

- n is the carrier concentration,
- e is the electron charge, and
- μ is the carrier mobility.

9.9.2 Metal-organic Frameworks

They are porous materials composed of metal ions coordinated with organic ligands. Their tunable porosity and electronic structure make them promising candidates for thermoelectric applications.

MOFs represent a versatile class of materials composed of metal nodes connected by organic linkers, forming a three-dimensional network stabilized by strong coordination bonds. The unique structure of MOFs allows for significant tunability of their properties, particularly through variations in the choice of metal ions and organic ligands. This tunability extends to their porosity and surface area, making them highly adaptable for a range of applications.

In terms of electrical properties, MOFs typically exhibit low electrical conductivity (σ) due to their insulating nature. However, this can be significantly enhanced by incorporating conductive linkers or guest molecules into the framework. This modification introduces pathways for charge transport, thereby improving the overall electrical performance of the material. Additionally, MOFs can exhibit high Seebeck coefficients (S), reaching up to 100 $\mu V/K$, primarily due to their tunable electronic structure. This tunability is a key advantage for optimizing thermoelectric performance.

MOFs have very low thermal conductivity (κ), largely due to their porous structure and the presence of numerous interfaces that scatter phonons. This low thermal conductivity is highly beneficial for thermoelectric applications, as it helps maintain a larger temperature gradient across the material, thereby enhancing energy conversion efficiency.

Enhancing the electrical conductivity through the introduction of conductive pathways, such as guest molecules or doping, plays a crucial role in improving the overall ZT value. Given these properties, MOFs are increasingly being explored for use in flexible thermoelectric composites. Their tunable electronic structure, low thermal conductivity, and ability to enhance electrical conductivity through functionalization make them highly attractive for applications in wearable electronics, energy harvesting, and other flexible technologies. By leveraging these characteristics, MOFs have the potential to significantly contribute to the development of next-generation thermoelectric materials.

9.9.3 2D Materials

Two-dimensional materials, including graphene, transition metal dichalcogenides (TMDs), and black phosphorus, have garnered significant interest in thermoelectric applications due to their unique combination of properties. These materials are composed of single or a few layers of atoms, with strong covalent bonds within the layers and weak van der Waals interactions between them. For instance, graphene consists of a single layer of carbon atoms arranged in a hexagonal lattice, offering exceptional mechanical strength and electrical properties.

Graphene, in particular, exhibits remarkably high electrical conductivity (σ), reaching up to 10^6 S/cm, primarily due to its high carrier mobility. However, its Seebeck coefficient (S) is typically low, ranging from 10 to 30 μV/K. This intrinsic limitation can be overcome through strategies such as doping or strain engineering, which can enhance the Seebeck coefficient and improve thermoelectric performance. Additionally, while graphene has high thermal conductivity, this property can be mitigated in composite materials through phonon scattering, thereby reducing thermal conductivity and enhancing the thermoelectric figure of merit.

Similarly, other 2D materials like TMDs and black phosphorus also offer tunable electronic properties, making them attractive for thermoelectric applications. These materials can be engineered through energy filtering or band engineering to optimize their Seebeck coefficients and overall thermoelectric performance.

Given these properties, 2D materials are increasingly being integrated into flexible thermoelectric composites. Their high electrical conductivity, combined with tunable Seebeck coefficients and reduced thermal conductivity in composite forms, makes them ideal candidates for applications in wearable electronics, flexible sensors, and energy-harvesting devices. By leveraging the unique attributes of 2D materials, researchers and engineers can develop next-generation thermoelectric systems that combine efficiency with mechanical flexibility and adaptability.

9.9.4 Self-healing and Stretchable Thermoelectric Materials

Self-healing materials can autonomously repair damage, such as cracks or scratches, extending the lifespan of thermoelectric devices. Self-healing is achieved through dynamic bonds, such as hydrogen bonds, ionic interactions, or reversible covalent bonds.

The self-healing efficiency (η) can be quantified as:

$$\eta = \frac{\text{Property after healing}}{\text{Original property}} \times 100\% \tag{9.14}$$

On the other hand, stretchable materials can withstand large deformations without losing functionality, making them ideal for wearable electronics [1].

Reference

1 Abera Waktole, D., Jia, B., Wang, W., Zuo, Z., and Ma, K. Integrating dual heat sources to enhance thermoelectric generator power output. *Appl. Therm. Eng.*, vol. 258 Jan. 2025. https://doi.org/10.1016/j.applthermaleng.2024.124500.

10

Integration of Si-based Micro-thermoelectric Generator Devices

Takanobu Watanabe

Waseda University, Tokyo, Japan

10.1 Introduction

Waste heat recovery is poised to play a pivotal role in the transition toward a more sustainable and carbon-neutral society. The total waste heat emissions in 2030 are estimated to account for 23–53% of the global input energy, with a potential recovery rate of 6–12% [1]. Among the various technologies available for heat recovery, such as the Rankine and Kalina cycles, solid-state thermoelectric generation, which utilizes the Seebeck effect of semiconductors, stands out owing to its simplicity and applicability to various heat resources of different temperatures.

In the late 1980s, a microsized thermoelectric generator (μ-TEG) utilizing silicon (Si) miniaturization technology was developed [2], although it did not gain widespread attention at that time. Interest in μ-TEG devices began to grow only after the release of wristwatches operated at body temperature by the Seiko Corporation in 1998 [3]. Strasser et al. [4] presented a μ-TEG device fabricated with the Si-CMOS process and demonstrated a high specific power-generation capacity (SPGC) of 0.06 μW$\text{cm}^{-2}\text{K}^{-2}$, where SPGC is the performance indicator of the output power of μ-TEG normalized to the device area and the square of the applied temperature difference:

$$\text{SPGC} = \frac{P_{\text{max}}}{A\Delta T^2}, \tag{10.1}$$

where P_{max}, A, and ΔT represent the maximum power, area, and applied temperature difference to the TEG device.

A significant breakthrough occurred in 2008 when Boukai et al. [5] and Hochbaum et al. [6] discovered the excellent thermoelectric properties of Si nanowires (NWs). They independently discovered that Si-NWs yield ZT values of 0.3–0.6 at room temperature, a significant improvement over the ZT of bulk

Si crystals, which was only approximately 0.01 and previously deemed inadequate for thermoelectric applications. The significant improvement in ZT can be attributed to the decrease in the lattice thermal conductivity of the nanometer-scaled Si crystals. This discovery has led to the development of various types of μ-TEGs utilizing Si as thermoelectric legs [7].

Conventional μ-TEGs have been designed mainly to achieve high energy conversion efficiency. The thermoelectric legs are typically suspended in a cavity space or filled with a material of higher specific thermal resistance to minimize heat flow outside the thermoelectric leg, as shown in Figure 10.1a. Meanwhile, the thermoelectric leg must have a high thermal resistance to establish a substantial temperature gradient across it. However, the potential benefits of miniaturizing TEG devices for overall performance improvement have not been fully recognized. The primary objective of developing μ-TEG devices was to explore new applications by integrating them with Si integrated circuits [7].

The author's research group proposed a novel design concept for μ-TEG devices [8,9] as shown in Figure 10.1b, where thermoelectric legs are directly placed on the substrate without the need for a cavity space underneath. The heat current flows perpendicular to the substrate, and the TEG is driven by a steep temperature gradient established around the heat inlet. By shortening the length of the thermoelectric legs, the thermoelectric performance is enhanced [10]. Performance here refers to the SPGC, and energy conversion efficiency is not the primary focus. In the proposed μ-TEG, the absence of cavities around the legs allows for increased output density through miniaturization and integration of the devices, similar to CMOS-integrated circuits. Suppose the fuel cost for heat energy is negligible. In that case, increasing

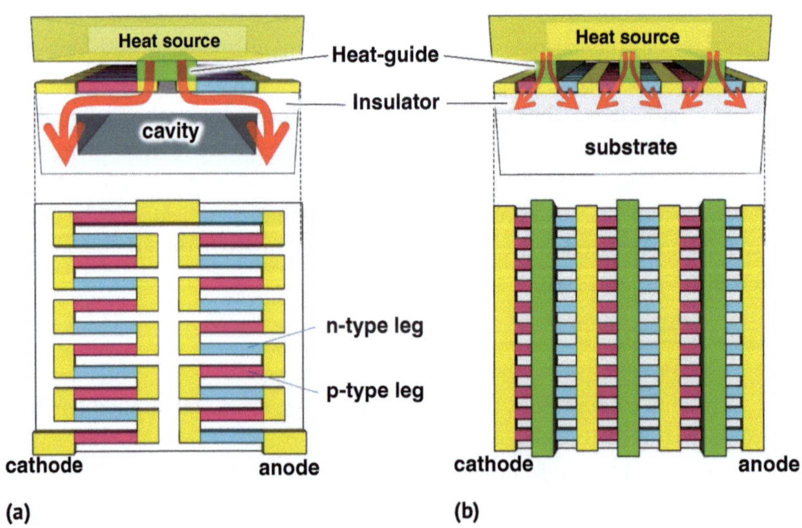

Figure 10.1 Schematic of (a) the typical structure of a conventional planar μ-TEG device. (b) Proposed μ-TEG device without a cavity. Red arrows show the heat flow.

the energy conversion efficiency may not be necessary. Rather, the total cost of power generation should be reduced if durable and high-power thermoelectric devices can be manufactured at a low cost [11]. As a preliminary experiment, a prototype of a single-stage, cavity-free TEG with Si-NW legs of 0.25 μm long was fabricated and recorded as an SPGC equivalent to 0.48 $\mu W cm^{-2} K^{-2}$ [12,13].

This study presents a series of device designs and experiments on large-scale, integrated, cavity-free μ-TEGs based on previous publications [14].

10.2 Device Design of Integrated TEG

10.2.1 Multistage Integration Schemes

Two types of multistage integration schemes exist for thermoelectric devices: unileg-type and bileg-type, as shown in Figure 10.2. The schematic of the unileg-type structure, which can be constructed using only n-type (or p-type) semiconductor for legs, is shown in Figure 10.2a. The output voltage of the device is increased by electrically connecting the hot- and cold-side sections with metallic wiring; however, the energy conversion efficiency is compromised owing to the inevitable heat leakage through this electrical wiring.

The bileg-type, shown in Figure 10.2b, comprises both p- and n-type semiconductors electrically connected in series, eliminating heat leakage through the wiring, as in the unileg-type. However, because p- and n-type semiconductors generally have different resistivities, the leg width ratio must be adjusted to optimize the power generation capacity per unit area when integrated in a planar manner.

In this study, both types of integrated TEGs were fabricated using a well-established and standard Si-CMOS process with multilayer wiring technology [14–20]. The optimal device layout was determined to maximize the SPGC as the objective function through finite element method (FEM) simulation, as elaborated hereafter.

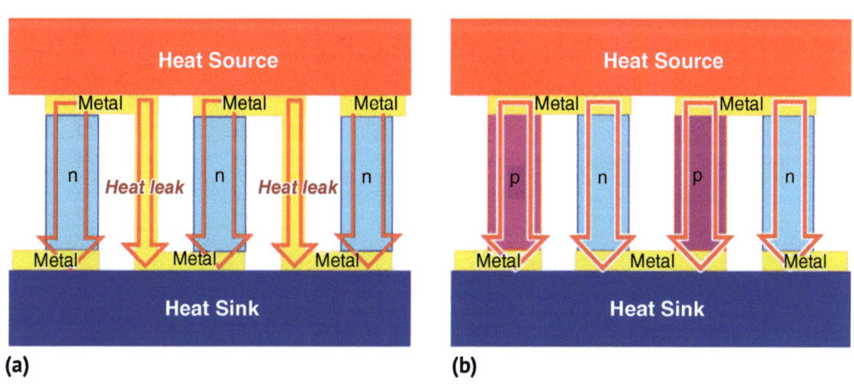

Figure 10.2 (a) Unileg and (b) bileg structures of multistage thermoelectric devices.

10.2.2 Unileg Device Layout

The unit cell of a unileg-type integrated device is shown in Figure 10.3. Adjacent unit cells are electrically connected through the first metal layer, and heat is transferred from an external heat source through the second metal layer to the hot side of each leg. The second metal layer utilized for heat transfer is referred to as a heat guide. The top view below the first metal layer, excluding the heat-guide layer, is shown in Figure 10.3c. The unit cell comprises two stages of leg bundles, where the hot-side electrodes are positioned adjacent to each other, and the hot- and cold-side electrodes are connected by the first metal layer with a meander shape. Based on FEM simulations, the temperature difference applied to the leg can be enhanced by increasing the heat injection pitch. This is achieved by placing the hot-side electrodes adjacent to each other, as shown in Figure 10.4a, rather than alternating the hot- and low-side electrodes, as shown in Figure 10.4b [15].

10.2.3 Bileg Device Layout

The unit cell of the bileg-type integrated device is shown in Figure 10.5. In this device, heat leakage through the metal wiring is not a concern, eliminating the need for special consideration of the first metal layer structure. However, the widths W_p

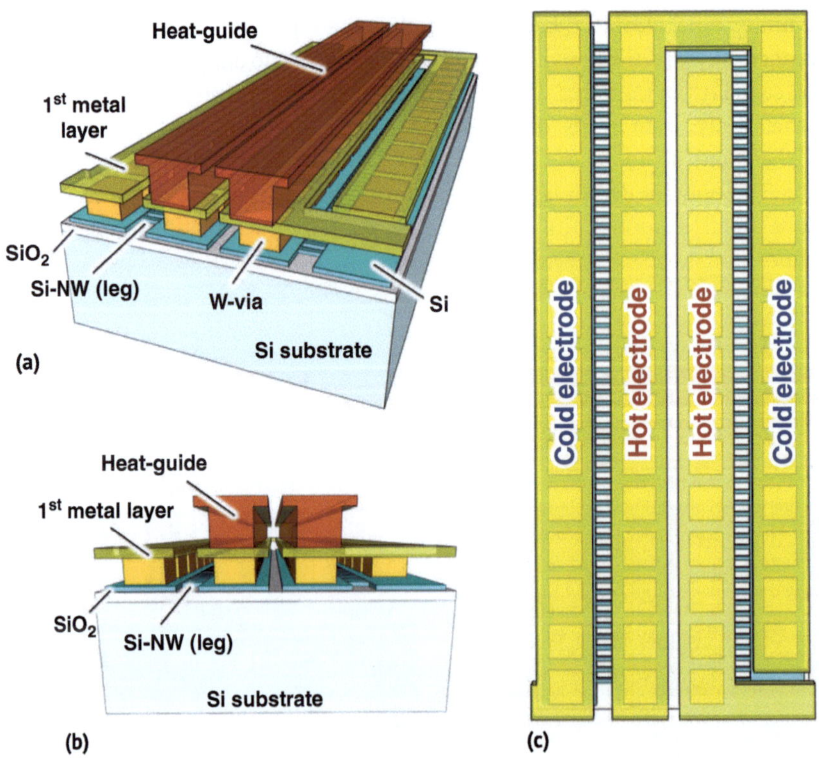

Figure 10.3 Unit structure of the unileg-type integrated thermoelectric device. (a) Bird's-eye view. (b) Cross-sectional view. (c) Top view without the heat-guide metal layer.

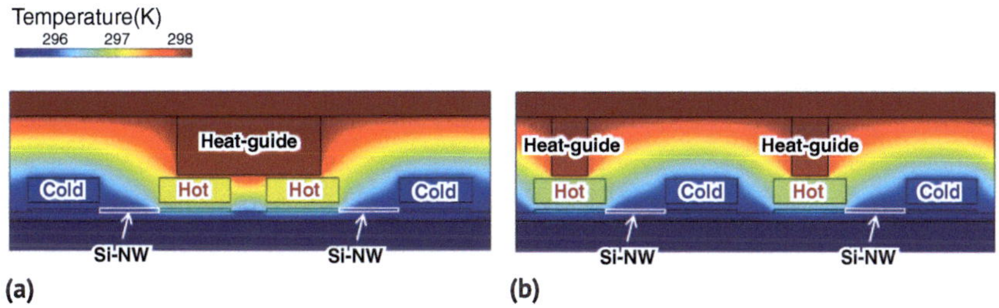

Figure 10.4 Cross-sectional temperature distributions simulated with FEM in the unileg-type integrated thermoelectric device [15]. (a) Hot sides of neighboring thermoelements are placed next to each other. (b) Hot- and cold-side electrodes are placed alternately.

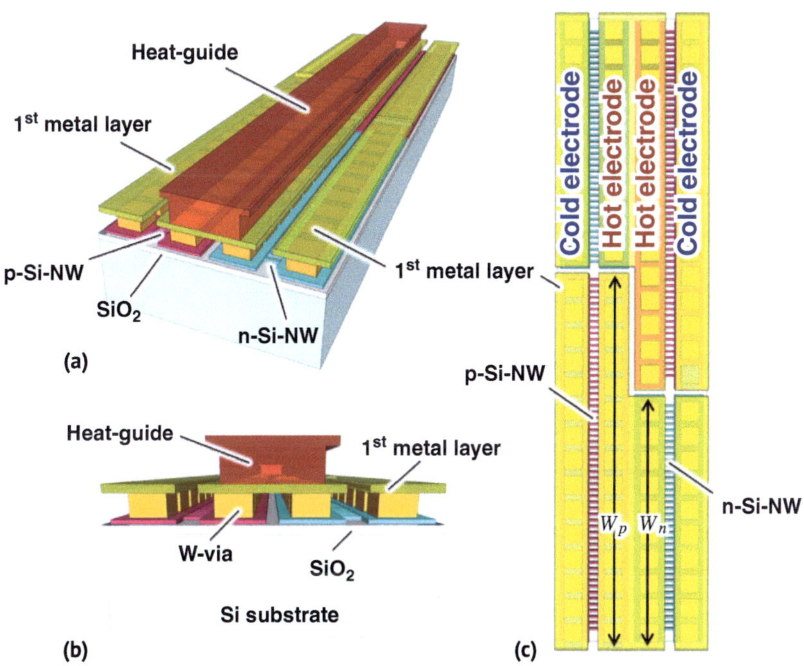

Figure 10.5 Unit structure of bileg-type integrated thermoelectric device. (a) Bird's-eye view. (b) Cross-sectional view. (c) Top view without the heat-guide metal layer.

of the p-type legs and W_n of the n-type legs must be adjusted to maximize the SPGC, and two alternating p–n leg pairs should be utilized as the unit structure, as shown in Figure 10.5c. The optimal width ratio can be determined using a simple thermoelectric equivalent circuit as follows:

$$\frac{W_p}{W_n} = \sqrt{\frac{\rho_p}{\rho_n}}, \tag{10.2}$$

where ρ_p and ρ_n represent the resistivity of p- and n-type legs, respectively [16]. The resistivity ratio ρ_p/ρ_n at which the power factor of Si reaches its maximum is approximately 2 to 3. Consequently, the optimum width ratio W_p/W_n befalls within the range of 1.4 to 1.7.

10.3 Experimental

10.3.1 Device Fabrication Process

The integrated thermoelectric device was fabricated using a 300 mm line in the superclean room of the Tsukuba Innovation Arena (TIA) at the National Institute of Advanced Industrial Science and Technology (AIST), Japan. The process flow for the device fabrication is shown in Figure 10.6. Si-NW legs with a width of less than 100 nm were formed on the top Si layer of a silicon-on-insulator (SOI) wafer through patterning using immersion ArF lithography and dry etching techniques. The SOI substrate comprised a 745 μm thick Si substrate, 145 nm thick buried-oxide layer, and 125 nm thick top Si layer. Subsequently, P^+ and BF^{2+} ions were implanted to form n- and p-type regions, with an acceleration voltage was 40 keV and doses of $7.5 \times 10^{14}\ cm^{-2}$ and $2.5 \times 10^{15}\ cm^{-2}$, respectively. The n- and p-type thermoelectric devices were composed of 120 and 180 Si-NWs per stage, respectively.

Next, a SiO_2 layer was deposited through chemical vapor deposition (CVD) as an interlayer dielectric, and 1.8×1.8 μm contact holes were formed by wet etching, with a 300 nm thick W plug subsequently deposited. Surface planarization was achieved through chemical mechanical polishing, after which TiN/AlCu/TiN/Ti (20 /30 /200 /50 nm) was deposited by sputtering to form the first metal layer.

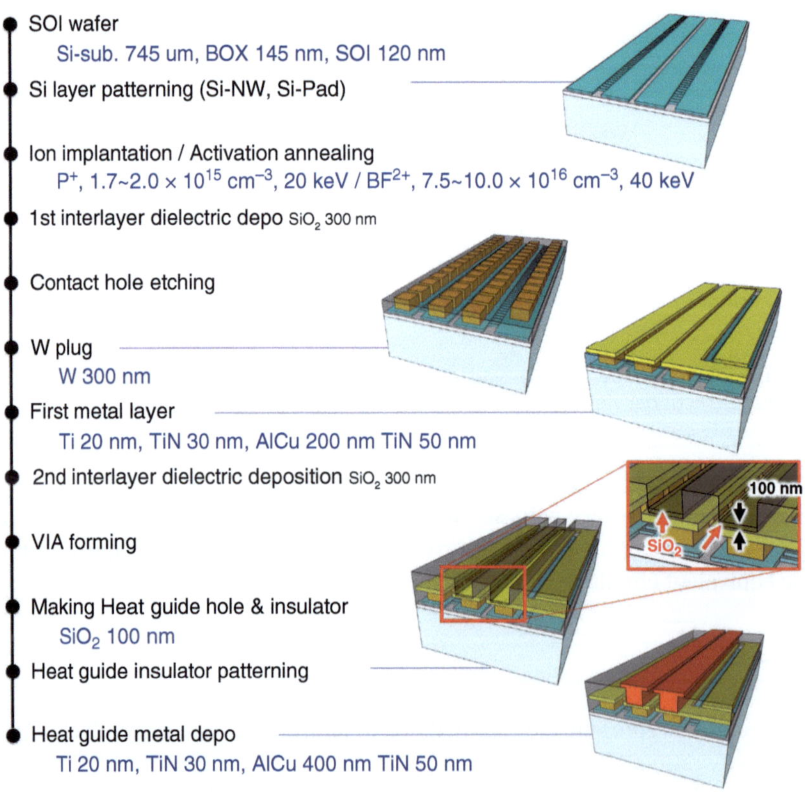

Figure 10.6 Process flow of integrated thermoelectric devices.

Next, a second interlayer of the SiO_2 film was deposited by CVD. The heat-guide section and contact area with the external electrodes were opened through wet etching. To insulate the heat guide and first wiring layer, a 100 nm thick SiO_2 layer was deposited onto the heat-guide section. Finally, Ti/TiN/AlCu/TiN (20 /30 /400 / 50 nm) was deposited as the heat-guide metal and patterned through dry etching.

Optical microscopy and cross-sectional electron microscopy images of the fabricated integrated μ-TEG device are shown in Figure 10.7.

10.3.2 Measurement Method

The thermoelectric performance of the μ-TEG device was characterized using a microprober system equipped with a temperature-controlled sample stage. The setup of the measurement is shown in Figure 10.8. The sample stage served as the cold sink, and the temperature was precisely controlled using a water chiller and Peltier device. A high thermal conductivity carbon sheet (SEKISUI POLYMATECH MANION50α) was affixed to the surface of the stage using silicone paste on which the sample was placed to minimize the contact thermal resistance. The thermoelectric performance was evaluated by applying a temperature difference to the sample devices in two ways: a custom-made microheater pressed against the sample surface (harvesting mode), as shown in Figure 10.8a, and a heat guide directly heated by Joule heating to establish a temperature gradient in the device (test mode), as shown in Figure 10.8b. The harvesting mode closely resembled actual operating conditions. However, controlling the contact thermal resistance with an external heat source can lead to significant measurement errors. However, the test mode allows for measurements with higher reproducibility as the contact thermal resistance with the high-temperature heat source is negligible. To considerably stabilize the contact thermal resistance with the heat source in the harvesting mode, silicone paste was applied to the tip of the microheater and wrapped in a carbon sheet. Electrical measurements were conducted using the four-probe method with a nanovoltmeter (KEITHLEY 2182A) and a microcurrent source (KEITHLEY 2401).

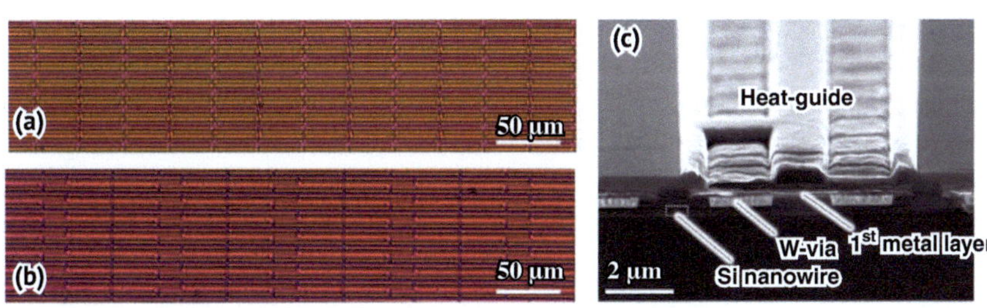

Figure 10.7 Optical microscopy images of (a) unileg-type and (b) bileg-type-integrated thermoelectric devices. (c) Cross-sectional scanning electron microscopy image of a bileg device.

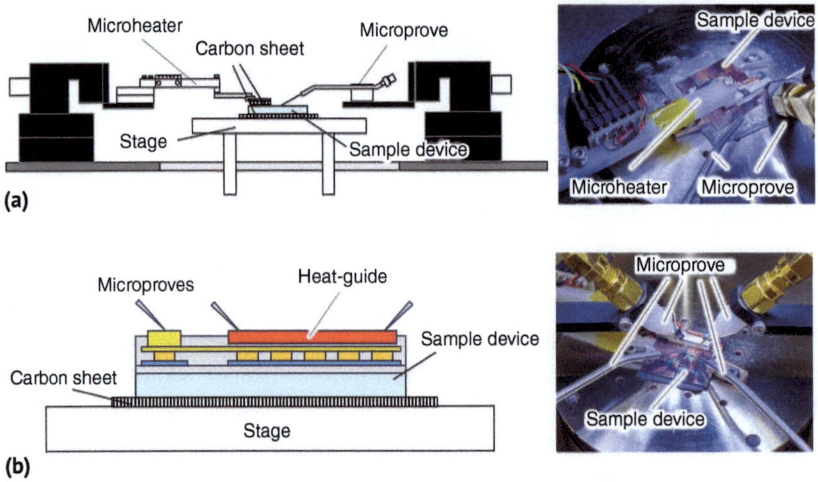

Figure 10.8 Experimental setup for measuring the thermoelectric characteristics. (a) Experimental setup of harvesting-mode characterization. (b) Setup for test-mode characterization.

10.4 Results and Discussion

The relationship between the maximum power-generation density and the length of Si-NW in unileg- and bileg-integrated TEGs is shown in Figure 10.9 [17]. These measurements were conducted in the harvesting mode using a microheater as the heat source, with a stage temperature of 20 °C and a microheater temperature of 25 °C. In both unileg and bileg devices, the thermoelectric power density increased as the Si-NW length decreased, indicating that the SPGC increased with miniaturization and integration. Despite the bileg devices having a lower integration density compared with unileg devices owing to the larger footprint of the p-type leg, the power density of bileg devices was slightly higher. This result indicates that the temperature difference applied to each stage of the leg is smaller in the unileged devices owing to heat leakage through the wiring [17].

The operating temperature dependences of the open-circuit voltage (V_{OC}), internal resistance (R), and maximum power ($P_{\max}$) on the operating temperature (T_{op}) for a unileg device comprising 144 stages with 0.25 μm long Si-NWs are shown in Figure 10.10 [18]. The evaluation was performed in the harvesting mode within a vacuum chamber to eliminate atmospheric effects. The heater temperature was maintained at $T_{op} + 3\,°C$ and the stage temperature at $T_{op} - 3\,°C$, resulting in a constant temperature difference of 6 °C. As shown in Figure 10.10, V_{OC} and $P_{\max}$ increased with T_{op}, indicating an enhancement in the effective Seebeck coefficient of the device. Higher temperatures led to improved power-generation performance within the device's temperature limit.

Further, the impact of the heat-guide structure on thermoelectric performance was investigated [18]. A comparison of the open-circuit voltages of the unileg devices (18786 stages) with varying second-layer dielectric thicknesses of 500 and

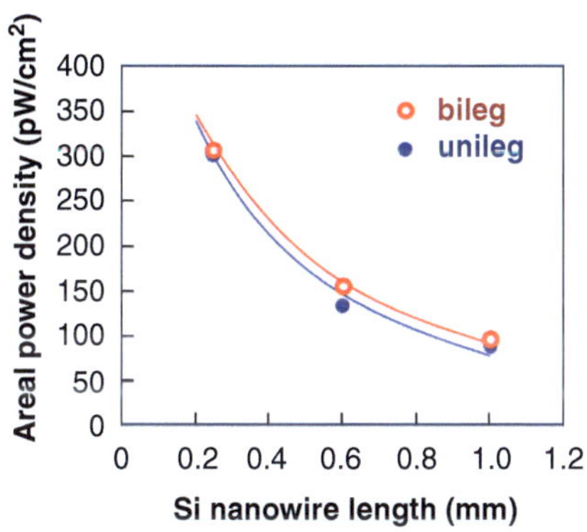

Figure 10.9 Maximum thermoelectric power normalized with the device versus Si-NW length measured in the harvesting mode. The external temperature difference was 5 °C. *Source*: [17] / with permission of IOP Publishing.

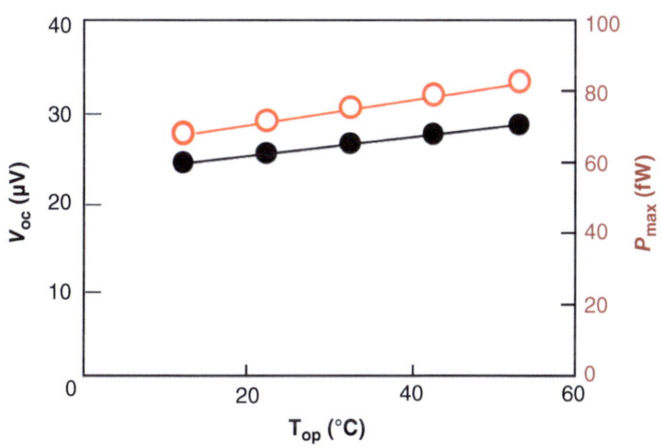

Figure 10.10 Operation temperature dependence of open-circuit voltage and maximum thermoelectric power from a unileg-integrated device with 144 stages. The external temperature difference was 6 °C. *Source*: Adapted from [18].

900 nm is shown in Figure 10.11. The measurement was performed in the harvesting mode with an applied temperature difference of 5 °C. The power density increased significantly with increasing thickness of the interlayer dielectric and heat-guide layers. This can be attributed to the increased temperature difference applied to the Si-NW part owing to the increased aspect ratio of the heat-guide layer. The performance of the fabricated devices showed a similar size dependence as anticipated, with the exception of a significant increase in the internal resistance observed in bileg-integrated devices, particularly at shorter leg lengths [19]. The relationship between internal resistance and leg length for bileg devices is shown in Figure 10.12a. While the experimental and FEM simulation results were closely aligned, the resistance was higher than predicted by the simple equivalent circuit model in the region of short leg lengths. As shown in Figure 10.12b, this discrepancy can be attributed to the spreading resistance effect, where the current density in the p-leg region, which extends beyond the n-leg width, decreases when the leg length

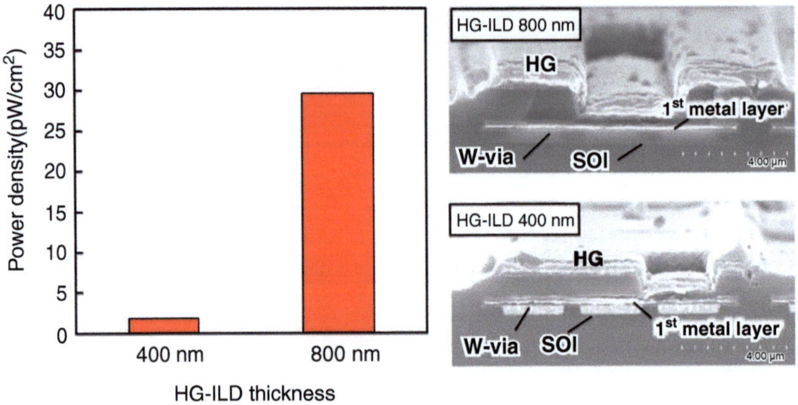

Figure 10.11 Impact of the height of the heat-guide on the thermoelectric power density. A larger temperature gradient was applied on the Si-NW part as the distance between the heater and the cold-side terminal increased. *Source*: [19] / with permission of IOP Publishing.

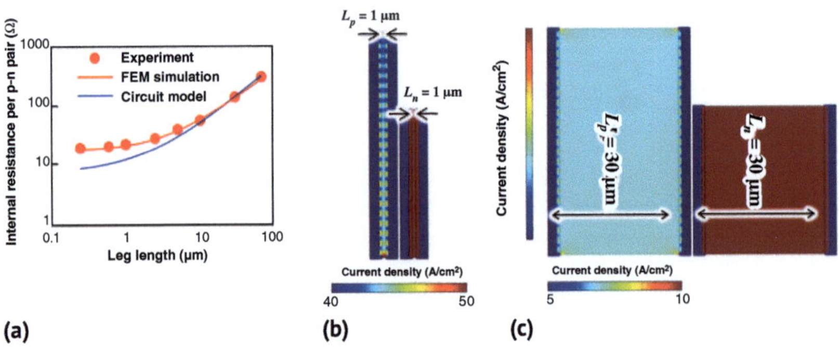

Figure 10.12 (a) Electrical internal resistance of bileg thermoelectric devices per one p–n pair. Simulated current density distributions at $\Delta T = 1$ K in the n- and p-type legs with (b) 1 µm long and (c) 30 µm long, respectively. *Source*: [20] / with permission of IOP Publishing.

is short. Once the leg length exceeds approximately 10 µm, the current is evenly distributed throughout the p-leg, rendering the spreading resistance negligible, as shown in Figure 10.12c. Therefore, for microdevices with leg lengths of less than a few micrometers, the spreading resistance becomes apparent at the point where the width changes abruptly, with $W_p/W_n = 1$ being the optimal choice.

The relationships between the areal power-generation density and applied temperature difference for various Si-based TEG devices reported to date are shown in Figure 10.13. The dashed lines in the figure represent the iso-SPCG lines that are proportional to the square of the applied temperature difference ΔT. Most of the data are aligned parallel to these lines, highlighting the significance of the SPGC as a performance indicator. The highest reported SPGC value is 373 µWcm^{-2}K^{-2} by Curtin et al. [21], followed by 28 µW m^{-2}K^{-2} by Hu et al. [22], both of which were achieved in the test mode for vertically oriented leg devices. Recently Yanagisawa et al. [23] obtained an SPGC of 1.3 µWcm^{-2}K^{-2}, setting a new record in harvesting-mode operation. In our study, the highest value obtained thus far is 0.9 µWcm^{-2}K^{-2}

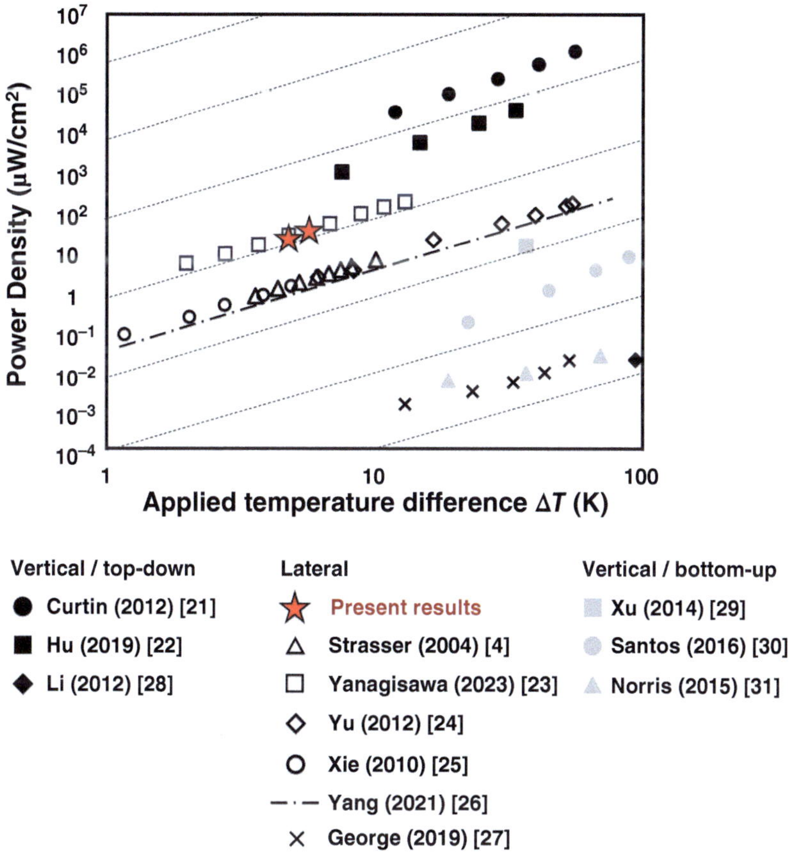

Figure 10.13 Relationships between areal power density and applied temperature for Si-based thermoelectric generators reported to date.

for a unileg device. This demonstrated that high-power-generation performance can be achieved through miniaturization alone without cavities.

10.5 Summary and Conclusion

This study developed and reviewed Si-based and cavity-free integrated thermoelectric devices. Utilizing Si-NWs as thermoelectric elements, multistage thermoelectric devices were fabricated using a standard Si-CMOS process. We demonstrated that the power-generation capacity of these devices increased with miniaturization and integration. Remarkably, a high SPGC of 0.9 $\mu Wcm^{-2}K^{-2}$ was achieved for a unileg device employing 0.25 μm-long Si-NWs. Furthermore, the power-generation capacity could be further improved by thickening the outermost metal layer for heat injection. The potential for even greater enhancement in the SPGC through continued miniaturization and optimization of device parameters is anticipated.

Acknowledgments

This research was supported by JST-CREST (JPMJCR15Q7 and JPMJCR19Q5) and a Grant-in-Aid for Scientific Research (22H01530). Micro-thermoelectric devices were fabricated in the TIA Super Clean Room at the National Institute of AIST and the Advanced Research Infrastructure for Materials and Nanotechnology at the National Institute for Materials Science (NIMS-ARIM) in Japan. The author expresses gratitude to collaborators Dr. Takeo Matsuki and Dr. Motohiro Tomita, as well as all the students who contributed to this study. The author would also like to extend his gratitude to Mr. Daisuke Kawaguchi of Hamamatsu Photonics K.K. for his assistance with the stealth dicing of the 300 mm wafers.

References

1 Firth, A., Zhang, B. and Yang, A. (2019) Quantification of global waste heat and its environmental effects. *Appl. Energy*, 235, 1314–34.

2 Rowe, D.M., Morgan, D.V. and Keily, J.H. (1989) Miniature low-power/high-voltage thermoelectric generator. *Electron. Lett.*, 25, 166–8.

3 Kishi, M., Nemoto, H., Hamao, T., Yamamoto, M., Sudou, S., Mandai, M. and Yamamoto, S. (1999) Micro-thermoelectric modules and their application to wrist-watches as an energy source. *Proceedings of IEEE 18th International Conference on Thermoelectrics*, 301–7.

4 Strasser, M., Aigner, R., Franosch, M. and Wachutka, G. (2002) Miniaturized thermoelectric generators based on poly-Si and ploy-SiGe surface micromachining. *Sens. Actuators A, Phys.*, 97-98, 535–42.

5 Boukai, A.I., Bunimovich, Y., Tahir-Kheli, J., Yu, J.K., Goddard III, W.A. and Heath, J.R. (2008) Silicon nanowires as efficient thermoelectric materials. *Nature*, 451, 168–71.

6 Hochbaum, A.I., Chen, R., Delgado, R.D., Liang, W., Garnett, E.C., Najarian, M., Majumdar, A. and Yang, P. (2008) Enhanced thermoelectric performance of rough silicon nanowires. *Nature*, 451, 163–7.

7 Watanabe, T. (2021) Microthermoelectric devices using Si nanowires. In: Roji, F. (ed.), Thermoelectric Energy Conversion, Elsevier, 503–20.

8 Watanabe, T., Asada, S., Xu, T., Hashimoto, S., Ohba, S., Himeda, Y., Yamato, R., Zhang, H., Tomita, M., Matsukawa, T., Kamakura, Y. and Ikeda, H. (2017) A scalable Si-based micro thermoelectric generator. *2017 IEEE Electron Devices Technology and Manufacturing Conference (EDTM) Proceedings of Technical Papers*, 86–7.

9 Zhang, H., Xu, T., Hashimoto, S. and Watanabe, T. (2018) The possibility of mW/cm^2-class on-chip power generation using ultrasmall Si nanowire-based thermoelectric generators. *IEEE Trans. Electron Devices*, 65, 2016–23.

10 Zhang, Q., Deng, K., Wilkenes, L., Reith, H. and Nielsch, K. (2022) Micro-thermoelectric devices. *Nat. Electron.*, 5, 333.

11 Min, G. (2006) Thermoelectric module design theories. In: Rowe, D.M. (ed.) Thermoelectrics Handbook, CRC Press, 11-1–14.

12 Tomita, M., Ohba, S., Himeda, Y., Yamato, R., Shima, K., Kumada, T., Xu, M., Takezawa, H., Mesaki, K., Tsuda, K., Hashimoto, S., Zhan, T., Zhang, H., Kamakuri, Y., Suzuki, Y., Inokawa, H., Ikeda, H., Matsukawa, T., Matsuki, T. and Watanabe, T. (2018) 10 μW/cm^2-class high power density silicon thermoelectric energy harvester compatible with CMOS-VLSI technology, *Symposium 2018 VLSI Technology Digest Technical Papers*, 93–4.

13 Tomita, M., Ohba, S., Himeda, Y., Yamato, R., Shima, K., Kumada, T., Xu, M., Takezawa, H., Mesaki, K., Tsuda, K., Hashimoto, S., Zhan, T., Zhang, H., Kamakuri, Y., Suzuki, Y., Inokawa, H., Ikeda, H., Matsukawa, T., Matsuki, T. and Watanabe, T. (2018) Modeling, simulation, fabrication and characterization of a 10 μw/cm^2 class Si-nanowire thermoelectric generator for IoT applications. *IEEE Trans. Electron Devices*, 65, 5180–8.

14 Watanabe, T. Large-scale integrated silicon thermoelectric device. *IEEJ Trans. Electron. Inf. Syst.*, 145, 371–376.

15 Abe, K., Oda, K., Tomita, M., Matsuki, T., Matsukawa, T. and Watanabe, T. (2020) Effect of unit-cell arrangement on performance of multi-stage-planar cavity-free unileg thermoelectric generator using silicon nanowires. *2020 International Conference on Simulation of Semiconductor Processes and Devices (SISPAD)*, 75–8.

16 Chen, J., Lin, B., Wang, H. and Lin, G. (2000) Optimal design of a multi-couple thermoelectric generator. *Semicond. Sci. Technol.*, 15, 184.

17 Arai, S., Miura, T., Mahfuz, M.M.H., Matsuki, T., Miyake, Y., Arayama, R. and Watanabe, T. (2024) Experimental demonstration of scalability in a cavity-free planar silicon-integrated thermoelectric device. *Jpn. J. Appl. Phys.*, 63, 02SP38.

18 Hoshina, T., Tomita, M., Matsuki, T. and Watanabe, T. (2024) Operating temperature dependency of power generation capacity in silicon planar integrated micro thermoelectric generators. *IEEE Trans. Electron Devices*, 71(4), 2624–2630.

19 Kuga, K., Mahfuz, M.M.H., Matsuki, T. and Watanabe, T. (2024) Impact of heat guide structure on the power generation performance of integrated cavity-free micro thermoelectric generators. *Jpn. J. Appl. Phys.*, 63, 02SP74.

20 Tomita, M., Kashizaki, T., Hoshina, T., Kurosaki, A., Arai, S., Matsuki, T. and Watanabe, T. (2023) Experimental demonstration of spreading resistance effect in a miniaturized bileg thermoelectric generator. *Jpn. J. Appl. Phys.*, 62, SC1066.

21 Curtin, B.M., Fang, E.W. and Bowers, J. (2012) Highly ordered vertical silicon nanowire array composite thin films for thermoelectric devices. *J. Electron. Mater.*, 41, 887–894.

22 Hu, G., Edwards, H. and Lee, M. (2019) Silicon integrated circuit thermoelectric generators with a high specific power generation capacity. *Nat. Electron.*, 2, 300–306.

23 Yanagisawa, R., Koike, S., Nawae, T., Tsujii, N., Wang, Y., Mori, T., Ruther, P., Paul, O., Yoshida, Y., Harashima, J., Kinumura, T., Inada, Y. and Nomura, Masahiro. (2024) Planar-type silicon thermoelectric generator with phononic nanostructures for 100 μW energy harvesting, *Mater. Today Phys.*, 45, 101452.

24 Yu, X., Wang, Y., Liu, Y., Li, T., Zhou, H., Gao, X., Feng, F., Roinila, T. and Wang, Y. (2012) CMOS MEMS-based thermoelectric generator with an efficient heat dissipation path. *J. Micromech. Microeng.*, 22, 105011.

25 Xie, J., Lee, C. and Feng, H. (2010) Design, fabrication, and characterization of CMOS MEMS-based thermoelectric power generators. *J. Microelectromech. Systems*, 19, 317.

26 Yang, S.M. and Wang, S. H. (2021) Development of a thermoelectric energy generator with double cavity by standard CMOS process. *IEEE Sens. J.*, 21, 250.

27 George, A., Yanagisawa, R., Anufriev, R., He, J., Yoshie, N., Tsujii, N., Guo, Q., Mori, T., Volz, S. and Nomura, M. (2019) Thermoelectric enhancement of silicon membranes by ultrathin amorphous films. *ACS Appl. Mater. Interfaces*, 11, 12027.

28 Li, Y., Buddharaju, K., Singh, N. and Lee, S.J. (2012) Top-down silicon nanowire-based thermoelectric generator: design and characterization. *J. Electron. Mater.*, 41, 989–92.

29 Xu, B., Khouri, W. and Fobelets, K. (2014) Two-sided silicon nanowire array/bulk thermoelectric power generator. *IEEE Electron Device Lett.*, 35, 596–8.

30 Santos, J.D., Salleras, M., Donmez, I., Gadea, G., Calaza, C., Morata, A., Tarancón, A. and Fonseca, L. (2016) Power response of a planar thermoelectric microgenerator based on silicon nanowires at different convection regimes. *Energy Harvest. Syst.*, 3, 335–42.

31 Norris, K.J., Garrett, M.P., Zhang, J., Coleman, E., Tompa, G.S. and Kobayashi, N.P. (2015) Silicon nanowire networks for multi-stage thermoelectric modules. *Energy Convers. Manag.* 96, 100–4.

11

Design of Heat Guide Layers in Micro-thermoelectric Generators

Tianzhuo Zhan

School of Integrated Circuit Science and Engineering, Beihang University, Beijing, China

11.1 Introduction

11.1.1 Thermoelectric Generators

Trillions of sensors will be deployed in the future Internet of Things (IoT) society for data collection and exchange. IoT sensors should be wireless for convenient installation and maintenance cost savings. For the power supply of a tremendous number of wireless sensor nodes, batteries in various forms, such as Li-ion battery, fuel cell, and supercapacitor, are traditional solutions [1]. However, batteries suffer from various limitations. For example, batteries must be replaced or recharged due to their limited lifetime, even though the performance of the battery is gradually improving. This should be avoided in many application scenarios, such as infrastructure health monitoring, industry condition monitoring, marine monitoring, military monitoring, animal tracking and control, and biomedical implants, where replacement of batteries is difficult and has a high cost. To overcome these limitations and meet the demands of many IoT wireless sensor applications, two solutions are promising. One is to make wireless sensors less power hungry. By optimizing the design of sensor systems, the power consumption of wireless sensors is continually dropping. The current power consumption of low-power wireless sensors is on the μW or mW level [1]. In some millimeter-scale ultra-low-power IoT sensors, standby power could be as low as nW level [2, 3]. The other solution is to use energy-harvesting devices to harvest energy from the environment for the power supply [4]. The energy harvester can recharge batteries to extend their lifetime or even replace them. Therefore, energy-harvesting technology will play an important role in realizing a perpetual self-powered wireless sensor network in the future IoT society.

Energy-harvesting technology can be primarily categorized into four types according to the energy sources available for harvesting, namely light energy, radio frequency (RF) radiation energy, thermal energy, and vibration energy [5]. By utilizing temperature differences in the environment, thermoelectric materials can convert thermal energy into electrical energy directly via the Seebeck effect [6]. Thermal energy harvesting can be achieved by thermoelectric generators (TEGs). TEGs have the advantages of direct energy conversion, no moving parts, high reliability, no maintenance, and long service lifetime [7]. Thus, compared with energy harvestings using kinetic energy sources such as piezoelectric, triboelectric, and magnetostrictive, TEG is very suitable for applications in harsh environments. For example, a radioisotope thermoelectric generator (RTG) converts the thermal energy released by nuclear fission into electricity, which has been successfully utilized as the power source of satellites and deep space probes [8]. Some RTGs have provided power for more than 40 years, showing their stability and reliability. Furthermore, benefiting from great progress achieved in microfabrication and VLSI circuit technology, millimeter-scale ultra-small sensor nodes have come into reality [2, 3]. The scalability in volume makes TEGs competent for perpetual power supply for these miniaturized ultra-low-power IoT wireless sensor nodes. However, miniaturized micro-TEGs (μTEG) have several limitations. One of the major drawbacks of μTEGs is the low output power compared with other types of power generators, which makes it difficult to meet the practical power requirements of most wireless sensors. One reason for the low output power of μTEGs is the low figure of merit (ZT) value, which is a dimensionless value that measures the efficiency of thermoelectric materials. Over the past decades, the ZT value has increased slowly with existing materials, with the realized value in the vicinity of 1. In recent years, nanostructured materials have been widely studied, and ZT values in the 2–3 range have been reported by lowering thermal conductivity while maintaining electrical conductivity [9, 10]. Another reason is the small temperature difference from the hot junction to the cold junction of the thermocouples in μTEGs. These reasons reflect the challenges to further improve the performance of μTEGs.

11.1.2 Parasitic Thermal Resistance in TEGs

Parasitic resistance is a common parasitic effect in semiconductor devices. Parasitic resistance refers to an unwanted electrical resistance that causes a voltage drop when current flows through the device, thus affecting the performance of the device. Correspondingly, parasitic thermal resistance refers to an undesired thermal resistance that causes a temperature gradient that has no contribution to the thermoelectric power generation when heat flows in TEGs, which should be diminished for power enhancement. Figure 11.1 shows the schematic diagram of a typical vertical-type TEG with a sandwich structure. Heat is absorbed from the hot source and ejected from the heat sink to form a temperature difference. Thermocouples consist of many pairs of n-type and p-type thermoelements, which are electrically connected in series and thermally in parallel. In planar TEGs, the direction of heat and current flow is different from that in vertical TEGs, but the basic working principles are the same. Figure 11.2 shows the equivalent thermal circuit of TEGs [11]. Q is the heat flux. A, L, σ, and κ are the cross-sectional area, the length, the electrical conductivity,

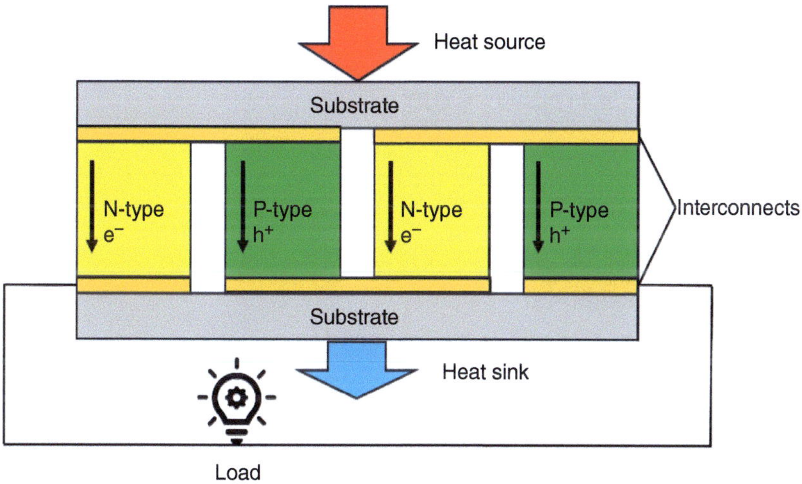

Figure 11.1 Schematic diagram of a typical vertical-type TEG with a sandwich structure.

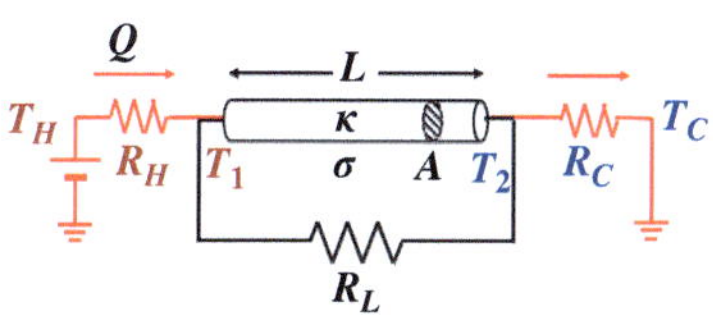

Figure 11.2 Equivalent thermal circuit of TEGs. Q is the heat flux. A, L, σ, and κ are the cross-sectional area, the length, the electrical conductivity, and the thermal conductivity of the thermoelements, respectively. T_H and T_C are the temperatures at the heat source and the heat sink, respectively. T_1 and T_2 are the temperatures at the hot junction and the cold junction of the thermoelements, respectively. R_H is the thermal resistance between the heat source and the hot junction of the TEG, and R_C is the thermal resistance between the cold junction of the TEG and the heat sink.

and the thermal conductivity of the thermoelements, respectively. T_H and T_C are the temperatures at the heat source and the heat sink, respectively. T_1 and T_2 are the temperatures at the hot junction and the cold junction of the thermoelements, respectively. R_H is the thermal resistance between the heat source and the hot junction of the TEG, and R_C is the thermal resistance between the cold junction of the TEG and the heat sink. Therefore, R $(R_H + R_C)$ is the total parasitic thermal resistance of the TEG. Furthermore, the thermal resistance of the package, which can be considered as a part of R, should also be considered in the structure optimization of a packaged TEG. When a certain temperature difference is applied to a packaged TEG, the ratio of the TEG thermal resistance to the total thermal resistance determines the temperature difference between the cold junction and the hot junction of the TEGs. Therefore, the performance of TEGs deteriorates with increasing resistance of the package.

According to the Seebeck effect, the open-circuit voltage V_{OC} is given by the following equation:

$$V_{OC} = S\,(T_H - T_C - QR)$$

where S is the Seebeck coefficient of thermoelements. The maximum output power $P_{\max}$ is given by the following equation:

$$P_{\max} = \frac{V_{OC}^2}{4R} = \frac{S^2 \sigma A \left(T_H - T_C - QR\right)^2}{4L}$$

The temperature difference between the heat source and the heat sink $T_H - T_C$ can be calculated by:

$$T_H - T_C = Q\left(R + \frac{L}{\kappa A}\right)$$

The maximum output power $P_{\max}$ can be derived as:

$$P_{\max} = \frac{S^2 \sigma A L \left(T_H - T_C\right)^2}{4\left(L + \kappa A R\right)^2}$$

When the parasitic thermal resistance R is much smaller than the intrinsic thermal resistance of the thermoelements, the maximum output power $P_{\max}$ can be written as:

$$P_{\max} = \frac{S^2 \sigma A \left(T_H - T_C\right)^2}{4L}$$

However, in miniaturized μTEGs, especially the feature size of which is in nanoscale, R could be significantly larger than the intrinsic thermal resistance of the thermoelements, and even the thermal boundary resistance may be larger than the intrinsic thermal resistance of the thermoelements. In this case, the maximum output power $P_{\max}$ can be written as:

$$P_{\max} = \frac{S^2 \sigma L \left(T_H - T_C\right)^2}{4A\kappa^2 R^2}$$

Thus, R plays a critically important role in the maximum output power, which should be suppressed as much as possible.

11.1.3 Silicon Nanowire Micro-TEGs

Bismuth telluride (Bi_2Te_3) and its alloys are commonly used thermoelectric materials for TEGs and refrigeration [12]. In addition, SnSe [13], half-Heusler [14], and skutterudite materials [15] have been found to have excellent thermoelectric properties. However, these thermoelectric materials are difficult to integrate with existing semiconductor chip systems, because they are incompatible with the semiconductor fabrication technology. Furthermore, in some application scenarios, such as energy harvesting for the power supply of low-power IoT wireless sensing nodes, the primary concern is not necessarily conversion efficiency. Trade-offs need to be made between the conversion efficiency and the fabrication cost of μTEGs [16]. Si is a competitive material with very low cost due to abundant raw material and mass production by manufacturing lines of the semiconductor industry. However, bulk Si is not suitable as a thermoelectric material because of its high thermal conductivity (~150 W/mK), leading to an extremely low ZT value of 0.01. Recently, Si nanowires (SiNWs) with nanoscale diameters have been reported with high thermoelectric efficiencies. The thermal conductivity has been reduced by nearly two orders of magnitude compared to bulk Si without significantly affecting the power factor, and a ZT value of 0.6 at room temperature has been achieved [17, 18].

According to structures, there are two different types of SiNW-μTEGs, namely, vertical type and planar type, depending on whether the direction of the temperature gradient is perpendicular or parallel to the surface of the substrate. In vertical-type SiNW-μTEGs, a large temperature difference can be retained across the vertically aligned SiNWs, and thus large output power. However, the fabrication cost is high due to the high requirement of technology such as a deep etching process. By comparison, the fabrication cost for planar-type SiNW-μTEGs is lower than that of vertical μTEGs because of its compatibility with IC planar technology. However, the use of substrates with a high thermal conductivity reduces the temperature difference across the SiNWs, resulting in poor performance. To solve this problem, the substrate needs to be etched away to form isolated SiNWs, leading to a high fabrication cost and fragile structure. In response to this issue, a miniaturized planar-type SiNW-μTEG has been proposed, which could be simply fabricated on commercially available silicon-on-insulator (SOI) substrates using the complementary metal-oxide semiconductor (CMOS)-compatible processes [19]. Figure 11.3 shows the schematic of the proposed planar-type SiNW-μTEG structure. In this SiNW-μTEG architecture without a thermal isolation cavity, the steep temperature gradient formed in close proximity to the hot junction of thermocouples is used for power generation. When the heat source is applied perpendicular to the substrate, an exuded thermal field is formed across the SiNWs. As a result, a steep temperature gradient is generated near the heat source, which decays exponentially along the length of the SiNWs away from the heat source. The exuded thermal field can be used for power generation. Therefore, there is no need to etch away the substrate to form suspended SiNWs, leading to a low fabrication cost and well-protected SiNWs.

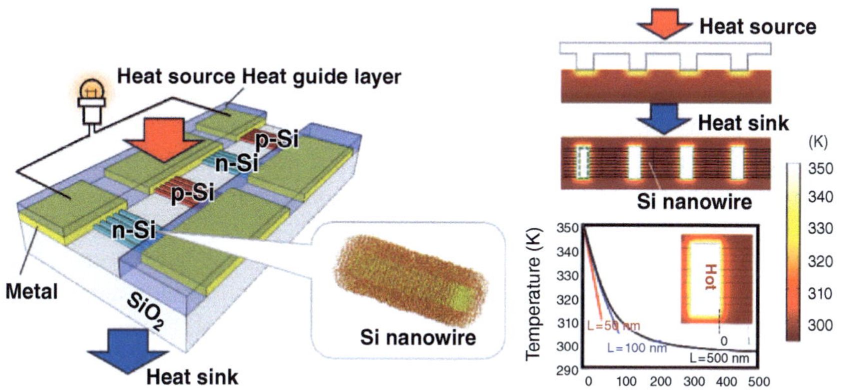

Figure 11.3 Schematic of the proposed planar-type SiNW-μTEG structure. The steep temperature gradient formed near the hot junction of thermocouples is used for power generation.

11.1.4 Power Enhancement in SiNW-μTEGs by Thinning Substrate

Tomita et al. have fabricated a prototype proposed planar SiNW-μTEG on SOI substrates [20]. The thicknesses of the Si substrate, the buried SiO_2 layer, and the top SOI layer are 745 μm, 145 nm, and 50 nm, respectively. SiNWs were patterned on the SOI layer by ArF immersion lithography and reactive-ion etching (RIE). Phosphorus ions with a dose of 5.0×10^{15} ions/cm^2 were implanted at an acceleration energy of 10 keV. The implantation was followed by activation annealing at 1000 °C. A single device is comprised of a 70-line SiNW bundle with a pitch of 500 nm. Both ends of the SiNWs were connected to separate Si pads. These SiNWs were wrapped by a 10-nm-thick SiO_2 layer. Next, this SiO_2 layer at the Si-pad regions was removed. A 400-nm-thick Al/TiN/Ti was deposited on the Si pad as the electrode, which is commonly used in multilayered metal wiring in CMOS devices. Then, a forming gas anneal was performed at 400 °C to suppress the contact resistance between the electrode and the Si-pad and to suppress the interface trap charges at the SiO_2/Si interface. To reduce the parasitic thermal resistance of the SiNW-μTEG, the thickness of the Si substrate was thinned from 745 to 50 μm by backside grinding. The thermoelectric properties of the SiNW-μTEGs were measured using a lab-built probe system equipped with a microthermostat comprising a resistive heater and a platinum thermometer on an aluminum nitride (AlN) ceramic plate. The electrical characterization of the SiNW-μTEGs was performed using a semiconductor-device analyzer (Agilent, B1500A, Keysight Technologies Inc., USA). The hot side of the electrode was heated using the microthermostat, which was maintained at 297–300 K. The base stage beneath the Si substrate was maintained at 290–292 K by a water chiller.

The thermoelectric power density was significantly enhanced by thinning the Si substrate. The applied temperature difference was 5 K; the length and the width of the SiNWs were 0.25 μm and 65 nm, respectively. The power density of the SiNW-μTEG with a 50-μm-thick Si substrate reached 12 μW/cm^2. Furthermore, V_{OC} was enhanced from 7.4 to 14.6 μV by thinning the substrate from 745 to 50 μm. This enhancement can be attributed to the decreased parasitic thermal resistance induced by the Si substrate. Furthermore, the TE performance of the fabricated devices was simulated by the finite-element method (FEM) using COMSOL Multiphysics. The heat conduction, the Poisson, and the charge continuity equations, together with the drift–diffusion current, the Seebeck, and the Peltier equations, were simultaneously solved. A 2D periodic boundary condition is adopted in the horizontal directions. The experimental results agree well with the FEM simulation results.

This study has demonstrated that by decreasing the parasitic thermal resistance introduced by the Si substrate, the power generation of the SiNW-μTEG can be significantly enhanced by one order of magnitude. However, in this study, a prototype unileg μTEG with only n-type SiNWs was fabricated for the test. For more energy harvesting to power IoT sensors, a large-scale bileg μTEG with many n-type and p-type SiNWs is necessary. Figure 11.4 shows the schematic diagram of a bileg SiNW-μTEG, in which a multilayered structure consisting of a heat guide layer, an interconnect layer, a thermocouple layer, a dielectric interlayer, and the underlying substrate is illustrated. The n-type and p-type SiNW thermocouples were connected

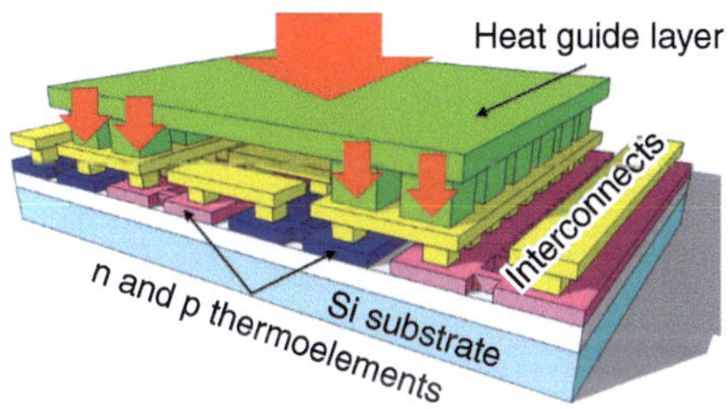

Figure 11.4 Schematic diagram of a bileg SiNW-μTEG with n-type and p-type thermocouples, in which a multilayered structure consisting of a heat guide layer, an interconnect layer, a thermocouple layer, a dielectric interlayer, and the underlying substrate is illustrated.

electrically in series by the interconnects and isolated electrically by the dielectric interlayer. The heat guide layer was used to guide the heat flow from the heat source to the hot junction of each thermocouple. The dielectric layer with low thermal conductivity can block the heat flow from the heat source to the cold junction of each thermocouple to maintain the temperature difference across thermocouples. The heat guide layer should be electrically insulative due to its direct contact with the interconnect metal layer and have a high thermal conductivity for heat guiding. In this SiNW-μTEG, parasitic thermal resistance is introduced not only by the Si substrate but also by the heat guide layer. Therefore, a heat guide layer with low thermal resistance needs to be designed for power enhancement.

11.2 AlN Heat Guide Layers in SiNW-μTEGs

AlN is a ceramic material with good electrical insulation properties and high thermal conductivity comparable to metallic aluminum. Furthermore, AlN has a thermal expansion coefficient that closely matches that of silicon. These properties make it suitable for thermal management in electronic and optoelectronic devices, such as power and RF semiconductors, laser diodes, and deep-ultraviolet (UV) photonics, where heat dissipation is critical during high-power and high-temperature operation. [21–23] Obviously, these properties also make AlN promising as heat guide layers in SiNW-μTEGs.

Zhan et al. have studied the use of AlN thin films as the heat guide layer in cavity-free planar SiNW-μTEGs [24]. A prototype SiNW-μTEG has been fabricated on an SOI substrate. Figure 11.5a schematically demonstrates the main steps of the process used to fabricate the SiNW-μTEGs. The thicknesses of the Si substrate, the BOX layer, and the top SOI layer are 745 μm, 145 nm, and 50 nm, respectively. SiNWs were patterned on the SOI layer by electron-beam (EB) lithography and RIE methods. An SAL-601 negative resist was used for the EB lithography. The RIE process was performed with inductively coupled plasma using $C_3F_8/O_2/Ar$ mixing gas. Next, thermal oxidation was performed under a dry O_2 atmosphere at 850 °C for 3 h, to form an SiO_2 layer with a thickness of 20 nm on the surface of the SiNWs for protection. Consequently, the thickness of the SiNWs decreased to approximately 35 nm. Phosphorus ions were then implanted at an acceleration energy of 25 keV, with a

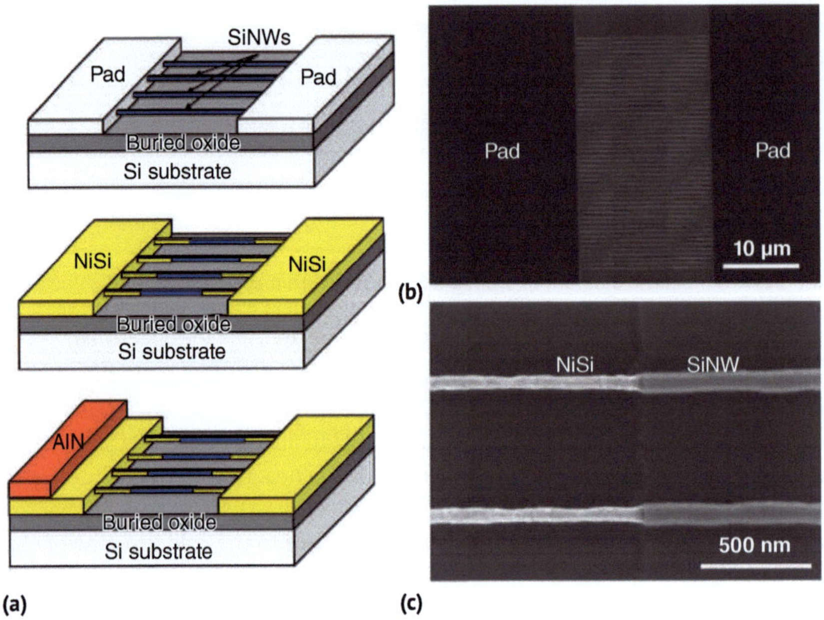

Figure 11.5 (a) Schematic illustration of the main steps of the process used to fabricate the SiNW-μTEG. (b) Top-view SEM image of one array with 50 SiNWs. (c) High-resolution SEM image of two individual SiNWs. *Source*: (a) Ref. [24] / Taylor & Francis / CC BY 4.0; (b, c) [24] / Taylor & Francis Group / CC BY 4.0.

dose of 1.0×10^{15} ions/cm^2. The implantation was followed by activation annealing at 950 °C under vacuum conditions for 10 min. Next, the SiO_2 layer at the pad regions was removed via photolithography and a wet-etching process using buffered hydrofluoric acid. Then, the Si pads and both ends of the SiNWs were nickelized by depositing a Ni film with a thickness of 20 nm using magnetron sputtering. This was followed by rapid thermal annealing at 410 °C for 20 min. Both ends of the SiNWs were connected to the NiSi pads. The purpose of the nickelidation of the Si pads and the SiNWs is to form low-resistive nickel silicides, which offer low contact resistance and few interfacial defects with Si.

Figure 11.5b and c shows the top-view scanning electron microscopy (SEM) images of a part of the fabricated SiNW-μTEG. The fabricated SiNW-μTEG consists of eight arrays of 400 legs. with a pitch of 500 nm. The SiNWs have a rectangular cross section. The thickness and width of the SiNWs were 35 and 100 nm, respectively. The length of the SiNWs ranged from 8 to 280 μm. Finally, an AlN film was deposited onto one of the NiSi pads via magnetron sputtering as the heat guide layer to guide the heat flow from the heat source to the hot junction of the SiNWs. Two deposition methods were utilized to grow the AlN films. One method involved the nonreactive RF sputtering method, where AlN and Ar are used as the sputtering target and sputtering gas, respectively. The other method involved reactive RF sputtering, where Al and an Ar/N_2 mixture are used as the sputtering target and sputtering gas, respectively. The flux ratio of Ar/N_2 mixing gas is 16:4. The total flow of the Ar/N_2 mixing gas was 20 sccm. The substrate temperature, sputtering power, and base pressure of the sputtering system were 25 °C, 300 W, and 1×10^{-5} Pa, respectively.

The microstructures of the AlN films deposited by the two methods were characterized by SEM and transmission electron microscopy (TEM). Figure 11.6a and b shows the cross-sectional SEM images of the AlN films deposited using the two methods. The reactively sputtered AlN film exhibits a polycrystalline structure, and the grain boundaries are aligned perpendicular to the surface of the Si substrate. The nonreactively sputtered AlN film exhibits an amorphous-like structure, and no crystal grains can be observed. The cross-sectional TEM images in Figure 11.6c and d show the microstructures of the AlN films more clearly with high resolution. The reactively sputtered AlN film possesses much larger grains than the nonreactively sputtered AlN film. The different diffraction patterns also confirm the different crystallinities of the AlN films deposited using the two methods.

X-ray diffraction (XRD) was employed to evaluate the crystalline quality of AlN films. The comparison of XRD patterns of the AlN films deposited by reactive and nonreactive sputtering methods is shown in Figure 11.7a. A sharp peak at $2\theta \approx 36°$can be observed in the case of reactively sputtered AlN films, corresponding to the (0002) diffraction peak of AlN. By comparison, the nonreactively sputtered AlN film shows a very weak and broad peak. The average grain size can be determined from the peak width according to the Scherrer equation. The full-width at half maximum values of the AlN (0002) peaks of the reactively sputtered and nonreactively sputtered AlN films are 0.3 and 2.1, respectively. Correspondingly, the average grain sizes of the reactively and nonreactively sputtered AlN films were calculated

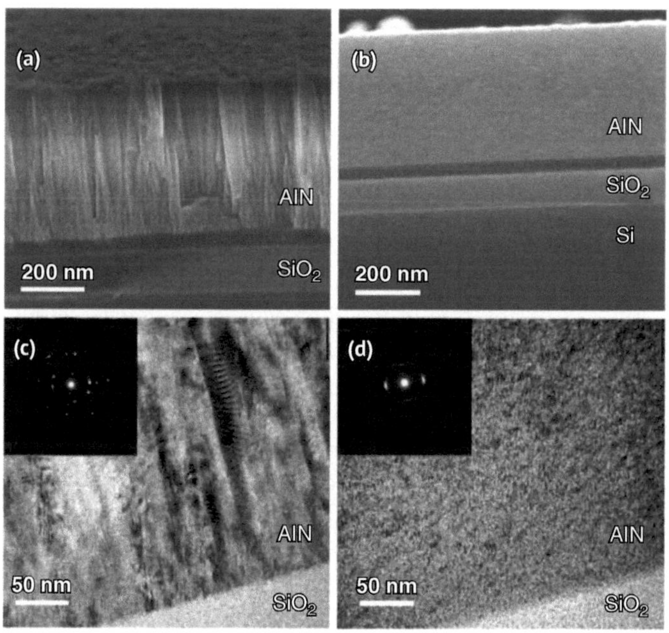

Figure 11.6 Cross-sectional SEM (a and b) and TEM (c and d) images of the AlN films deposited by reactive sputtering (a and c) and nonreactive sputtering (b and d). The insets show electron diffraction patterns. *Source*: Ref. [24] / Taylor & Francis Group / CC BY 4.0.

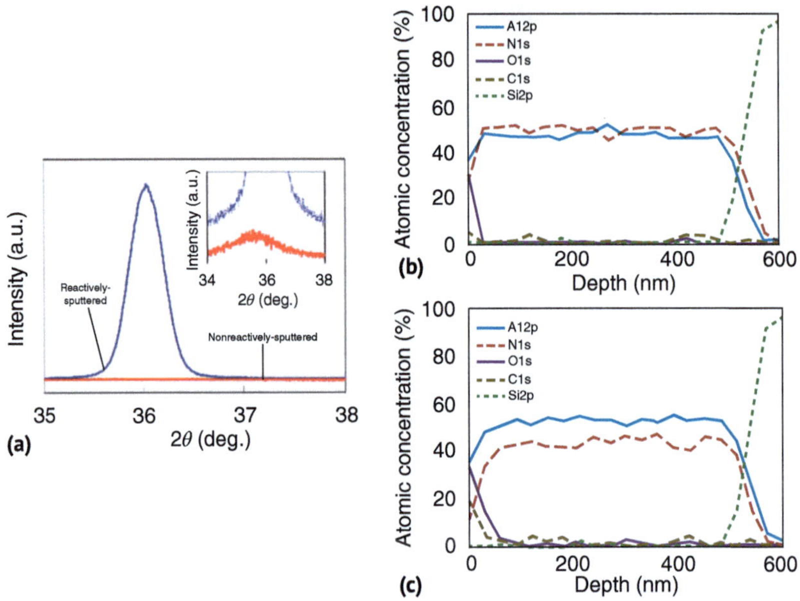

Figure 11.7 (a) Comparison of XRD patterns of AlN films deposited by reactive and nonreactive sputtering methods. Atomic concentration measured by XPS as a function of depth in AlN films deposited by reactive sputtering (b) and nonreactive sputtering (c). *Source*: Ref. [24] / Taylor & Francis / CC BY 4.0.

to be 25 and 3.7 nm, respectively. The SEM, TEM, and XRD characterization results indicate that the crystalline quality of the reactively sputtered AlN film is superior to that of the nonreactively sputtered AlN film.

The chemical composition of the two AlN films was characterized by X-ray photoelectron spectroscopy (XPS). Figure 11.7b and c shows the atomic concentration of the constituent elements as a function of depth for the two AlN films. In both AlN films, oxygen and carbon impurities can be observed near the top surface; this indicates contamination of the surfaces of the AlN films. A higher concentration of carbon impurity can be observed in the nonreactively sputtered AlN film. The atomic composition of the area below the contaminated surface layer primarily consists of Al and N. The reactively sputtered AlN film is stoichiometric, with an Al:N ratio of 1:1. However, in the case of the nonreactively sputtered AlN film, a non-stoichiometric compound with an Al:N ratio of approximately 4.5:5.5 was confirmed. Below the contaminated surface layers of both AlN films, the concentrations of the oxygen and carbon impurities are very low.

The thermal conductivities of the sputtered AlN films were measured using the ω method, which is based on the frequency-domain thermoreflectance technique [25]. T A thermoelectric cooling system was used to maintain the sample stage at 25 °C. The vacuum chamber was evacuated to ~0.1 Pa. The measuring principle is schematically shown in Figure 11.8a. The sample was heated with a pump laser, and its surface temperature was measured using a probe laser through the temperature-dependent reflectivity. Beams from the pump and probe lasers were focused coaxially on the surface of the film stack using an objective lens. The pump laser is a

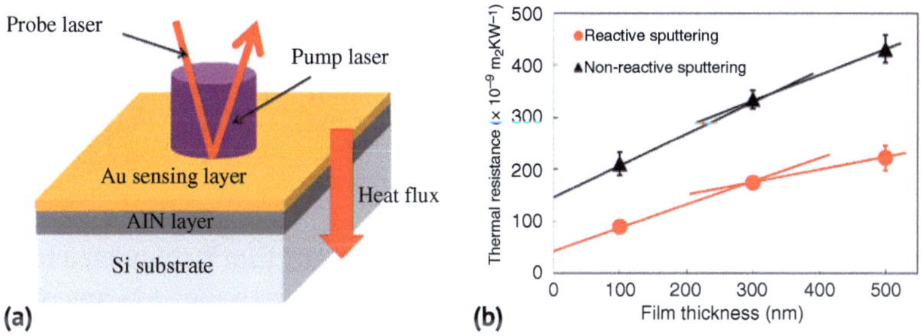

Figure 11.8 (a) Schematic of the Au/AlN/Si film stacks in the ω method. (b) Thermal resistances of Au/AlN/Si as a function of the AlN film thickness. The solid lines are marked by fitting the two data points (100 and 300 nm; 300 and 500 nm).

multimode diode laser with a wavelength of 405 nm and an output power of 360 mW. The probe laser is a single-mode diode laser with a wavelength of 635 nm and an output power of 5 mW. The pump laser was modulated at the frequencies of $\omega = 1$, 2, 4, and 8 kHz by the transistor–transistor logic signal from the function generator built in a DSP lock-in amplifier (Stanford SR8300). A 150-nm-thick Au film was then deposited on the AlN film via sputtering to function as a laser absorber and temperature sensor. The temperature response at the surface of the Au film was measured by the probe laser via the thermoreflectance technique. Because the diameter of the pump laser beam was much larger than that of the probe laser beam and the thermal diffusion length, a one-dimensional heat conduction in the normal-to-plane direction was assumed. The measurements were performed at three different locations on each sample; each measurement was performed three times to reduce errors. The in-phase temperature at the surface of the Au film, $T(0)$, can be expressed as a function of the properties of the film and the substrate, as shown here:

$$\frac{T\,(0)}{qd_{\mathrm{Au}}} = \frac{e^{-i\frac{\pi}{4}}}{\sqrt{2\lambda_S C_S}}\omega^{-\frac{1}{2}} + \left(1 - \frac{\lambda_{\mathrm{Au}} C_{\mathrm{Au}}}{\lambda_S C_S}\right)\frac{d_{\mathrm{Au}}}{\lambda_{\mathrm{Au}}} + R$$

where q is the heat flux, d is the thickness, λ is the thermal conductivity, C is the volumetric heat capacity, and R is the total thermal boundary resistance at the Au/AlN and AlN/Si interfaces. Subscript S denotes the substrate. A plot of $T(0)\,/\,(qd_{\mathrm{Au}})$ versus $\omega^{-1/2}$ gives a straight line, where the intercept is the sum of R and the thermal resistance of the Au film. By measuring R of samples with different AlN film thicknesses, the thermal conductivity of the AlN thin films can be determined from the slope of the straight line fitting the data points.

In this study, the thicknesses of the deposited AlN films were 100, 300, and 500 nm. Figure 11.8b shows the thermal resistances of Au/AlN/Si as a function of the thickness of the AlN films. For the reactively sputtered AlN films, the three data points cannot be plotted on a straight line, indicating that the thermal conductivity of the reactively sputtered AlN films is thickness dependent. By fitting the two data points of 300 and 500 nm, the thermal conductivity of the 500 nm-thick AlN film was determined to be 4.2 W/mK. From the cross-sectional TEM images, tens of nanometers of transition amorphous and nano-crystalline AlN are observed near its interface with the substrate. The small grains and defects near the interface region strongly scatter phonons, resulting in low thermal conductivity. Furthermore, phonons with long mean free paths (MFPs) over 0.3 μm contribute to ≈50% of the thermal conductivity in AlN [21]. This limits the theoretical upper limit of sub-300 nm thin AlN thermal conductivity. In the case of the nonreactively sputtered AlN films, the three data points are almost on a straight line, indicating that the thickness dependence of the thermal conductivity is weak. By fitting the two data points of 300 and 500 nm, the thermal conductivity of the nonreactively sputtered AlN film is calculated to be 2.1 W/mK. The thermal conductivity results correlate well with the SEM and XRD results, which indicates that the difference in the thermal conductivity is primarily induced by the differences in the crystalline quality of the two AlN films. The slight variation in the atomic composition should have a very weak influence on the thermal conductivity. However, the slightly higher carbon contamination level in the surface of the nonreactively sputtered AlN films may also lead to their lower thermal conductivity.

The parasitic thermal resistance of the SiNW-μTEGs introduced by the AlN heat guide layer comprises the intrinsic thermal resistance of the AlN film and the thermal boundary resistance (TBR) of the AlN/NiSi interface. According to the measurement principle of the ω method, the intercepts of the fitted straight lines give the total TBR values of the two interfaces (Au/AlN and AlN/Si) of the Au/AlN/Si samples. By fitting two data points (100 and 300 nm), we determined the TBR of the two interfaces. Figure 11.8b clearly shows that the TBR values of the two interfaces of the nonreactively sputtered samples (150 m^2K/GW) are much higher than those of the reactively sputtered samples (47 m^2K/GW). The large difference in the TBR of the two samples should be attributed to the difference in the crystalline quality of the two AlN films, which in turn significantly affects the interfacial properties, such as interfacial roughness, disorder, and bonding strength.

The thermoelectric properties of the SiNW-μTEGs were measured using a lab-built probe system equipped with a microthermostat comprising a resistive heater and a platinum thermometer on an AlN ceramic plate. The thermoelectric properties were measured by applying both a temperature difference and a loading voltage to the SiNW-μTEG. The thermostat was used as the heat source, which was placed on the AlN film and maintained at 45 °C. The base stage underlying the Si substrate was maintained at 20 °C by a water chiller as the heat sink. The electrical characterization of the SiNW-μTEGs was performed using a semiconductor-device analyzer (Agilent, B1500A, Keysight Technologies Inc., USA). The temperature difference across the SiNWs was measured using the IR thermography camera. Figure 11.9a–c shows a schematic of the measurement system, an optical photograph of the SiNW-μTEG, the electrical probes, and the thermostat during measurement, and the temperature distribution captured by the IR thermography camera during application of the heat source, respectively.

Figure 11.10a–d shows the measured maximum thermoelectric power, maximum power generation density, short-circuit current, and open-circuit voltage as a function of SiNW length, when two different AlN films were used as heat guide layers. The results show that an enhancement in the thermoelectric power of the SiNW-μTEGs has been achieved by using a reactively sputtered AlN heat guide layer.

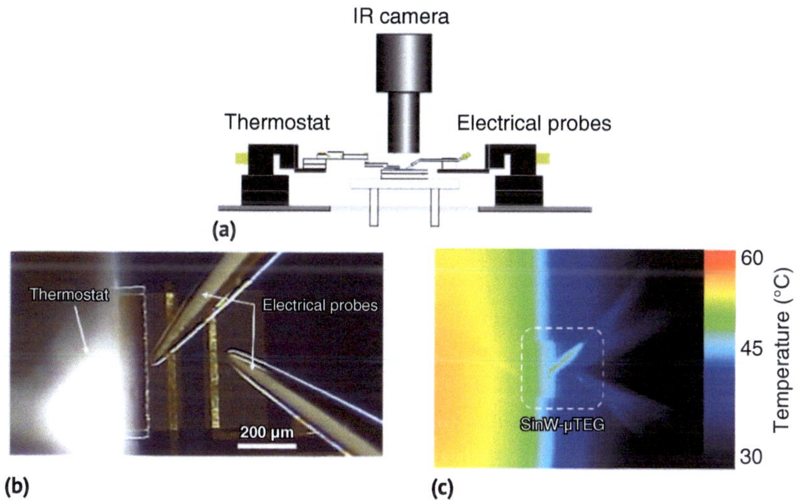

Figure 11.9 (a) Schematic of the measurement system. (b) Optical photograph of the SiNW-μTEG, the electrical probes, and the thermostat during measurement. (c) Temperature distribution captured by the IR thermography camera during application of the heat source. *Source*: (a) Ref. [24] / Taylor & Francis / CC BY 4.0; (b, c) Ref. [24] / Taylor & Francis Group / CC BY 4.0.

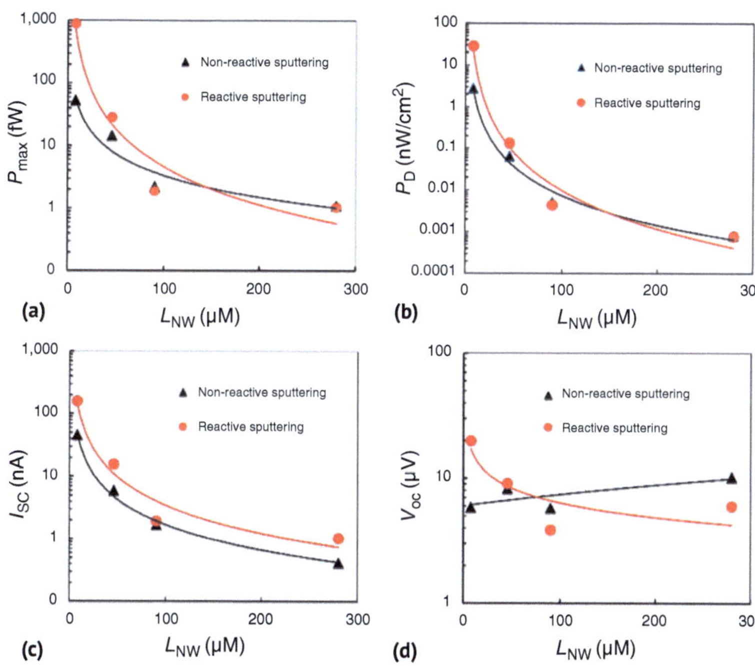

Figure 11.10 Measured maximum thermoelectric power (a), maximum power generation density (b), short-circuit current (c), and open-circuit voltage (d) as a function of SiNW length, when two different AlN films were used as heat guide layers. *Source*: Ref. [24] / Taylor & Francis / CC BY 4.0.

The enhancement should be attributed to the decrease of parasitic thermal resistance introduced by the AlN heat guide layer, which was caused by the improved thermal conductivity of the AlN films and the reduced TBR of the AlN/NiSi interface. However, the thermal conductivity of the reactively sputtered AlN film is still significantly lower than that of AlN in bulk form. Therefore, further improvement can be expected by improving the quality of AlN films.

Bulk-like thermal conductivities for AlN films have been reported. For example, Koh et al. have grown a series of AlN films on sapphire substrates ranging in thickness from 3 to 22 μm using metal-organic chemical vapor deposition (MOCVD) [26]. The measured thermal conductivities are 320 ± 42 W/mK at room temperature. However, the growth temperature in MOCVD was 1300 °C, which is incompatible with many requirements of CMOS processes, where deposition temperature less than 400 °C is typically required. Furthermore, AlN films in micrometer scale are not suitable for use in μTEGs. Zhang et al. have deposited AlN films with various thicknesses on silicon wafers by thermal ALD at 350 °C [27]. The thickness of AlN films was modified by performing 500, 1000, 1500, and 2000 ALD cycles. They found that when the number of ALD cycles was increased to 2000, the grains in AlN films grew larger and the crystal quality improved. A high thermal conductivity of 240.77 W/mK was obtained for the 206.8 nm-thick film deposited by 2000 cycles.

To overcome the drawbacks associated with high deposition temperature in various CVD methods, sputtering deposition at low temperature is still expected to deposit AlN film with high quality and high thermal conductivity. In the sputtering process, when the deposition temperature is fixed at a low temperature, other deposition parameters, such as base pressure, deposition temperature, sputtering power, and sputtering gas composition, could be modified to improve the purity and crystalline quality of the AlN films and the interfacial properties. Vaziri et al. have reported high cross-plane thermal conductivities approaching 90 W/mK for sub-300-nm-thick AlN films sputtered at low temperatures (<200 °C) [28]. In this work, AlN thin films were deposited using reactive DC magnetron sputtering of an Al target with mixed Ar/N_2 sputtering gas on 14-nm-thick thermally grown amorphous SiO_2 on Si substrates. AlN films with different structural and compositional properties were grown by modifying the sputtering parameters such as pressure, gas composition, DC power, and relative position of target and substrate. It is found that besides crystallite size and point defects, the alignment of AlN texture and structural homogeneity can greatly impact thermal conductivity, thermal boundary conductance, and thermal anisotropy. Perez et al. have deposited AlN films with thicknesses ranging from 100 nm to 1.7 μm at low temperature (<100 °C) by DC reactive balanced magnetron sputtering [29]. By controlling the sputtering gas composition of 25% Ar and 75% N_2 during deposition, a high thermal conductivity of 104 W/mK for 600 nm-thick AlN films has been achieved on c-Al_2O_3 substrates. Furthermore, the highest thermal conductivity values were found for depositions onto a cooled stage, which improved the film morphology. The previous two results have demonstrated the potential of AlN thin film as a heat-spreading layer for thermal management in 3D ICs due to the low-temperature back-end-of-line process-compatible deposition. Therefore, AlN thin films deposited by these techniques are also suitable as the heat guide layers in CMOS process-compatible SiNW-μTEGs.

11.3 Metal/Dielectric Heat Guide Layers in SiNW-μTEGs

In comparison to AlN film, metal films deposited by magnetron sputtering at room temperature demonstrate good crystallinity quality due to their relatively low melting point. Furthermore, the primary heat carrier in metals is free electrons. At room temperature, the typical value of the MFP of electrons in most metals is of the order of a few tens of nanometers [30]. Therefore, metal films with a thickness of hundreds of nanometers have bulk-like thermal conductivities [31]. If a thin dielectric layer is employed as the electrical insulation layer, metals such as Al and Cu films with thermal conductivities are promising candidates as the heat guide layer in μTEGs.

Zhan et al. have prepared various metal/adhesion/dielectric stacked heat guide layers using different combinations of metal, dielectric, and adhesion layers [32]. Figure 11.11a illustrates a schematic of the structure of the metal/adhesion/dielectric heat guide layers. The film stacks were deposited on SOI substrates by the magnetron sputtering method. The thicknesses of the base Si substrate, the BOX layer, and the top SOI layer were 750 μm, 145 nm, and 50 nm, respectively. The metal layers for heat conduction were Al and Cu. SiO_2 and AlN were used as the dielectric layer. Two transition metals, Ti and Cr, formed the adhesion layer between the metal and the dielectric layer to improve the adhesion of the metal/dielectric interface. Metal/dielectric samples with no adhesion layer were also prepared to serve as a reference for comparison. The thicknesses of the metal layer, adhesion layer, and dielectric layer were 300, 10, and 30 nm, respectively. The vacuum chamber of the sputtering system was evacuated to under a base pressure of 5.0×10^{-5} Pa. All thin films were deposited using a sputtering power of 300 W at a substrate temperature of 25 °C. The SiO_2 film was deposited by a nonreactive sputtering method, where SiO_2 was used as the sputtering target, while the AlN film was deposited using a reactive sputtering method with an Al target and an Ar/N_2 mixture gas with a flux ratio of 16:4. The total flow of the Ar/N_2 mixing gas was 20 sccm.

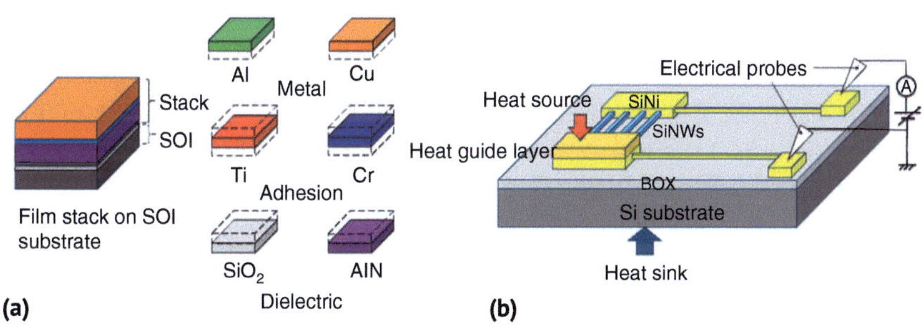

Figure 11.11 (a) Schematic of the structure of the metal/adhesion/dielectric stacked heat guide layers. The metal layers for heat conduction were Al and Cu. SiO_2 and AlN were used as the dielectric layer. Two transition metals, Ti and Cr, formed the adhesion layer between the metal and the dielectric layer to improve the adhesion of the metal/dielectric interface. (b) Schematics of the fabricated prototype SiNW-μTEG for evaluating the effect of TBR in the TC layers on power generation. *Source*: Ref. [32] / with permission of the American Chemical Society.

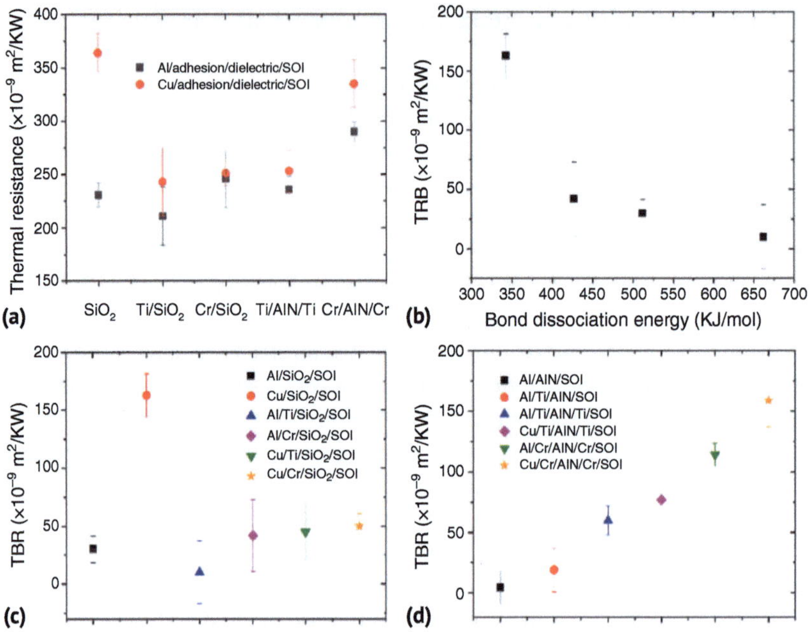

Figure 11.12 (a) Measured overall thermal resistance of all metal/adhesion/dielectric/SOI samples. (b) TBRs of the metal/SiO_2/SOI interfaces as a function of the bond dissociation energy of the metal–O bonds. (c) TBRs of metal/adhesion/dielectric/SOI samples when SiO_2 is used as the dielectric layer. (d) TBRs of metal/adhesion/dielectric/SOI samples when AlN is employed as the dielectric layer. *Source*: Reproduced from Ref. [32] / with permission of American Chemical Society.

The ω method was employed to measure the overall thermal resistance of the samples. It should be noted that the ω method cannot separate the thermal resistance of each constituent part. Figure 11.12a shows the measured overall thermal resistance of the stack films. The overall thermal resistances of Cu/adhesion/dielectric stacks are higher than their Al/adhesion/dielectric counterparts. The overall thermal resistance of the stack films consists of the intrinsic thermal resistance of each constituent layer and the TBRs of all interfaces. The thermal conductivities of the 30-nm-thick sputtered SiO_2 and AlN films were ~0.8 [33] and ~2.5 W/mK [24], respectively. This corresponds to the intrinsic thermal resistances of ~37.5 and ~12 m^2K/GW, respectively. The thermal resistances of the constituent metal and adhesion layers are on the order of 1 and 0.1 m^2K/GW, respectively, as calculated from the thermal conductivity and the thickness of each layer. The TBR of metal/metal interfaces is typically lower than 1 m^2K/GW. For example, the TBR of an Al/Cu interface was measured to be 0.25 m^2K/GW [34]. Therefore, the overall thermal resistance should be mainly contributed by the intrinsic thermal resistance of the dielectric layers and TBRs of the metal/dielectric and dielectric/SOI interfaces.

Figure 11.12c shows the TBRs of metal/adhesion/dielectric stacks when SiO_2 is used as the dielectric layer. The TBR of the Cu/SiO_2/SOI sample was significantly higher than that of the Al/SiO_2/SOI and Cu/SiO_2/SOI samples with an adhesion layer. All films were deposited under the same conditions, and interface properties such as interface roughness should be within the same order of magnitude for different samples. Thus, the extremely high TBR of the Cu/SiO_2 interface was attributed to the poor adhesion of the Cu/SiO_2 interface. It is known that Cu possesses a half-empty s shell and a tightly bound and filled *d* shell, therefore making it less chemically reactive than Al. Cu does not reduce SiO_2, and, thus, a graded oxide layer improving adhesion was unable to form [35]. In contrast, Al oxidizes well and strongly adheres to SiO_2 compared to Cu. A stronger interfacial bonding leads to a higher phonon transmission coefficient for both longitudinal and transverse modes [36–38]. This is considered the primary reason for the decreased TBR. To improve the poor bonding of the Cu/SiO_2 interface, a Ti and Cr transition metal layer was deposited prior to Cu film deposition. The results show that using a Ti or Cr adhesion layer significantly decreases the TBR of the Cu/SiO_2 interface, emphasizing the significance of interfacial adhesion in TBR. Furthermore, Ti decreased the TBR of the Cu/SiO_2 interface more than Cr. In contrast, the DMM predicted a higher TBR for the Ti/SiO_2 interface than the Cr/SiO_2 interface.

The bond dissociation energy can be used as a measure of the bonding strength of a chemical bond. The bond dissociation energies for Ti–O, Cr–O, and Cu–O bonds are 662, 427, and 343 kJ/mol, respectively. A stronger Ti–O bond than the Cr–O bond is one possible reason that Ti decreases the TBR of the Cu/SiO_2 interface more than Cr. The interfacial chemistry of the Ti/SiO_2 and Cr/SiO_2 interfaces was characterized using XPS. Figure 11.13a and b shows that Ti was only present as TiO_2, consisting of two Ti^{4+} peaks: Ti $2p_{1/2}$ (464.7 eV) and Ti $2p_{3/2}$ (458.6 eV), and no metallic Ti peak (454.0 eV) was observed. This indicates complete oxidation of Ti near the Ti/SiO_2 interface. For Cr, both Cr_2O_3 and metal Cr peaks were observed, including two Cr^{3+} peaks: Cr $2p_{1/2}$ (586.3 eV) and Cr $2p_{1/2}$ (576.6 eV). Two metal Cr peaks were also observed: Cr $2p_{1/2}$ (583.5 eV) and Cr $2p_{1/2}$ (574.1 eV). This indicates that Cr was not completely oxidized by SiO_2. The higher tendency of Ti to be oxidized with respect to Cr leads to the formation of a graded oxide layer and therefore a stronger adhesion of the Ti/SiO_2 interface than the Cr/SiO_2 interface. This presents another possible reason that the Ti decreases the TBR more than Cr. For the case of Al metal layer, using Ti and Cr as an adhesion layer decreases and increases the TBR of Al/SiO_2 interface, respectively. The bond dissociation energy of the Al–O bond was 512 kJ/mol, higher than the Cr–O bond but lower than the Ti–O bond. This may explain the underlying reason for the TBR of the Al/ SiO_2 interface to have decreased and increased when using the Ti and Cr adhesion layer, respectively. The results show that Ti is better than Cr as an adhesion layer in decreasing the TBR of the Al/SiO_2 and Cu/SiO_2 interfaces. The TBR of the metal/dielectric/SOI samples as a function of bond dissociation energy is plotted in Figure 11.12b, which shows that the TBR decreased with increasing bond dissociation energy of the metal–O bonds.

Figure 11.12d shows the TBRs of metal/adhesion/dielectric stacks when AlN was employed as the dielectric layer. The Al/AlN/SOI sample has the lowest TBR, followed by the Al/Ti/AlN/SOI sample. As opposed to the SiO_2 dielectric layer, the use of Ti as an adhesion layer slightly increased the TBR of the Al/AlN interface, although the bond dissociation energy of the Ti–N bond (464 kJ/mol) was higher than the Al–N bond (297 kJ/mol). It is known that Al atoms at the Al/AlN interface

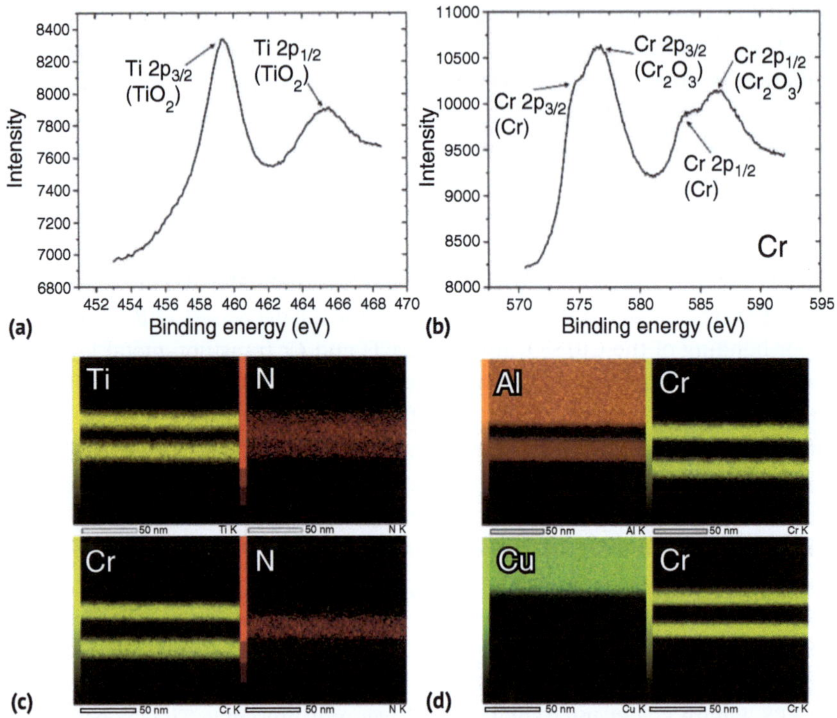

Figure 11.13 (a) XPS spectra near the Ti/SiO_2 interface. Ti is only present as TiO_2. (b) XPS spectra near the Cr/SiO_2 interface. Both Cr_2O_3 and metal Cr peaks are observed. (c) EDS elemental mappings near the Ti/AlN/Ti and Cr/AlN/Cr interfaces. N diffuses out of the AlN layer into the adjacent Ti adhesion layers but almost not into the Cr layers. (d) EDS elemental mapping near the Al/Cr/AlN/Cr and Cu/Cr/AlN/Cr interfaces. The occurrence of interdiffusion between the metal layer and the adhesion layer is observed. *Source*: (a, b) Reproduced from Ref. [32] / with permission of American Chemical Society; (c, d) Reproduced from Ref. [32] / with permission of American Chemical Society.

not only produce covalent bonds with N atoms in the AlN layer but also produce metallic bonds with other Al atoms in the AlN layer [39]. The Al—Al metallic bonds may facilitate the heat transport at the Al/AlN interface. Use of the two Ti layers above and below the AlN layer further increased TBR compared to the sample with the Ti layer only above. This may be attributed to the degradation of the adhesion of the AlN/SiO_2 interface by the insertion of the Ti layer between AlN and the native SiO_2, as the adhesion between AlN and native SiO_2 was strong due to the formation of silicon oxynitride [40]. Use of Cr increased the TBR of both the Al/AlN and the Cu/AlN interfaces more than that for Ti. One possible reason for this is that the bond dissociation energy of the Cr–N bond (378 kJ/mol) was weaker than that of the Ti–N bond (464 kJ/mol). We also characterized the EDS elemental mappings of the Al/Ti/AlN/Ti/SOI and Al/Cr/AlN/Cr/SOI samples, as shown in Figure 11.13c. The figure clearly demonstrates that N diffuses out of the AlN layer into the adjacent Ti adhesion layers; however, it almost does not diffuse into the Cr layers, indicating the lack of reaction between Cr and AlN. In comparison, the diffusion of N into the Ti layer leads to the formation of a graded TiN layer, facilitating adhesion and resulting in a stronger Ti/AlN bonding than Cr/AlN. This is another possible reason that the TBR in Al/Ti/AlN/Ti/SOI was lower than that for the Al/Cr/AlN/Cr/SOI samples. These results indicate that interfacial bonding strength plays a critical role in TBR. The TBRs of Cu/adhesion/dielectric/SOI samples are higher than their Al/adhesion/dielectric/SOI counterparts, despite sharing the same adhesion/dielectric interface. The EDS elemental mappings in Figure 11.13d show the occurrence of interdiffusion between the metal layer and the adhesion layer. The diffusion of Cu into the adhesion layer and the consequent formation of an alloy may weaken the adhesion strength of the adhesion/dielectric interface, resulting in higher TBR than the Al counterpart samples. This indicates that the interdiffusion of elements between the metal and the adhesion layers and the adhesion and the dielectric layers also affects TBR.

All results show that to obtain a low TBR of a metal/ adhesion/dielectric heat guide layer, Al is better than Cu as the metal layer, although Cu has a higher intrinsic thermal conductivity than Al. This is because Cu has poor adhesion to the dielectric layer, resulting in a high TBR of the metal/dielectric interface. Despite the use of an adhesion layer, diffusion of Cu into the adhesion layer may weaken the adhesion strength of the metal/dielectric interface. SiO_2 is better than AlN as a dielectric layer, although the AlN thin film has a higher intrinsic thermal conductivity than SiO_2. This is because metal–O chemical bonds have a higher bonding strength than their metal–N counterparts, resulting in a lower TBR of the metal/dielectric interface. However, the Al/AlN interface is an exception, which shows a lower TBR than the Al/SiO_2 interface, which may be attributed to the formation of Al–Al metallic bonds at the Al/AlN interface. Furthermore, Ti was better than Cr as an adhesion layer, as Ti has a stronger adhesion to oxide and nitride dielectric layers than Cr, enabling a lower TBR of the metal/dielectric interface. The stronger the chemical bond or the higher the tendency of the formation of a graded oxide or nitride layer at the interface, the lower the TBR.

To understand how the thermal resistance of the heat guide layer affects the power generation of μTEGs. Heat guide layers were deposited onto one pad via magnetron sputtering, which was schematically shown in Figure 11.11b. Five heat guide layers with different structures and thermal resistances were fabricated on SiNW-μTEGs. The structure and thickness of the five heat guide layers were as follows: AlN (300 nm), Cu/Ti/AlN/Ti (300/10/10/10 nm), Al/Ti/SiO_2 (300/10/10 nm), Al/Ti/SiO_2 (300/10/20 nm), and Al/Ti/SiO_2 (300/10/30 nm). Forming gas annealing at 400 °C for 30 min with 90% Ar + 10% H_2 was performed before the measurement of thermoelectric characteristics to decrease interface states at the Si/SiO_2 interface. A temperature difference of 25 °C was applied using a micro thermostat. Power generation was measured by loading a reverse current to the SiNW-μTEG using a direct current (DC) source instrument (Keithley 6220). The open-circuit voltage (V_{OC}) and short-circuit current (I_{SC}) were measured using a nanovolt/micro-Ohm meter (Keysight, 34420A, 7 1/2 Digit).

The measured Seebeck coefficient and electrical conductivity of the SiNWs were 206 μV/K and 148/Ω cm, respectively. This indicates that the thermoelectric power factor was 628 μW/mK^2. Figure 11.14a presents the current–voltage curves of the fabricated prototype SiNW-μTEGs with different heat guide layers. The slope of the

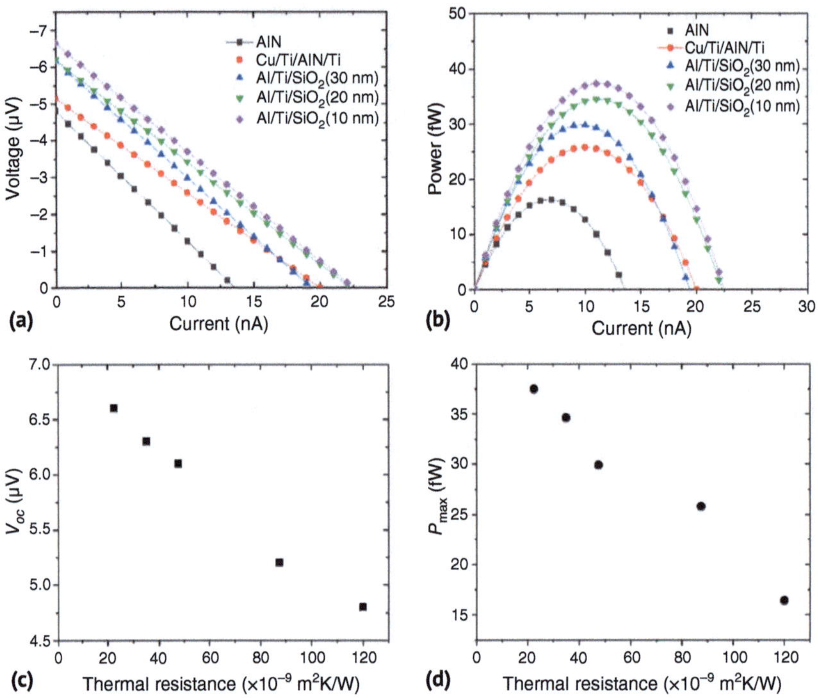

Figure 11.14 (a) Current–voltage curves of SiNW-μTEGs with different heat guide layers. (b) Current-power curves of SiNW-μTEGs with different heat guide layers. (c) Dependence of the V_{OC} of SiNW-μTEGs on the thermal resistance of the heat guide layers. (d) Dependence of P_{max} of SiNW-μTEGs on the thermal resistance of the heat guide layers. *Source:* Reproduced from Ref. [32] / with permission of American Chemical Society.

current–voltage curves denotes the internal electric resistance, which is the sum of the pad, nanowire, and contact resistances of the SiNW-μTEG. The different slope indicates that these SiNW-μTEGs have a different internal electric resistance. This should be attributed to the different nanowire or contact resistance due to processing variations or measuring conditions. V_{OC}, the intercept of the current–voltage curves, denotes the thermoelectric voltage of the SiNW-μTEG. The different V_{OC} should be attributed to the various temperature differences (ΔT) across the silicon nanowire in different devices. The different parasitic thermal resistance introduced by various heat guide layers in different devices leads to different across the silicon nanowire. The current–power curves of the SiNW-μTEGs with different heat guide layers are shown in Figure 11.14b. We measured a maximum thermoelectric voltage of 6.7 μV and a thermoelectric power (P_{max}) of 37.4 fW for the SiNW-μTEG when an Al/Ti/SiO_2 (300/10/10 nm) structure was employed as the heat guide layer. The minimum V_{OC} and P_{max} were attained for the SiNW-μTEG with an AlN heat guide layer (300 nm). Figure 11.14c and d plots the dependence of V_{OC} and P_{max} on the thermal resistance of the heat guide layers, respectively, showing that they increase with decreasing thermal resistance of the heat guide layers. As the thermal resistance of the SiNWs and the thermal resistance of the heat guide layer are connected in series, a relative decrease in the thermal resistance of the heat guide layer will result in an increase in ΔT across the SiNWs, leading to an increase in V_{OC} and P_{max}. It is anticipated that the thermoelectric power may be further enhanced if the thermal resistance of the heat guide layer can be reduced further. This may be realized by additional thinning of the dielectric layer or by lowering the TBR in the metal/dielectric heat guide layers.

11.4 Summary

μTEGs consist of many n-type and p-type thermocouples for high power generation. A heat guide layer is needed to guide the heat flow from the heat source to the hot junction of each thermocouple. The heat guide layer should be electrically insulative due to its direct contact with the interconnect layer and thermally conductive for heat guiding. AlN is a promising candidate as a heat guide material, which has good electrical insulation properties and high thermal conductivity. Furthermore, AlN has a thermal expansion coefficient that closely matches that of silicon. These properties make it suitable for thermal management in electronic and optoelectronic devices, such as power and RF semiconductors, laser diodes, and deep-UV photonics, where heat dissipation is critical during high-power and high-temperature operation. These properties also make AlN promising as heat guide layers in SiNW-μTEGs. However, AlN films in nanoscale thickness typically have lower thermal conductivity than bulk AlN. Furthermore, the fabrication of AlN heat guide layer on SiNW-μTEGs should be compatible with CMOS processes, where deposition temperature less than 400 °C is typically required. Therefore, it is difficult to grow high-quality AlN films using high-temperature CVD methods, causing no damage to the underlying device structures. In comparison, AlN films can be deposited using the sputtering method at low temperatures. Sputtering parameters such as pressure, gas composition, sputtering power, and relative position of target and substrate should be modified to optimize the deposition conditions to improve the quality of AlN films and reduce the defects at the interface. Metal/dielectric heat guide layers with a thermally conductive metal layer and an electrically insulative dielectric layer are alternatives as a heat guide layer. However, the TBR between the metal layer and dielectric layer should be suppressed. Because in nanoscale, TBR can dominate the total thermal resistance of the multilayered structures with interfaces. Interfacial properties such as roughness, interdiffusion, disorder, and chemical bonding should be optimized to decrease the TBR as much as possible.

References

1 Yan, J., Liao, X., Yan, D. and Chen, Y. (2018) Review of micro thermoelectric generator. *J. Microelectromechanical Syst.*, 27, 1–18.

2 Lee, Y., Bang, S., Lee, I., Kim, Y., Kim, G., Ghaed, M. H., Pannuto, P., Dutta, P., Sylvester, D. and Blaauw, D. (2013) A modular 1 mm^3 die-stacked sensing platform with low power I^2C inter-die communication and multi-modal energy harvesting. *IEEE J. Solid-State Circuits*, 48, 229–43.

3 Chen, G., Ghaed, H., Haque, R., Wieckowski, M., Kim, Y., Kim, G., Fick, D., Kim, D., Seok, M., Wise, K., Blaauw, D. and Sylvester, D. (2011) A cubic-millimeter energy-autonomous wireless intraocular pressure monitor. In *IEEE International Solid-State Circuits Conference*, San Francisco, CA, 310–312.

4 Zhou, G., Huang, L., Li, W. and Zhu, Z. (2014) Harvesting ambient environmental energy for wireless sensor networks: a survey. *J. Sensors*, 2014, 815467.

5 Harb A. (2011) Energy harvesting: state-of-the-art. *Renew. Energy*, 36, 2641–54.

6 Tritt, T.M. and Subramanian, M.A. (2006) Thermoelectric materials, phenomena, and applications: a bird's eye view. *MRS Bull.*, 31, 188–98.

7 Bell, L.E. (2008) Cooling, heating, generating power, and recovering waste heat with thermoelectric systems. *Science*, 321, 1457–61.

8 Yang, J. and Caillat, T. (2006) Thermoelectric materials for space and automotive power generation. *MRS Bull.*, 31, 224–9.

9 Zhao, L., Lo, S., Zhang, Y., Sun, H., Tan, G., Uher, C., Wolverton, C., Dravid, V.P. and Kanatzidis, M.G. (2014) Ultralow thermal conductivity and high thermoelectric figure of merit in SnSe crystals. *Nature*, 508, 373–7.

10 Snyder, G.J. and Toberer, E.S. (2008) Complex thermoelectric materials. *Nat. Mater.*, 7, 105–14.

11 Zhang, H., Xu, T., Hashimoto, S. and Watanabe, T. (2008) The possibility of mW/cm^2-class on-chip power generation using ultrasmall Si nanowire-based thermoelectric generators. *IEEE Trans. Electron Devices*, 65, 2016–23.

12 Venkatasubramanian, R., Siivola, E., Colpitts, T. and O'Quinn, B. (2001) Thin-film thermoelectric devices with high room temperature figures of merit. *Nature*, 413, 597–602.

13 Zhao, L., Tan, G., Hao, S., He, J., Pei, Y., Chi, H., Wang, H., Gong, S., Xu, H., Dravid, V., Uher, C., Snyder, G.J., Wolverton, C. and Kanatzidis, M.G. (2016) Ultrahigh power factor and thermoelectric performance in hole-doped single-crystal SnSe. *Science*, 351, 141–4.

14 Zhu, T., Fu, C., Xie, H., Liu, Y. and Zhao, X. (2015) High efficiency half-Heusler thermoelectric materials for energy harvesting. *Adv. Energy Mater.*, 5, 1500588.

15 Rogl, G. and Rogl, P. (2017) Skutterudites, a most promising group of thermoelectric materials. *Curr. Opin. Green Sustain. Chem.*, 4, 50–7.

16 LeBlanc, S., Yee, S.K., Scullin, M.L., Dames, C. and Goodson, K.E. (2014) Material and manufacturing cost considerations for thermoelectrics. *Renew. Sustain. Energy Rev.*, 32, 313—27.

17 Hochbaum, A., Chen, R., Delgado, R.D., Liang, W., Garnett, E.C., Najarian, M., Majumdar, A. and Yang, P. (2008) Enhanced thermoelectric performance of rough silicon nanowires. *Nature*, 451,163–7.

18 Boukai, A.I., Bunimovich, Y., Tahir-Kheli, J., Yu, J.-K., Goddard III, W.A. and Heath, J.R. (2008) Silicon nanowires as efficient thermoelectric materials. *Nature*, 451, 168–71.

19 Watanabe, T., Asada, S., Xu, T., Hashimoto, S., Oba, S., Himeda, Y., Yamato, R., Zhang, H., Tomita, M., Matsukawa, T., Kamakura, Y. and Ikeda, H. (2008) A scalable Si-based micro thermoelectric generator. In *2017 IEEE Electron Devices Technology and Manufacturing Conference (EDTM)*, 86–7.

20 Tomita, M., Oba, S., Himeda, Y., Yamato, R., Shima, K., Kumada, T., Xu, M., Takezawa, H., Mesaki, K., Tsuda, K., Hashimoto, S., Zhan, T., Zhang, H., Kamakura, Y., Suzuki, Y., Inokawa, H., Ikeda, H., Matsukawa, T., Matsuki, T. and Watanabe, T. (2018) Modeling, simulation, fabrication, and characterization of a 10-μW/cm^2 class Si-nanowire thermoelectric generator for IoT applications. *IEEE Trans. Electron Devices*, 65, 5180–8.

21 Xu, R.L., Rojo, M.M., Islam, S.M., Sood, A., Vareskic, B., Katre, A., Mingo, N., Goodson, K.E., Xing, H.G., Jena, D. and Pop, E. (2019) Thermal conductivity of crystalline AlN and the influence of atomic-scale defects. *J. Appl. Phys.*, 126, 185105.

22 Cheng, Z., Koh, Y. R., Mamun, A., Shi, J., Bai, T., Huynh, K., Yates, L., Liu, Z., Li, R., Lee, E., Liao, M.E., Wang, Y., Yu, H.M., Kushimoto, M., Luo, T., Goorsky, M.S., Hopkins, P.E., Amano, H., Khan, A. and Graham, S. (2020) Experimental observation of high intrinsic thermal conductivity of AlN. *Phys. Rev. Mater.*, 4, 044602.

23 Spina, L.L., Iborra, E., Schellevis, H., Clement, M., Olivares, J. and Nanver, L.K. (2008) Aluminum nitride for heat spreading in RF IC's. *Solid-State Electro.*, 52, 1359–63.

24 Zhan, T., Yamato, R., Hashimoto, S., Tomita, M., Oba, S., Himeda, Y., Mesaki, K., Takezawa, H., Yokogawa, R., Xu, Y., Matsukawa, T., Ogura, A., Kamakura, Y. and Watanabe, T. (2018) Miniaturized planar Si-nanowire micro-thermoelectric generator using exuded thermal field for power generation. *Sci. Technol. Adv. Mater.*, 19, 443–53.

25 Kato, R., Xu, Y. and Goto, M. (2011) Development of a frequency-domain method using completely optical techniques for measuring the interfacial thermal resistance between the metal film and the substrate. *Jpn. J. Appl. Phys.*, 50, 106602.

26 Koh, Y.R., Cheng, Z., Mamun, A., Hoque, M.S.B., Liu, Z., Bai, T., Hussain, K., Liao, M.E., Li, R., Gaskins, J.T., Giri, A., Tomko, J., Braun, J.L., Gaevski, M., Lee, E., Yates, L., Goorsky, M.S., Luo, T., Khan, A., Graham, S. and Hopkins, P.E. (2020) Bulk-like intrinsic phonon thermal conductivity of micrometer-thick AlN films. *ACS Appl. Mater. Interfaces*, 12, 29443—50.

27 Zhang, W., Li, J., Fang, J., Hui, L., Qin, L., Gong, T., Sun, F. and Feng, H. (2023) Atomic layer deposited high quality AlN thin films for efficient thermal management. *J. Mater. Chem. A*, 11, 21846.

28 Vaziri, S., Perez, C., Datye, I.M., Kwon, H., Hsu, C.-F., Chen, M.E., Noshin, M., Lee, T.-Y., Asheghi, M., Woon, W.-Y., Pop, E., Goodson, K.E., Liao, S.S. and Bao, X. (2024) AlN: an engineered thermal material for 3D integrated circuits. *Adv. Funct. Mater.*, 2024, 2402662.

29 Perez, C., McLeod, A.J., Chen, M.E., Yi, S.-I., Vaziri, S., Hood, R., Ueda, S.T., Bao, X., Asheghi, M., Park, W., Talin, A.A., Kumar, S., Pop, E., Kummel, A.C. and Goodson, K.E. (2023) High thermal conductivity of submicrometer aluminum nitride thin films sputter-deposited at low temperature. *ACS Nano*, 17, 21240—50.

30 Gall, D. (2016) Electron mean free path in elemental metals. *J. Appl. Phys.*, 119, 085101.

31 Avery, A.D., Mason, S.J., Bassett, D., Wesenberg, D. and Zink, B.L. (2015) Thermal and electrical conductivity of approximately 100-nm permalloy, Ni, Co, Al, and Cu films. *Phys. Rev. B*, 92, 214410.

32 Zhan, T., Ma, S., Jin, Z., Takezawa, H., Mesaki, K., Tomita, M., Wu, Y., Xu, Y., Matsukawa, T., Matsuki, T. and Watanabe, T. (2020) Effect of the thermal boundary resistance in metal/dielectric thermally conductive layers on power generation of silicon nanowire microthermoelectric generators. *ACS Appl. Mater. Interfaces*, 12, 34441–50.

33 Yamane, T., Nagai, N., Katayama, S. and Todoki, M. (2002) Measurement of thermal conductivity of silicon dioxide thin films using a 3ω method. *J. Appl. Phys.*, 91, 9772.

34 Gundrum, B.C., Cahill, D.G. and Averback, R.S. (2005) Thermal conductance of metal-metal interfaces. *Phys. Rev. B*, 72, 245426.

35 Nagao, K., Neaton, J.B. and Ashcroft, N.W. (2003) First-principles study of adhesion at CuO/SiO_2 interfaces. *Phys. Rev. B*, 68, 125403.

36 Losego, M.D., Grady, M.E., Sottos, N.R., Cahill, D.G. and Braun, P.V. (2012) Effects of chemical bonding on heat transport across interfaces. *Nat. Mater.*, 11, 502–6.

37 O'Brien, P.J., Shenogin, S., Liu, J., Chow, P.K., Laurencin, D., Mutin, P.H., Yamaguchi, M., Keblinski, P. and Ramanath, G. (2013) Bonding-induced thermal conductance enhancement at inorganic heterointerfaces using nanomolecular monolayers. *Nat. Mater.*, 12, 118–22.

38 Hu, M., Keblinski, P. and Schelling, P. K. (2009) Kapitza conductance of silicon–amorphous polyethylene interfaces by molecular dynamics simulations. *Phys. Rev. B*, 79, 104305.

39 Takashi, M. and Yoshihiro, K. (2018) Epitaxial Growth of III-Nitride Compounds: Computational Approach, Springer, Cham, Switzerland.

40 Lane, M., Dauskardt, R.H., Krishna, N. and Hashim, I. (2000) Adhesion and reliability of copper interconnects with Ta and TaN barrier layers. *J. Mater. Res.*, 15, 203–11.

12

Thermoelectric Thin-film Thermoelectric Generator

Koji Miyazaki

Department of Mechanical Engineering, Kyushu University, Fukuoka, Japan

12.1 Introduction

Electricity can be directly generated by heat flow through a solid material, a process called thermoelectric power generation [1, 2]. Since it has no moving parts, it is considered easy to miniaturize. The characteristics of thermoelectric materials are evaluated based on the Seebeck coefficient S (V/K), which represents the voltage generated per unit temperature difference, electrical conductivity σ (S/m), and thermal conductivity λ (W/(m K)). The efficiency η is defined as the ratio of heat input Q to power output W and is expressed by the following equations:

$$\eta = \frac{W}{Q} = \frac{T_h - T_c}{T_h} \times \frac{m_{\text{opt}} - 1}{m_{\text{opt}} + \frac{T_c}{T_h}}$$

$$m_{\text{opt}} = \sqrt{1 + \frac{1}{2}\frac{S^2\sigma}{\lambda}(T_h + T_c)} = \sqrt{1 + ZT} \tag{12.1}$$

where $T_h$is high temperature and T_c is low temperature at the both sides of the material. The dimensionless figure of merit ZT represents thermoelectric performance, with higher values indicating greater efficiency, as shown in Figure 12.1. The numerator of Z, $S^2\sigma$, is called the power factor. While both high values are desirable, they have a trade-off relationship: metals have a low Seebeck coefficient S, while insulators have low electrical conductivity σ, as shown in Figure 12.2. Therefore, semiconductors are used as thermoelectric materials, with various types such as Si and Si–Ge [3]. Covering all thermoelectric materials is challenging in the present section, so this overview focuses on thin-film fabrication of bismuth telluride (Bi_2Te_3), which exhibits the highest performance at room temperature, and its thermoelectric thin-film generator characteristics.

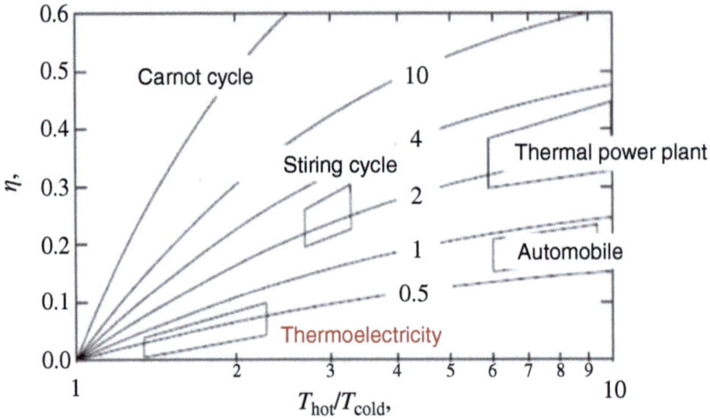

Figure 12.1 Theoretical efficiency of thermoelectric generators. The values on curves represent ZT. Higher ZT shows higher efficiency.

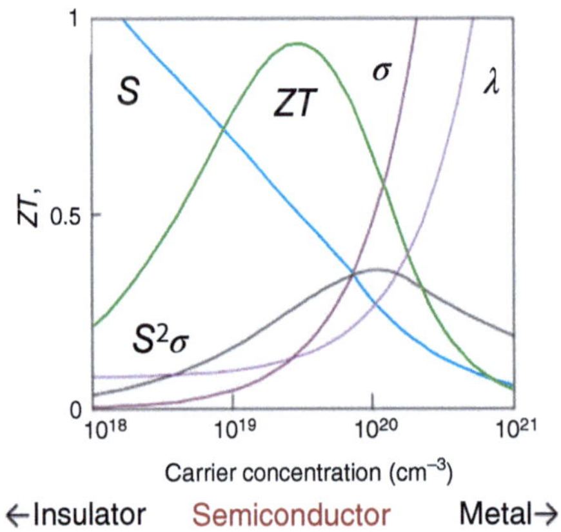

Figure 12.2 Schematic ZT curve with carrier concentration. The thermoelectric materials are typically heavily doped semiconductors with a carrier concentration ranging from 10^{19} to 10^{21} carriers per cm^3.

12.2 Bi_2Te_3 Thin-film Fabrication

Vacuum deposition methods are generally used for the fabrication of Bi_2Te_3 thin films [4]. The deposition process involves evacuating the chamber using a turbomolecular pump or an oil diffusion pump and then evaporating Bi_2Te_3 by some method to form a thin film on a substrate. Undoped Bi_2Te_3 exhibits n-type behavior, while doping with Sb produces p-type characteristics, and doping with Se enhances n-type behavior.

Bi_2Te_3 thin films cannot be formed by thermally evaporating Bi_2Te_3 due to stoichiometric deviation. In the crucible, an equilibrium state is reached, and Tellurium (Te), which has a high vapor pressure, is preferentially evaporated from the melt. Then, molecular beam epitaxy (MBE) is generally used to adjust the stoichiometry of the composite, in which Bismuth (Bi) and Tellurium (Te) are evaporated separately from two crucibles, allowing for the optimization of conditions to achieve the desired Bi_2Te_3 film [5]. When doping with Antimony (Sb) or Selenium (Se), a third crucible is used. To prevent composition deviation due to differences in vapor pressure at equilibrium, an alternative approach called pulsed laser deposition (PLD) is employed. In this method, Bi_2Te_3 is rapidly evaporated, causing the evaporated material to immediately form a thin film, thereby maintaining the correct composition under nonequilibrium conditions in a short period. PLD is widely used for fabricating thin films of materials with complex compositions, not just Bi_2Te_3 [6]. The method is also commonly used in the field of superconducting thin films, which require quaternary materials [7]. Based on a similar concept, another technique called flash evaporation deposition (FED) thermally evaporates Bi_2Te_3 instantaneously, as shown in Figure 12.3 [8, 9]. The raw material is pulverized into fine particles ranging from a few micrometers to several hundred micrometers in diameter, and small amounts are supplied onto a preheated tungsten boat. Upon contact with the tungsten boat, the fine particles evaporate within an extremely short time, enabling deposition onto the substrate while minimizing composition deviation.

Another method, arc plasma deposition, can also produce compositionally uniform Bi_2Te_3, illustrated in Figure 12.4 [10, 11]. In this technique, a high voltage is applied between electrodes and Bi_2Te_3, causing instantaneous evaporation. Since the voltage is applied in short pulses, the material evaporates instantly, maintaining its composition.

Regardless of the deposition method, post-deposition annealing is essential and is typically performed at around 250 °C in high-purity argon gas, argon with 5% hydrogen gas, and so on.

In addition to vacuum deposition techniques, various other methods have been proposed to obtain Bi_2Te_3 thin film, including metal-organic chemical vapor deposition (MOCVD) using chemical reactions [12, 13], solvothermal synthesis [14], where precursor materials are dissolved in a liquid and heated, and electrodeposition [15], where an electric current is applied to a solution to induce deposition. Since around 2010, spray printing [16] and screen printing [17–19] have also been reported.

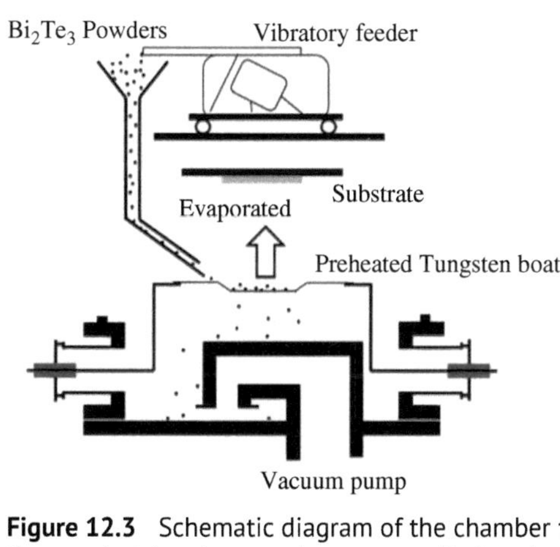

Figure 12.3 Schematic diagram of the chamber for the flash evaporation method. The thermoelectric micropowders are supplied to the preheated tungsten boat by a vibratory feeder.

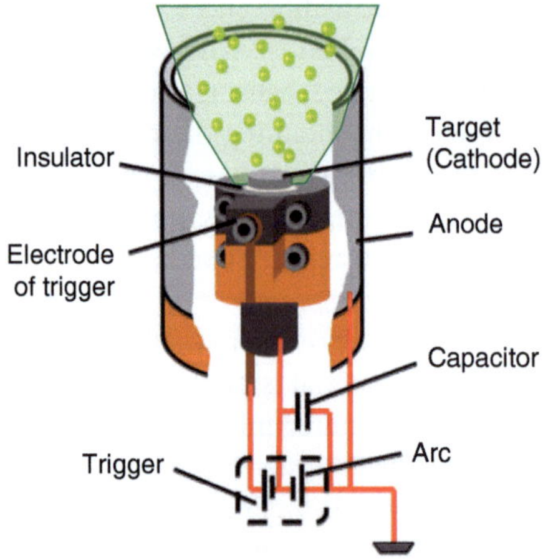

Figure 12.4 Schematic diagram of a coaxial-type vacuum arc evaporation source. Bi_2Te_3 is used as the cathode to generate Bi_2Te_3 plasma by applying a high voltage for a short duration at a frequency of 1 Hz.

Although these methods currently exhibit lower performance, they offer advantages in terms of low-cost fabrication and structural control.

12.3 Thermoelectric Generators with Bi_2Te_3 Thin Films

The standard clean room technique, such as the lift-off process, is generally used to make a thermoelectric microgenerator [20]. Also, electrodeposition has been reported for microgenerator fabrication to reduce the production costs [15, 21]. Moreover, it is difficult to treat Bi_2Te_3 as a non-Si material in the standard clean room. For example, the inkjet printing is used to make a microstructure for a Bi_2Te_3-based thermoelectric generator [22]. In our approach, since we successfully fabricated both p-type and n-type thin films using the FED method, we developed a mini-generator using a shadow mask [23]. The generator pattern can be created by thin-film deposition through shadow masks, as shown in Figure 12.5. The dimension of p- and n-type legs is 15 mm (length) × 1.0 mm (width) × 1.0 μm (thickness), and the spacing between legs is 0.6 mm. The overall size of the generator on the substrate is 22 mm long, 24 mm wide, and 1.0 mm thick, consisting of seven p/n couples (Figure 12.6). The maximum output power was 0.21 μW at a 30 °C temperature difference. By refining the shadow mask through microfabrication, we further advanced the fabrication of microgenerators [24]. We fabricated Si-based shadow masks to evaporate the materials with specific patterns. The shadow masks were created using standard microfabrication processes, as shown in Figure 12.7. We began with a silicon wafer and deposited a silicon nitride thin film on it using plasma-enhanced chemical vapor deposition. To pattern the configurations of each shadow mask, we spun photoresist onto the silicon nitride thin film. The photoresist was then exposed to ultraviolet light and developed to create the desired pattern. The silicon nitride thin films were etched into the desired patterns using reactive-ion etching. Finally, the remaining silicon in the patterns was etched by soaking the

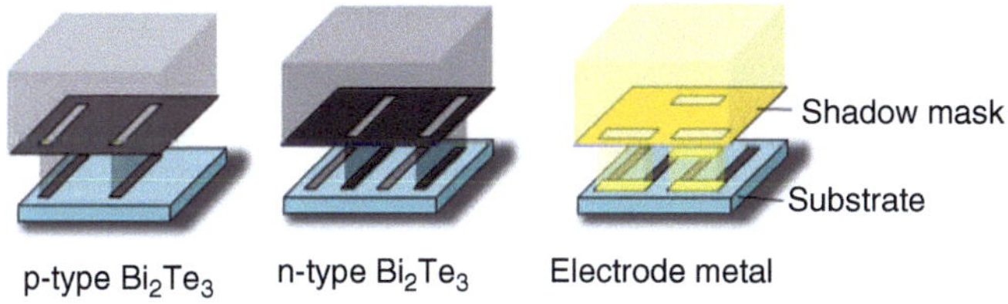

Figure 12.5 Schematic diagram of the thin-film patterning process by material deposition through shadow masks for the Bi_2Te_3-based thin-film thermoelectric generator.

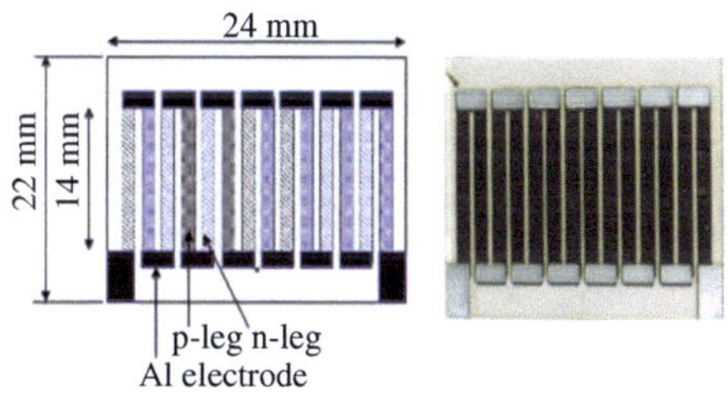

Figure 12.6 Schematic illustration and photograph of the Bi_2Te_3-based thin-film thermoelectric generator.

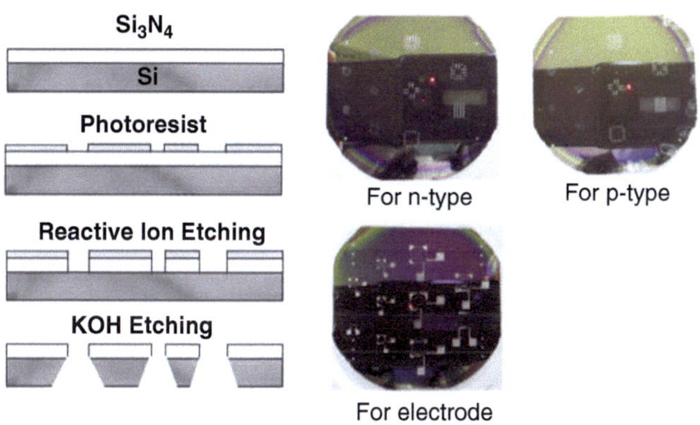

Figure 12.7 Schematic illustration and photograph of the microfabricated shadow masks.

wafer in KOH. Each of the fabricated shadow masks has patterns that are suitable for their respective p-type and n-type bismuth telluride materials and metal electrodes. To achieve a temperature difference in a small region simply by heating the bottom of the generator, we deposit Bi_2Te_3 through the shadow on an ultra-thin substrate of a silicon nitride (Si_3N_4) thin film, as shown in Figure 12.8. The silicon nitride–based thin film has a very small cross-sectional area in the in-plane direction, so it has a large thermal resistance. Therefore, if we heat at the bottom of the substrate, heat will conduct through the silicon nitride–based thin film from the silicon, with the result that a temperature distribution is obtained in the freestanding thin film. The processes for the fabrication of the substrate are almost the same as those for the shadow masks, as shown in Figure 12.7, but in the case of the substrate, we can fabricate the freestanding thin film by only etching the silicon substrate. The dimensions of the fabricated substrate are 4 mm × 4 mm with a 4 μm thickness.

The fabricated in-plane thermoelectric generators are shown in Figure 12.9. It is known that the output of the thermoelectric power generation is proportional to the volume of the thermoelectric material, as shown in Figure 12.10. As a result, the generated power was in the nW order, but we successfully confirmed the operation of the thermoelectric module.

Next, we fabricated a thermoelectric module composed of Bi_2Te_3 on a polyimide substrate to achieve a flexible module [24]. Since a lower thermal conductivity of the substrate allows for a greater temperature difference across the thermoelectric module, we introduced a porous structure in the substrate and deposited Bi_2Te_3 onto it.

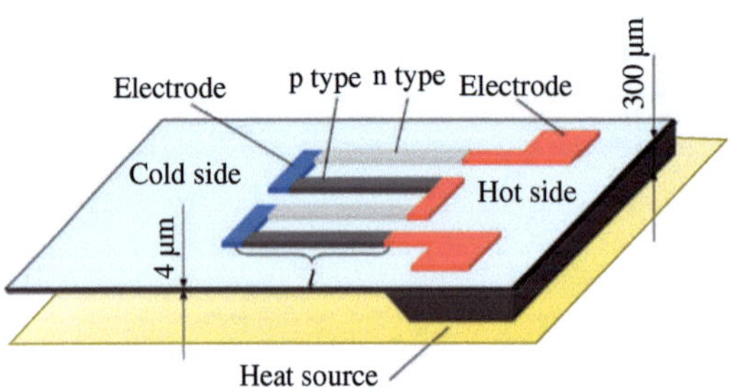

Figure 12.8 Schematic diagram of the in-plane thermoelectric device.

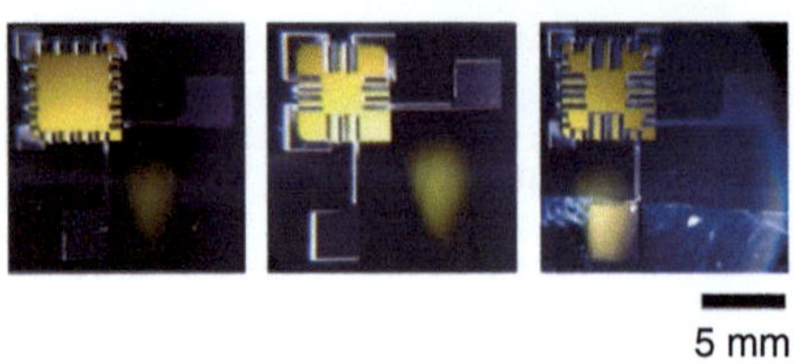

Figure 12.9 Photograph of the in-plane thermoelectric device.

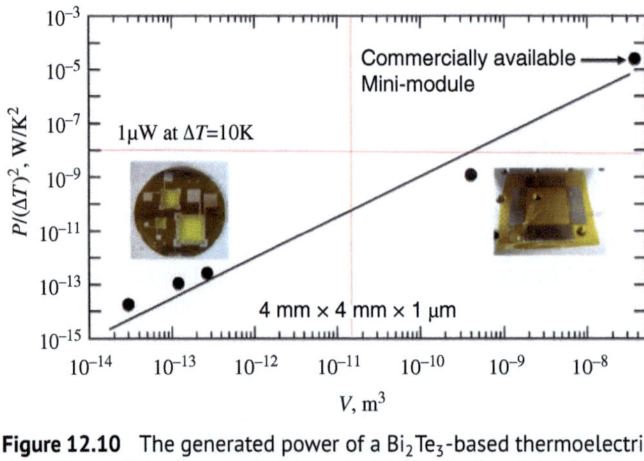

Figure 12.10 The generated power of a Bi_2Te_3-based thermoelectric generator depends on the volume of the thermoelectric material. The horizontal axis represents the volume, while the vertical axis represents the output power divided by the square of the temperature difference.

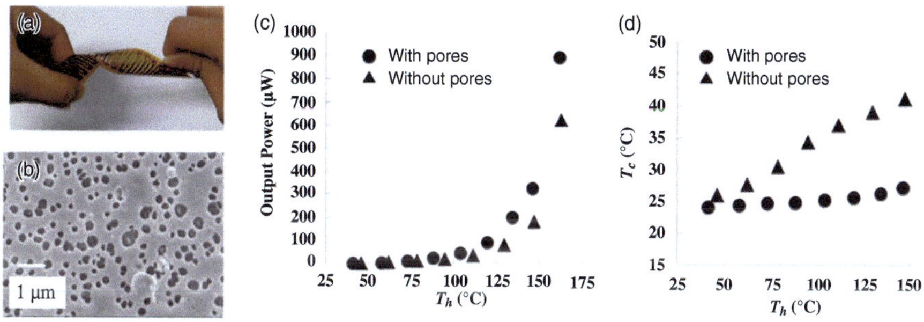

Figure 12.11 (a) Photograph of a flexible thermoelectric generator with microporous films. (b) The top view of the SEM image of the thermoelectric porous film. (c) The output power of a thermoelectric generator with porous films is higher than that of the films without pores. (d) The temperature at the cold side, T_c, is kept at a low temperature due to the high thermal resistance of the module.

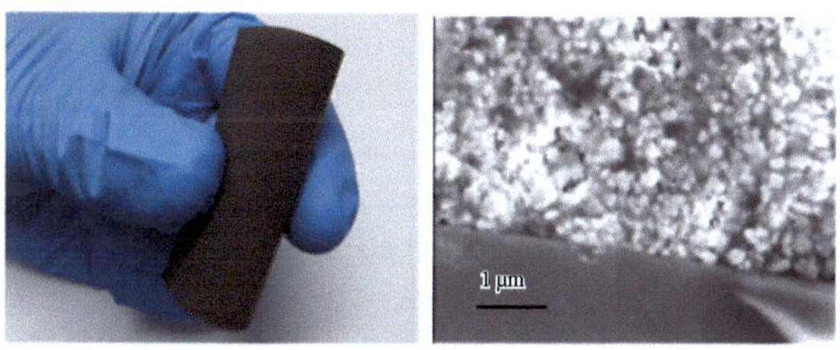

Figure 12.12 Left: Photograph of a flexible thermoelectric generator prepared by screen printing. Right: The cross-sectional view of the SEM image of the screen-printed Bi_2Te_3 film. *Source*: [19] / with permission of Springer Nature.

The output results of the thermoelectric module on the porous polymer substrate are shown in Figure 12.11. By reducing the apparent thermal conductivity of the module, we confirmed an increase in power generation.

12.4 Printed Thermoelectric Generators

A thermoelectric module is made by screen printing to reduce the production costs. Bi_2Te_3 fine particles are prepared by ball milling, and they are mixed with polyamic acid and an organic solvent for making a printing ink [18, 25]. The screen-printed film is shown in Figure 12.12. As previously mentioned, this method facilitates the fabrication of in-plane thermoelectric modules. However, since the film thickness can be adjusted to approximately 100 μm through printing, it is also possible to create conventional Π-type thermoelectric modules. When power generation is tested near room temperature, the module output power is up to 1 μW order. To improve the output of the thermoelectric module, increasing the packing density of the particle-based film to enhance its electrical conductivity is essential. A key challenge in development is determining how much the packing density can be increased through thermal processing without applying pressure in the fabrication process.

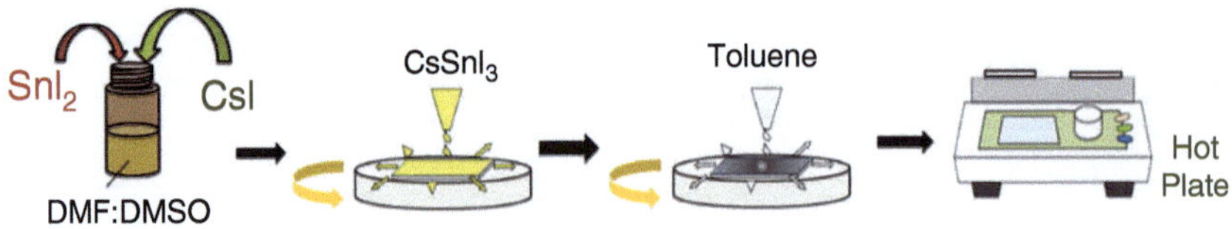

Figure 12.13 Halide perovskite–based thermoelectric film is made by spin-coating in a nitrogen environment glove box.

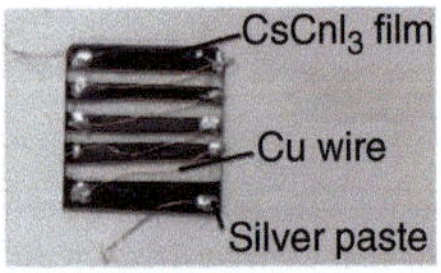

Figure 12.14 Photograph of a unileg module of a halide perovskite-based thermoelectric film using Cu wire legs.

Beyond the conventional development of Bi_2Te_3-based thermoelectric modules, thermoelectric modules using halide perovskites as printable thermoelectric materials have been carried out, such as PEDOT:PSS [26] and carbon nanotubes [27]. It has been demonstrated that $CsPbI_3$ operates as a highly efficient solar cell material [28, 29]. Sn has been proposed as a substitute for Pb for lead-free halide perovskite. $CsSnI_3$ thin films were prepared in a nitrogen environment glove box by the spin-coating technique on a glass substrate, as shown in Figure 12.13. The precursor solution of $CsSnI_3$ was prepared by mixing CsI and SnI_2.

We used toluene as an antisolvent after spin-coating the $CsSnI_3$ solution. The spin-coated films are annealed at 130 °C for 5 min. $CsSnI_3$ thin films turned black from yellow in color during the annealing process, which confirms the formation of $CsSnI_3$ crystals. By replacing CsI with MAI, $MASnI_3$ halide perovskites can also be synthesized [30]. $CsSnI_3$ exhibits significant self-doping due to the oxidation of Sn from Sn^{2+} to Sn^{4+}, resulting in high electrical conductivity. While this material does not achieve high efficiency as a solar cell, its high conductivity has led to its potential application in thermoelectric conversion [31]. However, since n-type halide perovskites have not yet been reliably synthesized, printed thermoelectric modules have not been fully commercialized. Nonetheless, a unileg thermoelectric module has been demonstrated as shown in Figure 12.14, with an output of 0.8 nW and a temperature difference of 5 °C [32].

12.5 Summary

Here, we introduced our progress in thermoelectric module development, focusing on Bi_2Te_3 over the past 10 years. Production costs become more important than power generation efficiency because advances in electronic circuits are reducing the required electric power for energy-harvesting applications. As a result, deposition techniques have transitioned from vacuum thin-film evaporation to low-cost printing in air. The screen-printing method is developed by making a viscous ink with Bi_2Te_3 powders in organic solvents. The present screen-printing method can be applied to other thermoelectric materials due to its simplicity. In recent years, not only Bi_2Te_3 but also printable thermoelectric materials have been reported, such as PEDOT:PSS, carbon nanotube, and halide perovskite. The exploration of new materials remains a crucial factor in advancing the practical use of thermoelectric power generation.

Symbols and Abbreviations

η	Efficiency
λ	Thermal conductivity, W/(m K)
σ	Electrical conductivity, S/m
m	Nondimensional parameter
S	Seebeck coefficient, V/K
T	Temperature, K
Q	Heat input, W
W	Electrical power, W
Z	Figure of merit, 1/K

References

1 Rowe, D.M. (1995) CRC Handbook of Thermoelectrics, CRC Press.

2 Snyder, G.J. and Toberer, E.S. (2008) Complex thermoelectric materials. *Nat. Mater.*, 7, 105–14.

3 Takashiri, M., Borca-Tasciuc, T., Jacquot, A., Miyazaki, K. and Chen, G. (2006) Structure and thermoelectric properties of boron doped nanocrystalline $Si_{0.8}Ge_{0.2}$ thin film. *J. Appl. Phys.*, 100, 054315.

4 Tang, X., Li, Z., Liu, W., Zhang, Q. and Uher, C. (2022) A comprehensive review on Bi_2Te_3-based thin films: thermoelectrics and beyond. *Interdiscipl. Mater.*, 1, 88–115.

5 Charles, E., Groubert, E. and Boyer, A. (1988) Structural and electrical properties of bismuth telluride films grown by the molecular beam technique. *J. Mater. Sci. Lett.*, 7, 575–7.

6 Saini, S., Mele, P., Miyazaki, K., and Tiwari, A. (2016) On-chip thermoelectric module comprised of oxide thin film legs. *Energy Convers. Manag.*, 114, 251–7.

7 Matsumoto, K. and Mele, P. (2010) Artificial pinning center technology to enhance vortex pinning in YBCO coated conductors. *Supercond. Sci. Technol.*, 23, 014001.

8 Völklein, F., Baier, V., Dillner, U. and Kessler, E. (1990) Transport properties of flash-evaporated $(Bi_{1-x}Sb_x)_2Te_3$ films I: optimization of film properties. *Thin Solid Films*, 187, 253–62.

9 Takashiri, M., Shirakawa, T., Miyazaki, K. and Tsukamoto, H. (2007) Fabrication and characterization of $Bi_{0.4}Te_{3.0}Sb_{1.6}$ thin films by flash evaporation method. *J. Alloys Compd.*, 441, 246–50.

10 Gasenkova, I.V. (1997) Bismuth telluride films prepared by vacuum-arc plasma method for thermoelectrical transducers. *Proceedings of 16th International Conference on Thermoelectrics*, 151–4.

11 Uchino, M., Kato, K., Hagino, H. and Miyazaki, K. (2013) Fabrication by coaxial-type vacuum arc evaporation method and characterization of bismuth telluride thin films. *J. Electron. Mater.*, 42, 1814–19.

12 Venkatasubramanian, R., Colpitts, T., Watko, E., Lamvik, M. and El-Marsry, N. (1997) MOCVD of Bi_2Te_3, Sb_2Te_3 and their superlattice structures for thin-film thermoelectric applications. *J. Cryst. Growth*, 170, 817–21.

13 Giani, A., Boulouz, A., Pascal-Delannoy, A., Foucaran, A., Charles, E. and Boyer, A. (1999) Growth of Bi_2Te_3 and Sb_2Te_3 thin films by MOCVD. *Mater. Sci. Eng. B*, 64, 19–24.

14 Hosokawa, Y., Wada, K., Tanaka, M., Tomita K. and Takashiri, M. (2018) Thermal annealing effect on structural and thermoelectric properties of hexagonal Bi_2Te_3 nanoplate thin films by drop-casting technique. *Jpn. J. Appl. Phys.*, 57, 02CC02.

15 Snyder, G.J., Lim, J.R., Huang, C.K. and Fleurial, J.P. (2003) Thermoelectric microdevice fabricated by a MEMS-like electrochemical process. *Nat. Mater.*, 2, 528–31.

16 Navone, C., Soulier, M., Testard, J., Simon J. and Caroff T. (2010) Optimization and fabrication of a thick printed thermoelectric device. *J. Electron. Mater.*, 40, 789–93.

17 Madan, D., Chen, A, Wright, P.K. and Evans, J.W. (2011) Dispenser printed composite thermoelectric thick films for thermoelectric generator applications. *J. Appl. Phys.*, 109, 034904.

18 Zhang, B., Sun, J., Katz, H.E., Fang, F. and Opila, R.L. (2010) Promising thermoelectric properties of commercial PEDOT: PSS materials and their Bi_2Te_3 powder composites. *ACS Appl. Mater. Interfaces*, 3, 3170–78.

19 Kato, K., Hagino, H. and Miyazaki, K. (2013) Fabrication of bismuth telluride thermoelectric films containing conductive polymer using a printing method. *J. Electron. Mater.*, 42, 1313–18.

20 Wang, Z., Leonov, V., Flourini, P. and Hoof, C.V. (2009) Realization of wearable miniaturized thermoelectric generator for human body applications. *Sens. Actuators A: Physical*, 156, 95–102.

21 Takahashi, M., Katou, Y., Nagata, K. and Furuta, S. (1994) The composition and conductivity of electrodeposited Bi-Te alloy films. *Thin Solid Films*, 240, 70–2.

22 Miyazaki, K., Iida, T. and Tsukamoto, H. (2003) Micro-fabrication of Bi_2Te_3 by using micro-jet, *Proceedings of the 22nd International Conference on Thermoelectrics*, 673–6.

23 Takashiri, M., Shirakawa, T., Miyazaki, K. and Tsukamoto, H. (2007) Fabrication and characterization of bismuth-telluride-based alloy thin film thermoelectric generators by flash evaporation method. *Sens. Actuators A*, 138, 329–34.

24 Kurosaki, J., Yamamoto, A., Tanaka, S., Cannon, J., Miyazaki, K. and Tsukamoto H. (2009) Fabrication and evaluation of a thermoelectric microdevice on a free-standing substrate. *J. Electron. Mater.*, 38, 1326–30.

25 Kato, K., Hatasako, Y., Kashiwagi, M., Hagino, H., Adachi, C. and Miyazaki, K. (2014) Fabrication of a flexible bismuth telluride power generation module using microporous polyimide films as substrates. *J. Electron. Mater.*, 43, 1733–9.

26 Kato, K., Kuriyama, K., Yabuki, T. and Miyazaki, K. (2020) Organic-inorganic thermoelectric material for a printed generator. *J. Phys. Conf. Ser.*, 1052, 012008.

27 Budnova, O. and Xavier, C. (2012) Towards polymer-based organic thermoelectric generators. *Energy Environ. Sci.*, 5, 9345–62.

28 Amma, Y., Miura, K., Nagata, S., Nishi, T., Miyake, S., Miyazaki, K. and Takashiri, M. (2022) Ultra-long air-stability of n-type carbon nanotube films with low thermal conductivity and all-carbon thermoelectric generators. *Sci. Rep.*, 12, 21603.

29 Baranwal, A.K., Nishimura, K., Liu, D., Kamarudin, M.A., Kapil, G., Saini, S., Yabuki, T., Iikubo, S., Minemoto, T., Yoshino, K., Miyazaki, K., Shen, Q. and Hayase S. (2022) Relationship between carrier density and precursor solution stirring for lead-free tin halide perovskite solar cells performance. *ACS Appl. Energy Mater.*, 5, 4002–7.

30 Saini, S., Matsumoto, I., Kishishita, S., Baranwal, A.K., Yabuki, T., Hayase, S. and Miyazaki, K. (2022) Use of anti-solvent to enhance thermoelectric response of hybrid halide perovskite thin films. *Jpn. J. Appl. Phys.*, 61, SE1019.

31 Saini, S., Baranwal A.K., Yabuki, T., Hayase, S. and Miyazaki, K. (2019) Growth of halide perovskites thin films for thermoelectric applications. *MRS Adv.*, 4, 1719–25.

32 Saini, S., Baranwal A.K., Yabuki T., Hayase, S. and Miyazaki, K. (2020) Unileg thermoelectric module comprised by coated halide-perovskite thin film. *J. Heat Transf.*, 142, 074502.

13

Frosting and Defrosting During Energy Harvesting from Air by Heat Pump

*Aye Aye Thet and Song Mengjie**

Department of Energy and Power Engineering, School of Mechanical Engineering, Beijing Institute of Technology, China

*Corresponding author: song@gmail.com

13.1 Background of Air-source Heat Pump

13.1.1 Heat Pump

A heat pump system is an efficient and ecofriendly solution for maintaining a comfortable indoor temperature, providing both heating and cooling with high efficiency [1]. In warmer months, it removes heat from indoors and transfers it outside, similar to how an air conditioner operates. During colder months, it extracts heat from an external source and delivers it to warm the indoor space. Globally, over 90% of the population lives in areas where heat pump units can effectively manage indoor climate control [1]. Studies have demonstrated that using heat pumps instead of traditional heating methods like coal or electricity can significantly reduce greenhouse gas emissions, particularly CO_2. As energy costs continue to rise, there is a growing interest in utilizing heat pump technology as an energy-efficient solution [1].

Figure 13.1 shows how CO_2 emissions changed for five energy sources from 2010 to 2022. Throughout the time frame, natural gas continues to be the leading contributor, highlighting its extensive use in energy generation [2]. Consistently high levels of indirect emissions are most likely due to carbon embedded in secondary processes or energy imported from elsewhere. Most notably, emissions from coal have decreased slightly, which is in keeping with the worldwide trend of eliminating coal as a result of climate change activities. The contributions from oil and nonrenewable waste have decreased, yet they remain consistent. These tendencies show how difficult it is to reduce overall emissions, especially those from fossil fuels, and to switch to greener energy sources.

Figure 13.2 notes that the heat pump sales corresponded to the world region from 2019 to 2023 [3]. This bar chart shows the sales of heat pumps in the United States, the European Union, China, Japan, and the rest of the world for the years 2019

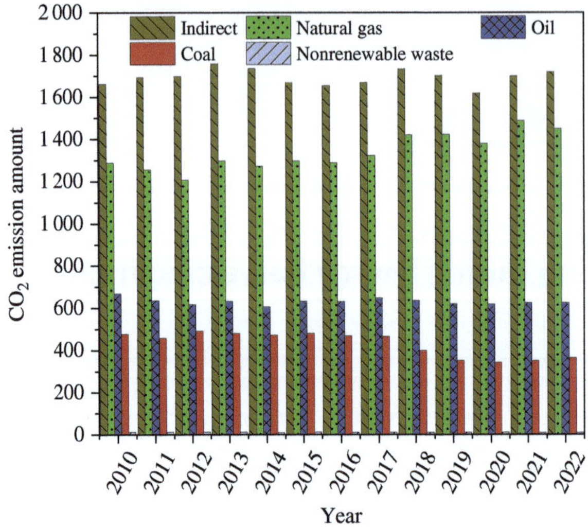

Figure 13.1 Yearly breakdown of CO_2 emissions by energy source (2010–2022).

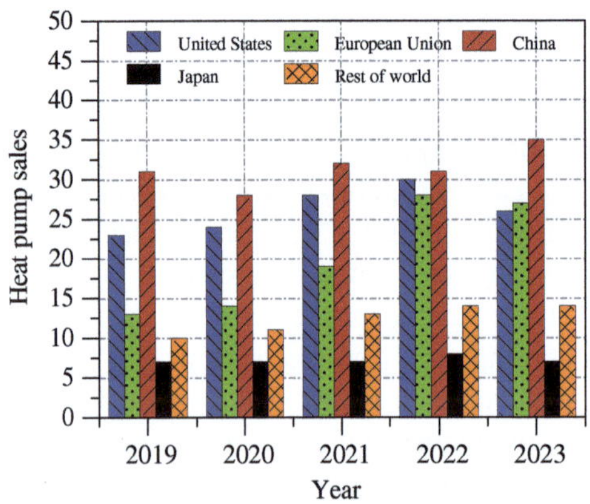

Figure 13.2 A comparative analysis of global markets for regional heat pump units (2019–2023).

through 2023. The grouped bar format, which displays the data, represents each zone with a different color for clarity. The horizontal axis displays 2019 through 2023, and the vertical axis indicates heat pump sales in gigawatts (GW). An explanation of the data, it is clear from the graphic that during the specified period, the European Union continuously had the greatest heat pump sales of any region. The European Union demonstrated its dominance and expanded the use of heat pump technology in 2019 by recording 31 GW in sales, which grew steadily to 35 GW by 2023. The second-largest market is the United States. Sales began at 23 GW in 2019, rose to 30 GW in 2022, and then slightly decreased to 26 GW in 2023. Although recent economic or regulatory changes may have slowed growth, this shows considerable growth initially, possibly spurred by federal subsidies for energy-efficient technologies. Sales in China are comparatively stable, peaking at 19 GW in 2021 after beginning at 13 GW

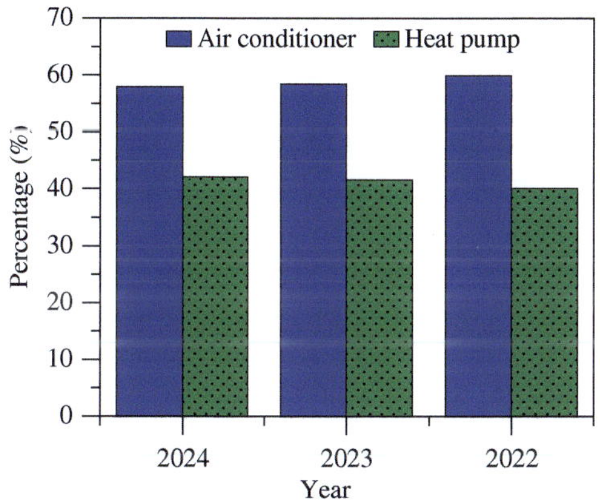

Figure 13.3 Year-to-date percentage distribution of air conditioners and heat pumps (2022–2024).

in 2019 and then leveling off at 14 GW in 2022 and 2023. This consistent pattern may be a result of the nation's emphasis on a variety of energy technologies and a sensible strategy for heat pumps within its framework for renewable energy. From 2019 to 2023, Japan consistently maintained a sales figure of 7 GW, the lowest of all the regions. This stability could indicate limited growth due to factors such as a smaller market, slower adoption, or reliance on alternative energy technology. The category for the rest of the world is growing steadily but somewhat. Sales grew steadily from 10 GW in 2019 to 14 GW in 2023. This probably indicates a rise in acceptance in developing markets or areas outside of China, the United States, and Europe, which are the main markets. All things considered, the graph gives a clear picture of the development of the heat pump market over time, highlighting regional variations in adoption and providing information on how market dynamics and regulations affect these patterns [3, 4].

Figure 13.3 illustrates the year-to-date percentage distribution of air conditioners and heat pumps from 2022 to 2024 [5]. The data compares the relative prevalence of these two technologies, air conditioners, and heat pumps, across three years. The graph highlights a consistent dominance of air conditioners in the market over three years, with heat pumps maintaining a steady but smaller share. This distribution underscores the need for further analysis to understand consumer preferences, technological advancements, and potential shifts driven by policy changes or environmental considerations [3]. There are three main types of heat pumps. These are air-source, water-source, and ground-source. Air-source heat pumps (ASHPs) are inexpensive to install and commonly used. They function well in moderate climates, using outside air as a medium for heat exchange. Water-source heat pumps dissipate heat by water instead of air. They require access to a well, lake, or other water source and are not used as commonly as air-source heat pumps. Ground-source, or geothermal, heat pumps utilize thermal energy stored underground, transferring heat similar to air-source heat pumps. They offer much more efficient operation with the constant temperature of the ground. Among these heat sources, air and water are the most common ones for space-heating heat pump units [6, 7]. Consequently, heat pump units like air-to-air, air-to-water, water-to-air, and water-to-water are frequently employed for buildings and industrial settings. ASHP units are popular due to their ease of installation and cost-effectiveness. These units operate on the principles of vapor compression refrigeration, using a refrigerant system with a compressor and condenser to transfer heat from one location to another [7]. Table 13.1 describes the comparative analysis of residential heating technologies: efficiency, sustainability, and application parameters.

One of the most effective heating options is an air-to-water heat pump, which transfers heat from the surrounding air to water for home hot water and space heating. Because these systems can transfer heat rather than create it, they are usually three to four times more efficient than conventional fossil fuel-based systems. For instance, employing these pumps can save energy expenses for households in Germany and France by 35% to 50% when compared to gas boilers, particularly in well-insulated homes [3]. Because the systems are frequently connected to storage tanks for water heating and distribution throughout the house, they also provide versatility. Some models are appropriate for different climates since they have extra cooling features. In many countries, government incentives have made them more accessible to consumers despite their higher initial costs [8].

The earth's constant temperature serves as a heat source for ground-source heat pumps. They are especially efficient in frigid climates where air-source systems may not function well, with efficiency three to five times higher than conventional systems. For residences with enough acreage for installation, they are an affordable solution due to their extended lifespan of 20 to 25 years and low operating costs [9]. However, only larger properties or new developments may use them due to their high initial installation costs and requirements for ground loops. Notwithstanding these obstacles, they are an essential component of the sustainable heating landscape due to their dependability and efficiency [10].

Air-to-air heat pumps may be used for both heating and cooling by transferring heat between the interior and outdoor air. Their dual-purpose usefulness and three to four times better efficiency than traditional systems make them particularly attractive in moderate climates [11]. However, because they do not provide hot water, heat pumps are less effective than air-to-water or ground-source systems. These systems are often preferred because they are easy to install and can be adapted to existing structures. With an average lifespan of 15 to 20 years and their ability to lower carbon footprints, these systems are popular as a more environmentally friendly alternative to conventional heating options [11]. Solar thermal heaters utilize rooftop collectors to capture solar energy for hot water production. As a zero-emission technology, they are ideal for sunny regions with ample sunlight available throughout the year. Their lifespan ranges from 20 to 30 years, and they have low running costs, making them a reliable and ecofriendly choice for residential hot water heating. However, their effectiveness is limited in areas with little to no solar exposure, and solar thermal systems typically only heat water, offering less versatility compared to heat pumps [12, 13].

District heating involves the centralized generation and distribution of heat through a network of underground pipes. These systems can be powered by low-carbon energy sources such as waste heat recovery and biomass, enhancing their

Table 13.1 Comparative analysis of residential heating technologies: efficiency, sustainability, and application parameters.

No.	Technology	Energy source	Efficiency	Applications	Lifespan	Environmental impact	Key insights
1	Air-to-water heat pumps	Outside air	Three to four times more efficient than traditional systems	Space heating, hot water	15–18 years	Low emissions (powered by electricity)	Highly energy-efficient and versatile; significant cost savings (e.g., 35% in Germany, 50% in France)
2	Ground-source heat pumps	Ground heat	Three to five times more efficient	Space heating, hot water	20–25 years	Very low emissions	Highly efficient but with higher installation costs; ideal for well-insulated homes
3	Air-to-air heat pumps	Outside air	Three to four times more efficient	Space heating, cooling	15–20 years	Low emissions	Great for dual heating/cooling systems; limited in providing hot water
4	Solar thermal heaters	Solar energy	Depends on sunlight availability	Hot water	20–30 years	Zero emissions	Best suited for sunny regions; dependent on weather conditions and solar panel placement
5	District heating	Centralized low-carbon sources	High efficiency (varies by system)	Space heating, hot water	Varies by infrastructure	Depends on the energy source	Efficient in urban areas; infrastructure-intensive; relies on renewable energy sources for impact
6	Biomass boilers	Biomass (wood, pellets, chips)	70–90% efficiency	Space heating, hot water	20 years	Moderate emissions	Sustainable if sourced properly but generates emissions during combustion
7	Electric radiators	Electricity	100% conversion of electricity to heat	Space heating	10–15 years	High emissions	High operational costs unless powered by renewable energy
8	Gas boilers and furnaces	Natural gas	90–98% efficient	Space heating, hot water	15–20 years	High emissions	Dominates in many markets but contributes to significant carbon emissions

efficiency. District heating is most effective in urban areas with high population density, as the infrastructure costs can be shared among a large user base. Integrating district heating with renewable energy sources has the potential to significantly reduce carbon emissions [14]. Biomass boilers use biological materials like wood pellets or logs to produce heat, achieving efficiencies of 70% to 90%, depending on the system design. Although biomass is renewable, it generates emissions during combustion, making it less sustainable than zero-emission technologies [15]. The sustainability of biomass boilers largely depends on the source of the biomass. When sourced responsibly, they can be a viable option in areas with abundant forestry resources [16].

Electric radiators convert electricity into heat with 100% efficiency. However, they are usually more expensive to operate unless powered by renewable energy. Their dependence on electricity from fossil-fuel-based grids results in higher emissions unless combined with renewable energy sources [17]. Electric radiators are best suited for small spaces or as supplementary heating sources. While they are easy to install and operate, they lack the efficiency and versatility of heat pumps [18]. Gas boilers and furnaces are market leaders in the global heating sector, boasting efficiencies between 90% and 98%. However, they contribute significantly to carbon emissions due to their reliance on fossil fuels. Although highly efficient, their market share is declining as governments strive for greener heating options. To meet global decarbonization goals, it is essential to phase out gas boilers and implement policies that support clean technologies like heat pumps [19, 20].

13.1.2 Air-source Heat Pump

An air-source heat pump can efficiently heat and cool your home. When properly installed, it can deliver up to four times more heat energy than the electrical energy it consumes. Unlike combustion heating systems, a heat pump transfers heat instead of converting it from fuel. ASHPs have been used for many years across the United States. Recent technological advances have made them a viable heating alternative even in regions with extended periods of subfreezing temperatures [6, 21]. ASHP units are used globally due to their high energy efficiency and ecofriendly nature. However, the outdoor coil surface of ASHP units is prone to frost deposition and accumulation, which can lead to significant negative effects [7].

The latest air-source heat pump technology has improved efficiency. The efficiency can be improved by using the additional device. These devices are (1) electronic and thermostatic expansion valves: these valves provide precise control over the refrigerant flow to the indoor coil, enhancing overall performance; (2) variable speed blowers: these blowers improve efficiency by adjusting airflow in part-load conditions, effectively addressing challenges such as restricted ducts, dirty filters, and contaminated coils; (3) enhanced coil design: thicker coils result in better dehumidification, contributing to improved indoor air quality; (4) advanced motor and compressor designs: inverter-driven systems can seamlessly adjust between low and high speeds, offering exceptional energy savings and superior humidity control [6]. Together, these innovative features elevate the effectiveness of air-source heat pumps, making them a more sustainable choice for heating and cooling needs.

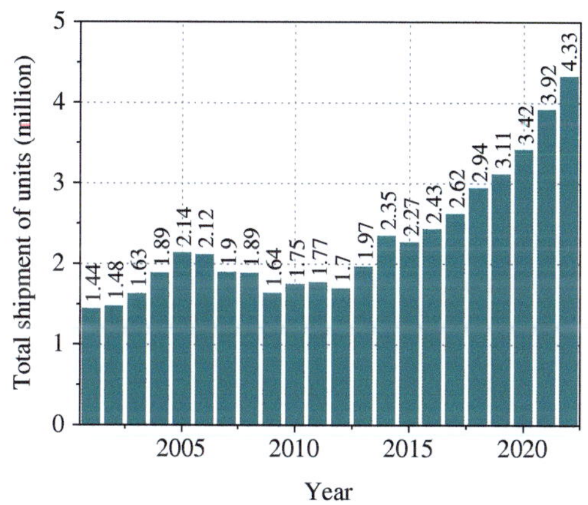

Figure 13.4 Annual shipments of air-source heat pumps in the United States (2001–2022).

Figure 13.4 illustrates the yearly shipments of air-source heat pumps made in the United States from 2001 to 2022 [3]. A discernible pattern of expansion can be seen in the data, which is depicted as blue square markers, throughout the two decades. The findings indicate that air-source heat pumps are becoming increasingly popular, particularly in recent years. This is most likely attributable to the increased need for energy-efficient heating and cooling solutions, as well as regulations and technologies that are supportive of this trend. The substantial increase that occurred after the year 2015 may be a result of shifting energy priorities or technological breakthroughs in heat pump technology.

13.2 Frosting Mechanism

A frosting mechanism refers to the physical process by which frost forms on a cold surface, essentially involving water vapor from the air condensing into tiny droplets on the surface and then freezing, creating a layer of ice crystals; this process is primarily driven by heat transfer and phase change, where the key element is a temperature difference between the cold surface and the surrounding humid air, causing water vapor to condense and freeze on the surface. This process is particularly relevant in various technical and engineering fields, such as refrigeration, air-source heat pumps, aircraft, wind power generation, and power grids [22].

Understanding the frosting mechanism is crucial for improving the efficiency and reliability of systems where frosting can occur [22]. Frosting is a complex process that involves the transmission of heat and mass, as well as a combination of phase transitions. Researchers have conducted extensive theoretical, experimental, and numerical research to gain a deeper understanding of the mechanism underlying icing. This section describes the most significant findings from the relevant studies. These findings encompass the frost creation process and morphology, the frost layer structure model, and the factors influencing frosting characteristics [22]. The frosting mechanism is a multifaceted process characterized by the accumulation of ice crystals during the freezing phase. This phenomenon can have a substantial effect on the

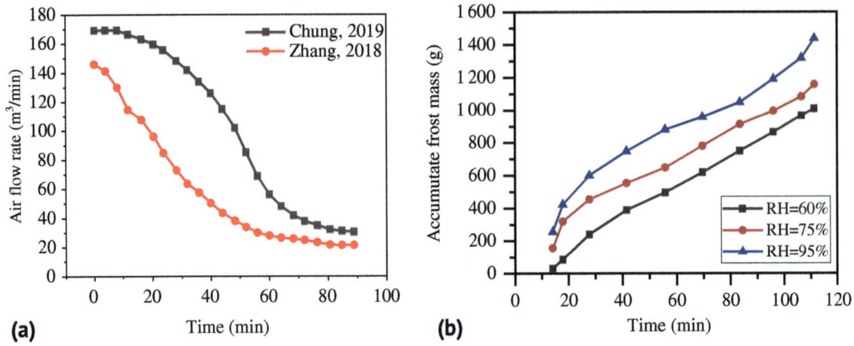

Figure 13.5 (a) Comparison of airflow performance. (b) Accumulated frost mass on time.

operational efficiency of heat exchangers, refrigeration units, and HVAC systems. This may result in diminished thermal efficiency, heightened energy consumption, and difficulties in measurement and modeling. The strong adherence of frost to surfaces complicates removal processes and poses risks to the underlying material. Environmental factors such as humidity, temperature, and airflow influence frost formation, resulting in variable outcomes in practical applications.

Figure 13.5a shows the comparison of airflow rate reduction over time during frost accumulation, based on data from Chung (2019) [23] and Zhang (2018) [24]. The graph illustrates the decline in airflow rate (m^3/min) as frost forms under different experimental conditions, with Chung's study showing a slower reduction than Zhang's. Figure 13.5b illustrates the relationship between accumulated frost mass (grams) and time (minutes) under three relative humidity (RH) conditions: 60%, 75%, and 95% [25]. This relationship provides insights into the effects of environmental humidity on frost formation dynamics. It demonstrates the pivotal role of RH in frost accumulation and highlights the engineering challenges associated with high-humidity environments. This data is essential for designing energy-efficient, frost-resilient systems in refrigeration, HVAC, and cryogenic applications [26].

13.2.1 Frost Formation Process and Stages

The formation of frost is a multistep physical process that occurs when water vapor in the air directly transforms into solid ice (a process known as deposition) on a surface that is below freezing. This phenomenon involves several distinct phases, each characterized by a specific combination of environmental factors and thermodynamic principles. When the surface temperature falls below the freezing point of water, which is 0 °C, the first stage in the development of frost takes place. Radiative cooling, ventilation, and air conditioning systems remove heat from the surface and simultaneously raise the air's dew point to the frost formation threshold [27].

The process of nucleation takes place when the temperature of the surface dips below the freezing and dew points. This makes it possible for water vapor molecules to bind to one another and create small ice crystals. Scratches and other surface defects, such as microroughness and scratches, can act as nucleation sites, which lead to the development of frost. The surface's qualities and the air's humidity levels influence the nucleation rate [28]. Once the nucleation process is complete, more water vapor molecules deposit onto the original ice nuclei, leading to the development of frost crystals. The continuation of this process is contingent upon the surface remaining cold and the availability of water vapor. During the development phase, the presence of humidity, air movement, and temperature gradients can produce different frost forms such as feathery hoarfrost or smooth glazing. The production of frost that is denser is often the consequence of conditions that involve higher humidity and calm air [29].

13.2.2 Solidification of Droplets on Cold Plate Surface

The heat and mass transfer properties of water droplets on chilled surfaces have been the subject of focused research. This includes the investigation of phenomena like delayed freezing, frost formation, ice adhesion, and ice nucleation dynamics. Song M studied the frosting process on a cold plate, which consists of three periods according to the growth timeline. Frost formation is a process that involves the phase transition of water vapor to solid ice, a phenomenon commonly observed in nature and industrial contexts. Understanding the steps of frost development can be beneficial in various fields, including meteorology, refrigeration, aviation, and agriculture, where frost accumulation can have significant effects on performance and safety [30]. In the initial stage, water droplets condense onto a surface. This occurs when the temperature of the surface drops below the dew point, causing water vapor in the air to condense into liquid droplets, which is necessary for the subsequent stages of frost formation. Condensation on a cold surface marks the initial phase in frost formation, as it introduces the water needed for ice nucleation when the temperature continues to decrease [30]. As the temperature decreases further, the surface cools to below freezing, initiating the second stage. At this stage, the liquid droplets begin to freeze, transitioning from a liquid to a solid state [31, 32]. Ice nucleation can occur through different mechanisms. For example, homogeneous nucleation happens within pure water droplets, whereas heterogeneous nucleation can occur around impurities or particles within the droplets, acting as a catalyst for ice formation [33]. The ice crystals form with a branching pattern, known as dendritic growth, a characteristic structure resulting from supercooling and the molecular structure of ice. The airflow over the surface promotes water vapor deposition onto the growing crystals, accelerating their development and leading to larger, more complex ice structures [30].

In the final stage, frost builds up on the surface as ice crystals accumulate and interlock, forming a continuous layer. The crystals expand vertically and laterally, creating a thicker, denser frost layer. This process is facilitated by the airflow, which supplies water vapor that condenses and freezes onto the frost layer. The final frost formation is a structured layer, distinguishable from the initial smooth droplets. This frost layer has a more crystalline appearance and a more substantial insulating effect due to its lower thermal conductivity compared to water, affecting the surface's thermal properties. In practical applications, this thick frost layer can lead to issues such as reduced thermal efficiency in refrigeration or increased aerodynamic drag on aircraft surfaces. In addition, during the droplet condensation period, water droplets initially form on the cold plate surface. These droplets coalesce due to surface energy, with larger droplets being more stable than smaller ones. At a critical time (t_c), these droplets solidify into ice particles or crystals [30]. The second period is the solidified liquid tip-growth period, which occurs from (t_c) to a transitional time (t_t). During this period, a singularity appears at the top of the ice crystal, forming a tip that grows vertically by absorbing vapor from the surrounding air. After (t_t), the frost layer growth period begins. A uniform porous frost layer forms, and small ice branches emerge around the vertically growing tip, creating a hollow frost structure. The final period is the frost layer's full growth period. In this period, the frost layer's density increases due to water vapor permeation and surface melting. This melting cycle, influenced by thermal resistance changes, continues until the frost layer surface remains below 0 °C, resulting in a thick frost or ice layer [30].

13.2.3 Frosting Models

Several types of frosting models are used to understand and predict the behavior of frost formation on cold plates during energy harvesting. These models can be broadly categorized into three main types [34]. There are three types of models: data-driven models (which include empirical and semiempirical models), analytical models, and numerical models. Table 13.2 shows the comparison of the four primary types of models. These are empirical, numerical, semiempirical, and analytical methods based on their foundational approaches, limitations, and typical applications.

13.2.3.1 Empirical Models

Empirical models for frosting are based on experimental data and observations, and they play a crucial role in predicting the behavior of frost formation and growth under different conditions. These models incorporate correlations for frost density

Table 13.2 Comparison of modeling approach for analysis of frost formation.

No.	Model type	Basis	Strengths	Limitations	Applications
1	Empirical	Experimental data	Simple, accurate within tested conditions	Limited applicability beyond tested conditions	Industrial quality control, stable conditions
2	Theoretical	First principles	Broad applicability, deep understanding	Computationally complex, may need simplifications	Research, new materials development
3	Semiempirical	Theory + experiment	Balanced accuracy and simplicity	Requires recalibration for different materials	Casting, welding, and industrial processes
4	Analytical	Physical laws	Exact solutions, useful for idealized conditions	Limited to simplified systems, real-world complexity	Academic research, early-stage modeling

and thermal conductivity, which are derived from experimental measurements and used to forecast changes in the frost layer's properties over time. Additionally, they estimate the frost growth rate based on factors such as temperature, humidity, and airflow. Empirical models also include correlations for heat and mass transfer processes during frosting, helping predict the heat transfer and moisture between the air and the frost layer. Furthermore, some empirical models incorporate correlations for defrosting processes, enabling the prediction of the energy required for defrosting and the duration of this cycle. While empirical models provide quick predictions based on observed data, they may not fully capture all the complexities of the frosting process, especially under varying conditions [35].

13.2.3.2 Analytical Models

Analytical models for frosting processing techniques are designed to predict the formation and growth of frost under various conditions. These models are based on fundamental heat and mass transfer principles, phase change, and fluid dynamics. These models and their key aspects are the following:

1. Heat and mass transfer equations
2. Phase-change dynamics
3. Frost growth rate
4. Thermal conductivity and density
5. Boundary conditions
6. Validation and calibration

The analytical models often begin with basic heat and mass transfer equations. These equations describe how to transfer heat and moisture between the air and the frost layer. The models account for the latent heat of phase change and sensible heat transfer [36]. Frost formation involves the phase change of water vapor to ice. Analytical models include equations that describe this phase-change process, considering factors such as temperature, humidity, and surface properties [37]. The frost growth rate is the rate at which frost grows is a critical parameter. Analytical models use equations to predict the frost growth rate based on environmental conditions and the properties of the surface on which frost forms [38]. The thermal conductivity and density of the frost layer change as it grows. Analytical models incorporate these changes to predict the thermal resistance of the frost layer accurately [39]. The boundary conditions are accurate the modeling requires proper boundary conditions. These include the temperature and humidity of the air, the temperature of the surface, and the initial conditions of the frost layer [40]. Analytical models are often validated and calibrated using experimental data. This ensures that the models can accurately predict frosting behavior under different conditions [36].

13.2.3.3 Numerical Models

Numerical models employ computational methods to simulate the frosting process. These models can account for complex interactions between heat transfer, mass transfer, and phase change [34, 41]. The three main frosting models can be classified as detailed stages in frosting models.

The initial stage is the water droplet condensation (WDC) stage; during this initial stage, as the temperature drops below the dew point, water droplets begin to form and grow on the cold plate surface through nucleation. Nucleation involves the formation of tiny water droplets due to the condensation nuclei, which are the tiny particles in the air that provide a surface for water vapor to condense onto. As the droplets continue to accumulate, they pave the way for the next stage of the frosting process [34]. The second stage is the water droplet freezing (WDF) stage.

In this stage, the water droplets that have condensed on the cold plate surface begin to transform as the temperature further decreases. The condensed water droplets freeze, starting from the surface and progressing inward. This process of freezing is initiated by the cold temperature of the plate, which causes the water droplets to transition from a liquid state to a solid state. Additionally, adjacent droplets can merge and freeze together, contributing to the formation of a solid ice layer [34]. The final stage is the frost layer growth stage. Once the water droplets have frozen, the next stage involves the growth of ice crystals from the frozen droplets. As more water vapor condenses and freezes, these ice crystals grow and interlock, forming a rough frost layer on the cold plate surface. The frost layer continues to thicken over time as the accumulation of frozen water droplets and ice crystals increases. This process of frost layer growth is influenced by factors such as temperature, humidity, and the duration of exposure to these conditions [30].

Figure 13.6 shows the relationship between frost density and effective thermal conductivity depending on data from two sets of studies: [42] and [30]. As frost density increases, both figures exhibit a rise in thermal conductivity, indicating that denser frost tends to conduct heat more effectively. However, the studies summarized by Zhang [42] show a significantly steeper increase, meaning that they report higher thermal conductivity values at comparable frost densities. This suggests that factors such as measurement techniques, frost structure, or environmental conditions in the Zhang [42] studies might have led to higher conductivity observations. The consistent gap between the two average lines indicates variability in the results across studies, emphasizing the need to consider different methodologies when analyzing frost's thermal properties. Understanding the detailed stages in frosting

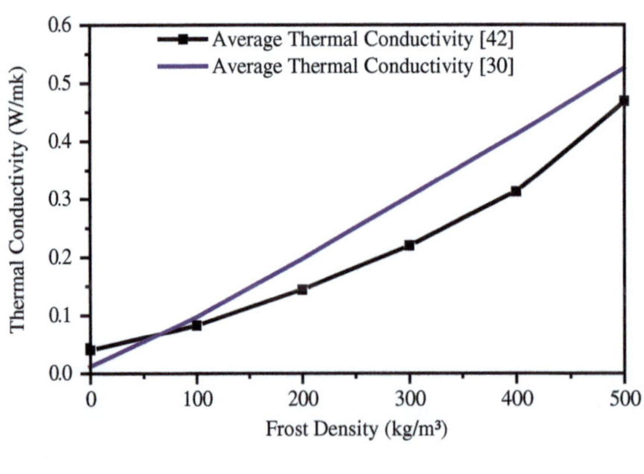

Figure 13.6 Relationship between the effective thermal conductivity and frost density.

models is essential for various applications, including refrigeration systems, cold storage, and air conditioning units. By comprehending the processes involved in frosting, engineers and designers can develop more efficient and effective systems to manage and control frosting, thereby improving the overall performance and longevity of equipment.

13.2.4 Limitations and Potential Use

One of the main drawbacks of icing mechanisms is their impact on energy efficiency. Frost development on heat exchangers, like those in air-source heat pumps, produces an insulating coating that significantly reduces heat transmission efficiency. To maintain appropriate temperatures, the insulating effect compels the system to use more energy, which raises operating costs and decreases overall efficiency [43]. The durability of the materials is another important restriction. Over time, material wear and deterioration may result from the cyclical process of icing and defrosting. Wear and tear on components subjected to repetitive temperature stressors may shorten the equipment's lifespan. This deterioration necessitates frequent maintenance and replacement of parts, thereby increasing operating expenses [44].

Additionally, frost accumulation threatens operational stability. System performance might fluctuate as a result of frost-blocking airflow and decreasing heat exchanger efficiency. These variations may disrupt normal operations, leading to frequent maintenance needs and potential system breakdowns. Excessive frost accumulation in difficult situations might result in total system failures, requiring prompt action. Environmental factors further complicate the efficiency of antifrosting techniques. Frost may build quickly and severely in areas with extremely low temperatures and high humidity, outpacing normal defrosting processes. This constraint necessitates the creation of more resilient and flexible strategies to control frost in a range of climates [35]. Despite these drawbacks, there are numerous ways to enhance system dependability and performance through the use of frosting techniques. Understanding and improving defrost cycles may result in considerable efficiency gains in energy systems, especially in HVAC and refrigeration applications. Controlling when and how defrosting occurs can keep systems running smoothly while saving energy [45]. Frosting and defrosting technologies are essential to the aerospace sector to guarantee aircraft performance and safety. By preventing frost accumulation on vital surfaces like wings and engines, efficient frost management systems improve operational dependability and flight safety. The integrity of airplanes in a variety of weather circumstances depends on these technologies [44].

Effective frost control is also advantageous for applications involving renewable energy. In colder climates, frost and ice buildup can hinder the functioning of solar panels and wind turbines. Sophisticated frosting processes can make these systems function well even in inclement weather, thereby increasing their dependability and energy output [43]. The food preservation sector employs controlled freezing methods to maintain the safety and quality of frozen goods. The food business can guarantee that items stay at ideal temperatures, maintaining their freshness, and prolonging shelf life by controlling the production of frost in storage and transportation systems. Reducing waste and upholding food safety regulations depend heavily on this application [69]. In conclusion, frosting mechanisms have a number of drawbacks, especially when it comes to operating stability, material durability, and energy efficiency, but their prospective applications across a range of sectors underscore their significance. By overcoming these constraints with creative designs and cutting-edge materials, we can optimize the advantages of frosting mechanisms, leading to more dependable and effective systems for a variety of uses.

13.2.5 Summary

In summary, the stages in frosting models, including WDC, WDF, and frost layer growth, play a crucial role in the formation and development of frost on cold surfaces. These stages are governed by physical principles related to temperature, humidity, and the behavior of water vapor, and they are fundamental to the understanding and management of frosting in diverse industrial and commercial settings [30].

13.3 Frosting of Air-source Heat Pump

Frost deposition and accumulation on the surface of the outdoor coil in an ASHP unit are inevitable and always have significant negative effects. To accurately predict and control a frosting defrosting cycle, the interrelated heat, mass, and momentum transport phenomena within frost and melted frost and at the air–frost interface, a moving boundary condition, should be clearly understood [7]. ASHPs are widely used for heating and cooling in residential and commercial buildings due to their energy efficiency and environmental benefits. However, one of the challenges associated with ASHPs is the formation of frost on the heat exchanger, which can significantly impact their performance. This chapter provides a detailed overview of the frosting phenomenon in ASHPs, including when and where frosting can occur, and the implications for system efficiency and maintenance [7]. The operation of air-source heat pumps in mild/cold environments, typically ranging from −15 °C to 6 °C, poses challenges related to frost formation during operation. When the surface temperature of the fins drops below the dew point temperature, water vapor in the air condenses and freezes, leading to the formation of frost. This layer of frost acts as insulation over the heat exchange area, hindering heat transfer from the air to the refrigerant. Studies have shown that increasing frost thickness results in higher heat transfer resistance and decreased air flow, leading to lower convective heat transfer and reduced operational performance, ultimately decreasing the coefficient of performance (COP) of the heat pump [46, 47]. An experimental analysis of frost thickness over time is shown in Figure 13.7. The graph demonstrates significant variations in frost thickness over time across different experimental conditions, with Ehsan's study showing the most rapid increase, while Lenic's shows a more gradual frost growth [48, 49, 50]. This comparison provides insight into how different conditions or methodologies might impact frost development rates in experimental setups.

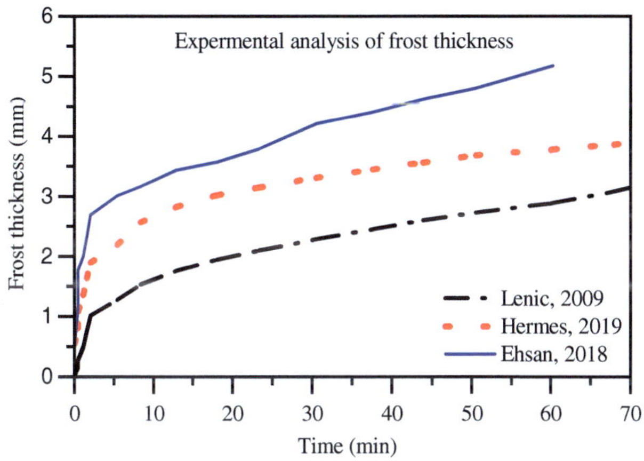

Figure 13.7 Experimental analysis of frost thickness.

Various factors, such as increased humidity, air temperature, surface temperature, and ambient air velocity, contribute to frost formation. To address the impact of frost on heat pump efficiency and prevent malfunctions, extensive research is being conducted on the subject of frosting on air-sourced heat exchangers. Several reports have examined the frost growth process on flat surfaces and described the different stages involved in frost formation. The growth of frost can also be explained by finned tube heat exchangers (FTHXs), where droplets condensed onto the fins and tubes freeze into ice crystals, forming a porous frost layer and eventually a denser frost layer. Gravity plays a role in the initial growth period, and the frost forms faster between fins than on a flat cold surface. The heat transfer coefficient initially increases, possibly due to the rough ice surface increasing the heat exchange area, but further examination is required to determine the cause [47].

Predicting frost growth is crucial for optimizing heat pump operation and avoiding efficiency losses. However, accurately describing the frost in models is challenging due to the unpredictable and uneven initial stage of drop formations and crystallization. The operational performance of heat pumps during frosting conditions has been modeled to understand the impact of frost on heat pump parameters. Studies have shown that the heating capacity is more sensitive to accumulated frost than the COP, and the loss of convective heat transfer due to decreased airflow through the evaporator plays a significant role in performance drop during later frosting stages. Finned heat exchangers are essential components in various thermal systems, enhancing heat transfer efficiency by increasing the surface area available for heat exchange. The design and configuration of the fins significantly influence their performance, particularly in terms of frost formation and solidification [26].

Figure 13.8 illustrates how the COP varies over time across three distinct humidity levels: 60%, 75%, and 95% [26]. This data, collected over a span of 120 minutes, effectively underscores the relationship between ambient humidity and the operational efficiency of heating systems. The graphic indicates that higher levels of humidity can significantly impair system performance. It is evident that systems consistently

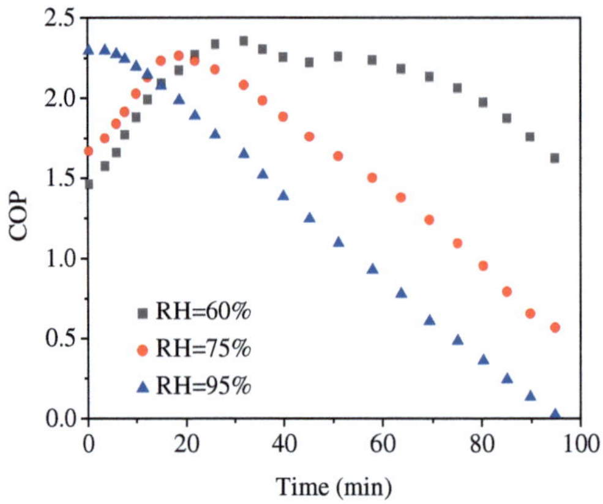

Figure 13.8 Variation of COP over time at different relative humidities.

operating in environments characterized by elevated RH are likely to experience performance degradation. Consequently, it might be necessary to implement optimized designs specifically tailored to mitigate these adverse effects. Such enhancements could include improved dehumidification processes or the integration of advanced heat exchanger technologies. The findings of this study highlight the critical importance of adapting both system designs and operational strategies to accommodate varying humidity conditions. This adaptation is essential for improving long-term performance and achieving greater energy efficiency. Furthermore, current research efforts are increasingly focused on enhancing the efficiency and functionality of air-source heat pump units, especially in colder climates.

13.3.1 Frosting on Tube-Fin Heat Exchangers

Tube-fin heat exchangers are commonly used in ASHPs due to their high heat transfer efficiency. When these heat exchangers are exposed to humid air, water vapor condenses on their surfaces and subsequently freezes, forming a layer of frost. This frost accumulation can have detrimental effects on the heat transfer characteristics of the heat exchangers, leading to decreased efficiency and increased energy consumption. Understanding the mechanisms of frosting on tube-fin heat exchangers and developing strategies to mitigate its effects are crucial for optimizing the performance of air-source heat pump units. The process of frosting on tube-fin heat exchangers begins with the condensation of water vapor on the fin surfaces. As the temperature of the heat exchanger drops below the dew point of the surrounding air, water droplets form on the fins. These droplets can coalesce and grow, eventually freezing as the temperature continues to decrease [7].

The formation of frost is influenced by several factors, including the temperature of the heat exchanger, the humidity of the air, and the airflow over the fins. The initial stage of frosting involves the nucleation and growth of ice crystals, which can create a rough and uneven surface on the fins. This roughness can significantly increase the heat transfer resistance, making it more challenging for the heat exchanger to maintain efficient energy harvesting. As the frost layer continues to grow, it can disrupt the airflow over the fins, further reducing their effectiveness in transferring heat. The presence of frost on tube-fin heat exchangers can lead to a decrease in the overall efficiency of the air-source heat pump unit. The increased thermal resistance caused by the frost layer can reduce the heat transfer rate, leading to higher energy consumption and decreased performance [7].

Additionally, the accumulation of frost on the fins can obstruct airflow, affecting the overall functionality of the heat exchanger. In some cases, the frost layer can become so thick that it completely blocks the airflow, causing the heat pump unit to fail [7]. One of the primary challenges in managing frosting on tube-fin heat exchangers is predicting when and where it will occur. The formation of frost is a dynamic process that can vary depending on the operating conditions of the heat pump unit. Changes in temperature, humidity, and airflow can all influence the rate and extent of frosting, making it essential to develop accurate predictive models to anticipate and mitigate its effects. Several strategies have been proposed to reduce the impact of frosting on tube-fin heat exchangers in air-source heat pump units. One approach is to incorporate advanced surface coatings or treatments that can inhibit the formation of frost. These coatings can create a hydrophobic surface that repels water, preventing the initial condensation of water droplets and delaying the onset of frost formation [7].

An alternate option is to implement active defrosting methods, which involve using additional heat sources or mechanical methods to remove frost accumulation from the heat exchanger fins. The energy consumption of the heat pump unit is likely to grow as a result of these measures, despite the fact that they are successful in preserving the efficiency of the system. In light of this, it is of the utmost importance to find a middle ground between the advantages of defrosting and the additional expenses incurred for energy. Additionally, continuing research is concentrating on the development of novel materials and creative designs for tube-fin heat exchangers with the goal of reducing the amount of icing that occurs. Some research has investigated different materials that have improved anti-frosting qualities, while others have focused on improving the geometric designs of the heat exchanger fins in order to limit the amount of frost that accumulates [7].

Figure 13.9 illustrates the success of a numerical model in predicting frost density [51, 52]. It also displays the challenges that the model faces. During the early phases of frost development, the model demonstrates a high degree of agreement with the experimental data, which indicates that it is accurate regarding the process. Nevertheless, as the formation of frost continues, a disagreement emerges, with the model having a tendency to overestimate the frost density in comparison to the measurements made under experimental conditions. This divergence might be attributable to simplifications or assumptions made within the computational model that do not adequately account for the complicated dynamics of frost formation that occur in the actual world. FTHXs have undergone significant optimization to reduce icing and improve thermal performance. Additionally, Yang et al. presented optimal design parameters for a FTHX in home refrigerators in 2006. Their study intended to improve the efficiency of heat transmission and to lengthen the amount of time that the system could operate under icing conditions. By optimizing the design of the

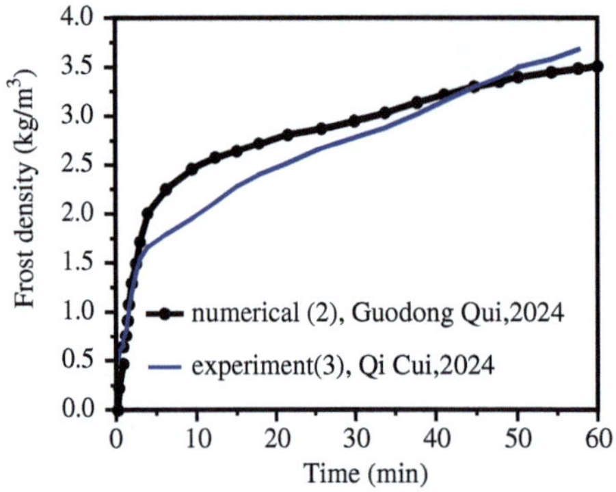

Figure 13.9 Comparison of numerical and experimental analysis for frost density (RH = 75%).

fins and tubes, they were able to obtain a 6.3% increase in the average heat transfer rate as well as a 12.9% extension in the amount of time necessary for operation [7].

A comprehensive investigation of the air-side heat transfer properties of flat FTHXs was conducted by Lee et al (2010). The investigation was conducted under trying circumstances. In the course of their trials, they experimented with a variety of important elements, including fin pitch, the number of tube rows, and tube alignment. Furthermore, their findings demonstrated that both the fin pitch and the staggered tube alignment had a substantial impact on the lowering of airflow, which, in turn, played an essential part in preventing the accumulation of frost [7]. More recently, the frosting behaviors and thermal performance of louvered fins with unequal louver pitch were investigated by Park et al. Their study demonstrated that using an unequal louver pitch design delayed frost blocking in the spaces between louvers at the front side, resulting in a 21% improvement in thermal performance. Additionally, they found that a design where the louver pitch successively decreased from the air inlet region to the redirection region provided more uniform frost growth and further enhanced thermal performance [53].

Table 13.3 provides an overview of various types of finned tubes utilized in heat exchange systems, particularly focusing on frost development and its implications for surface solidification within cooling systems. Each finned tube design has unique structural traits that influence the formation and accumulation of frost, which in turn impacts the overall heat transmission efficiency of the system. In heat exchangers commonly found in refrigeration and air conditioning applications, finned tubes serve to increase surface area, thereby enhancing heat transfer effectiveness. However, when temperatures dip below freezing, ice may accumulate on the surfaces of the tubes, resulting in diminished heat transmission efficiency due to increased thermal resistance. The specific fin design can either alleviate or exacerbate the accumulation of frost. The fundamental design of a plain finned tube features linear fins affixed to a central tube. The impact of frost on this design results in a uniform accumulation on the fin surfaces, which gradually increases thermal resistance and lowers pressure. While this uniform frost distribution may seem beneficial at first glance, it ultimately undermines the effectiveness of the heat exchange process

Table 13.3 Heat exchanger types and contributions to frost and solidification.

No.	Type	Description	Contribution to frost and solidification	References
1	Plain finned tube	Basic design with straight fins.	Frost forms uniformly, leading to a gradual increase in thermal resistance and pressure drop.	[54]
2	Spiral-finned tube	Fins are wound in a spiral around the tube.	Enhanced turbulence reduces frost thickness but increases frost density, affecting heat transfer efficiency.	[55]
3	Longitudinal finned tube	Fins run parallel to the tube length.	Directional airflow can lead to uneven frost distribution, impacting overall performance.	[56]
4	Perforated finned tube	Fins have perforations to increase turbulence.	Perforations enhance air mixing, reducing frost buildup but increasing pressure drop.	[57]
5	Louvered finned tube	Fins have small slits or louvers.	Louvers disrupt airflow, reducing frost formation but complicating defrosting processes.	[58]
6	Wavy finned tube	Fins have a wavy pattern.	Increased surface area and turbulence lead to higher frost accumulation but better heat transfer.	[59]
7	Offset Strip finned tube	Fins are offset to create turbulence.	Offset strips enhance heat transfer but can lead to non-uniform frost growth.	[60]

as the frost layer builds up over time. Although this design is advantageous in systems requiring uniform heat transfer, its effectiveness diminishes in scenarios where frost accumulation is problematic [54]. In a spiral-finned tube design, the fins are arranged in a spiral configuration around the tube. The spiral shape promotes turbulence in the airflow, reducing the thickness of the frost layer that develops. However, this increased turbulence can also lead to a denser frost layer, which, despite accumulating more slowly than in standard finned tubes, can still negatively impact heat transfer efficiency. The primary benefit of spiral-finned tubes lies in their ability to mitigate frost accumulation while maintaining effective heat transmission, although the increased density of frost may complicate defrosting procedures [55].

The design features of the longitudinal finned tube that are parallel to the length of the tube. The resultant directional airflow can lead to irregular frost distribution during solidification, with some areas of the tube experiencing greater frost buildup than others. Such variations in frost accumulation can result in inconsistent heat transfer efficiency, as regions with increased frost will exhibit higher thermal resistance. While the longitudinal fin design may enhance the specific airflow patterns, it complicates maintaining steady heat transmission due to the irregular nature of frost accumulation [56]. Perforated finned tube design incorporates fins with perforations and either holes or slits to enhance turbulence and promote air mixing.

The perforations facilitate improved air circulation, which helps to reduce frost accumulation on the surface. However, this airflow disruption may result in a higher pressure drop. Overall, reduced frost buildup can enhance heat transmission efficiency over time, making this design suitable for applications where minimizing frost accumulation is essential. Nonetheless, the accompanying pressure drop could impact energy efficiency [57]. Louvered finned tube: In this design, fins are equipped with small openings known as louvers that disrupt airflow across the surface [61]. By impeding airflow, the louvers effectively reduce frost development. While this helps lower frost accumulation, it creates challenges for defrosting, as airflow obstruction can hinder the removal of ice. Consequently, while louvered finned tubes can be beneficial for controlling frost, systems must integrate defrosting mechanisms to manage the inherent complications of this design [58].

The features of a wavy finned tube are the fins with a wavy configuration that enhances the surface area and induces turbulence. The undulating fins increase surface area, potentially leading to more substantial frost accumulation due to the additional area available for frost formation. Despite this risk, the turbulence generated by the wavy fins can enhance heat transfer efficiency. This design strikes a balance between increased frost buildup and effective heat transmission, making it suitable for systems capable of tolerating significant frost while still requiring efficient heat transfer [59]. The offset strip finned tube configuration features staggered fins that enhance airflow turbulence. While these offset fins improve heat transfer efficiency, they may also contribute to uneven frost formation, resulting in certain areas accumulating more frost than others. This irregular frost distribution can impair the overall effectiveness of the system, particularly in applications where consistent heat transmission is critical. This design is thus most appropriate for situations where effective heat transfer is essential and where some irregularity in frost distribution can be tolerated [60]. The configuration of finned tubes plays a crucial role in determining the frost formation, density, and distribution on their surfaces. Designs like plain finned tubes tend to produce uniform frost accumulation, thereby increasing thermal resistance and diminishing heat transmission efficiency. In contrast, alternatives such as perforated or louvered finned tubes help manage frost accumulation, although they may introduce increased pressure drops or complicate the defrosting process. Ultimately, the objective of the finned tube designs is to optimize heat transfer efficiency. While designs like wavy finned tubes enhance heat conduction by increasing surface area and turbulence albeit at the cost of higher frost accumulation, other designs, such as spiral or perforated finned tubes, facilitate frost management and prolonged efficiency. However, these designs may also introduce operational challenges, including increased pressure loss.

13.3.2 Frosting on Microchannel Heat Exchanger

ASHPs are increasingly using microchannel heat exchangers (MCHXs) due to their compact dimensions and superior heat transfer efficiency. Nonetheless, they have a greater susceptibility to icing than conventional tube-fin heat exchangers. Because MCHXs have more surface area and closer-spaced fins, frost forms faster when cold, damp air flows through the heat exchanger. This causes frost to build up on the fins, which greatly reduces the heat pump's efficiency. The formation of ice can obstruct airflow pathways, elevate pressure drops, and diminish heat transfer efficiency, eventually leading to increased energy consumption and reduced heating

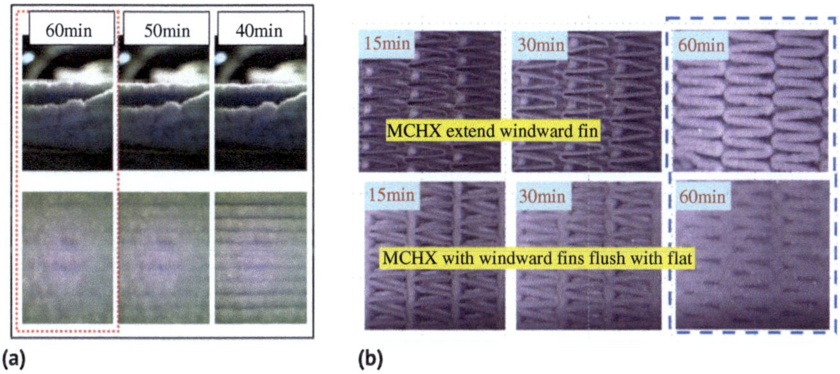

Figure 13.10 Comparison of frost condition on FTHX and MCHX heat exchangers. (a) Photo of the frosted FTHX. (b) Image of the frost on two tested of MCHX.

capacity. Moreover, frost accumulation on MCHXs is irregular, exhibiting greater thickness and density at the air input and reduced thickness toward the outflow, thereby affecting the overall efficacy of the heat exchanger [62]. The accumulation of frost on MCHXs markedly impacts heat transfer efficiency and overall system performance [63]. The small configuration of MCHX and the characteristics of microchannels lead to expedited icing, resulting in increased pressure drops and reduced heat transfer efficiency. Comprehending the determinants of ice, including channel geometry, airflow, and surface treatments, enables engineers to enhance the design of MCHX systems to mitigate the adverse impacts of frost, thereby improving long-term performance and efficiency.

The comparative analysis of the images reveals that the frosting condition and blockage in MCHX and FTHX types of heat exchangers are shown in Figure 13.10. The highlighted area in this figure focuses on frost formation overtime on two types of heat exchangers, FTHX and MCHX with different fin configurations [24, 64]. Comparative analysis of the images reveals that the frost blockage in MCHX with extend windward fin was less severe than in MCHX with windward fins flush with flat and FTHX. After 30 minutes, MCHX with windward fins flush with flat, and FTHX experienced significantly more severe frost blockage compared with MCHX extended windward fin. By 60 minutes, the extended fins of the MCHX windward fin had accumulated a substantial amount of frost, yet the airflow passage remained partially open. In contrast, the airflow passage in MCHX with windward fins flush with the flat was nearly completely obstructed by frost [64].

13.3.3 Characteristics of Heat Pumps During Frosting

Several significant factors negatively impact the performance of air-source heat pumps during icing conditions. Initially, the accumulation of ice forms an insulating barrier, hindering the effective transfer of heat between the air and the refrigerant. This thermal resistance reduces the system's heating capacity, thereby increasing energy consumption. This is because the heat pump has to work harder in order to keep the temperature at the correct level. Additionally, the accumulation of frost causes an increase in the pressure drop across the air-side heat exchanger during the process. This increase in pressure drop restricts airflow, exacerbating the decline in heat transfer efficiency and diminishing the system's overall performance [65, 66]. As a result, the overall performance of the ASHP units is significantly affected during frosting conditions. Moreover, ASHP units are equipped with defrost cycles to manage frost buildup. However, frequent defrost cycles can lead to a reduction in the overall efficiency of the system and an increase in wear and tear on system components. These defrost cycles are essential for melting the frost and maintaining the optimal functioning of the ASHP units, but the frequent activation of these cycles can have detrimental effects on the efficiency and longevity of the system components [64].

Excessive frost buildup can also lead to system malfunctions, including reduced airflow, increased pressure drops, and potential damage to heat exchanger components. These malfunctions can compromise the overall performance and reliability of the ASHP units, necessitating maintenance and repairs to restore the system to optimal working condition [67].

Additionally, the temperature of the heat pump components fluctuates during frosting and defrosting cycles, which further affects the overall performance and efficiency of the system. These temperature fluctuations can impact the operational stability of the ASHP units and contribute to variations in system performance during frost accumulation and defrosting processes [35]. In conclusion, the characteristics of heat pumps during frost conditions significantly impact the performance and efficiency of air-source heat pumps. Reduced heat transfer efficiency, increased pressure drops, frequent defrost cycles, system malfunctions, and temperature fluctuations all pose challenges associated with frosting. Understanding these factors is essential for effectively managing and maintaining ASHP units in environments prone to frost buildup [67].

13.3.4 Summary

A significant challenge for air-source heat pumps is the buildup of frost, which is especially problematic in environments that are both cold and damp. To ensure the effectiveness and reliability of the ASHP units, it is essential to understand frosting situations and implement solutions that mitigate icing. It is possible to considerably improve the performance of ASHP units by identifying the important elements that lead to frost formation, implementing new technologies and design enhancements, and so limiting the influence that frosting has on the operations of the system.

13.4 Anti-frosting Technology of Air-source Heat Pump

ASHPs are highly efficient for both heating and cooling, but they encounter challenges with frosting, particularly in cold and humid environments. To address this issue, various anti-frosting technologies have been developed, each with its own mechanisms and applications. Different strategies are used to minimize or eliminate frost formation and to remove any frost that does form, ensuring the efficient operation of the heat pump [7].

13.4.1 Preheating Method

Preheating the inlet air to an outdoor coil stands out as a simple yet highly effective technique for mitigating or preventing frosting in the ASHP units. The findings revealed a remarkable 38.0% increase in heating capacity and 57.0% enhancement in COP when the electric heater was employed to heat the inlet air at outdoor conditions below the 2 °C/1 °C (dry-bulb/wet-bulb temperature). However, it is important to acknowledge that the significant drawback of this approach lies in the high energy consumption associated with preheating in very cold regions. Comparative analyses of various frost-retarding measures underscored that preheating the inlet air is not economically viable in the regions characterized by prolonged periods of very low outdoor temperatures ranging from −54 to 10 °C [7]. Consequently, to bolster the economic feasibility of ASHP units, the energy used for preheating the inlet air must be derived from waste heat sources, such as that recovered from exhausted indoor air or wastewater [7]. In contrast, this method involves preheating the air before it reaches the heat exchanger, reducing the likelihood of frost formation by maintaining the surface temperature above the frost point.

13.4.2 Dehumidity Method

Dehumidification plays a crucial role in optimizing the performance of air-source heat pumps by minimizing frost formation, enhancing heat exchange efficiency, extending system lifespan, and reducing energy consumption. The dehumidity method to prevent frost formation on an air-source heat pump, a common method for anti-frosting technology, involves using a desiccant to decrease the moisture content of the incoming air before it reaches the evaporator coil. This can reduce the availability of water vapor to freeze into frost and can be accomplished by adding a dedicated dehumidification unit in the air intake path of the heat pump system [68, 69]. By lowering ambient humidity, dehumidification facilitates improved heat transfer and lessens wear on essential components. Moreover, it decreases the frequency of defrost cycles, which contributes to lower operational costs. Effective strategies for humidity reduction include utilizing advanced heat exchangers with hydrophilic coatings, implementing pre-dehumidification units, and optimizing refrigerant temperature and flow rates. These methods not only mitigate frost buildup but also enhance moisture management [70].

However, the existing challenges are particularly in low-temperature environments where dehumidification efficiency may decline. Additionally, some methods, such as desiccant systems, could lead to increased energy consumption [71]. Figure 13.11 illustrates the relationship between heat transfer (in kW) and time (in minutes) for different RH levels [26]. The goal of this graph is to demonstrate how varying levels of humidity affect heat transfer efficiency over time. The benefits of efficient dehumidification are manifold. It stabilizes heating performance, reduces maintenance needs, improves indoor air quality, and lowers long-term operational expenses. In various contexts, dehumidification is crucial, particularly in commercial settings such as greenhouses and industrial facilities, where accurate

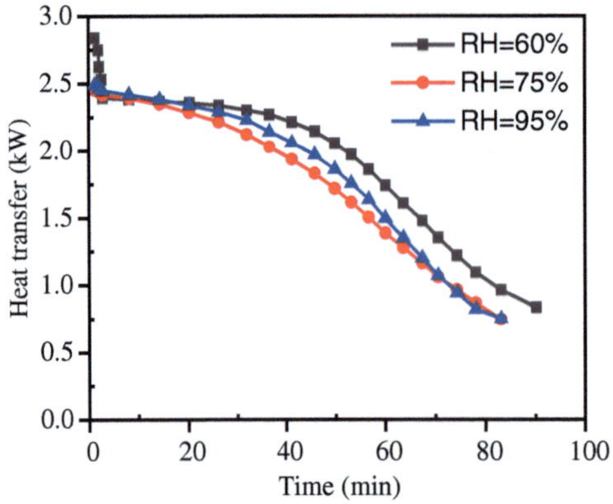

Figure 13.11 Effect of relative humidity on heat transfer over time.

temperature and humidity control are essential. Ultimately, effective dehumidification is key to achieving optimal energy efficiency and occupant comfort in both residential and commercial heating applications [72]. Dehumidification in anti-frosting ASHP units is essential for enhancing efficiency, reliability, and energy savings. By mitigating frost formation through innovative techniques, these systems ensure consistent performance in a variety of conditions, making them a cornerstone of sustainable heating solutions for residential and commercial applications.

13.4.3 Surface Treatment Method

Several studies have investigated the effects of fin surface coatings on the frosting and defrosting performance of outdoor coils. Okoroafor and Newborough's 2000 study revealed that cross-linked hydrophilic polymeric coatings notably decreased frost growth on cold surfaces exposed to warm, humid air, reducing frost thickness by 10–30% compared to uncoated metallic surfaces. In 2001, Wu and Webb demonstrated the preference for hydrophilic coatings in both frosting and defrosting processes due to their ability to result in less frost and retain water [7]. The application of surface coatings on fins in outdoor coils, such as those in HVAC systems, can significantly impact frosting and defrosting performance. Frost formation on coils can reduce heat transfer efficiency, increase energy consumption, and cause operational interruptions. Therefore, specialized coatings are applied to mitigate these effects, aiming to delay frost formation, facilitate defrosting, and improve overall thermal efficiency.

Table 13.4 shows a detailed discussion of how different coatings affect frosting and defrosting. Applying coatings to the heat exchanger fins can significantly impact frosting and defrosting performance. Hydrophilic, hydrophobic, and hygroscopic coatings have been shown to reduce frost growth, delay frost formation, and make the frost layer easier to remove. These treatments enhance the efficiency of the defrosting process and improve overall system performance [35]. Surface treatment methods effectively reduce icing on the heat exchangers of air-source heat pump units. Modifying the characteristics of the heat exchanger surfaces can reduce the rate of frost accumulation and enhance the system's overall efficiency.

Table 13.4 Coating types and their influence on frosting/defrosting performance metrics.

No.	Coating type	Material examples	Effects on frosting	Effects on defrosting	Key parameters for analysis	References
1	Hydrophobic coatings	Fluoropolymers, silanes	Delays frost formation, reduces water adhesion	Facilitates ice shedding, shortens defrost time	Water contact angle, adhesion force, defrost rate	[73, 74]
2	Superhydrophobic coatings	Silica, Teflon	Significantly delayed frosting prevents droplet adherence	Very effective in quick ice removal	Contact angle hysteresis, droplet roll-off angle	[75]
3	Hydrophilic coatings	Titanium dioxide (TiO_2)	Reduces droplet formation, promotes thin water film	Provides uniform defrosting, reduces frost buildup over cycles	Water spreading coefficient, freezing delay time	[76, 77]
4	Anti-icing coatings	Phase-change materials (PCMs)	Delays frosting by storing thermal energy	Maintains temperature for defrosting, reduces energy consumption	Phase-change temperature, thermal conductivity	[78, 79]
5	Ice phobic coatings	Silicone, Elastomers	Limited effect on frost delay, prevents strong ice adhesion	Enhances defrosting by reducing ice adhesion strength	Ice adhesion strength, mechanical durability	[80]

The following surface treatment techniques have proven effective in anti-frosting technology. Hydrophobic coatings are commonly used to reduce frost development on surfaces. These coatings are designed to repel water, preventing moisture from adhering to the surface of heat exchangers. By reducing water accumulation, hydrophobic coatings slow down the condensation process and minimize frost formation. They are particularly effective in high-humidity environments, significantly delaying frost formation and reducing its growth rate. Superhydrophobic coatings take hydrophobicity a step further by creating an even more water-repellent surface. These coatings typically feature nanoscale roughness, which enhances the contact angle between water droplets and the surface, further impeding frost formation [81, 82]. The superior water-repellent properties of superhydrophobic surfaces help to reduce frost accumulation and condensation, thereby maintaining heat transfer efficiency for longer periods [83].

Anti-frosting and anti-icing coatings are designed to prevent the accumulation of frost and reduce the adhesion of ice crystals after frost forms. These coatings typically consist of low surface energy compounds or polymers that impede the freezing process. In some cases, they can also help speed up the removal of ice once it has formed, thus improving the defrosting process [84]. By using these coatings, the negative effects of frost on heat pump efficiency can be mitigated, as they prevent ice or frost from obstructing airflow or impairing heat transfer. Electrostatic treatments involve applying an electric field to the surface of the heat exchangerto alter the behavior of water droplets. This method encourages droplet detachment and minimizes water retention on the surface, thereby decreasing the likelihood of frost formation. Such treatments can significantly reduce water adhesion to surfaces, leading to lower frost accumulation and enhanced heat exchanger efficiency, especially in changing weather conditions.

Nano-coatings involve the deposition of nanoparticles onto the surfaces of heat exchangers, creating a rough or porous texture that inhibits frost formation. These coatings increase the surface area and alter the microstructure, which prevents uniform frost development. The increased surface roughness allows for faster shedding of frost, preventing obstruction of the airflow through the heat exchanger [85]. Additionally, nano-coatings enhance the long-term durability of the heat exchanger, providing superior anti-frosting performance over time. Surface texturing refers to the mechanical or chemical modification of heat exchanger surfaces to create patterns or microstructures that inhibit frost development [86]. Textured surfaces promote irregular frost formation, thus reducing the likelihood of significant frost buildup in any one area. This approach can also aid in the removal of ice once it has formed, decreasing the need for repetitive defrost cycles and improving the overall efficiency of the heat pump.

There are various types of surface treatment techniques such as hydrophobic and superhydrophobic coatings, anti-frosting and anti-icing technologies, electrostatic treatments, nano-coatings, and surface texturing. These techniques effectively reduce frost accumulation on the heat exchangers of air-source heat pump units. These solutions enhance the performance of the heat pump system by preventing frost buildup and minimizing the need for frequent defrosting. Implementing these treatments allows ASHP units to operate more efficiently in cold and humid conditions, resulting in improved energy savings and extended operating lifespans.

13.4.4 Summary

By utilizing advanced technologies, air-source heat pumps can perform more effectively in colder climates, reducing problems related to frost and ensuring peak efficiency. The implementation of the anti-frosting technology is vital for improving the effectiveness, dependability, and versatility of ASHP units in the low-temperature areas. Addressing frost challenges through these innovations encourages the increased use of ASHP units as a sustainable and the energy-efficient heating option.

13.5 Defrosting of Air-source Heat Pump

Defrosting in an air-source heat pump system is an essential process that removes frost from the evaporator, ensuring optimal efficiency and proper operation. The following subsections provide a detailed description of the defrosting system based on the specified parameters.

13.5.1 Types of Defrosting Air-source Heat Pump

Defrosting is an essential process in air-source heat pump systems, particularly in colder climates where frost or ice can accumulate on the evaporator coil. Efficient defrosting not only helps maintain the heat pump's heating capacity but also reduces energy consumption. Table 13.5 outlines the main types of defrosting methods used in ASHP units, along with their key features and advantages. The effectiveness of air-source heat pump systems is dependent on a variety of defrosting techniques to keep them operational in cold weather conditions. Frost can build on the evaporator coils of a heat pump as the temperature lowers, which can result in a decrease in the heat pump's overall effectiveness. By utilizing several different defrosting procedures to combat this issue, each with its own set of processes, advantages, and limits. The following provides an overview of the nine basic defrosting processes used in ASHP units [35]. The process of defrosting through the use of the reverse cycle entails reversing the refrigeration cycle of the heat pump, which causes the system to briefly function in the cooling mode. This process melts the ice and effectively warms the exterior coil. This makes the reverse cycle an effective method of clearing ice, as it utilizes the compressor and refrigerant already present in the system, thereby eliminating the need for additional energy sources. Heating by electrical resistance: This technique uses electric resistance heaters to directly heat the evaporator coil, which subsequently melts the ice. Despite its ability to produce the desired results, its dependence on electricity, which may not be as effective of cost as using the heat pump's refrigerant, makes it less energy efficient than other methods [35].

The process of hot-gas defrosting involves redirecting hot refrigerant gas from the compressor to the evaporator coil, thereby bypassing the defrosting process. The hot gas bypasses the expansion valve and directly enters the coil, assisting in melting the previously collected frost. While this technology is efficient and eliminates the need for external energy sources, it may significantly reduce the system's heating capacity during defrost cycles. Passive defrosting methods rely on natural processes, like changes in the temperature of the air outside, to melt the frost without the need for any kind of assistance from an outside source. Although this method is energy efficient, it may not be sufficient in situations where the temperature is extremely low or where frost accumulates quickly. Ultrasonic defrosting is a method of defrosting that involves the use of high-frequency sound waves to shake the ice that is on the evaporator coils. This causes the ice to break apart and fall off. Although this approach is still in development and has the potential to be highly energy efficient, commercial systems currently employ it far less frequently than traditional methods [87].

Defrosting through the use of specialized chemicals can treat the evaporator coils in certain systems, lowering the freezing point of water and preventing frost accumulation. While this technology uses less energy, it raises concerns about the maintenance and refilling of chemical agents, as well as the potential long-term environmental impact. Some heat pumps employ sophisticated coatings on the evaporator coils to prevent frost from clinging to the surface in the first place. These coatings aim to enhance the system's efficiency by decreasing the accumulation of frost and the frequency of defrosting cycles. One of the disadvantages of these coatings is that they can be relatively expensive and may lose effectiveness over time. Defrosting through the use of heat recovery: The process of heat recovery defrosting includes melting the frost by utilizing waste heat from the system's operation, such as the heat created by the compressor. Because this approach recycles energy inside the system, it is not only ecologically beneficial but also efficient in terms of energy use. However, if the system fails to produce enough waste heat to effectively manage the defrosting process, it may not be successful [35, 88].

Assistance with defrosting utilizing phase-change materials (PCMs) has the ability to absorb and store significant quantities of heat throughout the process of transitioning from a liquid to a solid state and then release this heat when they form a solid state. It is possible to melt ice more effectively by including PCMs in the heat pump units. This is because PCMs absorb system heat and gradually release it to thaw coils. This technique has the potential to enhance efficiency; nevertheless, it necessitates the proper selection of materials in order to guarantee compliance with the thermal dynamics of the system. Every defrosting technique has its own advantages and disadvantages, and a variety of factors, such as the system's design, the surrounding environment's temperature, the approach's energy efficiency, and running cost considerations, frequently influence the choice of method. For instance, people extensively utilize reverse cycle defrosting due to its effectiveness. On the other hand, techniques such as the utilization of ultrasonic waves or phase-change materials are still in the process of development and may offer potential energy savings in the future. Ultimately, the optimal defrosting method for each air-source heat pump system will vary, as it depends on the specific needs and conditions of the system [87].

13.5.2 Efficiency

The efficiency of defrosting an air-source heat pump depends on the method used, environmental conditions, and the specific design of the heat pump system. Air-source heat pumps rely on defrosting techniques to maintain their operational effectiveness and reliability in frost-prone environments. Frost buildup significantly impacts the efficiency of heat transfer, lowering heating capacity and increasing energy usage [89]. Several defrosting methods are commonly used in ASHP systems, including hot-gas bypass defrosting (HGBD), reverse cycle defrosting (RCD), electric heating defrosting (EHD), and compressor shutdown defrosting (CSD). Table 13.5 summarizes the efficiency, energy penalty, and downtime impact of different defrosting methods for air-source heat pumps. It presents an efficiency analysis of defrosting methods used in air-source heat pump systems [87, 90–92].

Table 13.5 Efficiency analysis of defrosting methods in ASHP units.

No.	Defrosting method	Efficiency (%)	Energy penalty (%)	Downtime impact (%)
1	Compressor shutdown	20	80	90
2	Electric heating defrosting	50	50	50
3	Reverse cycle defrosting	70	30	20
4	Hot-gas bypass defrosting	90	10	5

Table 13.6 Summary of defrosting efficiency by climate.

No.	Defrosting method	Cold climate	Moderate climate	Warm climate
1	Compressor shutdown	Low efficiency	Moderate efficiency	High efficiency
2	Electric heating	Moderate efficiency	Moderate efficiency	Low efficiency
3	Reverse cycle	Moderate efficiency	High efficiency	High efficiency
4	Hot-gas bypass	High efficiency	High efficiency	Moderate efficiency

Table 13.7 Comparison of control strategies.

No.	Strategy	Efficiency	Energy savings	Complexity	Suitability	References
1	Time-based	Low	Low	Simple	Predictable frost areas	[87]
2	Sensor-based	Moderate	Moderate	Moderate	Variable climates	[90]
3	Predictive	High	High	Complex	Advanced systems	[91]
4	Adaptive	Very high	Very high	High	Extreme climates	
5	On-demand	Maximum	Maximum	High	Modern ASHP units	[92]
6	IoT-based smart	Very high	High	Very high	Commercial systems	[93]

Efficient defrosting methods are crucial for maintaining the performance and reliability of ASHP systems in frost-prone environments. Continued research and innovation in defrosting technologies will further enhance the feasibility and efficiency of ASHP units across diverse climatic conditions. Table 13.6 describes a summary of the efficiency of different defrosting methods across various climate conditions (cold, moderate, and warm climates) [87, 90–92].

13.5.3 Defrosting Control Strategies for Air-source Heat Pumps

The performance and efficiency of air-source heat pumps depend on efficient defrosting management techniques. To limit energy loss, guarantee continuous heating, and lessen the negative consequences of frost formation, these techniques control the start, length, and frequency of defrosting cycles. In this section, we provide a thorough explanation of the various defrosting control techniques. Table 13.7 compares different control strategies used for managing systems like defrosting or temperature regulation. In contrast, the defrosting management techniques significantly impact the effectiveness and efficiency of ASHP units. The time-based management is still a simple and inexpensive solution, but it is ineffective in conditions that fluctuate. Modern ASHP units can benefit from sensor-based and adaptive control systems since they are more sensitive and energy efficient. The future of defrosting control lies in predictive and Internet of Things (IoT)–based smart systems, which use cutting-edge technology to maximize efficiency and lower operating expenses. These tactics fit in nicely with the growing need for ecological and energy-efficient heating systems.

Table 13.8 Comparison of temperature drops across methods.

No.	Defrosting method	Temperature drop (°C)	Remarks
1	CSD	3–6	Long defrost times cause significant cooling effects indoors.
2	EHD	1–2	Minimal fluctuation; fast defrosting ensures better thermal comfort.
3	RCD	2–5	Noticeable cooling due to heat extraction from the indoor space.
4	HGBD	<1	Negligible temperature fluctuation; partial heating ensures indoor comfort.

13.5.4 Indoor Air Temperature Fluctuation During Defrosting

Indoor air temperature fluctuation during defrosting is a critical aspect of air-source heat pump units, as it directly impacts occupant comfort and the overall perception of the system's performance. Different defrosting methods result in varying degrees of temperature drops inside the heated space due to changes in the heat pump's operation during frost removal. The fluctuation of the indoor temperature is influenced by the defrosting technique, system design, and external environmental conditions. The extent of indoor air temperature fluctuation during defrosting depends heavily on the defrosting method employed [90]. Hot-gas bypass defrosting stands out as the most effective solution for maintaining thermal comfort, while reverse cycle defrosting and electric heating defrosting offer balanced approaches depending on the climate and system design. In contrast, compressor shutdown is the least efficient and causes the most significant temperature fluctuations, making it unsuitable for modern applications. Improved defrosting technologies and better building insulation can further mitigate temperature fluctuations, enhancing user comfort and system efficiency. Table 13.8 compares the temperature drops and related remarks for different defrosting methods. It highlights the variations in indoor temperature impact during the defrosting process for various techniques [87, 90–92]. Each approach equilibrates the efficacy of the defrosting procedure with its effect on interior thermal comfort, with certain ways emphasizing rapid defrosting and others mitigating temperature variations.

13.5.5 Defrosting Models for Air-source Heat Pumps

Defrosting models are mathematical and computational frameworks used to simulate and optimize the process of frost accumulation and removal in air-source heat pumps [94]. These models help in understanding the dynamics of frost formation, predicting defrosting requirements, and designing control strategies to improve system efficiency and performance.

Table 13.9 compares defrosting models, categorized by their model types, focus areas, strengths, and limitations [22, 47, 69]. Defrosting models are crucial in advancing ASHP technology by improving efficiency and reducing operational costs. Frost

Table 13.9 Comparison of defrosting models.

No.	Model type	Focus	Strengths	Limitations
1	Frost formation models	Frost growth dynamics	Accurate frost prediction	Limited to frost accumulation
2	Heat transfer models	Energy exchange during defrosting	Optimizes defrost energy use	Does not predict frost onset
3	Dynamic defrosting models	Time-dependent frost dynamics	Captures transient behaviors	Computationally intensive
4	Integrated control models	Real-time defrost optimization	Combines prediction and adaptation	Requires advanced sensors

formation and heat transfer models are foundational, while dynamic and integrated control models represent state-of-the-art solutions for real-time defrosting optimization. Future research can focus on hybrid models combining predictive and adaptive capabilities for enhanced system performance.

Table 13.10 shows the most commonly used methods to manage frost formation on ASHP units, each with its own set of benefits and drawbacks. Among them, reverse cycle defrosting is widely utilized in air-source heat pumps to manage frost accumulation on the outdoor heat exchanger. Frost formation is a common issue in ASHP units, particularly in cold and humid climates, as the outdoor unit operates at temperatures below the freezing point. This frost can significantly impair the heat transfer efficiency of the system, necessitating effective defrosting methods to maintain optimal performance [7]. The mechanics of reverse cycle defrosting are relatively straightforward. When the heat pump unit detects that frost has built up to a certain level, it initiates the defrost cycle. The reversing valve, a critical component in the heat pump units, changes the direction of the refrigerant flow. As a result, the outdoor coil, which normally acts as an evaporator in heating mode, becomes a condenser. During the defrost cycle, the heat pump unit does not provide heat to the indoor space, as it operates in cooling mode in colder climates [7].

Additionally, the energy used to melt frost does not contribute to heating, which can cause a temporary drop in indoor temperature. This interruption may be noticeable in indoor spaces and can reduce the overall efficiency of the system. Another important factor to consider is the frequency and duration of defrost cycles. Frequent defrosting can lead to increased wear and tear on system components, particularly the reversing valve and compressor. It can also result in higher energy consumption, as the system must work harder to maintain the desired indoor temperature. Therefore, optimizing the timing and duration of defrost cycles is essential to balance effective frost removal with energy efficiency.

Despite these challenges, reverse cycle defrosting remains a popular choice due to its simplicity and effectiveness. Advances in heat pump technology have improved defrost cycle management, including more precise control algorithms and sensors that better detect frost accumulation. These advancements help minimize the impact on indoor comfort and the operating efficiency of heat pump units.

Table 13.10 The most commonly used method to manage frost formation on ASHP units [7].

No.	Defrosting Method	Mechanism	Advantages	Disadvantages
1	Reverse cycle defrosting	Utilizing an indoor coil as the evaporator	Efficient, uses existing components	Temporary heating interruption, indoor temperature drop
2	Hot-gas bypass defrosting	Bypass hot refrigerant gas to the outdoor coil	Quick, minimal indoor temperature impact	Requires additional components, refrigerant loss
3	Electric heating defrosting	Uses electric heaters to melt frost	Simple, effective in severe frost	High energy consumption, operational cost
4	Hot water spray defrosting	Spray hot water onto the outdoor coil	Effective, quick	Required water supply, potential freezing issues
5	Compressor shutdown defrosting	Shuts down the compressor, allowing the coil to warm up	Simple, low-cost	Long defrosting time, significant heating interruption

13.5.6 Summary

In summary, reverse cycle defrosting is a common and effective method used in air-source heat pumps to manage frost accumulation on the outdoor heat exchanger. By temporarily switching to cooling mode and directing hot refrigerant vapor to the outdoor coil, the system can melt the frost and restore heat transfer efficiency. While this method can reduce overall system efficiency during the defrost cycle and temporarily affect indoor comfort, ongoing technological advancements continue to improve its performance and reliability.

13.6 Conclusion and Outlook

The examination of frosting during energy harvesting from the air using a cold plate has unveiled several crucial findings.

1. Frost formation dynamics: Frost initially develops at the base of the fins due to lower temperatures and gradually extends to the tips as they cool down. This progression significantly impacts the efficiency of the heat exchanger over time.
2. Impact on airflow: Frost accumulation has a notable effect on airflow, leading to severe blockage in certain configurations. This obstruction can result in decreased efficiency and increased energy consumption.
3. Comparative performance: Various heat exchanger designs demonstrate different levels of frost blockage. For example, MCHX-A exhibited less severe frost blockage compared to MCHX-B, highlighting the significance of design in addressing frost-related issues. Future research should prioritize the development of advanced anti-frosting technologies, such as surface treatments and coatings that can delay or prevent frost formation. Enhancing defrosting

mechanisms, such as reverse cycle defrosting, has the potential to improve the efficiency and reliability of heat exchangers operating in cold environments.

There is a need to explore new materials with superior thermal properties and frost resistance to facilitate the development of more efficient energy harvesting systems. Customizing heat exchanger designs to suit specific environmental conditions can significantly optimize performance and minimize the impact of frosting in diverse climates. These advancements will play a critical role in improving the performance and sustainability of energy harvesting systems utilizing cold plates.

Nomenclatures

Variable	Description	Unit
t_c	Critical time	min
t_t	Transitional time	min

Abbreviations

ASHP	Air-source heat pump
COP	Coefficient of performance
CO_2	Carbon dioxide
CSD	Compressor shutdown
°C	Degree Celsius
EHD	Electric heating defrosting
FFASHP	Frost-free air-source heat pump unit
FTHX	Fin and tube heat exchanger
GW	Gigawatts
HGBD	Hot-gas bypass defrosting
HVAC	Heating, ventilation, and air conditioning
MCHX	Microchannel heat exchangers
PCMs	Phase-change materials
RCD	Reverse cycle defrosting
RH	Relative humidity
T_iO_2	Titanium dioxide
WDC	Water droplet condensation
WDF	Water droplet freezing
YTD	Year-to-date

References

1 Wei, P.S., Huang, C.C. and Lee, K.W. (2003). Nucleation of bubbles on a solidification front-experiment and analysis. *Metall. Mater. Trans. B.*, 34(3), 321–32.

2 IEA (2023). CO_2 emissions in 2022, IEA, Paris. https://www.iea.org/reports/co2-emissions-in-2022, Licence: CC BY 4.0.

3 IEA (2023). Global heat pump sales continue double-digit growth, IEA, Paris. https://www.iea.org/commentaries/global-heat-pump-sales-continue-double-digit-growth,Licence: CC BY 4.0.

4 Energy Savings Trust on air-to-air heat pumps. https://www.ehpa.org/news-and-resources/publications/european-heat-pump-market-and-statistics-report-2023/

5 IEA (2024). Heat pump sales by country or region, 2019-2023, IEA, Paris. https://www.iea.org/data-and-statistics/charts/heat-pump-sales-by-country-or-region-2019-2023, Licence: CC BY 4.0.

6 Department of Energy. (n.d.) Air source heat pumps. https://www.energy.gov/energysaver/air-source-heat-pumps

7 Mengjie, S., Shiming, D., Chaobin, D., Ning, M. and Zhihua, W. (2018). Review on improvement for air source heat pump units during frosting and defrosting. *Appl. Energy*, 221, 1150–70.

8 IEA (2024). Levelised cost of heating and cooling of residential heat pumps and alternatives in hot summer-cold winter climates in Chinese urban areas, IEA, Paris. https://www.iea.org/data-and-statistics/charts/levelised-cost-of-heating-and-cooling-of-residential-heat-pumps-and-alternatives-in-hot-summer-cold-winter-climates-in-chinese-urban-areas, Licence: CC BY 4.0

9 Ground-Source Heat Pumps: Overview of Market Status, Barriers to Adoption, and Options for Overcoming Barriers. U.S. Department of Energy Efficiency and Renewable Energy Geothermal Technologies Program Navigant Consulting, Inc. February 3, 2009.

10 U.S. Energy Information Administration. (n.d.). Geothermal heat pumps. Retrieved from https://www.eia.gov/energyexplained/geothermal/heat-pumps.php

11 International Energy Agency (IEA). (2023). Heating. IEA, Paris. https://www.iea.org/energy-system/buildings/heating

12 Solar Energy Industries Association (SEIA). Solar Thermal: Energy for a Sustainable Future. Retrieved from https://www.seia.org/initiatives/solar-heating-cooling

13 International Renewable Energy Agency (IRENA) reports on district heating networks: IRENA. Renewable Energy in District Heating and Cooling: A Sector Roadmap. https://www.irena.org/publications

14 Studies on district heating's role in urban decarbonization: IRENA and partner studies on the integration of district heating into sustainable urban infrastructure. Specific academic references may vary by study. https://www.irena.org/publications

15 Biomass Energy Centre's analysis of boiler efficiencies: Forest Research UK, Biomass Boilers and their In-Situ Performance. https://www.forestresearch.gov.uk/

16 Sustainable biomass sourcing policies: UK Department for Business, Energy & Industrial Strategy, "Sustainability Standards for Biomass." https://www.gov.uk/

17 European Commission reports on renewable energy integration: European Commission, "Renewable Energy Progress Report." https://energy.ec.europa.eu/

18 UK Government's guide on electric heating systems: UK Department for Energy Security and Net Zero, Electric Heating Systems Explained. https://www.gov.uk/

19 IEA analysis on gas boiler market trends: International Energy Agency (IEA), The Future of Heat: A Transition to Cleaner Heating. https://www.iea.org/

20 Climate action reports on the phase-out of fossil fuel heating systems: Reports and initiatives from Climate Action Network and relevant IEA publications. https://climatenetwork.org/

21 Wilson, J.H., Munankarmi, P., Less, B.D., Reyna, J.L. and Rothgeb, S. (2024). Heat pumps for all? Distributions of the costs and benefits of residential air source heat pumps in the United States. Elsevier Inc. https://doi.org/10.1016/j.joule.2024.01.022

22 Gu, T., Zhao, Y., Liu, Y. and Wang, D. (2023). A review of condensation frosting mechanisms and promising solutions. *Crystals*, 13, 493. https://doi.org/10.3390/cryst13030493.

23 Chung, Y., Yoo, J.W., Kim, G.T. and Kim, M.S. (2019). Prediction of the frost growth and performance change of air source heat pump system under various frosting conditions. *Appl. Therm. Eng.*, 147, 410–20.

24 Zhang, L., Jiang, Y., Dong, J., Yao, Y. and Deng, S. (2018). An experimental study of frost distribution and growth on finned tube heat exchangers used in air source heat pump units. *Appl. Therm. Eng.*, 132, 38–51.

25 Huang, L.Z., Tian, Y.Y., Song, M.J. and Zhang, L. (2024). Experimental study of relative humidity effect on the edge effect of frosting characteristics on a vertical cold plate surface. *Appl. Therm. Eng.*, 236, 121537.

26 van Belois, B.C. (2024). The effects of frost formation on the performance of a compact residential air-to-water heat pump. Master thesis. Eindhoven University of Technology.

27 Lamb, D. and Verlinde, J. (2011). Physics and Chemistry of Clouds, Cambridge University Press.

28 Gonda, T. (2004). Nucleation of ice crystals in supercooled conditions. *J. Cryst. Growth*, 74(1), 131–9.

29 Libbrecht, K.G. (2005). The physics of ice crystal growth. *Rep. Prog. Phys.*, 68(4), 855–95.

30 Song, M. and Dang, C. (2018). Review on the measurement and calculation of frost characteristics. *Int. J. Heat Mass Transf.*, 124, 586–614.

31 Chaudhary, G. and Li, R. (2014). Freezing of water droplets on solid surfaces: an experimental and numerical study. *Exp. Therm. Fluid Sci.*, 57, 86–93.

32 Hindmarsh, J.P., Russell, A.B. and Chen, X.D. (2003). Experimental and numerical analysis of the temperature transition of a suspended freezing water droplet. *Int. J. Heat Mass Transf.*, 46, 1199–213.

33 Pruppacher, H.R. and Klett, J.D. (1997). Microphysics of Clouds and Precipitation, Kluwer Academic Publishers.

34 Zhang, L., Song, M., Zhan, T., Shen, J. and Dang, C. (2022). Cold plate temperature effect on droplet and frost crystal behaviors at the early condensation frosting stage considering plate edge effect. *Eng. Sci.*, 21(2), 801.

35 Lu, Z., Huang, R. and Woods, J. (2024). A review of modeling approaches for predicting frost growth and defrosting on tube-fin heat exchangers. In *20th International Refrigeration and Air Conditioning Conference at Purdue*, West Lafayette, IN.

36 Janssen, D.D. (2013). Experimental Strategies for Frost Analysis, University of Minnesota.

37 Shahane, S., Shen, Y. and Wang, S. (2021). Numerical Simulations of Frost Growth Using Mixture Model on Surfaces with Different Wettability, University of Illinois at Urbana-Champaign.

38 Xia, Y., Hrnjak, P.S. and Jacobi, A.M. (2006). An Analytical and Empirical Study of Frost Accumulation Effects on Louvered-Fin, Microchannel Heat Exchangers, IDEALS.

39 Pegallapati, A.S. and Ramgopal, M. (2020). Modeling of frosting on fin-and-tube heat exchanger of a domestic refrigerator. In *Advances in Air Conditioning and Refrigeration: Select Proceedings of RAAR 2019* (pp. 235–242). Springer, Singapore.

40 Miettinen, J. (1997). Calculation of solidification-related thermophysical properties for steels. *Metall. Mater. Trans., B, Process Metall. and Mater. Process. Sci.*, 28(2), 281–97, https://doi.org/10.1007/s11663-997-0095-2.

41 Kim, C., Lee, J. and Lee, K.-S. (2017). Numerical modeling of frost growth and densification on a cold plate using frost formation resistance. *Int. J. Heat Mass Transf.*, 115, 1055–63.

42 Zhang, L., Mengjie, S., Shiming, D., Jun, S. and Chaobin, D. (2022). Frosting mechanism and behaviors on surfaces with simple geometries: a state-of-the-art literature review. *Appl. Therm. Eng.*, 215, 118984.

43 Mittal, K.L. and Choi, C.H. (eds.) (2020). Ice Adhesion: Mechanism, Measurement, and Mitigation. John Wiley & Sons.

44 Zhao, Z., Li, X., Li, W., Liu, M., Hu, Z., Jiang, T., Wang, H. and Zhao, Y. (2024). Progress in mechanism design of functional composites for anti-ice/deicing materials. *Surf. Sci. Tech.*, 2, 2.

45 Huang, S., Yu, H., Zhang, M., Qu, H., Wang, L., Zhang, C., Yuan, Y. and Zhang, X. (2023). Advances, challenges and outlooks in frost-free air source heat pumps: a comprehensive review from materials, components to systems. *Appl. Therm. Eng.*, 234, 121163.

46 Frosting and Defrosting on Air Source Heat Pump Evaporators, DiVA portal, CAMILLA SANDSTRÖM.

47 Song, M., Dang, C., Higashi, T. and Hihara, E. (2020). Review of experimental data associated with the solidification characteristics of water droplets on a cold plate surface at the early frosting stage. *Energy Build.*, 223, 110103.

48 Lenic, K., Trp, A. and Frankovic, B. (2009). Transient two-dimensional model of frost formation on a fin-and-tube heat exchanger. *Int. J. Heat Mass Transf.*, 52 (1–2), 22–32.

49 Hermes, C.J.L., Piucco, R.O., Barbosa, J.R., and Melo, C. (2019). A study of frost growth and densification on flat surfaces. *Exp. Therm. Fluid Sci.*, 33(2), 371–9.

50 Ehsan, A., Oleg, I., Stefano, L. and Carlo, I. (2023). Numerical simulation of frost formation on a plate-fin evaporator. In *Proceedings of the 3rd World Congress on Momentum, Heat and Mass Transfer (MHMT'18), Budapest, Hungary–April 12–14, 2018, ICMFHT 125*, https://doi.org/10.11159/icmfht18.125

51 Qiu, G., Li, S., Shi, Y., Fang, X. and Cai, W. (2024). Development and validation of a numerical model for frost growth based on nucleation theory. *Int. J. Heat Mass Transf.*, 221, 125137.

52 Cui, Q., Liu, J., Shi, W., Wang, Y. and Yuan, T. (2024). Research progress of evaporator surface frost suppression technology. *Acad. J. Sci. Technol.*, 9(2), 288–95.

53 Park, J.S., Kim, D.R. and Lee, K.S. (2016). Frosting behaviors and thermal performance of louvered fins with unequal louver pitch. *Int. J. Heat Mass Transf.*, 95, 499–505.

54 Jones, M.D. and Smith, A.K. (2015). Heat transfer in plain finned tube systems: theory and applications. *Int. J. Therm. Sci.*, 34(5), 451–65.

55 Liu, W. and Zhao, Y. (2017). Enhancing heat transfer through spiral fin design: effects on frost formation and density. *Int. J. Heat Mass Transf.*, 42(2), 101–12.

56 Thompson, E. and Green, R.P. (2018). Directional airflow and frost patterns in longitudinal finned tubes. *Appl. Therm. Eng.*, 56(7), 620–35.

57 Choi, S.J. and Park, J. (2019). Perforated finned tubes: impact of perforations on turbulence, frost control, and pressure drop. *J. Mech. Eng. Sci.*, 49(9), 870–83.

58 Hernandez, P. and Lee, T.H. (2020). Analysis of louvered fin designs for optimized frost control in heat exchangers. *Int. J. Refrig.*, 78(12), 745–59.

59 Zhao, H. and Cheng, Y. (2021). Wavy fin patterns: enhancing heat transfer and frost accumulation in HVAC systems. *J. Therm. Sci. Appl.*, 61(4), 310–25.

60 Wang, D. and Kim, J. (2022). Effect of offset strip fins on turbulence and frost growth in heat exchangers. *Trans. ASME*, 67(3), 520–32.

61 Xia, Y. and Jacobi, A.M. (2010). A model for predicting the thermal-hydraulic performance of louvered-fin, flat-tube heat exchangers under frosting conditions. *Int. J. Refrig.*, 33(2), 321–33.

62 Jones, K. and Steven M. (2013). Characteristics of Frost Growth On Micro-Channel Heat Exchangers.

63 Song, S., Bullard, C.W. and Hrnjak, P.S. (2002). Frost deposition and refrigerant distribution in microchannel heat exchangers. *ASHRAE Trans.*, 108(2), 944–53.

64 Xiong, T., Liu, G. and Yan, G. (2023). Performance improvement of microchannel heat exchangers with modified louver fins under frosting conditions. *Appl. Sci.*, 13, 6378. https://doi.org/10.3390/app13116378.

65 Li, Z., Qiao, H. and Aute, V. (2019). Tube-fin heat exchanger circuitry optimization for improved performance under frosting conditions. In *Proceedings of the 13th International Modelica Conference, Regensburg, Germany, March 4–6, 2019*, https://doi.org/10.3384/ecp19157259

66 Ye, H.Y. and Lee, K.S. (2013). Performance prediction of a fin-and-tube heat exchanger considering air-flow reduction due to the frost accumulation. *Int J Heat Mass Tran.*, 67, 225–233.

67 Rauscher, S. (2020). Heat pumps in winter: evaporator frosting. *TLK Energy* (27 April).

68 Fricke, B. (2021). A Critical Literature Review of Defrost Technologies for Heat Pumps and Refrigeration Systems Kashif Nawaz ORNL, United States of America, nawazk@ornl.gov.

69 Zhang, Y., Zhang, G., Zhang, A., Jin, Y., Ru, R. and Tian, M. (2018). Frosting Phenomenon and frost-free technology of outdoor air heat exchanger for an air source heat pump system in China: an analysis and review. *Energies*, 11(10), 2642. https://doi.org/10.3390/en11102642.

70 Wang, F., Wang, Z., Zheng, Y., Lin, Z., Hao, P., Huan, C. and Wang, T. (2015). Performance investigation of a novel frost-free air-source heat pump water heater combined with energy storage and dehumidification. *Appl. Energy*, 139, 212–19.

71 Wang, F., Liang, C. and Zhang, X. (2018). Research of anti-frosting technology in refrigeration and air conditioning fields: a review. *Renew. Sustain. Energy Rev.*, 81, 707–22.

72 Narayanan, R., Sethuvenkatraman, S. and Pippia, R. (2022). Energy and comfort evaluation of fresh air-based hybrid cooling system in hot and humid climates. *Energies*, 15(20), 7537.

73 Li, K., Xia, D., Luo, S., Zhao, Y., Tu, R., Zhou, X., Zhang, H. and Su, L. (2022). An experimental investigation on the frosting and defrosting process of an outdoor heat exchanger in an air conditioning heat pump system for electric vehicles. *Appl. Therm. Eng.*, 201, 117766.

74 Lee, S.H., Kim, J., Kim, D.W., Lee, J. and Joo, J.B. (2026). FeSO4–Palmitic acid chemical bath deposition for practical superhydrophobic coatings with frost suppression performance in heat pump systems. Available at SSRN 6718340.

75 Wang, F., Zhou, Y., Yang, W., Ni, M., Zhang, X. and Liang, C. (2020). Anti-frosting performance of sprayable superhydrophobic coating suitable for outdoor coil of air source heat pump. *Appl. Therm. Eng.*, 169, 114967. https://doi.org/10.1016/j.applthermaleng.2020.114967.

76 Chen, J.S., Chao, S.Y. and Chen, C.C. (2023). The effect of hydrophilic surface coating of fins on the performance of fin-and-tube heat exchangers. *Appl. Sci.*, 13(18), 10450.

77 Gu, Y., He, G., Li, S., Ding, W., Li, H. and Duan, J. (2022). Study on frost-suppression characteristics of superhydrophobic aluminium surface heat exchanger applied in air source heat pump. *Sustainability*, 14(4), 1954.

78 Li, S., Tan, Y., Li, S. and Xiao, A. (2024). A multifunctional superhydrophobic phase-change photothermal coating for concrete anti-icing. Polymer International.

79 Wang, C., Jia, X., Zhao, P., Wang, L. and Shi, L. (2024). A review of the application of solid-liquid phase change materials in water heaters. *Therm. Sci.*, 28(1 Part B), 409–24.

80 Saeidiharzand, S. (2023). Anti-Icing Functional Biphilic Surfaces And Functionalized Multiscale Metal Organic Framework-Based Coatings (Doctoral dissertation). Doctoral dissertation, Sabancı University.

81 Oberli, L., Caruso, D., Hall, C., Fabretto, M. and Evans, D. (2014). Condensation and freezing of droplets on superhydrophobic surfaces. *Adv. Colloid Interface Sci.*, 210, 47–57

82 Chang, S.N., Zhou, S. and Zheng, H.K. (2018). Freezing characteristics of water drop on different wettability surface. In *5th International Conference on Experimental Fluid Mechanics - ICEFM 2018-38*, July 2–4.

83 Lee, H., Yang, J.B. and Kim, D.R. (2024). Anti-frosting characteristics of superhydrophobic-hydrophilic wettability switchable surfaces. *Int. J. Heat Mass Transf.*, 221, 125035. https://doi.org/10.1016/j.ijheatmasstransfer.2023.125035.

84 Jhee, S., Lee, K.-S. and Kim, W.-S. (2002). Effect of surface treatments on the frosting/defrosting behavior of a fin-tube heat exchanger. *Int. J. Refrig.*, 25(8), 1047–53. http://doi.org/10.1016/S0140-7007(02)00008-7.

85 Venkateswaran, N., Subbaiyan, N., Punniyakotti Varadharajan, G., Vellingiri, S., Asary, A.R., Giri, J.M. and Petchimuthu, P. (2023). A comprehensive review of the energy efficiency on nano coated fin and tube condenser. *Environ. Qual. Manag.*, 33(1), 293–301.

86 Chen, T., Chen, Y., Xu, J., Ren, L., Cong, Q., Jin, J. and Choy, K.L. (2022). Influence of wing scales of *Catocala electa Borkhason* on the frosting process. *J. Taibah Univ. Sci.,* 16(1), 240–49.

87 Song, M., Dong, J., Wu, C., Jiang, Y. and Qu, M. (2017). Improving the frosting and defrosting performance of air source heat pump units: review and outlook, *HKIE Trans.*, 24(2), 88–98. https://doi.org/10.1080/1023697X.2017.1313134.

88 Liu, D., Zhao, F.Y. and Tang, G.F. (2007). Frosting of heat pump with heat recovery facility. *Renew. Energy (in Chinese)*, 32(6), 1228–42.

89 Song, M.J. and Mao, N. (2020). Experimental study on the melted frost influence on the metal energy storage during an air source heat pump defrosting. *Energy Build.*, 214, 109809.

90 Minglu, Q., Liang, X., Deng, S. and Yiqiang, J. (2010). Improved indoor thermal comfort during defrost with a novel reverse-cycle defrosting method for air source heat pumps. *Build. Environ.,* 45(11), 2354–61.

91 Chung, Y., Na, S.I., Yoo, J.W. and Kim, M.S. (2021). A determination method of defrosting start time with frost accumulation amount tracking in air source heat pump systems. *Appl. Therm. Eng.*, 184, 116405.

92 Wu, C., Wang, Z., Li, X., Xu, Z., Xu, C., Yang, Y., Zhao, W. and Wang, Y. (2023). Field experimental research on retarding frosting and improving defrosting performance of air source heat pumps through variable frequency technology. *J. Build. Eng.*, 64, 105628.

93 Rahman, H.U., Mehmood, M.U. and Lazoglu, I. (2024). A centralized frost detection and estimation scheme for internet-connected domestic refrigerators. *Int. J. Refrig.*, 169, 194–203.

94 Song, M.J., Deng, S.M. and Xia, L. (2014). A semi-empirical modeling study on the defrosting performance for an air source heat pump unit with local drainage of melted frost from its three-circuit outdoor coil. *Appl. Energy*, 13631, 537–47.

14

Magnetostrictive Materials and Composites for Energy-harvesting Applications

Hiroki Kurita

Graduate School of Environmental Studies, Tohoku University, Sendai, Miyagi, Japan

14.1 Introduction

The Internet of Things is revolutionizing the entire world. Sensors are installed in facilities and equipment, and analog processes are visualized as digital data, which is then stored and analyzed on the cloud via the Internet. It enables improvements in quality, efficiency, and productivity, as well as the creation of new values and services, and is about to revolutionize how business is conducted in various fields. The number of sensors is expected to increase at an explosive rate and reach one trillion by 2030 [1]. The power required to drive sensors and communicate data is several microwatts. However, using batteries for a trillion sensors poses an extremely large social problem regarding the environment, resources, and cost. Energy harvesting, which recovers power from unused energy (vibration, heat, light, radio waves, etc.) that exists widely in the natural environment, has attracted attention and is expected to be used as power for sensor drive and data communication [2].

Piezoelectric ceramics, promising energy harvesting materials, are brittle insulators, so even if the voltage obtained from vibration is large, the current is small, and large power cannot be obtained. Lead zirconate titanate (PZT) piezoelectric ceramics, which enable vibration energy harvesting, contain hazardous substances (lead), so there is a demand for the development of lead-free piezoelectric materials due to environmental concerns [3]. On the other hand, Tb–Dy–Fe alloys (Terfenol-D) and Fe–Ga alloys (Galfenol), which exhibit super magnetostrictive properties, have attracted attention as energy harvesting materials and are being actively studied [4, 5]. In addition, Fe–Co alloy thin films [6] have been developed by sputtering, and Fe–Co alloys [7] by forging and cold working, and the effects of heat treatment and other factors on magnetostrictive properties have been investigated. Furthermore, Fe–Ga/Fe–Co alloy multilayer materials [8], AlN/Fe–Co multilayer materials [9], Tb–Dy–Fe/Fe–Co alloy multilayer materials [10], and Fe–Co alloy coatings [11] have been developed for sensor applications. Research has also been conducted on stress-induced magnetic wall movement of Fe–Co alloy microwires [12].

Composite materials combine two or more materials according to the purpose of use to provide properties that cannot be obtained alone and are used in various fields. Recently, research and development of multifunctional composites, in which composites with lightweight and high strength and stiffness are combined with completely new functions (e.g., sensors, actuators, energy harvesting and storage, self-healing, and morphing) to achieve multiple functions simultaneously, has been promoted. Research trends are evolving toward developing new applications and market expansion for composite materials. This chapter addresses the fundamentals of energy harvesting for functional composites made of Fe–Co alloys and provides an overview of the evaluation of reverse magnetostrictive properties. Recent research on vibration and impact energy-harvesting properties is also mentioned.

14.2 Magnetic Field-dynamic Field Mutual Interference and Basic Equation

In a Cartesian coordinate system$(\mathrm{O} - x_1x_2x_3)$, consider an elongated magnetostrictive round bar of length (L) and cross-sectional area (A). The easy axis of magnetization is in the x_3 direction, coinciding with the length direction. Suppose the length of the round bar is extremely large compared to its diameter. In that case, the problem is reduced to one dimension, and the following equations describe the constitutive behavior of the magnetostrictive material [13].

$$\varepsilon_{33} = s_{33}\sigma_{33} + (d_{33} + m_{33}H_3)\,H_3 \tag{14.1}$$

$$B_3 = (d_{33} + m_{33}H_3)\,\sigma_{33} + \mu_{33}H_3 \tag{14.2}$$

where σ_{33} and ε_{33} are components of the stress tensor and strain tensor, respectively. B_3 and H_3 are components of the magnetic flux density vector and magnetic field strength vector, respectively. s_{33} is the elastic compliance coefficient. d_{33} is the magnetic field constant. μ_{33} is the magnetic permeability. m_{33} is the second-order magnetoelastic constants, which depend on the stress [14, 15].

Considering the case where a uniform compressive load $(\sigma_0\,(t)\,A)$, which depends on time (t), acts on a magnetostrictive round bar in the direction of the easy axis of magnetization, that is, the x_3 direction, compressive stress $(\sigma_{33} = \sigma_0(t))$ is generated in the round bar, resulting in a change in the magnetic flux density (B_3) inside the round bar, according to Eq. (14.2). The voltage (V_{out}) generated in the magnetostrictive round bar through the coil with n turns per unit length can be obtained from Faraday's law as follows.

$$V_{\mathrm{out}} = -\mathrm{nLA}\frac{\mathrm{d}B_3}{\mathrm{d}t} \tag{14.3}$$

Substituting Eq. (14.2) into Eq. (14.3), the output voltage depends on the number of coil turns (n), the volume of magnetostrictive material (LA), the piezomagnetic constant (d_{33}) and the load speed $(\mathrm{d}\sigma_{33}/\mathrm{d}t)$.

14.3 Fe–Co Alloy Fiber Integrated Composites

Fe–Co alloys have excellent workability and strength, good magnetization and magnetostriction properties, and can be embedded in resin as wire. This chapter outlines the characteristics of Fe–Co alloy fiber composites and recent results elucidating their inverse magnetostrictive properties. Also, it shows an example of impact energy-harvesting experiments applied to the heel of pumps.

Magnetostrictive fiber composites have been designed and developed by exploiting the property of magnetic flux to penetrate easily in the fiber's length direction and devising a manufacturing process for embedding in epoxy resin. Figure 14.1 shows various magnetostrictive materials.

Galfenol bulk and Fe–Co alloy wire-embedded epoxy resin matrix composite have almost the same volume and weight (about 12.8 cm^3 and 2.14 g, respectively), respectively (0.625 g for the wire alone), lighter than the bulk material and can be integrated into a structure, allowing for greater design flexibility. It also offers a high degree of freedom in design because it can be molded into the structure. When a coil is placed outside this composite material, the compressive stress ($\sigma_{33} = \sigma_0 (t)$) changes the magnetic flux density (B_3) and generates a voltage (V_{out}). The output voltage per unit volume increases with increasing load speed and varies depending on the fiber diameter [13], tensile residual stress [16], initial bias magnetic field [17], and heat treatment temperature [17].

Epoxy resin composites have been fabricated using woven materials of Fe–Co alloy wires and polyester fibers, and changes in magnetic flux density under compression and three-point bending loading have been measured [18]. Since Fe–Co alloy wires have high magnetic anisotropy and a longitudinal elastic modulus significantly different from that of epoxy resin, the composite material can be used when an external force is applied to the composite material. A large stress is generated only in the Fe–Co alloy wire [13]. Therefore, the apparent inverse magnetostrictive

Figure 14.1 (a) Terfenol-D bulk, (b) Galfenol bulk, (c) 0.2 mm diameter Fe–Co alloy wire, (d) Fe–Co alloy wire integrated polyester fabric, (e) Fe–Co alloy wire-embedded epoxy resin matrix composite, and (f) Fe–Co alloy wire embedded carbon fiber reinforced polymer.

properties of the Fe–Co alloy wire composite are comparable to those of Terfenol-D. The reverse magnetostrictive properties of CFRP laminates embedded with Fe–Co alloy wires have also been investigated similarly [19]. Furthermore, Fe–Co alloy wire-embedded aluminum (Al) alloy composites (see Figure 14.2) have been fabricated, and a power output of approximately 5.5 mW has been obtained by compressive loading [20].

A prototype Fe–Co alloy wire-embedded shoe for women has been fabricated, and the output power due to foot stomping during walking exercises has been evaluated [21]. Figure 14.3a shows a Fe–Co alloy wire-embedded epoxy resin energy-harvesting device (10 mm in diameter and 35 mm in length) with a coil. As shown in Figure 14.3b, it was mounted on the heel of a shoe for women, and light-emitting diode (LED) lighting was confirmed at an impact load of 100 to 200 N. As shown in Figure 14.4, there was variation in the output power during walking. Although variations were observed, more than 100 μW of power was obtained from the impact during walking. Although the results are not shown, the output power varies depending on the habit of walking, not on the human's body weight.

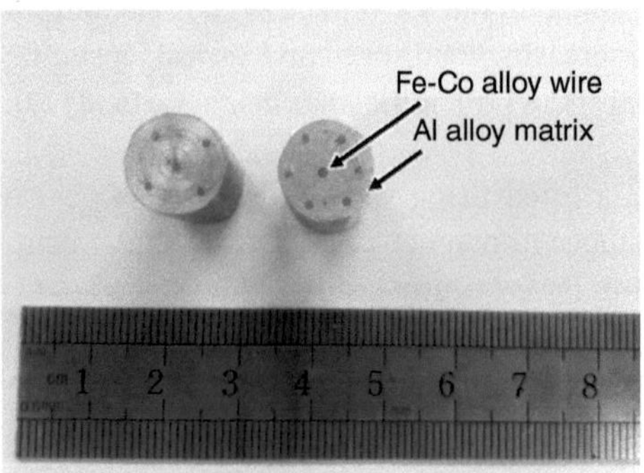

Figure 14.2 Fe–Co alloy wire-embedded Al alloy composites. *Source*: [20] / with permission of Elsevier.

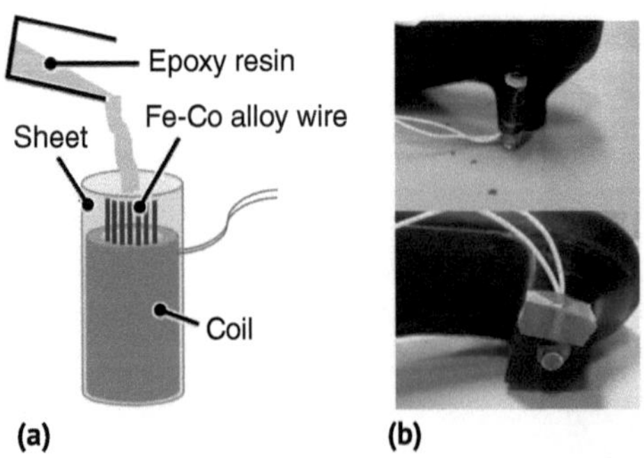

Figure 14.3 (a) Illustration of Fe–Co alloy wire-embedded epoxy resin energy-harvesting device and (b) embedding in a shoe for women. *Source*: [21] / MDPI / CC BY 4.0.

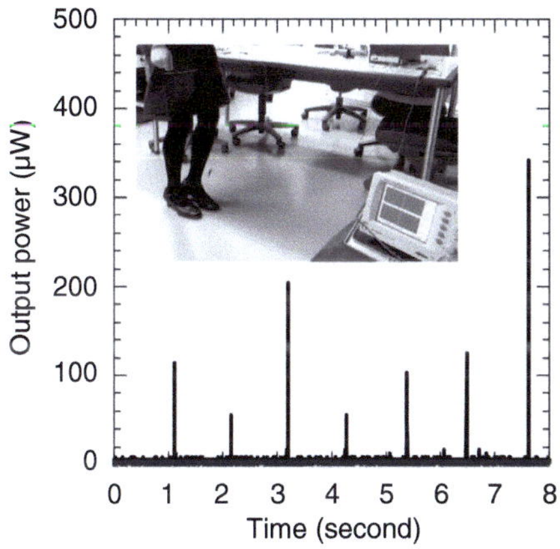

Figure 14.4 Impact output power of the embedded Fe–Co alloy wire-embedded shoe for women.

14.4 Fe–Co Alloy Clad Plate

When a magnetostrictive plate is subjected to a bending load and deformed so that the lower side is convex, compressive stress is generated on the upper side of the plate, and tensile stress on the lower side. The boundary between the area subjected to compressive stress and the area subjected to tensile stress is called the neutral plane, and the positive and negative values of stress (σ_{33}) differ between the area above and below the neutral plane. As a result, the magnetic flux density (B_3) in Eq. (14.2), is not uniform, hindering the magnetostrictive plate's high output power. Therefore, to increase the output power by bending vibration as much as possible, some ingenuity is required to design magnetostrictive materials. Here, the vibration energy-harvesting characteristics of Fe–Co alloy clad plates and the recent research results are described.

A Fe–Co alloy clad plate fabricated by thermal diffusion bonding a cold-rolled plate of Fe or Ni with a cold-rolled plate of Fe–Co alloy produces a greater vibration power output when subjected to bending vibration than a Fe–Co alloy plate alone [22]. Figure 14.5 shows a cantilevered beam of Fe–Co/Ni clad plate and Fe–Co alloy plate subjected to a concentrated load at the free end, with a length of 65 mm, a thickness of 1 mm, and a width of 5 mm. A finite element analysis with 0.05 mm forced deflection applied at the free end and considering the mutual interference of the magnetic fields shows that the vertical stress distribution in the length direction in both materials is about the same, as shown in Figure 14.6a. The difference in the magnetic flux density distribution between the two materials can be understood as the reason why the Fe–Co/Ni clad plate has a larger vibration energy-harvesting characteristic despite the equal stress generated in the materials. As shown in Figure 14.6b, the magnetic flux density of the Fe–Co/Ni clad plate has the same direction above and below the neutral plane, while the direction is opposite above and below the neutral plane for the Fe–Co alloy plate. In other words, for the Fe–Co/Ni clad plate, the magnetic fluxes above and below the neutral plane change in opposite directions due to bending

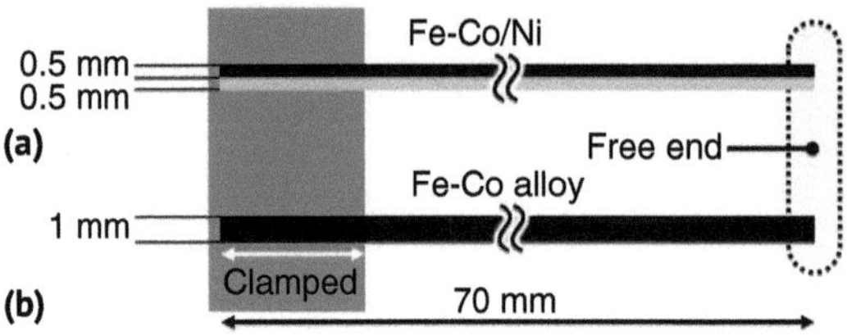

Figure 14.5 Schematic illustration of cantilever beams of (a) Fe–Co/Ni clad plate and (b) Fe–Co alloy plate.

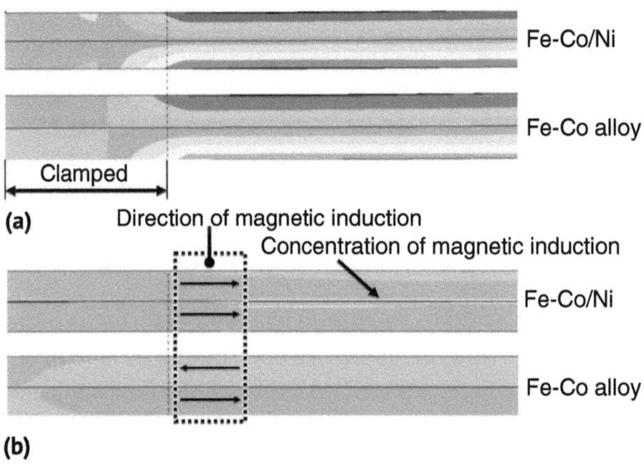

Figure 14.6 (a) Stress distribution and (b) flux density distribution of a cantilever under static bending.

vibration, resulting in a smaller dB_3/dt in Eq. (14.3). On the other hand, in the case of a clad steel plate with Fe and Fe–Co alloy, the magnetic flux changes only for Fe–Co alloy due to bending vibration, and in the case of a clad plate with Ni, which has a negative coercive constant, instead of Fe, the magnetic flux of Ni and Fe–Co alloy changes in the same direction. In the case of a clad plate using Ni with a negative magnetic constant instead of Fe, the magnetic fluxes of Ni and Fe–Co alloy change in the same direction, increasing the output power.

Furthermore, although stress concentration is not observed in both materials, magnetic flux density concentration occurs at the interface of the Fe–Co/Ni clad plate. This flux density concentration is another reason for the increase in output power. It seems to be due to the continuity of the normal direction component of the magnetic flux density vector and the tangential direction component of the magnetic field strength vector at the interface between Fe–Co alloy and Ni.

Figure 14.7 shows an example of the results of a bending vibration energy-harvesting experiment using a vibration energy-harvesting device [23]. The number of coil turns is 64 000. Figure 14.7a and b show the relationship between the power density output by bending vibration and the load resistance and acceleration of a Fe–Co/Ni clad plate and a PZT-based piezoelectric ceramic plate (commercial product), respectively, at a vibration frequency of 50 Hz. The Fe–Co/Ni clad plate can

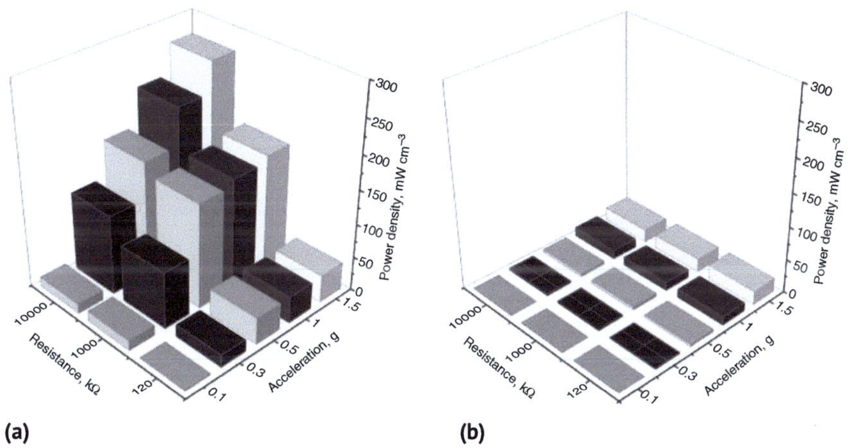

Figure 14.7 Relationship between resistance, acceleration, and power density of (a) Fe–Co/Ni clad plate and (b) PZT-based piezoelectric ceramic plate.

generate power from micro-vibrations, which cannot be achieved with PZT-based piezoelectric ceramic plates and is expected to have various applications.

14.5 Nitriding

Surface treatment methods are widely used to improve the performance and serviceability of engineering components, and nitriding is currently one of the most versatile surface treatment methods. These methods allow fine-tuning surface quality-dependent properties such as fatigue, wear, and corrosion resistance. Nakajima et al. nitrided the rolled Fe–Co/Ni clad plates at 580 °C for 30 minutes and evaluated the effects of nitriding on their magnetostriction, magnetic flux density change, and output voltage by three-point bending tests under a magnetic field [24].

Figure 14.8 shows the magnetostriction curve of the nitrided and as-rolled Fe–Co/Ni clad plates. The magnetostriction rate (initial slope) of the Fe–Co/Ni clad plates was larger than that of the Fe–Co alloy. The nitrided Fe–Co/Ni clad plates showed higher values than the as-rolled ones. This is attributed to the alignment of the crystallographic orientation in the <001> direction by nitriding treatment [24].

Figure 14.9 shows the relationship between the magnetic flux density change and load for nitrided and as-rolled Fe–Co/Ni clad plates obtained by three-point bending. The magnetic flux density change increased almost linearly with load. This is due to the magnetic wall migration in the Ni layer, which is caused by tensile stress. The slope of the initial magnetic flux density change (piezomagnetic constant) of the nitrided Fe–Co/Ni clad plates was larger than that of the as-rolled ones. This result suggests that the nitrided Fe–Co/Ni clad plate has superior energy-harvesting performance. The output voltage of the nitrided Fe–Co/Ni clad plates (110 mV) was higher than that of the as-rolled ones (85 mV), as shown in Figure 14.10. Hence, nitriding is effective in improving the magnetostrictive properties of Fe–Co alloys, that is, sensor and energy-harvesting performance.

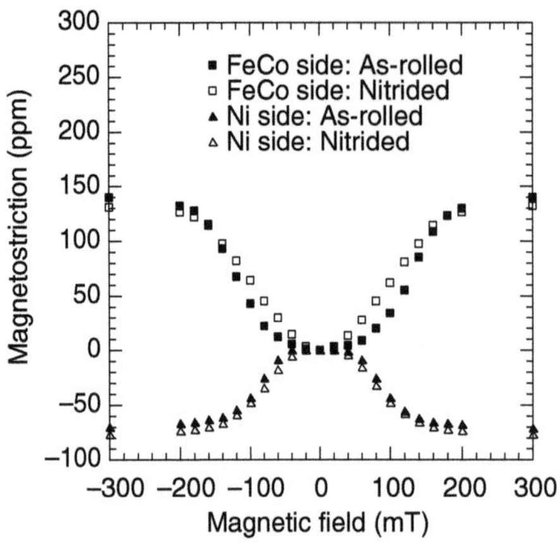

Figure 14.8 Magnetostriction curves for the Fe–Co/Ni clad plate.

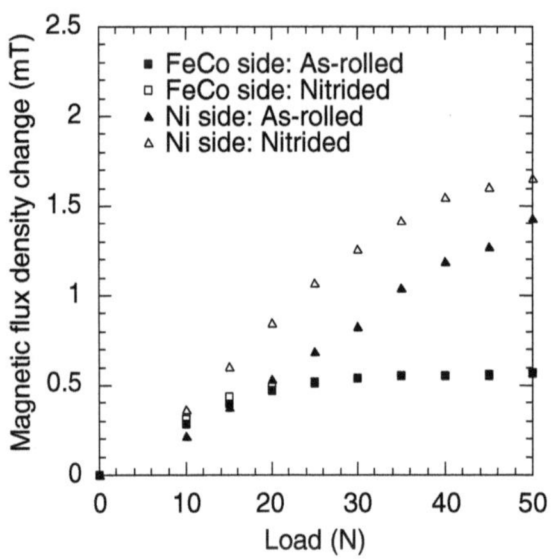

Figure 14.9 Magnetic flux density change for the Fe–Co/Ni clad plate.

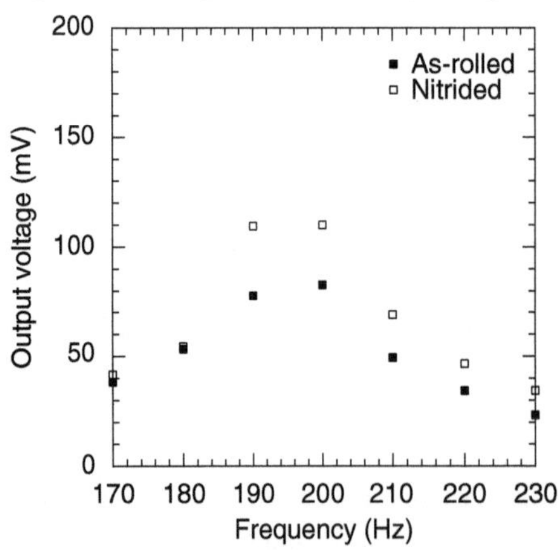

Figure 14.10 Output voltage for the Fe–Co/Ni clad plate.

14.6 Notch Introduction

Yang et al. introduced notches of 0, 0.5, and 1.0 mm into Fe–Co alloys to investigate the change in magnetic flux density under compressive stress loading and output power when subjected to impact [25]. Figure 14.11 shows the specimen dimensions. Figures 14.12 and 14.13 show a schematic diagram and a view of the experiment, respectively. The bias magnetic field (B_0) was 350 mT. The load resistance was set to 10 kΩ. The compressive stress was varied from 0 to 100 MPa. The nominal stress (σ_0) was calculated by dividing the static compressive load (P_0) by the cross-sectional area of the unnotched specimen. The cross-section of the outer magnet was aligned with that of the specimen. A Tesla meter was used to measure the magnetic flux density change caused by the applied compressive load (see Figure 14.12a for probe location). The impact test was then performed, as shown in Figure 14.12b. The specimen was enclosed in a coil with an inductance of 12.6 mH and a length of 15 mm. The load P was increased from 0 to the maximum load (P_{max}). Real-time output power and applied stress were measured simultaneously using a general-purpose oscilloscope.

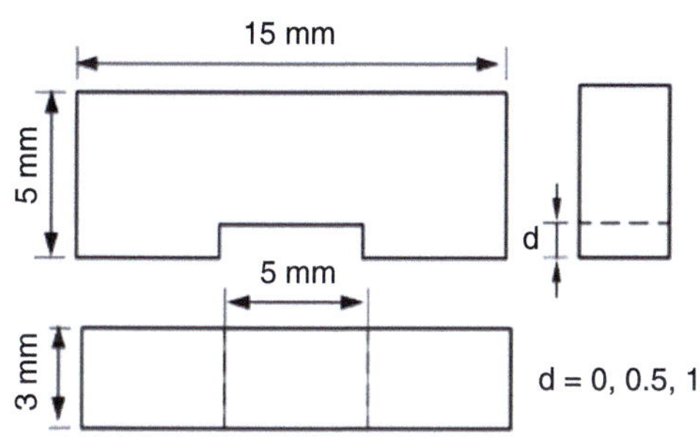

Figure 14.11 Dimension of the notch-introduced specimens.

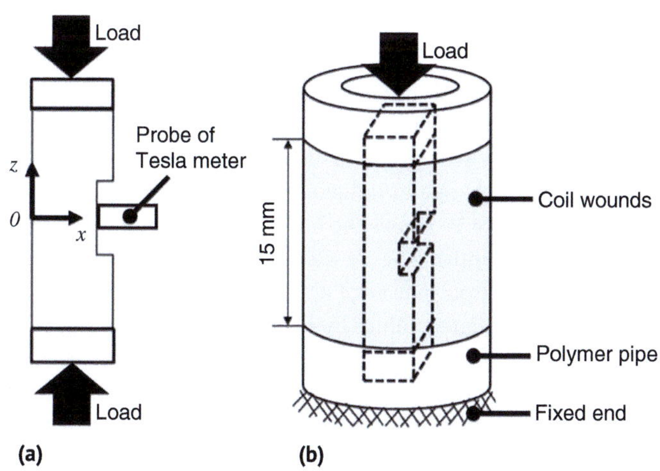

Figure 14.12 Schematic diagram of (a) compressive and (b) before-impact tests.

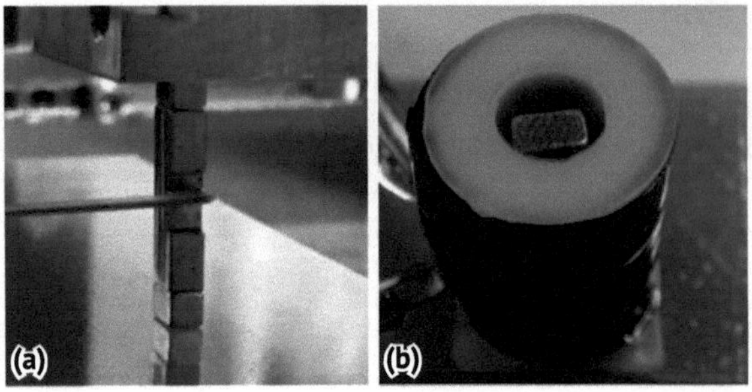

Figure 14.13 View of (a) compressive test and (b) before-impact test.

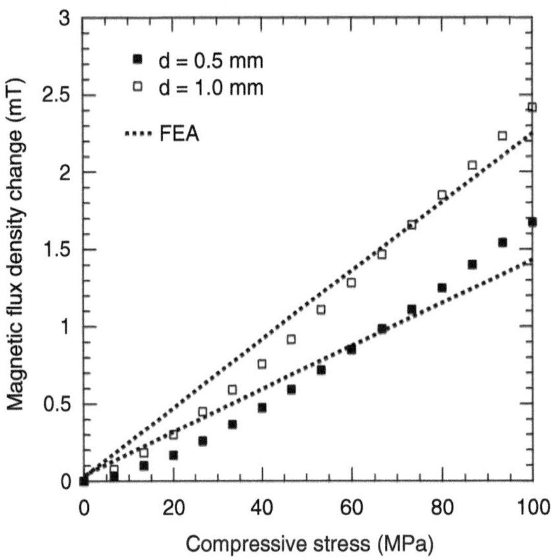

Figure 14.14 Magnetic flux density change during compressive stress loading on Fe–Co alloys with different notch depths.

Figure 14.14 shows the magnetic flux density change during compressive stress loading on Fe–Co alloys with different notch depths. The magnetic flux density changes for these two Fe–Co alloys with different notches were proportional to the applied compressive stress. The change in induced flux increased dramatically with increasing notch depth. Figure 14.14 shows the output power when impacting Fe–Co alloys with different notch depths. The Fe–Co alloy with the 1 mm notch introduced showed an increase in output power of approximately 400% compared to that without a notch, and the calculated values by FEA were in close agreement with the experimental values.

Similarly, the output power was proportional to the notch depth in the impact experiments. Figure 14.15 shows the stress distribution around the notch calculated by FEA. The stress distribution is concentrated around the notch. This stress concentration greatly improved the material's local reverse magnetostrictive properties

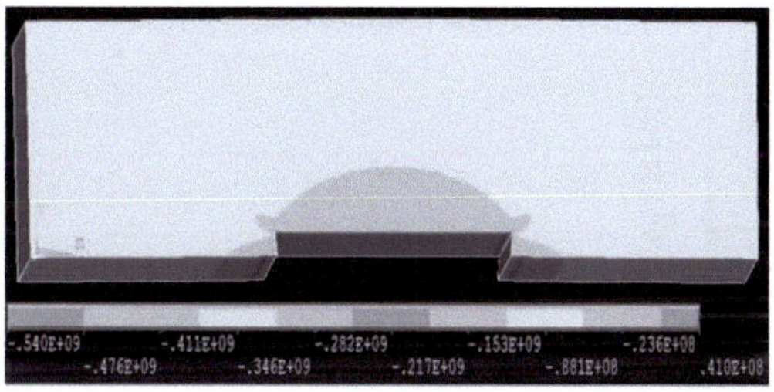

Figure 14.15 Stress distribution around the notch calculated by FEA.

and energy-harvesting performance. It should be noted, however, that the introduction of notches may significantly reduce the strength and fatigue life of the material.

14.7 Directed Energy Deposition

Fe–Co alloy cubes ($1 \times 1 \times 1$ cm^3) were built on a stainless-steel plate by directed energy deposition (DED) under a controlled atmosphere ($O_2 < 10$ ppm, $H_2O < 150$ ppm) [26]. Building conditions were a hatch space of 0.56 mm, a scanning speed of 1000 mm/min, and a layer thickness of 0.2 mm. Laser power was set to 200, 250, or 300 W, corresponding to an energy density of 107.1, 133.9, or 160.7 J/mm^3, respectively. The scanning and stacking directions were y- and z-directions, respectively.

The magnetostrictive properties of the DEDed Fe–Co alloy were measured during the vibrations that were applied perpendicular to the direction of the magnetic field using a biaxial strain gauge. Figure 14.16 shows the magnetostriction versus magnetic field curve of the DEDed Fe–Co alloy and the hot-rolled one. The slope represents the piezomagnetic constant, a parameter directly related to the energy-harvesting performance. The DEDed Fe–Co alloy exhibited magnetostriction three times greater than the conventional hot-rolled Fe–Co alloy. Regardless of the direction of the applied magnetic field, the initial slope of the Fe–Co alloy became smaller with increasing laser power. However, the Fe–Co alloy showed a large tilt (i.e., piezomagnetic constant) with increasing laser power.

Figure 14.17 shows the kernel average misorientation (KAM; left half) and inverse pole figure (IPF; right half) maps obtained from the electron backscatter diffraction of the DEDed Fe–Co alloy. The KAM and IPF maps show that due to the large hatch space between the two subsequent layers at the interface, the voids are preferentially in the y-direction (i.e., scanning direction). The alignment indicates that the deformation under the magnetic field in the x-direction was smaller, as indicated by the presence of voids in the y-direction. Thus, the piezomagnetic constant under the magnetic field in the x-direction was larger than in the y-direction. These voids contribute to the magnetostriction in the x-direction. The degree of magnetostriction in the x-direction was like that of the hot-rolled Fe–Co alloy. KAM maps showed strong distortion near grain boundaries in the y-z plane.

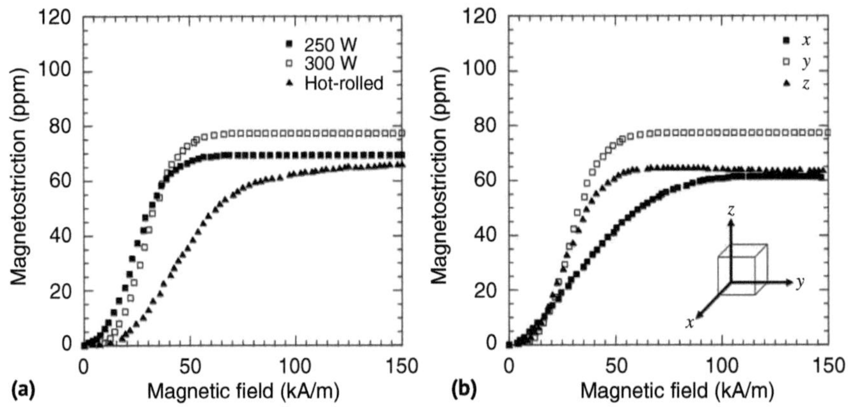

Figure 14.16 Magnetostriction versus magnetic field curve of (a) DEDed Fe–Co alloy in y-direction and hot-rolled one and (b) DEDed Fe–Co alloy built by a laser power of 300 W in each direction.

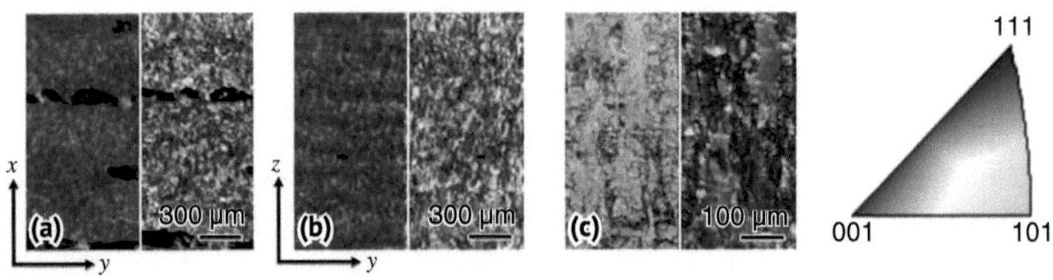

Figure 14.17 KAM (left half) and IPF (right half) maps obtained from the electron backscatter diffraction of (a) $x - y$ plane, (b) $y - z$ plane of the DEDed Fe–Co alloy, and (c) hot-rolled Fe–Co alloy. The black part in (a) $x - y$ plane is void. *Source*: [26] / MDPI / CC BY 4.0.

Consequently, DED significantly increases the freedom in modeling Fe–Co alloys and improves the magnetostrictive properties by affecting the internal microstructure through laser scanning. However, further experimentation is essential to accurately understand the building conditions and Fe–Co alloys.

14.8 Laser Powder Bed Fusion Process (LPBF)

Regarding the results of the notch introduction to the Fe–Co alloy plate, the Fe–Co alloy plates ($70 \times 5 \times 1.6$ cm^3) with the fully dense and honeycomb structures were manufactured by a laser powder bed fusion process (LPBF) with a laser power of 250 W, a scanning speed of 1000 mm/s, and a hatch space of 0.08 mm [27]. The wall thickness of the honeycomb structure was controlled to 250 μm, and the cell width to 2.5 mm. The power density was measured as the output power divided by the volume of the alloy. Figure 14.18 shows a schematic diagram of the vibration and impact energy-harvesting performance tests. The acceleration was fixed at 20 m/s^2. Sinusoidal vibration was employed. With 25 mm from the end fixed to the fixture (i.e., 45 mm effective length), the output voltage was obtained during vibration or impact at 200 to 600 Hz with a sampling frequency of 50 kHz and resistance of 1 MΩ.

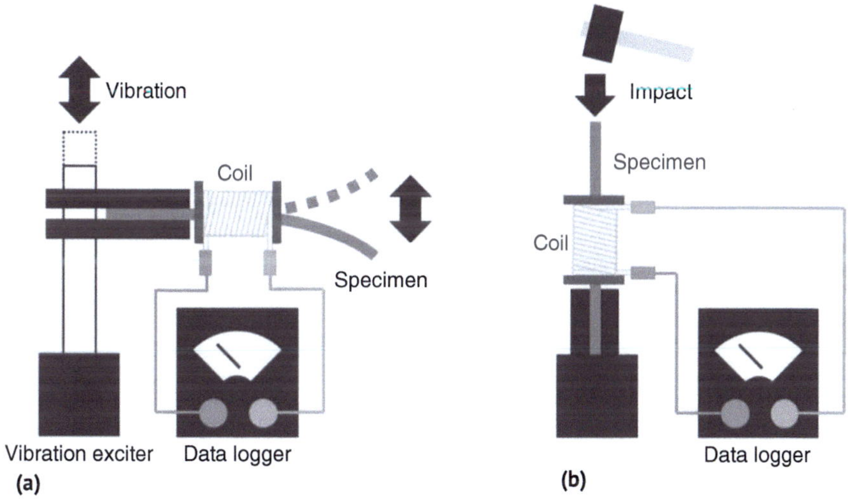

Figure 14.18 Schematic diagram of the vibration and impact energy-harvesting performance tests.

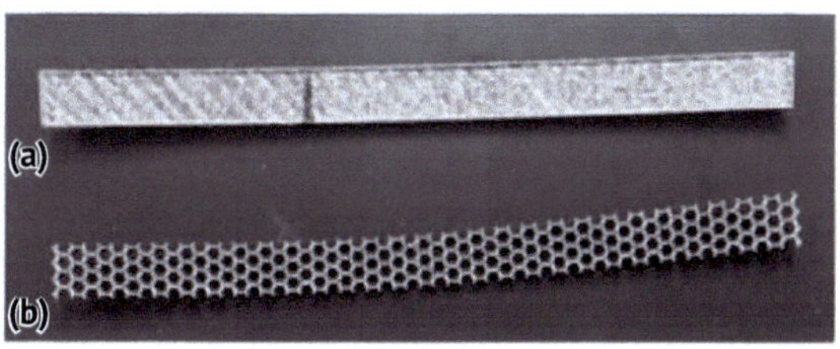

Figure 14.19 Appearance of the Fe–Co alloy plate with fully dense and honeycomb structures. *Source*: [27] / with permission of Elsevier.

The coil resistance was 11.42 kΩ, the load resistance was 11.72 kΩ, the number of coil turns was 28,000, and the coil diameter was 0.05 mm.

Figure 14.19 shows the appearance of the Fe–Co alloy plate with the fully dense and honeycomb structures. Figure 14.20a shows the output voltage versus frequency of the Fe–Co alloy plate in the vibration energy-harvesting test. The resonant frequencies of the fully dense Fe–Co alloy plate and the honeycomb structure Fe–Co alloy plate were 487 and 293 Hz, respectively. This result indicates that the resonance frequency shifts with structural changes in the Fe–Co alloy plate and that the honeycomb structure has a lower resonance frequency than the fully dense structure. Since the vibration frequencies in daily life tend to be low, vibration energy-harvesting devices should have low resonance frequencies. The relationship between power density and resistance at the resonant frequencies was investigated (see Figure 14.20b). In the vibration test, the honeycomb structure exhibited a power density of 4.7 times greater than the fully dense one. Thus, the honeycomb structure plate works effectively for energy harvesting regarding resonant frequency and power density. However, the honeycomb structure plate is damaged when subjected to vibration cycles of several minutes. Therefore, the honeycomb structure needs to be improved. Alternatively, other structures may be best suited to harness vibration energy.

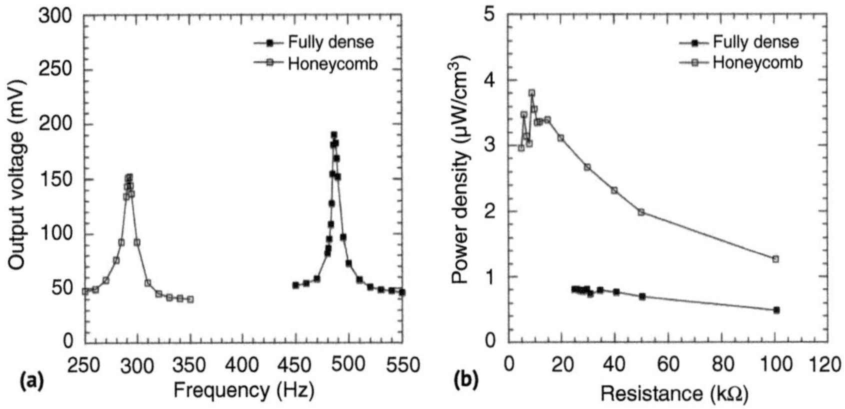

Figure 14.20 (a) Output voltage versus frequency of the Fe–Co alloy plate in the vibration energy-harvesting test and (b) Relationship between power density and resistance at the resonant frequencies.

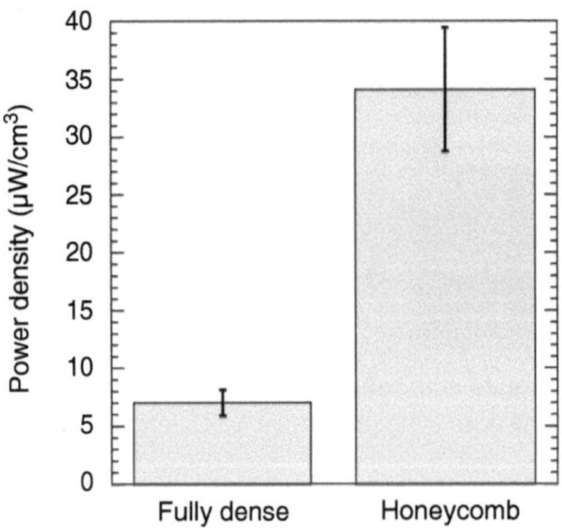

Figure 14.21 Power density of the Fe–Co alloy plate in the impact energy-harvesting test.

Figure 14.21 shows the power density of the Fe–Co alloy plate in the impact energy-harvesting test. The honeycomb structure exhibited a power density 4.9 times greater than the fully dense one in the impact test. The honeycomb structure of the Fe–Co alloy plate did not break during the impact test. It is well known that magnetostrictive materials can also be used as particulate matter sensors by utilizing shifts in resonant frequency or output voltage. Their weight governs the sensitivity of magnetostrictive particulate matter sensors. Therefore, such sensors must be lightweight to achieve high sensitivity. Therefore, using honeycomb or other designed structures will likely provide high-energy-harvesting performance and sensitivity as particulate matter sensors.

14.9 Conclusions

Considering the balance between magnetostrictive properties, mechanical properties, workability, and price, Fe–Co alloys have the greatest potential for practical use among magnetostrictive materials. Compared with piezoelectric materials, magnetostrictive materials have advantages in large power output and superior toughness. However, magnetostrictive materials are heavy; for example, the density of Fe–Co alloy is about 7.8 g/cm^3. In addition, the output power depends on the size of the coil, and the installation of large coils is essential for high-energy harvesting. The density of copper wire, the main material used in the coils, is 8.96 g/cm^3, making it heavy. Considering that the generated magnetic flux density changes have adverse effects on surrounding electronic devices, it is thought that applications such as energy harvesters in factories are more suitable than portable small devices and aviation applications. This chapter describes composite fabrication, cladding, nitriding, and structural design by additive manufacturing to reduce weight and improve performance. Stress-concentrated structure design effectively enhances the inverse magnetostrictive properties of magnetostrictive materials (i.e., energy-harvesting performance), but there is concern that it may reduce mechanical properties (especially fatigue properties). Future research on magnetostrictive energy harvesters will focus on energy-harvesting performance and durability.

References

1 Bryzek, J. (2013). TSensors Summit for Trillion Sensor Roadmap. 151.
2 Narita, F. and Fox, M. (2018). A review on piezoelectric, magnetostrictive, and magnetoelectric materials and device technologies for energy harvesting applications. *Adv. Eng. Mater.*, 20, 1700743.
3 Wang, Z. and Narita, F. (2019). Fabrication of potassium sodium niobate nano-particle/polymer composites with piezoelectric stability and their application to unsteady wind energy harvesters. *J. Appl. Phys.*, 126, 224501.
4 Atulasimha, J. and Flatau, A.B. (2011). A review of magnetostrictive iron–gallium alloys. *Smart Mater. Struct.*, 20, 043001.
5 Deng, Z. and Dapino, M.J. (2017). Review of magnetostrictive vibration energy harvesters. *Smart Mater. Struct.*, 26, 103001.
6 Nakajima, T., Takeuchi, T., Yuito, I., Kato, K., Saito, M., Abe, K., Sasaki, T., Sekiguchi, T., and Yamaura, S. (2014). Effect of annealing on magnetostrictive properties of Fe–Co alloy thin films. *Matl. Trans.*, 55(3), 556–60.
7 Yamaura, S., Nakajima, T., Satoh, T., Ebata, T., and Furuya, Y. (2015). Magnetostriction of heavily deformed Fe–Co binary alloys prepared by forging and cold rolling. *Mater. Sci. Eng. B*, 193, 121–9.
8 Liu, L., Zhan, Q., Yang, H., Li, H., Zhang, S., Liu, Y., Wang, B., Tan, X., and Li, R.W. (2016). Magnetostrictive GMR spin valves with composite FeGa/FeCo free layers. *AIP Adv.*, 6, 035206.

9 Bennett, S.P., Baldwin, J.W., Staruch, M., Matis, B.R., LaComb, J., Jvan'tErve, O.M., Bussmann, K., Metzler, M., Gottron, N., Zappone, W., LaComb, R., and Finkel, P. (2017). Magnetic field response of doubly clamped magnetoelectric microelectromechanical AlN-FeCo resonators, *Appl. Phys. Lett.*, 111, 252903.

10 Zhu, L., Li, K., Luo, Y., Yu, D., Wang, Z., Wu, G., Xie, J., and Tang, Z. (2019). Magnetostrictive properties and detection efficiency of TbDyFe/FeCo composite materials for nondestructive testing. *J. Rare Earths*, 37(2), 166–70.

11 Wang, W., Jia, Y., Xue, X., Liang, Y., and Du, Z. (2018). Magnetostrictive effect in micro-dotted FeCo film coated surface acoustic wave devices. *Smart Mater. Struct.*, 27, 105040.

12 Bhatti, S., Ma, C., Liu, X., and Piramanayagam, S.N. (2019). Stress-induced domain wall motion in FeCo-based magnetic microwires for realization of energy harvesting. *Adv. Electron. Mater.*, 5, 1800467.

13 Narita, F. and Katabira, K. (2017). Stress-rate dependent output voltage for $Fe_{29}Co_{71}$ magnetostrictive fiber/polymer composites: fabrication, experimental observation and theoretical prediction. *Mater. Trans.*, 58, 302–4.

14 Wan, Y., Fang, D., and Hwang, K.C. (2003). Non-linear constitutive relations for magnetostrictive materials. *Int. J. Nonlin. Mech.*, 38, 1053–65.

15 Jia, Z., Liu, W., Zhang, Y., Wang, F., and Guo, D. (2006). A nonlinear magnetomechanical coupling model of giant magnetostrictive thin films at low magnetic fields. *Sens. Actuators A: Physical*, 128, 158–64.

16 Narita, F. (2017). Inverse magnetostrictive effect in $Fe_{29}Co_{71}$ wire/polymer composites. *Adv. Eng. Mater.*, 19, 1600586.

17 Katabira, K., Kimura, N., Yamazaki, T., and Narita, F. (2019). Evaluation of the inverse magnetostrictive effect in heat-treated Fe–Co wire/polymer composites. *Mater. Des. Process. Commun.*, 1, e81.

18 Katabira, K., Yoshida, Y., Masuda, A., Watanabe, A., and Narita, F. (2018). Fabrication of Fe–Co magnetostrictive fiber reinforced plastic composites and their sensor performance evaluation. *Materials*, 11, 406.

19 Katabira, K., Kurita, H., Yoshida, Y., and Narita, F. (2019). Fabrication and Characterization of carbon fiber reinforced plastics containing magnetostrictive Fe–Co fibers with damage self-detection capability. *Sensors*, 19, 4984.

20 Yang, Z., Nakajima, K., Jiang, L., Kurita, H., Murasawa, G., and Narita, F. (2019). Design, fabrication and evaluation of metal-matrix lightweight magnetostrictive fiber composites. *Mater. Des.*, 175, 107803.

21 Kurita, H., Katabira, K., Yoshida, Y., and Narita, F. (2019). Footstep energy harvesting with the magnetostrictive fiber integrated shoes. *Materials*, 12, 2055.

22 Yang, Z., Nakajima, K., Onodera, R., Tayama, T., Chiba, D., and Narita, F. (2018). Magnetostrictive clad steel plates for high-performance vibration energy harvesting. *Appl. Phys. Lett.*, 112, 073902.

23 Yang, Z., Kurita, H., Onodera, R., Tayama, T., Chiba, D., and Narita, F. (2019). Evaluation of vibration energy harvesting using a magnetostrictive iron–cobalt/nickel-clad plate. *Smart Mater. Struct.*, 28, 034001.

24 Nakajima, K., Kurita, H., and Narita, F. (2019). Performance boost of Co-rich Fe–Co based alloy magnetostrictive sensors via nitrogen treatment. *Sens. Actuators A: Physical*, 295, 75–83.

25 Yang, Z., Kurita, H., Takeuchi, H., Nakajima, K., and Narita, F. (2019). Effect of notches on magnetic induction variation for magnetostrictive clad plate. *Sens. Mater.*, 31(11), 3681–90.

26 Nakajima, K., Leparoux, M., Kurita, H., Lanfant, B., Cui, D., Watanabe, M., Sato, T., and Narita, F. (2022). Additive manufacturing of magnetostrictive Fe–Co alloys. *Materials*, 15(3), 709.

27 Kurita, H., Lohmuller, P., Laheurte, P., Nakajima, K., and Narita, F. (2022). Additive manufacturing and energy-harvesting performance of honeycomb-structured magnetostrictive Fe_{52}-Co_{48} alloys. *Addit. Manuf.*, 54, 102741.

15

Mechanics of Magnetostrictive Materials and Composites

Fumio Narita

Graduate School of Environmental Studies, Tohoku University, Sendai, Japan

15.1 Introduction

In response to the threat of emerging and re-emerging infectious diseases, including new coronaviruses, there is growing interest in the Internet of Things (IoT) and digital transformation (DX). These technologies connect various objects to the Internet, leveraging digital advancements to drive disruptive innovation globally. It is expected that, by 2030, the number of sensors for IoT and DX will surpass one trillion, indicating the onset of a trillion-sensor society. Conversely, the operation of sensors and data communication requires power, and relying on batteries for a trillion sensors annually presents a considerable environmental, resource, and cost challenge. Consequently, energy harvesting—an approach that captures electrical energy from abundant, untapped kinetic and thermal energy in the natural environment—is drawing attention [1, 2]. This method is expected to serve as a self-sustaining power source for sensors and data communication.

Magnetostrictive Fe–Co alloy is a promising energy-harvesting material with excellent mechanical properties, high saturation magnetization, and low cost. Recently, we proposed the cobalt ferrite–spattered Fe–Co alloy plate [3], Fe–Co wire–dispersed epoxy composite stud bolt [4], and Fe–Co fiber–embedded glass fiber–reinforced polymer composites [5].

To enhance the performance of the previously mentioned magnetostrictive materials and composites, repeated trial fabrication and evaluation can be costly and time-consuming. Therefore, computer-based numerical material mechanics studies and investigations are a more effective approach.

Considering the aforementioned information, this chapter explores the fundamentals of the mechanics of magnetostrictive materials and composites and provides an overview that emphasizes the equations pertinent to electromagnetic and mechanical fields. It also discusses fundamental theories for the design, development, and evaluation of energy-harvesting devices and sensors, including actuators.

15.2 Interaction Between the Electromagnetic and Mechanical Fields

When a magnetic field is applied to a ferromagnetic material, the material's length undergoes a change. This phenomenon is known as the magnetostrictive effect and was first described by Joule in 1842 [6]. Figure 15.1a illustrates a schematic diagram of the magnetostrictive materials. The interior of the material is divided into regions referred to as magnetic domains, each exhibiting spontaneous magnetization. As the external magnetic field increases, the direction of spontaneous magnetization in each domain aligns with the magnetic field, causing the material to strain (either stretch or shrink) in response to the rotation of the magnetization. This phenomenon is termed magnetostriction.

Subsequently, Villari discovered that the magnetic flux changed when an external force was applied to a ferromagnetic material exhibiting the magnetostrictive effect; he termed this phenomenon the inverse magnetostrictive effect. As illustrated in Figure 15.1b, when a magnetized material experiences a compressive load, the magnetization in the material rotates, leading to a leakage flux from the material's surface. It is important to note that the magnetic flux per unit area is referred to as magnetic flux density.

Figure 15.2 illustrates the general relationship between the magnetic field and strain for magnetostrictive materials. Following the discovery of magnetostriction in ferromagnetic materials such as Fe and Ni, as well as various alloys, multiple macroscopic models have been developed to describe this phenomenon. This section

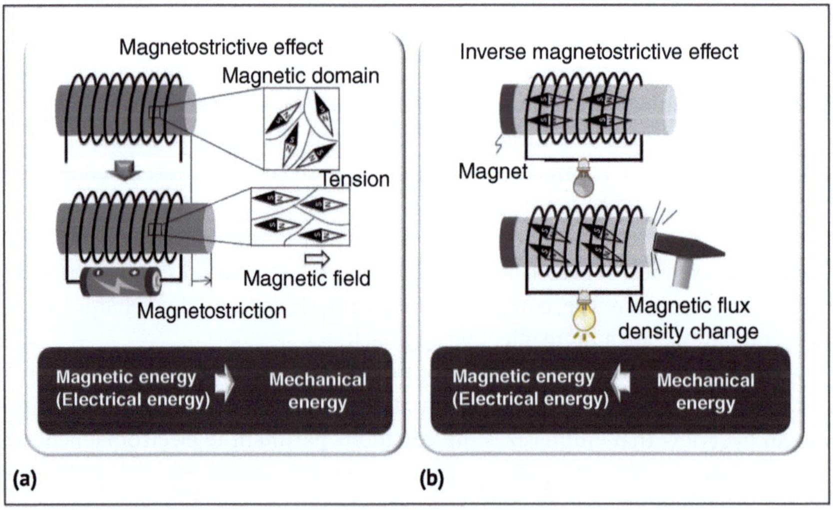

Figure 15.1 (a) Magnetostrictive and (b) inverse magnetostrictive effects.

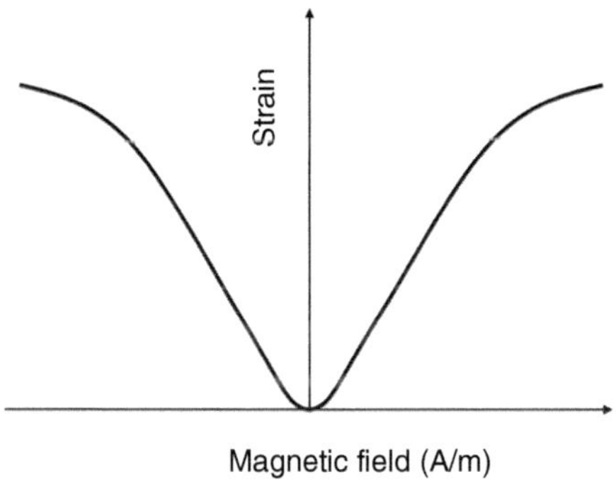

Figure 15.2 Magnetic field–strain curve.

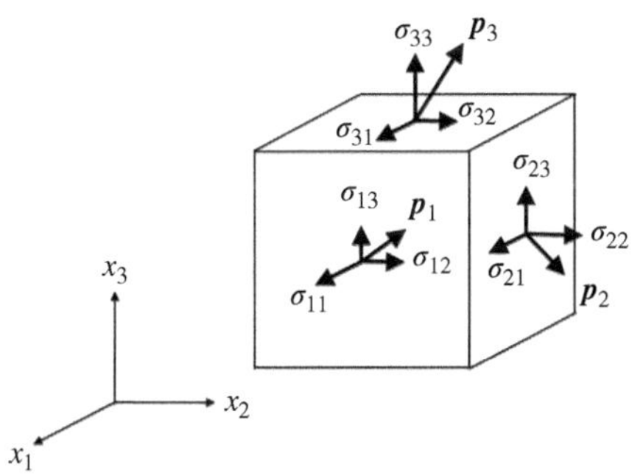

Figure 15.3 Stress vector and stress components.

presents the simplest mechanical model [7], which assumes a linear relationship between the magnetic field and the strain, considering a relatively small magnetic field. The model neglects hysteresis and is mathematically manageable.

15.2.1 Stress and Strain

To understand the magnetic deformation in magnetostrictive materials, it is beneficial to use stress and strain [8]. In the Cartesian coordinate system O-x_1, x_2, x_3 depicted in Figure 15.3, we consider a deformable body with a plane perpendicular to the coordinate axes x_1, x_2, and x_3. The stress acting on the plane perpendicular to the x_i-axis (the x_i plane) is a vector defined by both magnitude and direction, represented as $\boldsymbol{p}_i$. The stress vector $\boldsymbol{p}_i$ can be formulated as $\boldsymbol{p}_i = \sigma_{i1}\mathbf{e}_1 + \sigma_{i2}\mathbf{e}_2 + \sigma_{i3}\mathbf{e}_3 = \sigma_{ij}\mathbf{e}_j$, where the unit basis vectors $(\mathbf{e}_1, \mathbf{e}_2, \mathbf{e}_3)$ align with the coordinate axes x_1, x_2, x_3, using the summation convention for the stress components σ_{ij}. Here, $\boldsymbol{p}_i$ and $\mathbf{e}_i$ are not necessarily aligned. When the stress components are organized into a matrix, the result is as follows:

$$[\sigma_{ij}] = \begin{bmatrix} \sigma_{11} & \sigma_{12} & \sigma_{13} \\ \sigma_{21} & \sigma_{22} & \sigma_{23} \\ \sigma_{31} & \sigma_{32} & \sigma_{33} \end{bmatrix} \tag{15.1}$$

Here, $\sigma_{11}, \sigma_{22}, \sigma_{33}$ are the components perpendicular to the plane and represent the normal stress. $\sigma_{12}, \sigma_{13}, \sigma_{21}, \sigma_{23}, \sigma_{31}, \sigma_{32}$ are the tangential components of the surface and represent the shear stress. σ_{ij} is a second-order tensor with magnitude and two directions, and it is called a stress tensor.

The equilibrium or balance equations are given by

$$\begin{aligned}
&\frac{\partial \sigma_{11}}{\partial x_1} + \frac{\partial \sigma_{21}}{\partial x_2} + \frac{\partial \sigma_{31}}{\partial x_3} + F_1 = 0 \\
&\frac{\partial \sigma_{12}}{\partial x_1} + \frac{\partial \sigma_{22}}{\partial x_2} + \frac{\partial \sigma_{32}}{\partial x_3} + F_2 = 0 \\
&\frac{\partial \sigma_{13}}{\partial x_1} + \frac{\partial \sigma_{23}}{\partial x_2} + \frac{\partial \sigma_{33}}{\partial x_3} + F_3 = 0
\end{aligned} \tag{15.2}$$

where F_1, F_2, F_3 are the components of the body force vector per unit volume. By applying the summation convention, we can express it as:

$$\frac{\partial \sigma_{ji}}{\partial x_j} + F_i = 0 \tag{15.3}$$

In general,

$$\sigma_{ij} = \sigma_{ji} \tag{15.4}$$

is a statement of the symmetry of the stress tensor. Consequently, the stress state in the object is characterized by nine stress components; however, there are only six independent components, indicating that the stress tensor is symmetric.

If the displacement vector is u_1, u_2, u_3, the six strain components can be expressed by:

$$\begin{aligned}
&\varepsilon_{11} = \frac{\partial u_1}{\partial x_1}, \varepsilon_{22} = \frac{\partial u_2}{\partial x_2}, \varepsilon_{33} = \frac{\partial u_3}{\partial x_3} \\
&\varepsilon_{23} = \frac{1}{2}\left(\frac{\partial u_2}{\partial x_3} + \frac{\partial u_3}{\partial x_2}\right), \varepsilon_{31} = \frac{1}{2}\left(\frac{\partial u_3}{\partial x_1} + \frac{\partial u_1}{\partial x_3}\right), \varepsilon_{12} = \frac{1}{2}\left(\frac{\partial u_1}{\partial x_2} + \frac{\partial u_2}{\partial x_1}\right)
\end{aligned} \tag{15.5}$$

or

$$\varepsilon_{ij} = \frac{1}{2}\left(\frac{\partial u_i}{\partial x_j} + \frac{\partial u_j}{\partial x_i}\right) \tag{15.6}$$

15.2.2 Energy and Work

When magnetostrictive materials undergo deformation owing to an external force, they store elastic strain energy internally. The elastic strain energy per unit volume is referred to as the elastic strain energy density. When a strain $\mathrm{d}\varepsilon_{ij}$ is applied to the magnetostrictive material, the elastic strain energy density can be expressed as follows:

$$\sigma_{ij}\mathrm{d}\varepsilon_{ij} = \sigma_{11}\mathrm{d}\varepsilon_{11} + \sigma_{22}\mathrm{d}\varepsilon_{22} + \sigma_{33}\mathrm{d}\varepsilon_{33} + 2\left(\sigma_{12}\mathrm{d}\varepsilon_{12} + \sigma_{23}\mathrm{d}\varepsilon_{23} + \sigma_{31}\mathrm{d}\varepsilon_{31}\right) \tag{15.7}$$

Conversely, when the magnetic flux density of a magnetostrictive material is varied by $\mathrm{d}B_m$, the material stores magnetic work

$$\mathrm{d}W = H_m\mathrm{d}B_m = H_1\mathrm{d}B_1 + H_2\mathrm{d}B_2 + H_3\mathrm{d}B_3 \tag{15.8}$$

where H_m is a component of the magnetic field strength vector. Consequently, the change in internal energy U resulting from the strain $\mathrm{d}\varepsilon_{ij}$ and the flux density $\mathrm{d}B_m$ is

$$\mathrm{d}U = \sigma_{ij}\mathrm{d}\varepsilon_{ij} + H_m\mathrm{d}B_m \tag{15.9}$$

Conversely, the Gibbs free energy can be written as:

$$G = U - \sigma_{ij}\varepsilon_{ij} - H_mB_m \tag{15.10}$$

The total derivative of Eq. (15.10) is:

$$\mathrm{d}G = \mathrm{d}U - \mathrm{d}\sigma_{ij}\varepsilon_{ij} - \mathrm{d}H_mB_m - \sigma_{ij}\mathrm{d}\varepsilon_{ij} - H_m\mathrm{d}B_m \tag{15.11}$$

Considering Eq. (15.9),

$$\mathrm{d}G = -\varepsilon_{ij}\mathrm{d}\sigma_{ij} - B_m\mathrm{d}H_m \tag{15.12}$$

becomes

$$\varepsilon_{ij} = -\left(\frac{\partial G}{\partial \sigma_{ij}}\right)_H, B_m = -\left(\frac{\partial G}{\partial H_m}\right)_\sigma \tag{15.13}$$

The partial differentiation of the strain ε_{ij} with respect to the magnetic field strength H_m and the magnetic flux density B_m with respect to the stress σ_{ij} results in

$$\left(\frac{\partial \varepsilon_{ij}}{\partial H_m}\right)_\sigma = \left(\frac{\partial B_m}{\partial \sigma_{ij}}\right)_H = d_{mij} \tag{15.14}$$

where d_{mij} is the piezomagnetic constant, representing the initial slope of the curve in the linear range shown in Figure 15.2.

When a magnetic field is applied to the magnetostrictive material along the x_3 direction, it undergoes distortion in the same direction. This operating mode is known as the longitudinal mode, and, in this case, Eq. (15.14) is expressed as:

$$\left(\frac{\partial \varepsilon_{33}}{\partial H_3}\right)_\sigma = \left(\frac{\partial B_3}{\partial \sigma_{33}}\right)_H = d_{333} \tag{15.15}$$

15.2.3 Constitutive Equation

Magnetostrictive materials generate strain when subjected to stress and when exposed to a magnetic field. Additionally, magnetization is influenced both by the application of a magnetic field and by stress. These interactions can be mathematically represented by assuming that the material demonstrates reversible properties. From Eqs. (15.12) and (15.13), we obtain:

$$\mathrm{d}\varepsilon_{ij} = \left(\frac{\partial \varepsilon_{ij}}{\partial \sigma_{kl}}\right)_H \mathrm{d}\sigma_{kl} + \left(\frac{\partial \varepsilon_{ij}}{\partial H_k}\right)_\sigma \mathrm{d}H_k \tag{15.16}$$

$$\mathrm{d}B_i = \left(\frac{\partial B_i}{\partial \sigma_{jk}}\right)_H \mathrm{d}\sigma_{jk} + \left(\frac{\partial B_i}{\partial H_k}\right)_\sigma \mathrm{d}H_k \tag{15.17}$$

Given by

$$\left(\frac{\partial \varepsilon_{ij}}{\partial \sigma_{kl}}\right)_H = s^{\mathrm{H}}_{ijkl}, \left(\frac{\partial B_i}{\partial H_k}\right)_\sigma = \mu^{\mathrm{T}}_{ik} \tag{15.18}$$

the constitutive equations

$$\varepsilon_{ij} = s^{\mathrm{H}}_{ijkl}\sigma_{kl} + d_{kij}H_k \tag{15.19}$$

$$B_i = d_{jki}\sigma_{jk} + \mu^{\mathrm{T}}_{ik}H_k \tag{15.20}$$

are obtained if $d\sigma_{ij}$ and dH_k are small. Here s^{H}_{ijkl} is the elastic compliance coefficient under a constant magnetic field, and μ^{T}_{ik} is the magnetic permeability under constant stress.

In general, the constitutive equations can be simplified as shown in Table 15.1, using p or q instead of $ijkl$ when the subscripts i and j are interchangeable and k and l are interchangeable [9]. Considering Table 15.1, Eqs. (15.19) and (15.20) are as follows:

$$\begin{Bmatrix} \varepsilon_{11} \\ \varepsilon_{22} \\ \varepsilon_{33} \\ 2\varepsilon_{23} \\ 2\varepsilon_{31} \\ 2\varepsilon_{12} \end{Bmatrix} = \begin{bmatrix} s^{\mathrm{H}}_{11} s^{\mathrm{H}}_{12} s^{\mathrm{H}}_{13} s^{\mathrm{H}}_{14} s^{\mathrm{H}}_{15} s^{\mathrm{H}}_{16} \\ s^{\mathrm{H}}_{12} s^{\mathrm{H}}_{22} s^{\mathrm{H}}_{23} s^{\mathrm{H}}_{24} s^{\mathrm{H}}_{25} s^{\mathrm{H}}_{26} \\ s^{\mathrm{H}}_{13} s^{\mathrm{H}}_{23} s^{\mathrm{H}}_{33} s^{\mathrm{H}}_{34} s^{\mathrm{H}}_{35} s^{\mathrm{H}}_{36} \\ s^{\mathrm{H}}_{14} s^{\mathrm{H}}_{24} s^{\mathrm{H}}_{34} s^{\mathrm{H}}_{44} s^{\mathrm{H}}_{45} s^{\mathrm{H}}_{46} \\ s^{\mathrm{H}}_{15} s^{\mathrm{H}}_{25} s^{\mathrm{H}}_{35} s^{\mathrm{H}}_{45} s^{\mathrm{H}}_{55} s^{\mathrm{H}}_{56} \\ s^{\mathrm{H}}_{16} s^{\mathrm{H}}_{26} s^{\mathrm{H}}_{36} s^{\mathrm{H}}_{46} s^{\mathrm{H}}_{56} s^{\mathrm{H}}_{66} \end{bmatrix} \begin{Bmatrix} \sigma_{11} \\ \sigma_{22} \\ \sigma_{33} \\ \sigma_{23} \\ \sigma_{31} \\ \sigma_{12} \end{Bmatrix} + \begin{bmatrix} d_{11} d_{21} d_{31} \\ d_{12} d_{22} d_{32} \\ d_{13} d_{23} d_{33} \\ d_{14} d_{24} d_{34} \\ d_{15} d_{25} d_{35} \\ d_{16} d_{26} d_{36} \end{bmatrix} \begin{Bmatrix} H_1 \\ H_2 \\ H_3 \end{Bmatrix} \tag{15.21}$$

$$\begin{Bmatrix} B_1 \\ B_2 \\ B_3 \end{Bmatrix} = \begin{bmatrix} d_{11} d_{12} d_{13} d_{14} d_{15} d_{16} \\ d_{21} d_{22} d_{23} d_{24} d_{25} d_{26} \\ d_{31} d_{32} d_{33} d_{34} d_{35} d_{36} \end{bmatrix} \begin{Bmatrix} \sigma_{11} \\ \sigma_{22} \\ \sigma_{33} \\ \sigma_{23} \\ \sigma_{31} \\ \sigma_{12} \end{Bmatrix} + \begin{bmatrix} \mu^{\mathrm{T}}_{11} \mu^{\mathrm{T}}_{12} \mu^{\mathrm{T}}_{13} \\ \mu^{\mathrm{T}}_{12} \mu^{\mathrm{T}}_{22} \mu^{\mathrm{T}}_{23} \\ \mu^{\mathrm{T}}_{13} \mu^{\mathrm{T}}_{23} \mu^{\mathrm{T}}_{33} \end{bmatrix} \begin{Bmatrix} H_1 \\ H_2 \\ H_3 \end{Bmatrix} \tag{15.22}$$

For example, d_{333} in Eq. (15.15) can be rewritten as d_{33}. When the directions of magnetic polarization (magnetic bias) and prestress (mechanical bias) are aligned

Table 15.1 Tensor and matrix notation.

ij* or *kl	***p* or *q***
11	1
22	2
33	3
23 or 32	4
31 or 13	5
12 or 21	6

with the x_3-axis (the easy axis of magnetization), the constitutive Eqs. (15.21) and (15.22) for a polycrystalline ferromagnetic material are presented as follows [10]:

$$\begin{Bmatrix} \varepsilon_{11} \\ \varepsilon_{22} \\ \varepsilon_{33} \\ 2\varepsilon_{23} \\ 2\varepsilon_{31} \\ 2\varepsilon_{12} \end{Bmatrix} = \begin{bmatrix} s_{11}^{H} & s_{12}^{H} & s_{13}^{H} & 0 & 0 & 0 \\ s_{12}^{H} & s_{11}^{H} & s_{13}^{H} & 0 & 0 & 0 \\ s_{13}^{H} & s_{13}^{H} & s_{33}^{H} & 0 & 0 & 0 \\ 0 & 0 & 0 & s_{44}^{H} & 0 & 0 \\ 0 & 0 & 0 & 0 & s_{44}^{H} & 0 \\ 0 & 0 & 0 & 0 & 0 & s_{66}^{H} \end{bmatrix} \begin{Bmatrix} \sigma_{11} \\ \sigma_{22} \\ \sigma_{33} \\ \sigma_{23} \\ \sigma_{31} \\ \sigma_{12} \end{Bmatrix} + \begin{bmatrix} 0 & 0 & d_{31} \\ 0 & 0 & d_{31} \\ 0 & 0 & d_{33} \\ 0 & d_{15} & 0 \\ d_{15} & 0 & 0 \\ 0 & 0 & 0 \end{bmatrix} \begin{Bmatrix} H_1 \\ H_2 \\ H_3 \end{Bmatrix} \tag{15.23}$$

$$\begin{Bmatrix} B_1 \\ B_2 \\ B_3 \end{Bmatrix} = \begin{bmatrix} 0 & 0 & 0 & 0 & d_{15} & 0 \\ 0 & 0 & 0 & d_{15} & 0 & 0 \\ d_{31} & d_{31} & d_{33} & 0 & 0 & 0 \end{bmatrix} \begin{Bmatrix} \sigma_{11} \\ \sigma_{22} \\ \sigma_{33} \\ \sigma_{23} \\ \sigma_{31} \\ \sigma_{12} \end{Bmatrix} + \begin{bmatrix} \mu_{11}^{T} & 0 & 0 \\ 0 & \mu_{11}^{T} & 0 \\ 0 & 0 & \mu_{33}^{T} \end{bmatrix} \begin{Bmatrix} H_1 \\ H_2 \\ H_3 \end{Bmatrix} \tag{15.24}$$

15.2.4 One-dimensional Constitutive Equation

In the Cartesian coordinate system O-x_1, x_2, x_3, consider an elongated magnetostrictive rod oriented along the x_3 direction, with both the magnetic field and the direction of elongation aligned with the x_3-axis. When the length of the rod is considerably larger than its diameter, the problem can be treated as one dimensional. Consequently, the only relevant physical quantities in the constitutive Eqs. (15.23) and (15.24) are stress σ_{33}, strain ε_{33}, magnetic field strength H_3, and magnetic flux density B_3. Thus, the one-dimensional constitutive equation can be expressed as follows:

$$\varepsilon_{33} = s_{33}^{H}\sigma_{33} + d_{33}H_3 \tag{15.25}$$

$$B_3 = d_{33}\sigma_{33} + \mu_{33}^{T}H_3 \tag{15.26}$$

Eliminating the magnetic field H_3 from Eqs. (15.25) and (15.26) yields:

$$\varepsilon_{33} = s_{33}^{B}\sigma_{33} + \frac{d_{33}}{\mu_{33}^{T}}B_3 \tag{15.27}$$

Similarly, eliminating the stress σ_{33} yields

$$B_3 = \frac{d_{33}}{s_{33}^{H}}\varepsilon_{33} + \mu_{33}^{S}H_3 \tag{15.28}$$

The following is established:

$$s_{33}^{B} = \left(\frac{d\varepsilon_{33}}{d\sigma_{33}}\right)_B = s_{33}^{H}\left(1 - \frac{d_{33}^2}{s_{33}^{H}\mu_{33}^{T}}\right) = s_{33}^{H}\left(1 - k_{33}^2\right) \tag{15.29}$$

$$\mu_{33}^{\mathrm{S}} = \left(\frac{\mathrm{d}B_3}{\mathrm{d}H_3}\right)_{\varepsilon} = \mu_{33}^{\mathrm{T}}\left(1 - \frac{d_{33}^2}{s_{33}^{\mathrm{H}}\mu_{33}^{\mathrm{T}}}\right) = \mu_{33}^{\mathrm{T}}\left(1 - k_{33}^2\right) \tag{15.30}$$

$k_{33}^2 = d_{33}^2/s_{33}^{\mathrm{H}}\mu_{33}^{\mathrm{T}}$ is the magnetomechanical coupling coefficient, indicating the efficiency of conversion between mechanical (elastic strain) energy and magnetic energy, as shown in Figure 15.1.

The constitutive Eqs. (15.27) and (15.28) describe the stress and magnetic field:

$$\sigma_{33} = \frac{1}{s_{33}^{\mathrm{B}}}\varepsilon_{33} - \lambda_{33}B_3 \tag{15.31}$$

$$H_3 = -\lambda_{33}\varepsilon_{33} + \frac{1}{\mu_{33}^{\mathrm{S}}}B_3 \tag{15.32}$$

The coefficient λ_{33} is given as follows:

$$\lambda_{33} = \frac{d_{33}}{\mu_{33}^{\mathrm{T}}s_{33}^{\mathrm{B}}} = \frac{d_{33}}{\mu_{33}^{\mathrm{S}}s_{33}^{\mathrm{H}}} \tag{15.33}$$

15.2.5 Coupled Electromagnetic and Mechanical Field Equations for the Magnetostrictive System Design

Now, in the Cartesian coordinate system O-x_1, x_2, x_3 shown in Figure 15.4, consider a magnetostrictive cylindrical rod whose length l is much larger than its diameter. The length direction coincides with the x_3-axis (easy axis of magnetization), and the cross-sectional area is A. In addition, the magnetostrictive rod is harmonically excited at a frequency ω such that elastic wave propagation does not occur inside [11].

Among the physical quantities in the constitutive Eqs. (15.31) and (15.32)—stress $\sigma_{33}\exp(j\omega t)$, strain $\varepsilon_{33}\exp(j\omega t)$, magnetic field strength $H_3\exp(j\omega t)$, and magnetic flux density $B_3\exp(j\omega t)$—two can be freely chosen as input values while the other two serve as output values [11]. Here, t is the time, and $j^2 = -1$. In this section, voltage V and current I are selected as the input values, replacing magnetic field strength and flux density, while force F_3 and phase velocity v are designated as the output values, replacing stress and strain. For clarity, the term $\exp(j\omega t)$ will be omitted hereafter, which results in a multiplication by $1/j\omega$ instead of an integration with respect to time and a multiplication by $j\omega$ instead of a differentiation with respect to time.

If the number of turns of the coil is N, the magnetic field of the solenoid is given by:

$$H_3 = \frac{NI}{l} = nI \tag{15.34}$$

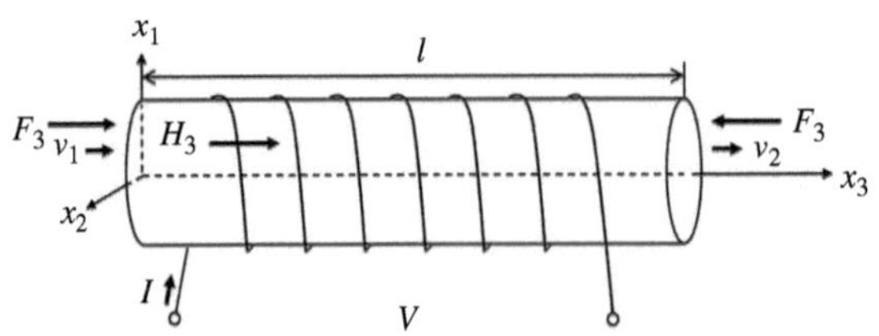

Figure 15.4 Magnetostrictive rod and coil.

where $n = N/l$ is the number of coil turns per unit length. Because Faraday's law is given by:

$$V = -NA\frac{dB_3}{dt} \tag{15.35}$$

the magnetic flux density in the solenoid with cross section A is

$$B_3 = -\frac{V}{j\omega NA} \tag{15.36}$$

As illustrated in Figure 15.4, let the phase velocities at $x_3 = 0, l$ be v_1 and v_2, respectively. Because displacement is derived by integrating the phase velocity over time, the strain ε_{33} can be expressed in terms of the phase velocity as follows.

$$\varepsilon_{33} = \frac{v_2 - v_1}{j\omega l} \tag{15.37}$$

Therefore, by substituting Eqs. (15.36) and (15.37) into Eq. (15.31), we obtain

$$F_3 = -\sigma_{33}A = -\frac{A}{s_{33}^{B}}\frac{v_2 - v_1}{j\omega l} - \lambda_{33}\frac{V}{j\omega N} \tag{15.38}$$

The minus sign in σ_{33} indicates that the force F_3 is positive for compressive stress. By substituting Eqs. (15.34), (15.36), and (15.37) into Eq. (15.32), we obtain:

$$I = -\lambda_{33}\frac{v_2 - v_1}{j\omega N} - \frac{l}{\mu_{33}^{S}}\frac{V}{j\omega N^2 A} \tag{15.39}$$

If the inductance at zero strain, or the fixed state, is set to:

$$L^{S} = \frac{\mu_{33}^{S}N^2 A}{l} \tag{15.40}$$

and Eq. (15.33) is considered, Eqs. (15.38) and (15.39) can be rewritten as follows:

$$F_3 = -\frac{A}{j\omega l s_{33}^{B}}(v_2 - v_1) - \frac{d_{33}}{j\omega N \mu_{33}^{T} s_{33}^{B}}V \tag{15.41}$$

$$I = -\frac{d_{33}}{j\omega N \mu_{33}^{T} s_{33}^{B}}(v_2 - v_1) - \frac{1}{j\omega L^{S}}V \tag{15.42}$$

The aforementioned equation is the basic equation called the impedance representation of magnetostriction [12].

15.3 Fiber Composite Materials

As illustrated in Figure 15.1b, the application of an external force to a magnetostrictive material generates a leakage flux from its surface. This leads to a change in the magnetic flux that penetrates the coil placed outside the material, enabling voltage generation. Fe–Co fibers possess characteristics such as a demagnetizing field coefficient and high magnetic anisotropy. When dispersed in polymers, they can be used to develop composite materials with enhanced magnetostrictive properties. Furthermore, polymer composites offer numerous advantages, including their lightweight nature and the potential for prestressing effects during the manufacturing process. In this section, we focus on magnetostrictive composite rods and explore their applications in impact energy harvesting and stress sensing, emphasizing the inverse magnetostriction effect (Figure 15.1b) or the leakage flux effect.

15.3.1 Composite Bar

Consider a magnetostrictive composite rod of length l, consisting of a magnetostrictive cylinder fully bonded to an elastic cylinder in the Cartesian coordinate system O-x_1, x_2, x_3, as illustrated in Figure 15.5. The length of the rod is aligned along the x_3-axis, which serves as the easy magnetization direction. Let A^{f} and A^{m} be the cross-sectional areas of the magnetostrictive and elastic cylinders, respectively, with the total cross-sectional area given by $A = A^{\mathrm{f}} + A^{\mathrm{m}}$. Here, the volume fraction of the magnetostrictive cylinder is defined as $v^{\mathrm{f}} = A^{\mathrm{f}}/A$. The superscripts f and m denote the magnetostrictive and elastic cylinders, respectively. The magnetostrictive composite rod is subjected to a uniform compressive load F_3 and a bias magnetic field H_0.

When the length of the composite rod is considerably greater than its diameter, the problem can be simplified to a one-dimensional analysis. From Eqs. (15.25) and (15.26), the constitutive equation for the magnetostrictive cylinder is given as follows:

$$\varepsilon_{33}^{\mathrm{f}} = s_{33}^{\mathrm{Hf}}\sigma_{33}^{\mathrm{f}} + d_{33}^{\mathrm{f}}H_3^{\mathrm{f}} \tag{15.43}$$

$$B_3^{\mathrm{f}} = d_{33}^{\mathrm{f}}\sigma_{33}^{\mathrm{f}} + \mu_{33}^{\mathrm{Tf}}H_3^{\mathrm{f}} \tag{15.44}$$

Assuming that the strength of the magnetic field induced by the inverse magnetostriction effect is sufficiently small compared to the bias field H_0, Eqs. (15.43) and (15.44) can be expressed using H_0 as follows:

$$\varepsilon_{33}^{\mathrm{f}} = s_{33}^{\mathrm{Hf}}\sigma_{33}^{\mathrm{f}} + d_{33}^{\mathrm{f}}H_0 \tag{15.45}$$

$$B_3^{\mathrm{f}} = d_{33}^{\mathrm{f}}\sigma_{33}^{\mathrm{f}} + \mu_{33}^{\mathrm{Tf}}H_0 \tag{15.46}$$

The strain $\varepsilon_{33}^{\mathrm{m}}$ and flux density B_3^{m} of the elastic cylinder are given by:

$$\varepsilon_{33}^{\mathrm{m}} = s_{33}^{\mathrm{m}}\sigma_{33}^{\mathrm{m}} \tag{15.47}$$

$$B_3^{\mathrm{m}} = \mu_0 H_3^{\mathrm{m}} \tag{15.48}$$

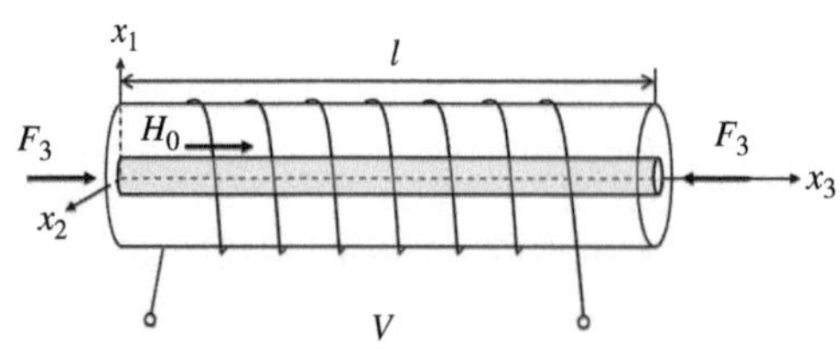

Figure 15.5 Magnetostrictive composite rod and coil.

where $\mu_0 = 4\pi \times 10^{-7}$ H^{m} is the magnetic permeability in free space.

Assuming a perfect joint between the magnetostrictive cylinder and the elastic cylinder ($\varepsilon_{33}^{\mathrm{f}} = \varepsilon_{33}^{\mathrm{m}}$), we obtain:

$$s_{33}^{\mathrm{Hf}}\sigma_{33}^{\mathrm{f}} + d_{33}^{\mathrm{f}}H_0 = s_{33}^{\mathrm{m}}\sigma_{33}^{\mathrm{m}} \tag{15.49}$$

and from the balance of forces, the following equation holds:

$$F_3 = A^{\mathrm{f}}\sigma_{33}^{\mathrm{f}} + A^{\mathrm{m}}\sigma_{33}^{\mathrm{m}} \tag{15.50}$$

From Eqs. (15.49) and (15.50), the stresses

$$\sigma_{33}^{\mathrm{f}} = \frac{s_{33}^{\mathrm{m}}(F_3/A) - d_{33}^{\mathrm{f}}\left(1 - \upsilon^{\mathrm{f}}\right)H_0}{s_{33}^{\mathrm{m}}\upsilon^{\mathrm{f}} + s_{33}^{\mathrm{Hf}}\left(1 - \upsilon^{\mathrm{f}}\right)} \tag{15.51}$$

$$\sigma_{33}^{\mathrm{m}} = \frac{s_{33}^{\mathrm{Hf}}(F_3/A) + d_{33}^{\mathrm{f}}\upsilon^{\mathrm{f}}H_0}{s_{33}^{\mathrm{m}}\upsilon^{\mathrm{f}} + s_{33}^{\mathrm{Hf}}\left(1 - \upsilon^{\mathrm{f}}\right)} \tag{15.52}$$

can be obtained. Substituting Eq. (15.51) into Eq. (15.46) and using Eq. (15.35), the output voltage

$$V = -NA^{\mathrm{f}}\frac{s_{33}^{\mathrm{m}}d_{33}^{\mathrm{f}}}{s_{33}^{\mathrm{m}}\upsilon^{\mathrm{f}} + s_{33}^{\mathrm{Hf}}\left(1 - \upsilon^{\mathrm{f}}\right)}\frac{\mathrm{d}\sigma_{33}^{0}}{\mathrm{d}t} \tag{15.53}$$

of the magnetostrictive composite rod owing to the compressive load F_3 is obtained. Here $\sigma_{33}^{0} = F_3/A = \sigma_{33}^{\mathrm{f}}\upsilon^{\mathrm{f}} + \sigma_{33}^{\mathrm{m}}\left(1 - \upsilon^{\mathrm{f}}\right)$ is the average normal stress acting on the composite bar.

15.3.2 Prototype and Compression Shock Power Generation Experiments

Magnetostrictive composite rods were produced by embedding 125 Fe–Co fibers, each with a diameter of 0.2 mm and a length of approximately 19 mm, into bisphenol-F epoxy resin. The rods were cured at room temperature for 24 h and subsequently heated to 80 °C for 3 h in a temperature-controlled chamber [13]. The fiber volume fraction is $\upsilon^{\mathrm{f}} = 0.0565$, and the composite rod's cross-sectional area is $A = 69.5$ mm^2. After fabrication, experiments to generate power from compressive stress were performed. The load was applied using an Autograph (SHIMAZU AG-50kNXD), and the output voltage V was recorded using a coil with a resistance of 106 Ω. This coil consists of 2500 turns over a length of 36 mm.

Figure 15.6 shows the correlation between the output voltage density $V/Al\upsilon^{\mathrm{f}}$ and the stress rate $\mathrm{d}\sigma_{33}^{0}/\mathrm{d}t$ during the unloading of the fabricated magnetostrictive composite bar, subjected to a maximum average stress of 8 MPa. The physical properties used are $s_{33}^{\mathrm{Hf}} = 5.5 \times 10^{-12}$ m^2/N, $d_{33}^{\mathrm{f}} = 0.125 \times 10^{-9}$ m/A, and $s_{33}^{\mathrm{m}} = 400 \times 10^{-12}$ m^2/N. Although the experimentally obtained voltage is slightly higher than that predicted by Eq. (15.53), the overall results are satisfactory. The output voltage rose with an increasing stress rate, demonstrating that the magnetostrictive composite rod produces power when subjected to compressive stress. Additionally, while the results are not presented, it is interesting to note that the output voltage is also influenced by the maximum average stress, allowing for the measurement of stress based on the output voltage value.

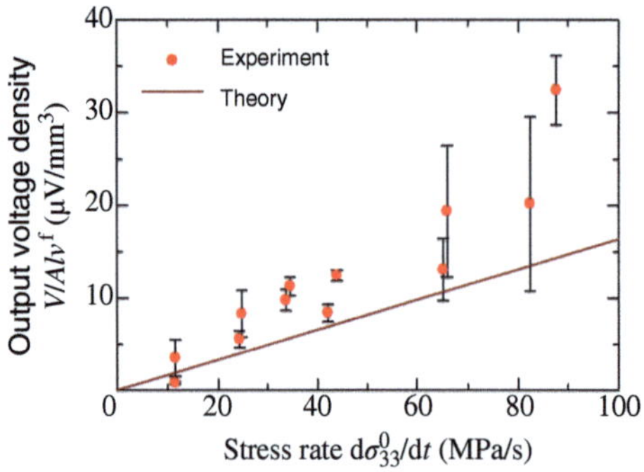

Figure 15.6 Relationship between the output voltage and the stress rate.

15.4 Laminated Plate

In the previous section, we examined the inverse magnetostriction effect (Figure 15.1b) and discussed impact energy harvesting and stress sensing using Eq. (15.26). Conversely, according to Eq. (15.25), when a magnetic field is applied along the easy axis of magnetization (x_3-axis), the magnetostrictive material will either stretch or shrink. For instance, as illustrated in Figure 15.7, a Fe–Co/Ni laminate featuring varying positive and negative values of the piezomagnetic constant d_{33} in the top and bottom layers exhibits bending as a whole; this occurs when one layer is stretched by the magnetic field while the other is contracted. To theoretically predict the deformation behavior of magnetostrictive laminated plates under the influence of magnetic fields, laminated beam theory proves valuable; however, it is essential to account for the interaction between electromagnetic and mechanical fields. In this section, we focus on the magnetostrictive effect (Figure 15.1a), formulate a foundational problem related to the bending of magnetostrictive laminated beams, and perform an analysis to explore the actuator response.

15.4.1 Laminated Beam

As illustrated in Figure 15.8, let us consider a magnetostrictive laminated cantilever beam, with one end fixed and the other end free, in the Cartesian coordinate system O-x_1, x_3. The length direction of the beam aligns with the x_3-axis, which represents the easy axis of magnetization. The dimensions of the beam are defined as follows: length l, width b, and thickness h. Additionally, the laminated beams are subjected to a uniform bending moment M^{M} resulting from the static magnetic field loading.

When the x_1-axis is oriented vertically downward, the vertical displacement w at any position x_3 is referred to as the deflection. The differential equation governing the deflection curve is expressed as follows:

$$\frac{\mathrm{d}^2 w}{\mathrm{d}x_3^2} = -\frac{M^{\mathrm{M}}}{D} \tag{15.54}$$

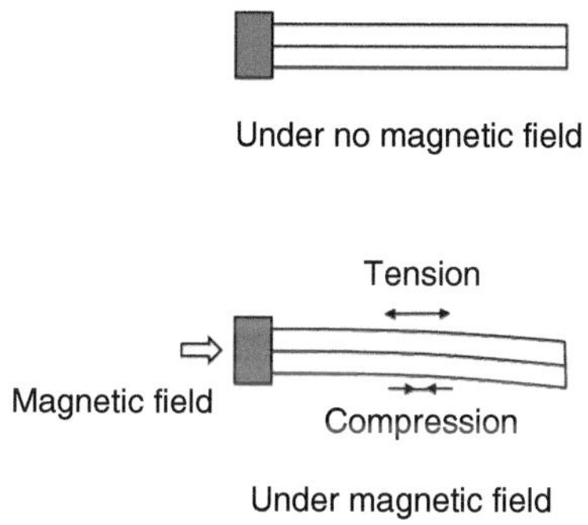

Figure 15.7 Deflection of magnetostrictive laminate in response to magnetic fields.

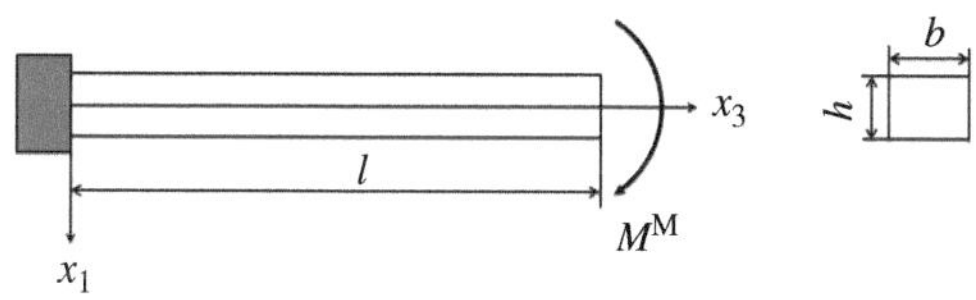

Figure 15.8 Magnetostrictive laminates under bending moment.

where D is the bending stiffness of the laminated beam. By integrating Eq. (15.54) twice and considering the boundary condition ($x_3 = 0$; $w = \mathrm{d}w/\mathrm{d}x_3 = 0$) for the cantilever beam, the deflection is obtained as:

$$w = -\frac{M^{\mathrm{M}}}{2D} x_3^2 \tag{15.55}$$

with the maximum deflection occurring at the tip ($x_3 = l$).

The bending moment M^{M} acting on the cross-section of the laminate is given by

$$M^{\mathrm{M}} = b \int_{-h/2}^{h/2} \sigma_{33} x_1 \mathrm{d}x_1 \tag{15.56}$$

Considering only the linear deformation and substituting Eq. (15.25) into Eq. (15.56) while taking Eq. (15.34) into account, we can derive the bending moment induced by the static magnetic field as follows:

$$M^{\mathrm{M}} = -\frac{bh^2}{8}\left[-\frac{(d_{33})_1}{\left(s_{33}^{\mathrm{H}}\right)_1} + \frac{(d_{33})_2}{\left(s_{33}^{\mathrm{H}}\right)_2}\right] nI \tag{15.57}$$

Here, subscripts 1 and 2 denote the upper and lower layers, respectively. Similarly, the bending stiffness of the laminate is given by:

$$D = b \int_{-h/2}^{h/2} \frac{1}{s_{33}^{\mathrm{H}}} x_1^2 \mathrm{d}x_1 = \frac{bh^3}{24}\left[\frac{1}{\left(s_{33}^{\mathrm{H}}\right)_1} + \frac{1}{\left(s_{33}^{\mathrm{H}}\right)_2}\right] \tag{15.58}$$

Substituting Eqs. (15.57) and (15.58) into Eq. (15.55), we obtain the relationship between deflection w and current I.

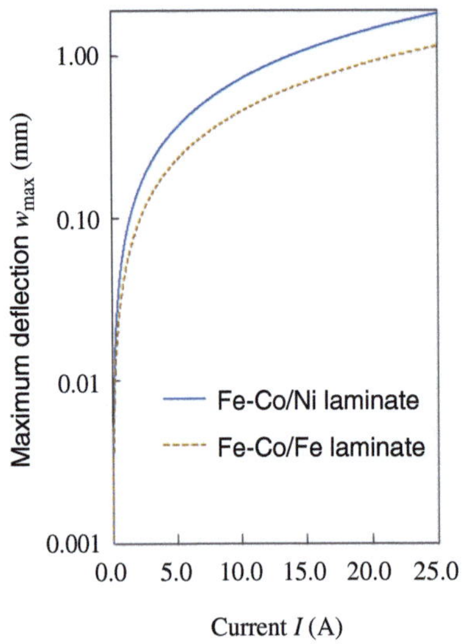

Figure 15.9 Relationship between maximum tip deflection and current.

15.4.2 Numerical Analysis

Figure 15.9 shows the relationship between maximum tip deflection and current for the Fe–Co/Ni and Fe–Co/Fe laminates. In this analysis, the parameters assumed were $l = 70$ mm, $b = 10$ mm, $h = 1$ mm, and $n = 2500$ turns/mm. The physical properties used were $s_{33}^{H} = 5 \times 10^{-12}$ m^2/N and $d_{33} = 0.125 \times 10^{-9}$ m/A for Fe–Co, while for Ni and Fe, the properties were $s_{33}^{H} = 5 \times 10^{-12}$ m^2/N, $d_{33} = -0.0736 \times 10^{-9}$ m/A and $s_{33}^{H} = 5 \times 10^{-12}$ m^2/N, $d_{33} = 0$ m/A, respectively. For the Fe–Co/Ni laminate, a deflection of approximately 1 mm can be achieved at approximately 14 A.

Currently, piezoelectric ceramic bimorphs are the most widely used actuators that use flexural displacement [14]. On the other hand, magnetostrictive laminates, made from metal, are known for their durability and ability to generate considerable force, making them promising candidates for future applications as bending actuators.

15.5 Conclusion

This chapter provides a concise overview of the mechanics involved in the design, development, and evaluation of magnetostrictive materials, highlighting the current status of the research on magnetostrictive composite materials. The application of the mechanics of magnetostrictive materials and composites is crucial for the optimal design of magnetostrictive devices that demonstrate interactions between electromagnetic and mechanical fields. Furthermore, given that magnetostrictive devices operate noncontact and are influenced by the surrounding magnetic field, it is essential to perform numerical simulations that also incorporate the electromagnetic field in free space.

Advancements in the field of mechanics of magnetostrictive materials and composites are expected in the future. While our discussion has primarily focused on applications in energy-harvesting devices, sensors, and actuators, this research is also expected to prove invaluable for crack detection and fatigue life prediction, both of which require considerations of fracture mechanics.

Acknowledgment

The author greatly acknowledge the support of this work by Japan Society for the Promotion of Science (JSPS), Core-to-Core Program grant number: JPJSCCA20200005.

References

1 Deng, Z. and Dapino, M.J. (2017) Review of magnetostrictive vibration energy harvester. *Smart Mater. Struct.*, 26, 103001.

2 Narita, F. and Fox, M.A. (2018) Review on piezoelectric, magnetostrictive, and magnetoelectric materials and device technologies for energy harvesting applications. *Adv. Eng. Mater.*, 20, 1700743.

3 Kurita, H., Fakhruddin, S.M.B., Inoue, K.Y., Nakaki, T., Kuroda, S., Wang, Z., Araki, W., Shiku, H. and Narita, F. (2023) Energy-harvesting and mass sensor performances of magnetostrictive cobalt ferrite-spattered Fe–Co alloy plate. *J. Alloys Compd.*, 951, 169844.

4 Mori, K., Narita, F. and Iriya, N. (2023) Magnetostrictive composite bolt fastening characteristics and internal axial force. *Adv. Eng. Mater.*, 25, 2300771.

5 Miyashita, T., Katabira, K., Kurita, H. and Narita, F. (2024) Measurement of magnetic flux density changes in mode I interlaminar fracture in magnetostrictive fiber–embedded glass fiber-reinforced polymer composites. *J. Compos. Sci.*, 8, 8.

6 Tremolet de Lacheisserie, E.D. (1993) Magnetostriction: Theory and Applications of Magnetoelasticity, CRC Press, Boca Raton.

7 Butterworth, S. and Smith, F.D. (1931) The equivalent circuit of the magnetostriction oscillator. *Proc. Phys. Soc.*, 43, 166–85.

8 Nye, J.F. (1957) Physical Properties of Crystals, Oxford University Press, Oxford.

9 Tiersten, F. (1969) Linear Piezoelectric Plate Vibrations, Plenum Press, New York.

10 IEEE Std 319-1990. (1991) IEEE Standard on Magnetostrictive Materials: Piezomagnetic Nomenclature, New York.

11 Engdahl, G. (2000) Handbook of Giant Magnetostrictive Materials, Academic Press, San Diego.

12 Butler, J.L. (1986) Highly Active Magnetostrictive Transducers, Image Acoustics Inc., Massachusetts.

13 Narita, F. and Katabira, K. (2017) Stress-rate dependent output voltage for $Fe_{29}Co_{71}$ magnetostrictive fiber/polymer composites: fabrication, experimental observation and theoretical prediction. *Mater. Trans.*, 58, 302–4.

14 Narita, F. and Wang, Z. (2025) Piezoelectric Materials, Composites, and Devices, Fundamentals, Mechanics, and Applications, Elsevier, London.

16

Additive Manufacturing of Magnetostrictive Alloys: Current Trends and Perspectives

Laheurte Pascal[1] *and Lohmuller Paul*[1,2]

[1] *Laboratoire d'Etude des Microstructures et de Mécanique des Matériaux, CNRS-Université de Lorraine, France*
[2] *Institut Jean Lamour, IJL, CNRS-Université de Lorraine, Campus ARTEM, Nancy Cedex, France*

16.1 Introduction

Additive manufacturing (AM) processes encompass all manufacturing processes based on layer-by-layer creation of a 3D volume. Generally, these methods can be divided into four main stages: numerical modeling of the 3D volume, slicing into layers of a given thickness and assigning a layer-filling strategy, layer-by-layer manufacturing itself, and post-processing stages (removal of manufacturing supports, surface finishing, heat treatment, etc.). Layer-by-layer volume processing offers many advantages over conventional processes, whether subtractive (machining, electrochemical, etc.) or forming (casting, rolling, etc.). These include:

- Production of complex 3D geometries (internal cooling channels, lattice structures, etc.)
- Optimization of part weight/performance ratios (topological optimization schemes, mesostructured materials)
- Reducing the number of parts in an assembly (simultaneous manufacture of embedded parts)
- Reduced manufacturing wastes (reuse of initial materials, absence of swarf, etc.)

Although initially developed for photosensitive materials (stereolithography apparatus (SLA) [1]), diversification of additive technologies today enables the manufacture of a wide range of polymeric, metallic, ceramic, and composite materials. As magnetostrictive materials are mainly metallics (metals or alloys), ceramics (ferrites), or metallic glasses, not all additive processes can be used for their production.

First, we will look at the processes and strategies used for the additive production of these materials. We will then look at the benefits that can be derived from the use of additive processes for the production of magnetostrictive materials. Finally, we'll look at related applications and future developments.

16.2 Magnetostrictive Materials Processing by AM

The predominantly metallic nature of magnetostrictive materials strongly determines AM processes that can be employed for their production. Two main strategies can be distinguished: direct production of material or alloy or composite production form using particles in a matrix (often polymers).

Direct processing is therefore based on the additive processes commonly used for the production of metallic materials, such as powder bed melting (PBF) and direct energy deposition (DED). The PBF process consists of two stages: a powder bed deposition step and a selective melting step of the area corresponding to the manufactured layers of the 3D volume. These two steps are repeated until the complete volume is obtained. The process uses metallic or ceramic powders, generally with a particle size distribution between 10 and 60 μm. The thermal energy required to melt the powder is either supplied by a laser or an electron beam, which is, respectively, referred to as laser powder bed fusion (LPBF) or electron beam melting (Figure 16.1a). It should be noted that alloying can be carried out on the basis of a prealloyed powder or an in situ alloy with a mixture of elementary powders, but it is not possible to adjust the alloy composition during the manufacture, and multimaterial manufacturing is complex to implement. This process offers the best resolutions (of the order of a few tens of microns), enabling the production of thin structures, particularly architectural structures (lattice structures, triply periodic minimal surfaces (TPMS), stochastic scaffolds). The main drawback of this process is that it requires a large stock of initial powder, making it unsuitable for the development of new alloys.

DED processes are also based on the use of powders, mainly metallic, with a particle size distribution generally larger than LPBF, between 40 and 100 μm. The main difference is that the powder is fed by injection nozzles directly to the thermal source, usually a laser (Figure 16.1b). The laser source/injection nozzle pair is mounted on a robotic arm, as is the manufacturing substrate, enabling three- to five-axis manufacturing. The DED process therefore provides access to larger manufacturing volumes, enabling the production of large parts, although its resolution is lower, limiting its viability for the production of thin structures. The main advantage of DED processes is the possibility of using several injection nozzles, each associated with a different metal or alloy. This enables multimaterial parts to be produced or composition gradients to be achieved. Another advantage of DED processes is that they can be used to refill preexisting parts (repairing, local functionalization).

The ultrasonic additive manufacturing (UAM) process is also used to produce magnetostrictive materials. It consists of friction welding of sheets together, followed by a milling step corresponding to the cross section of the desired part (Figure 16.1c). The UAM process can also be used to produce multimaterial parts but is more limited in terms of accessible geometry and resolution than the PBF or DED processes.

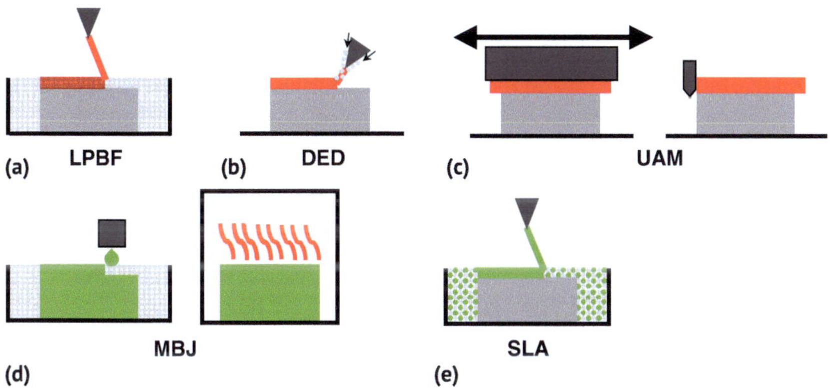

Figure 16.1 (a) Schematic representation of AM process, laser powder bed fusion (LPBF), (b) direct energy deposition (DED), (c) ultrasonic additive manufacturing (UAM), (d) metal binder jetting (MBJ), (e) stereolithography apparatus (SLA).

Metal binder jetting (MBJ) technology can also be used to manufacture parts from magnetostrictive materials. MBJ processes also employ a metal powder bed, but this is consolidated by a selectively deposited binder. A debinding and sintering step then produces a fully metallic part (Figure 16.1d). The main advantage of MBJ technology is that it is faster and less costly than LPBF technology and enables complex structures to be produced with interesting resolution without the need for support structures. However, the sintering step induces important shrinkage, which needs to be anticipated.

Methods for the additive production of indirect magnetostrictive materials are more aimed at the production of a composite in which the active material is often present in particulate form. This approach can be found in selective laser sintering (SLS) processes (Figure 16.1e). SLS processes are based on the use of preheated polymer powders, with a laser source providing the local energy required to melt the particles. In this context, the use of a mixture of polymer powder and magnetostrictive powder enables the production of a composite without the need to melt the metal particle. This strategy can also be applied to SLA processes, where the metal powder is mixed with the photosensitive resin.

There are just a few studies about AM of magnetostrictive alloys, and magnetostrictive properties as such are often not investigated to the detriment of the magnetic properties. However, as with any manufacturing process, a non-negligible part of new material development involves optimizing manufacturing conditions. In FA, the main aim of these parametric studies is to ensure material integrity (absence of cracks, lowest porosity rate). Second, the influence of the set of parameters on microstructures and, therefore, material properties is studied. Parametric studies of this type have been carried out on FeCo alloys [1–3], FeCoV [4, 5], FeGa [6], and FeSi [7, 8]. MBJ technology eliminates the need to optimize processing parameters, as consolidation is based on resin properties and the sintering step is mainly a function of the melting temperature of the magnetostrictive material concerned. Palmer et al. and Lima et al. have, respectively, investigated the magnetostrictive and magnetoelectric properties of $CoFe_2O_4$ ferrites obtained by an MBJ process [9, 10], and

Ahmed et al. have studied the magnetostrictive properties of a 17/4ph stainless steel [11]. Qi et al. are also interested in $CoFe_2O_4$ in composite form obtained by SLS in a PLLA matrix but focus mainly on the magnetic properties [12]. Paul et al. have studied the magnetostrictive and mechanical properties of an epoxy/Terfenol-D composite obtained by direct ink writing [13]. The possibility of in situ alloying offered by DED technology also makes it possible to study complex compositions such as the magnetic Fe–Si–B–Cu–Nb alloy [14].

16.3 Benefits of AM for the Production of Magnetostrictive Materials

AM of a magnetostrictive material can be of interest in terms of the material itself, its microstructure, and its properties. On the other hand, the design aspects and design freedom are associated with the AM process.

The development of new material grades is a long and costly process, based on both numerical and experimental aspects. DED processes are particularly well suited to this type of development. In particular, they enable the production of materials with complex compositions of up to 10 elements, with relatively modest powder feedstock (much lower than LPBF processes) [14]. DED processes can also be used to modify the composition of the deposited material during the manufacturing, enabling the production of chemical composition gradients [15]. This approach, used in the work of Geng et al., enables the exploration of the Fe–Co binary system across the whole spectrum of composition, structural phases, and magnetic properties. It also enables the identification of new compositions and their determination. This process, combined with physicochemical characterization, enables rapid database acquisition for exploring the links between chemical composition/microstructures/properties. It is also suitable for introducing alloying elements in low proportions. Finally, it can also be used to explore rare-earth systems at low cost.

The properties of magnetostrictive materials are known to be highly dependent on their microstructures and, therefore, on the associated manufacturing process. In fact, for a given composition, grain size, morphology, and orientation have their own influence on the magnetostrictive properties of the material. Indeed, for metallic processes requiring particle melting, the solidification rate is responsible for grain size and morphology. In general, the higher the solidification speed, the more columnar the grain shape. This induces anisotropy of properties depending on the direction of stress on a sample [5]. Solidification speed may also be responsible for the presence of amorphous phases. The adjustment and optimization of processing parameters (laser power, laser scanning speed, etc.) influences the fraction of amorphous phase in relation to the solid solution and therefore the associated properties [8]. Texture is also known to have a strong influence on magnetostrictive properties, as shown, for example, by the study of Downing et al. on Fe–Ga and Fe–Al alloys [16] or the work of Embadi et al. [17]. LPBF and DED processes are known to generate microstructures that are strongly oriented along the manufacturing direction, enabling these properties to be greatly improved in this direction. This has been highlighted in the work of Gao et al. [18] and Na et al. [19]. More conventionally, heat treatments can be used to control microstructures and thus adapt material properties. Riipinen et al. [4] have shown a significant influence of heat treatments on the mechanical properties of an FeCoV alloy produced by LPBF. It is also important to note that the texture produced during the manufacturing process can be retained after heat treatment, as demonstrated by Gao et al. [18].

Magnetostrictive materials are notoriously known as brittle or having a low manufacturability; they are commonly used in thin sheet form, requiring numerous rolling/annealing steps. The layer-by-layer production of a part greatly reduces the number of manufacturing steps between the initial material and the finished product. In fact, apart from the finishing stages, the entire shaping process takes place in the same operation. Once the processing parameters have been optimized, it becomes possible to shape materials with a reputation for brittleness, such as Fe–Si alloys or Terfenol-D [2, 7].

One of the major advantages of additive processes is their ability to produce complex geometries that are unattainable by conventional processes. This freedom of design, combined with the resolution of certain FA processes (such as LPBF), enables the development of mesostructured or architecturally structured materials. By analogy with "conventional" foams such as metal foams, FA allows complete control over the shape, scale, and distribution of pores in a given volume. This results in several families of architecturally structured materials, such as lattice structures (periodic organization of a beam network in a unit cell, enabling the realization of a network by tiling), TPMS structures (which are a mathematical description of a surface in a volume also used as a unit cell), or stochastic networks. These materials are therefore ideal not only for optimizing the weight/performance ratio but also for all phenomena favored by a large exchange surface (thermal dissipation, electrochemical synthesis, osseointegration, filtration support, etc.). In the case of magnetostrictive materials, eddy currents induce a loss of energy due to the Joule effect. One way of counteracting this phenomenon is to use a stack of metal sheets insulated from each other. The use of an architectural material acts in the same way. Kurita et al. have highlighted the influence of architecture on the energy harvesting of a magnetostrictive material. The use of a structured material enables energy harvesting approximately five times greater than that of a solid material inscribed in an equivalent volume. Ahmed et al. are also interested in the influence of architecture, using MBJ technology to highlight the role of the fraction of magnetostrictive material on magnetostriction, with a deformation two times greater in the case of material after sintering [11]. In addition, they highlight scale effects, with an increase in properties and a decrease in element size.

Additive processing of magnetostrictive materials is relatively recent (the first paper around 2016), so industrial applications are still very limited. Most of the work undertaken is still at the laboratory level, and there are still many of them. Currently, the number of alloy grades available is very small and is concentrated around FeCo, FeGa, $CoFe_2O_4$, and FeSi alloys. The Fe–Ni (amorphous or crystalline), Mn–Zn, and Ni–Zn families of alloys may also offer advantages for FA processing, particularly in view of their brittle nature. On the other hand, the use of the DED process can enable the identification of new compositions, as well as the alloying optimization of preexisting compositions.

Magnetostrictive materials are mainly used for applications such as sensors, implants, or energy harvesting. Gao et al. have studied the magnetostrictive properties and biodegradability of an FeGa alloy produced by LPBF for implantology applications [20]. Qi et al. propose a wireless sensor for implantology based on a polymer/magnetostrictive particle composite [12]. Lima et al. propose a multimaterial coupling with a magnetostrictive screen for capturer/activator applications. We can also cite the work of Schleider and Dapino, who propose a FeGa/Al 6061 composite produced by UAM for vibration-damping applications [21]. In terms of energy harvesting, pioneering work has been carried out on the influence of material architecture [2, 3]. Further studies are underway, but the coupling of magnetostrictive materials and FA seems promising and could lead to an important increase in the amount of energy involved.

References

1 Jones, N.J., Yoo, J., Ott, R., Lambert, P.K., Petculescu, G., Simsek, E., Schlagbel, D. and Lograsso, T.A. (2018) Magnetostrictive performance of additively manufactured. *AIP Adv.*, 8, 056403.

2 Nakajima, K., Leparoux, M., Kurita, H., Lanfant, B., Cui, D., Watanabe, M., Sato, T. and Narita, F. (2022) Additive manufacturing of magnetostrictive Fe–Co alloys. *Materials*, 15, 709.

3 Kurita, H., Lohmuller, P., Laheurte, P., Nakajima, K. and Narita, F. (2022) Additive manufacturing and energy-harvesting performance of honeycomb structured magnetostrictive Fe52–Co48 alloys. *Addit. Manuf.*, 54, 102741.

4 Riipinen, T., Metsä-Kortelainen, S., Lindroos, T., Keränen, J.S., Manninen, A. and Pippuri-Mäkeläinen, J. (2019) Properties of soft magnetic Fe-Co-V alloy produced by laser powder bed fusion. *Rapid Prototyp. J.*, 25, 699–707.

5 Kustas, A.B., Susan, D.F., Johnson, K.L., Whetten, S.R., Rodriguez, M.A., Dagel, D.J., Michael, J.R., Keicher, D.M. and Argibay, N. (2018) Characterization of the Fe-Co-1.5V soft ferromagnetic alloy processed by Laser Engineered Net Shaping (LENS). *Addit. Manuf.*, 21, 41–52.

6 Yoa, X., Deng, W., Wang, L., Gao, C. and Shuai, C. (2024) Laser-beam powder bed fusion of magnetostrictive Fe81Ga19 alloys: parameter optimization, microstructural evolution and magnetostrictive properties. *Phys. Scr.*, 99, 064936.

7 Özden, M.G. and Morley, N.A. (2023) Optimizing laser additive manufacturing process for Fe-based nano-crystalline magnetic materials. *J. Alloys Compd.*, 960, 170644.

8 Garibaldi, M., Ashcroft, I., Hillier, N., Harmon, S.A.C. and Hague, R. (2018) Relationship between laser energy input, microstructures and magnetic properties of selective laser melted Fe-6.9%wt Si soft magnets. *Mater. Charact.*, 143, 144–51.

9 Palmer, S. and Deng, Z. (2020) Additive manufacturing of magnetostrictive thin film sensors, In *Sensors and Smart Structures Technologies for Civil, Mechanical, and Aerospace Systems, Chez Proceedings SPIE Volume* 11379.

10 Lima, A.C., Pereira, N., Policia, R., Ribeiro, C., Correia, V., Lanceros-Mendez, S. and Martins, P. (2019) All-printed multilayer materials with improved magnetoelectric response. *J. Mater. Chem. C*, 7, 5394.

11 Ahmed, N., Deffley, R., Kundys, B. and Morley, N.A. (2023) 3D printing of magnetostrictive property in 17/4 ph stainless steel. *J. Magn. Magn. Mater.*, 585, 171115.

12 Qi, F., Gao, X., Shuai, Y., Peng, S., Deng, Y., Yang, S., Yang, Y. and Shuai, C. (2022) Magnetic-driven wireless electrical stimulation in a scaffold. *Compos. Part B*, 237, 109864.

13 Paul, A.A., Nguyen, Q.H. and Shen, W. (2024) 3D printed magnetostrictive polymer composites (MCPs) for wireless stress sensing. *Mater. Des.*, 247, 113402.

14 Conteri, R., Borkar, T., Nag, S., Jaeger, D., Chen, X., Ramanujan, R.V. and Banerjee, R. (2017) Laser additive processing of Fe-Si-B-Cu-Nb magnetic alloys. *J. Manuf. Process.*, 29, 175–81.

15 Geng, J., Nlebedim, I.C., Besser, M.F., Simsek, E. and Ott, R.T. (2016) Bulk combinatorial synthesis and high throughput characterization for rapid assessment of magnetic materials: application of Laser Engineered Net Shaping (LENS™). *JOM*, 68, 1972–7.

16 Downing, J.R., Na, S. and Flatau, A.B. (2017) Compressive pre-stress effects on magnetostrictive behaviors of highly textured Galfenol and Alfenol thin sheets. *J. Magn. Magn. Mater.*, 7, 056420.

17 Gao, C., Yao, W., Deng, Y., Pan, H. and Shuai, C. (2023) Laser-beam powder bed fusion followed by annealing with stress: a promising route for magnetostrictive improvement of polycrystalline Fe81Ga19 alloys. *Addit. Manuf.*, 68, 103516.

18 Embadi, A.A., Nedjad, S.H. and Ghavifekr, H.B. (2014) Effect of solidification texture on the magnetostrictive behavior of Galfenol. *Metall. Mater. Trans. A*, 45, 906–10.

19 Na, S., Galuardi, J. and Flatau, A.B. (2017) Consolidation of (001)-oriented Fe–Ga flakes for 3-D printing of magnetostrictive powder materials. *IEEE Trans. Magn.*, 53, 2503204.

20 Gao, C., Zeng, Z., Peng, S. and Shuai, C. (2022) Magnetostrictive bulk Fe-Ga alloys prepared by selective laser melting for biodegradable implant applications. *Mater. Des.*, 220, 110861.

21 J. Schleider and Dapino, M.J., (2014) Stiffness tuning of FeGa structures manufactured by ultrasonic additive manufacturing, In *Industrial and Commercial Applications of Smart Structures Technologies, Chez Proceedings Volume 9059*, San Diego.

17

Energy Harvesting with Spin–Orbit Torque: From Magnetization Switching to Emerging Spintronics

Shutaro Karube

Institute for Chemical Research, Kyoto University, Kyoto, Japan

17.1 Introduction

Spintronics is a new field of electronics that actively utilizes the spin degree of freedom, in addition to the charge degree of freedom of electrons and holes, which have conventionally been handled. The opening physical phenomenon of spintronics is nothing other than the giant magnetoresistance (GMR) effect [1, 2], which was the subject of the Nobel Prize in Physics in 2007. This phenomenon appears in multilayer structures composed of ferromagnets and nonmagnetic layers with nanoscale thickness, where spin-dependent electron scattering gives rise to magnetoresistance. The nonmagnetic layer plays the role of interlayer exchange coupling mediated by conduction electrons, and, depending on its thickness, it is possible to form an antiferromagnetic coupling state [3, 4]. As a conventional magnetoresistive effect, the anisotropic magnetoresistance (AMR) effect observed in a single ferromagnetic layer has long been known, but its magnetoresistance ratio was at most around 1–2% [5]. In contrast, the GMR effect exhibits magnetoresistance ratios of several tens of percent, making it not only a new discovery in fundamental research but also a physical phenomenon specialized for applications in the electronics industry. Since this phenomenon allows highly sensitive detection of magnetic fields, it could be applied as a magnetic sensor and was used in magnetic read heads of hard disk drives [6, 7]. As a result, it became possible to sensitively read the stray magnetic field from each information bit in magnetic disks, greatly contributing to high-density information storage, and therefore it became the subject of the Nobel Prize. In addition, a similar phenomenon known as the tunneling magnetoresistance (TMR) effect is also well known in spintronics as a phenomenon that exhibits a large magnetoresistance ratio. This phenomenon is realized in magnetic tunnel junction (MTJ) devices, in which a thin insulating layer of about 1–2 nm thickness is sandwiched between ferromagnetic layers. Similar to the GMR effect, this magnetoresistive effect arises from the difference between parallel and antiparallel magnetization alignments of the ferromagnetic layers and originates from spin-dependent tunneling conduction. The two ferromagnetic layers of this device are called the pinned layer and the free layer, and by reversing the magnetization of the free layer, parallel and antiparallel configurations of magnetization are realized. The pinned layer is coupled with an antiferromagnetic layer, and its magnetization is fixed by exchange bias. In 1995, independent groups led by Miyazaki and Moodera realized about 20% TMR ratio at room temperature by using amorphous insulating layers (aluminum oxide) [8, 9]. Furthermore, by optimizing the selection of ferromagnetic layers and the fabrication of tunnel barriers, the magnetoresistance ratio improved to about 80% but showed a saturation tendency. Subsequently, in 2001, first-principles calculations of epitaxial tunnel junctions theoretically predicted a TMR ratio exceeding 1000% [10], and, in 2004, independent groups led by Yuasa and Parkin discovered a giant room-temperature TMR effect [11, 12]. This phenomenon originates from band-selective tunneling and achieves a very high TMR ratio due to strong spin filtering. At present, TMR ratios exceeding 600% have been obtained using epitaxial MgO barriers [13], and they are also used commercially as MgO-TMR read heads [14]. In this way, the discoveries of the GMR effect and the TMR effect have led to the development of highly sensitive magnetic sensing technology in the field of spintronics. The MTJ device in which the TMR effect is realized can be utilized not only for sensing but also as a nonvolatile information storage device. This device is called magnetic random access memory (MRAM). By reversing the magnetization of the free layer in the MTJ part of MRAM, the binary state of 0 or 1 for one bit can be realized, and sufficient sensing is possible through the TMR effect. Depending on the difference in the method of magnetization reversal, there are several types of MRAM. Mainly in spintronics, there are spin-transfer torque MRAM (STT-MRAM), spin–orbit torque MRAM (SOT-MRAM) [15], and voltage-controlled MRAM (VC-MRAM) [16]. In VC-MRAM, a voltage is applied to the ferromagnet of the free layer, and by changing the electronic orbitals near the interface, the magnetic anisotropy is modulated, and the magnetization reversal barrier is reduced, thereby inducing magnetization reversal through thermal excitation or an assisting magnetic field [16]. While VC-MRAM induces magnetization reversal by voltage, the other two types of MRAM induce magnetization reversal by current. In STT-MRAM, spin-polarized current in the ferromagnetic layer exerts a spin-transfer torque on the magnetization of the free layer, while, in SOT-MRAM, spin current generated by applying current to a nonmagnetic layer gives rise to SOT [17]. Figure 17.1 shows the MTJ unit cell. Since STT-MRAM requires tunneling of a sufficient amount of spin-polarized current to induce magnetization reversal, the insulating layer can be damaged, and the device suffers from poor endurance. In contrast, in SOT-MRAM, the large current required to generate spin current does not pass through the tunnel barrier, so damage to the insulating layer is almost negligible. In this way, SOT-MRAM has clearly separated write and read paths and boasts high endurance, and the probability of read disturbance is also very low. Furthermore, magnetization reversal by SOT can be induced on a sub-nanosecond timescale, allowing ultrafast information rewriting, which is comparable to the speed of conventional information devices such as DRAM and SRAM [18]. From these perspectives, nonvolatile MRAM driven by SOT is regarded as a "killer

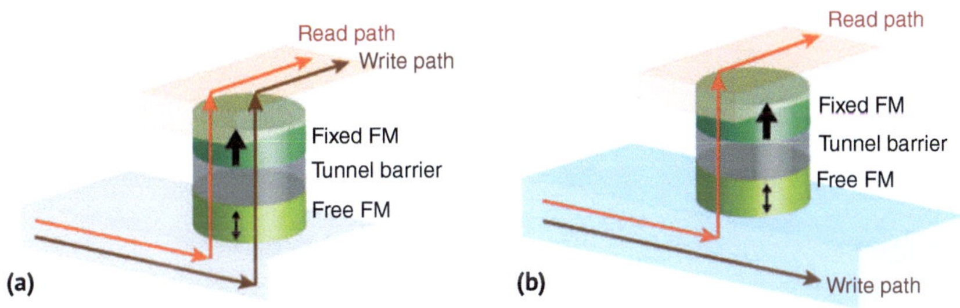

Figure 17.1 Schematic diagram of the MTJ unit cell in MRAM. (a) STT-MRAM, (b) SOT-MRAM, showing the details of the read path and write path. *Source*: [15] / with permission of John Wiley & Sons.

device" for efficiently processing the enormous amount of information brought by today's rapidly advancing information society. Research on the underlying phenomena, namely, spin-current generation, SOT, and magnetization switching, has therefore become increasingly important. This chapter introduces these physical phenomena from the perspective of fundamental research.

17.2 Spin-current and Spin–Orbit Torque

The spin current refers to a flow that carries only pure spin angular momentum without any net charge current. It can be generated by applying a charge current to various materials through spin–orbit interaction in solids. The most fundamental principle of spin-current generation is the spin Hall effect. This phenomenon was theoretically predicted in 1971 by D'yakonov and Perel as the generation of spin current in semiconductors [20], and in 2004 it was experimentally observed for the first time in GaAs, where up-spins and down-spins accumulate at the left and right edges of a current channel, respectively, as detected by the magneto-optical Kerr effect [21]. In addition to semiconductors, Hirsch reconsidered in 1999 that the spin Hall effect could also occur in metals [22], and in 2006 and 2007, independent groups led by Valenzuela, Saitoh, and Otani demonstrated its existence in metals [23–25]. Interestingly, the current direction $\mathbf{J}_\mathrm{C}$, the spin polarization direction $\boldsymbol{\sigma}$ and the spin-current direction $\mathbf{J}_\mathrm{S}$ are mutually orthogonal, expressed as $\mathbf{J}_\mathrm{S} \propto \mathbf{J}_\mathrm{C} \times \boldsymbol{\sigma}$ (Figure 17.2). As mentioned above, because spin–orbit interaction is the origin, finite spin currents can be generated in transition metals with large atomic numbers such as Pt, Ta, and W. One of the key parameters in the spin Hall effect is the spin Hall angle, which corresponds to the conversion efficiency from charge current to spin current. For SOT-MRAM applications, materials with a larger spin Hall angle are highly important, since they enable spin-current generation sufficient for magnetization switching at lower charge currents. Detection methods of spin currents include direct observation of spin accumulation, as in the GaAs example described earlier, but more quantitatively, a ferromagnetic layer is usually attached to the nonmagnetic layer generating spin current, and the SOT exerted on the magnetization of the ferromagnet is detected via magnetoresistance or rectification effects. Although some

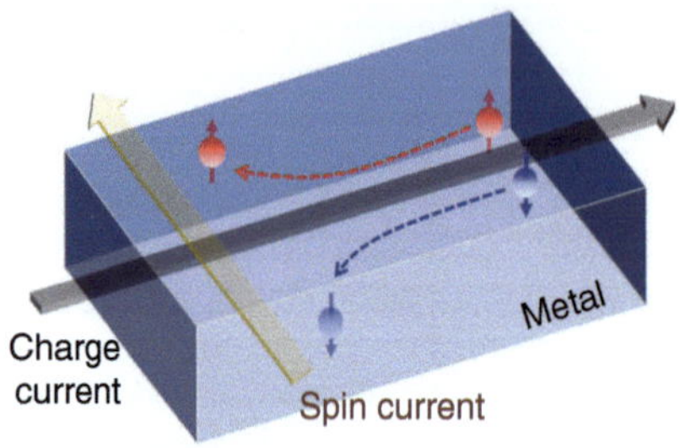

Figure 17.2 Schematic illustration of the spin Hall effect. When a charge current is applied to a heavy metal, a transverse spin current is generated, with the spin polarization direction perpendicular to both the charge and spin current directions.

variation exists depending on the method, it is generally known that materials such as Pt and Ta exhibit spin Hall angles of a few percent [26]. The physical origin of the spin Hall effect is broadly classified into two categories: an intrinsic mechanism arising from the electronic band structure, and an extrinsic mechanism caused by impurity scattering. The extrinsic mechanism, independent of the band structure, produces spin current through spin-dependent asymmetric scattering of electrons by impurity potentials such as heavy elements. It is mainly divided into skew scattering and side-jump mechanisms [27]. In fact, reports show that large spin Hall angles can be obtained by adding small amounts of Bi or Ir impurities to Cu, which has weak spin–orbit interaction [28, 29]. Furthermore, it has been reported that oxidation of transition metals such as Cu [30], Re [31], and W [32] can enhance the spin Hall angle. Thus, the extrinsic mechanism plays an important role in achieving large spin Hall angles.

Next, regarding the intrinsic mechanism, the Berry phase—a geometric phase of the electronic wavefunction—is deeply involved [33–35]. According to Bloch's theorem, electrons in a crystal are described by functions $u_{n\mathbf{k}}(\mathbf{r})$ arising from the periodicity of the lattice,

$$\psi_{n\mathbf{k}}(\mathbf{r}) = e^{i\mathbf{k}\cdot\mathbf{r}} u_{n\mathbf{k}}(\mathbf{r}) \tag{17.1}$$

where $n, \mathbf{k}$ denote the band index and the wavevector, respectively. The Berry connection $\mathbf{A}_n(\mathbf{k})$, corresponding to a gauge field in momentum space, is defined as

$$\mathbf{A}_n(\mathbf{k}) = i \langle u_{n\mathbf{k}} |\nabla_{\mathbf{k}}| u_{n\mathbf{k}} \rangle \tag{17.2}$$

which can be interpreted as the rate of phase change with respect to $\mathbf{k}$. The phase acquired by an electron moving adiabatically along a closed curve C is the Berry phase γ_n, given by

$$\gamma_n = \oint_C \mathbf{A}_n(\mathbf{k}) \cdot d\mathbf{k} \tag{17.3}$$

The Berry connection behaves similarly to a vector potential, and the corresponding fictitious magnetic field in momentum space, the Berry curvature $\mathbf{\Omega}_n(\mathbf{k})$, is defined as

$$\mathbf{\Omega}_n(\mathbf{k}) = \nabla_{\mathbf{k}} \times \mathbf{A}_n(\mathbf{k}) \tag{17.4}$$

This Berry curvature is the origin of the intrinsic spin Hall effect. The semiclassical equation of motion for an electron is

$$\frac{d\mathbf{r}(t)}{dt} = \frac{1}{\hbar}\nabla_{\mathbf{k}}\varepsilon_n(\mathbf{k}) - \frac{d\mathbf{k}}{dt} \times \Omega_n(\mathbf{k}) \tag{17.5}$$

where the first term represents the group velocity determined by the band dispersion, and the second term is the anomalous velocity arising from the Berry curvature. Here, $\mathbf{r}(t)$ is the position of the electron wave packet, and $\varepsilon_n(\mathbf{k})$ denotes the energy dispersion. This anomalous velocity in the second term produces a transverse velocity perpendicular to the applied electric field, which is the origin of spin-current generation. Interestingly, the same Berry curvature underlies the anomalous Hall effect observed in materials with spontaneous magnetization, such as ferromagnets. Both the spin Hall effect and the anomalous Hall effect originate from spin–orbit interaction. According to the Kubo formula, the spin Hall conductivity can be obtained by summing the Berry curvature over momentum space, enabling first-principles prediction of the spin Hall effect magnitude for each material.

The spin Hall effect introduced so far represents the prototype for spin-current generation, and in SOT-MRAM devices, SOT generated via the spin Hall effect in heavy metals is the mainstream approach. However, in recent years, it has been found that spin currents can also be generated based on a completely different principle. This phenomenon occurs through surface or interfacial states in topological insulators (TIs) [36], or via Rashba and Dresselhaus spin–orbit interactions that split electronic bands [37], and is referred to as the Edelstein effect (Figure 17.3). In 1990, Edelstein theoretically pointed out that in two-dimensional electron

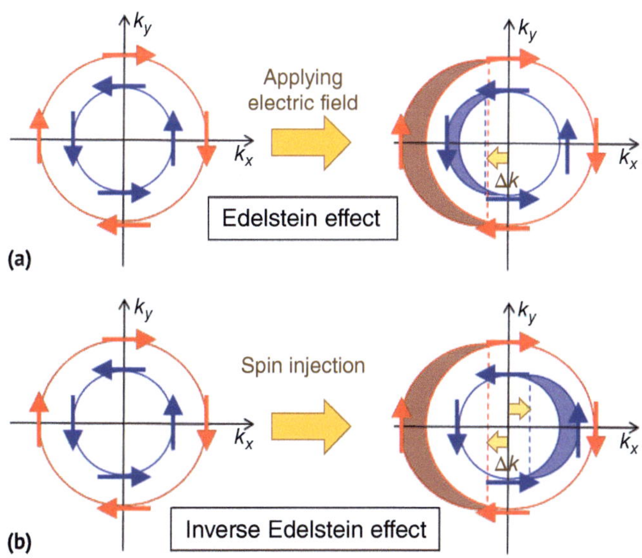

Figure 17.3 Schematic illustrations of (a) the Edelstein effect and (b) the inverse Edelstein effect in a Rashba system. In the Edelstein effect, application of an electric field shifts the Fermi contour, thereby inducing spin accumulation in the system. In the inverse Edelstein effect, spin injection causes shifts of the Fermi contour corresponding to the spin polarization direction, resulting in the induction of an electric field (or electric current).

systems with broken spatial inversion symmetry, an applied electric field induces spin polarization of the electrons [38]. Later, in 2000, Ganichev and collaborators experimentally observed spin polarization for the first time in a two-dimensional electron system in III–V compound semiconductor n-GaAs/AlGaAs quantum wells using optical methods [39]. Subsequently, in 2013, Rojas-Sánchez et al. reported a large inverse Edelstein effect at a Rashba interface in Ag/Bi [40]. In their study, spin currents were injected into the Rashba interface via a method called spin pumping [41], and the rectified charge current was measured as a voltage, marking a pioneering demonstration of spin conversion phenomena. Following this, large Edelstein effects utilizing the surface states of TIs such as Bi_2Se_3 and the interface at nonmagnetic metal/Bi_2O_3 have been widely reported [42–45], leading to a surge of research on spin-current generation in two-dimensional surface and interface systems. Although the Edelstein effect is a spin accumulation phenomenon in two-dimensional systems, when a metallic layer is adjacent to the spin-split interface, the nonequilibrium spin accumulation diffuses, resulting in a spin current. Like the spin Hall effect, the Edelstein effect generates spin currents upon applying a charge current, and thus can also be regarded as a type of spin conversion phenomenon, though it is qualitatively distinct from the spin Hall effect. Based on the spin Hall effect and the Edelstein effect described above, spin currents can be generated in a variety of material systems. As mentioned earlier, when these spin currents are injected into a ferromagnet, they exert a finite torque on the magnetization. This is known as SOT. SOT is mainly classified into two types: field-like torque and damping-like torque (Figure 17.4). These correspond to the precessional torque and damping torque described by the original Landau–Lifshitz–Gilbert (LLG) equation. The LLG equation, including these SOTs, can be expressed as follows.

$$\frac{\mathrm{d}\mathbf{m}}{\mathrm{d}t} = \mathbf{m} \times \mathbf{H}_{\mathrm{eff}} + \alpha \mathbf{m} \times \frac{\mathrm{d}\mathbf{m}}{\mathrm{d}t} + \frac{\gamma}{M_{\mathrm{S}}} [\tau_{\mathrm{FL}} \mathbf{m} \times \boldsymbol{\sigma} + \tau_{\mathrm{DL}} \mathbf{m} \times \mathbf{m} \times \boldsymbol{\sigma}] \tag{17.6}$$

Here, $\mathbf{m}, \mathbf{H}_{\mathrm{eff}}, \boldsymbol{\sigma}, \alpha, \gamma, M_{\mathrm{S}}, \tau_{\mathrm{FL}}, \tau_{\mathrm{DL}}$ represent the magnetization, effective magnetic field, spin, Gilbert damping constant, electron gyromagnetic ratio, saturation magnetization, field-like torque coefficient, and damping-like torque coefficient, respectively. The first term corresponds to the precessional motion, the second term to damping, and the third term to the SOT. As illustrated in Figure 17.4, consider a bilayer system composed of a heavy metal layer and a ferromagnetic layer. When a charge current is applied to the heavy metal layer, such as Pt, spin current is injected into the ferromagnetic layer above. The spin current, primarily generated via the spin Hall effect, acts on the magnetization as a damping-like torque. The field-like torque is believed to be induced by interfacial spin accumulation due to the Edelstein effect (sometimes simply referred to as the Rashba effect). As mentioned earlier, the Edelstein effect can also behave like a spin current generated by the conventional spin Hall effect through spin diffusion, and it has been shown that it can also generate damping-like torque [46].

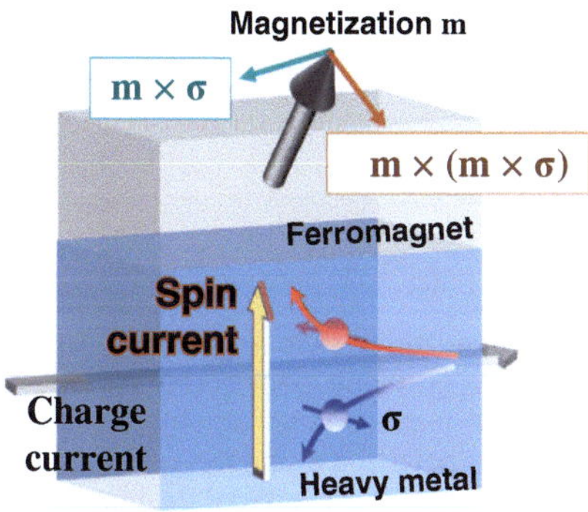

Figure 17.4 Illustration of SOT generated in a ferromagnetic layer by applying a charge current to a heavy metal/ferromagnet bilayer. The torques acting on the magnetization $\mathbf{m}$ are shown: $\mathbf{m} \times \boldsymbol{\sigma}$ represents the field-like torque, and $\mathbf{m} \times (\mathbf{m} \times \boldsymbol{\sigma})$ represents the damping-like torque.

17.3 Magnetization Switching Driven by Spin–Orbit Torque

By utilizing the spin current (or SOT) described in Section 17.2, we can induce magnetization switching. Depending on the magnetization direction (magnetic anisotropy) of the ferromagnet, there are mainly three types of magnetization switching schemes [47] (Figure 17.5). Due to the junction between the nonmagnetic layer, such as Pt, which generates the spin current, and the ferromagnetic layer where the magnetization switching occurs, the spin current naturally propagates along the z-direction (out-of-plane). If the applied charge current flows along the x-direction, the spin polarization direction of the spin current is automatically determined along the y-direction due to the orthogonality of the spin Hall effect. Therefore, in configurations such as Type-Y, where the magnetization is parallel to the spin polarization direction, magnetization switching can be achieved solely by the SOT. In contrast, for Type-Z and Type-X configurations, when using SOT generated by the spin Hall effect, it is necessary to apply an external magnetic field to break the symmetry of the magnetization switching. This is because, when the torque acts on both the magnetization-up and magnetization-down states, the torque itself is present, but it acts equally on both states, resulting in no energetic difference and thus failing to deterministically induce switching. Consequently, in most cases, external magnetic fields are considered necessary for SOT-induced magnetization switching. Configurations like Type-Y, which do not require an external magnetic field, are of great practical importance and are referred to as field-free magnetization switching (although Type-Y SOT-MRAM has in-plane magnetization, making it unsuitable for high-density storage). The historical background of SOT-induced magnetization switching traces back to 2009, when it was first demonstrated in a ferromagnetic semiconductor system, GaMnAs [48]. Shortly thereafter, from 2011 onward, numerous reports of this switching phenomenon in metallic systems were published. Metallic systems are particularly significant because they allow room-temperature operation, providing the foundational

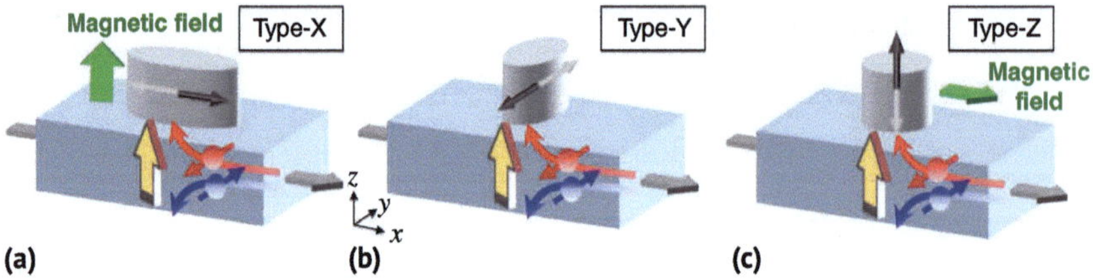

Figure 17.5 Schematic illustration of SOT-induced magnetization switching in a nonmagnetic layer/ferromagnetic dot structure. (a) Type-X (in-plane magnetization), (b) Type-Y (in-plane magnetization), and (c) Type-Z (out-of-plane magnetization) unit cells. Since the spin current generated by the spin Hall effect is polarized along the y-direction (current: x-direction, spin current: z-direction), the application of an external magnetic field is required to assist switching for configurations other than Type-Y.

knowledge for SOT-MRAM operation. In 2012, Liu et al. fabricated an MTJ device corresponding to an MRAM unit cell, successfully demonstrating SOT-induced magnetization switching via the TMR effect [49].

17.4 Field-free Magnetization Switching

As mentioned earlier, SOT-induced magnetization switching using the spin Hall effect requires the assistance of an external magnetic field, which has been a significant obstacle for MRAM applications. Therefore, demonstrating field-free magnetization switching has been a major challenge in spintronics. In particular, field-free switching in perpendicular-magnetized films (Type-Z), which are advantageous for high-density MRAM, is critically important. Here, several approaches are introduced. They mainly involve: (1) operations that modulate the symmetry of the magnetization itself, (2) operations that break the mirror symmetry of the spin-current generator, and (3) operations that break the time-reversal symmetry of the spin-current generator. While many other attempts exist, we focus here on these three.

First, we discuss operations that modulate the magnetization. As described in Section 17.3, Type-Z SOT-induced magnetization switching required an external magnetic field in the x-direction to create an energy difference between the up and down magnetization states, thereby enabling deterministic switching. This external field can be replaced by coupling a perpendicular-magnetized ferromagnet with an antiferromagnet, which induces an effective in-plane field via exchange bias (Figure 17.6). This drives field-free magnetization switching [50].

Next, we discuss operations that break the mirror symmetry of the spin-current generator. This operation allows the generation of spin currents with out-of-plane (z-direction) spin polarization, which cannot be induced by the conventional spin Hall effect alone, thereby enabling field-free magnetization switching. Mirror symmetry, as the term suggests, refers to the symmetry with respect to reflection: whether an object appears the same when mirrored. A well-known example is the human hand, which breaks mirror symmetry; no matter how you translate or rotate it, the left and right hands never coincide. This property is called chirality, and other

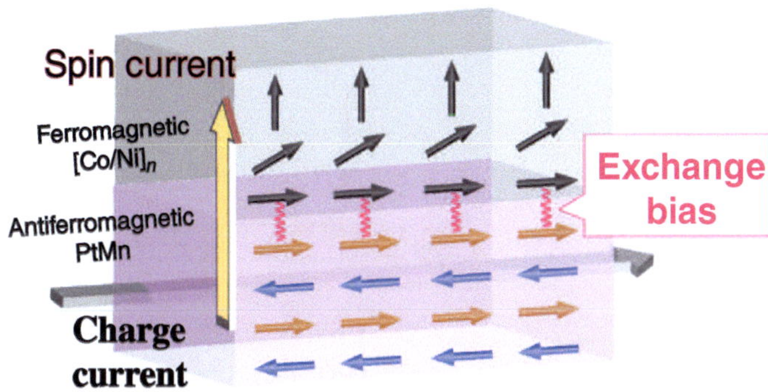

Figure 17.6 Junction structure of the antiferromagnetic PtMn/ferromagnetic multilayer $[Co/Ni]_n$. The $[Co/Ni]_n$ perpendicular magnetization cants under the exchange bias from PtMn, effectively creating a situation equivalent to an in-plane magnetic field, enabling field-free magnetization switching.

examples include helical structures such as DNA. Chirality has recently become an important concept in spintronics. Typically, heavy metals such as Pt, which are useful for the spin Hall effect, have cubic crystal structures with high symmetry. Thus, the structure appears identical when mirrored along the x, y, or z axes. The orthogonality of the charge current direction, spin-current direction, and spin polarization in the spin Hall effect is a consequence of this crystal symmetry. On the other hand, if mirror symmetry is broken by some operation or design, out-of-plane spin polarization becomes allowed. For a simple example, consider the zx-plane as a mirror plane. The y-axis is the normal direction to this mirror plane, and, under reflection, the y-component of a vector changes sign. If a charge current flows along x and the spin current propagates along y, only the spin-current direction changes sign; accordingly, the spin polarization must also reverse, due to the orthogonality condition in the spin Hall effect. For spins, which are axial vectors (while charge current is a polar vector), mirror symmetry in highly crystalline systems like Pt forbids out-of-plane spin components. Conversely, if mirror symmetry is broken, out-of-plane spin components can appear. Therefore, by generating spin currents in systems where mirror symmetry is broken, one can obtain spin currents with out-of-plane spin components capable of switching perpendicular magnetization. Based on this principle, two main approaches exist: making the sample structure asymmetric or choosing a low-symmetry crystal. Using structural asymmetry, Wang et al. in 2014 successfully generated out-of-plane spin polarization in a Ta/CoFeB/TaOx system (Figure 17.7a) by introducing in-plane structural asymmetry (Figure 17.7b)—wedge-shaped TaOx and controlled CoFeB oxidation—and demonstrated field-free magnetization switching (Figure 17.7b) [51]. Using low-symmetry crystals, Ralph et al. experimentally discovered out-of-plane spin-polarized spin currents in WTe_2 in 2017 [52]. Figure 17.7c and d shows the WTe_2 crystal structure: applying charge current along the high-symmetry b-axis preserves mirror symmetry (Figure 17.7c), yielding conventional spin Hall behavior, whereas charge current along the a-axis (Figure 17.7d), where mirror symmetry is broken, allows out-of-plane spin polarization, as confirmed by spin-current detection experiments. Based on this discovery, Chen

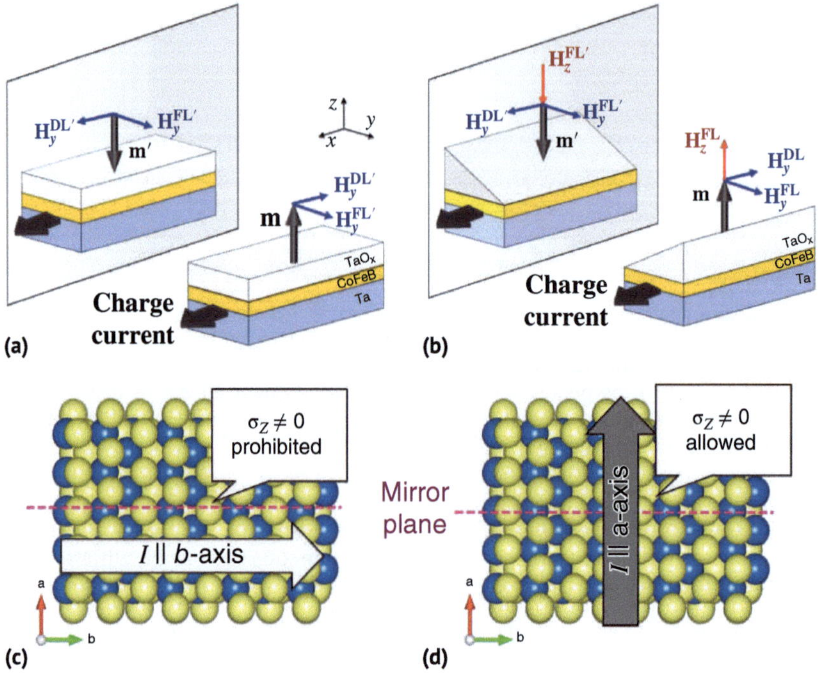

Figure 17.7 Upper panels: Schematic illustrations of structures (a) without and (b) with in-plane structural asymmetry. In conventional structures without asymmetry, mirror symmetry is preserved, which forbids the generation of an out-of-plane effective magnetic field; as a result, the symmetry required for magnetization reversal cannot be broken, and magnetization switching does not occur. In contrast, when in-plane asymmetry is present, mirror symmetry is broken, allowing the generation of an out-of-plane effective magnetic field, as indicated by the red arrows, which breaks the symmetry of magnetization reversal and enables field-free magnetization switching. Lower panels: Schematic illustrations for WTe_2 when an electric current is applied (c) along the *b*-axis and (d) along the *a*-axis. When the current flows along the *a*-axis, where mirror symmetry is broken, out-of-plane spin polarization (out-of-plane effective magnetic field) can be generated, and field-free magnetization switching is realized in this system as well.

et al. in 2021 demonstrated field-free magnetization switching in a $SrRuO_3/WTe_2$ junction, highlighting the usefulness of mirror-symmetry-breaking materials such as WTe_2 [53].

Third, we discuss operations that break the time-reversal symmetry of the spin-current generator. Simply put, this refers to spin-current generation phenomena in ferromagnets or antiferromagnets, which inherently break time-reversal symmetry. In these systems, the constraints on the spin polarization direction seen in the conventional spin Hall effect do not apply, and the spin polarization is determined by the magnetic order within the material. In other words, when the polarity of the magnetization or Néel vector is reversed, the sign of the generated spin polarization is also reversed—an extremely interesting phenomenon. Spin-current generation in ferromagnets is called the spin anomalous Hall effect. It was theoretically predicted by Taniguchi et al. in 2015 [54] and experimentally discovered in a CoFeB/Cu/Py system by Iihama et al. in 2018 [55]. Normally, ferromagnets are used to detect spin currents, so using a ferromagnetic spin-current generator might seem difficult. However, they employed two ferromagnets with different saturation magnetizations and detected spin currents via ferromagnetic resonance, successfully measuring at the resonance points of each ferromagnet. They confirmed that spin currents comparable to those generated by conventional spin Hall materials like Pt were produced, affecting the magnetization dynamics. Next, we discuss spin-current generation in antiferromagnets. As of 2025, research on spin-current generation using antiferromagnets has been thriving, and a variety of phenomena have been discovered. Importantly, all of these phenomena have successfully achieved field-free magnetization switching, making them highly promising for SOT-MRAM applications. The starting point of this research was the magnetic spin Hall effect, discovered in the noncollinear antiferromagnet Mn_3Sn. Mn_3Sn, although an antiferromagnet, exhibits large anomalous Hall and anomalous Nernst effects comparable to ferromagnets and displays intriguing properties in spin-current generation. In conventional spin Hall effects, the sign of the spin Hall angle is determined by the electronic band structure and does not change for the same material: Pt is positive, Ta and W are negative. In contrast, in Mn_3Sn, it was discovered in 2019 [56] that reversing the polarity of the magnetic octupole, characteristic of its noncollinear antiferromagnetism, reverses the sign of the spin Hall angle. By reversing the polarity of the magnetic octupole with an external magnetic field B, the polarity of the spin accumulation signal appearing at the surface via the magnetic spin Hall effect is also reversed. Based on this discovery, by carefully choosing the crystal plane, magnetic octupole direction, and current direction in Mn_3Sn, one can generate spin currents with out-of-plane spin polarization. Using this principle, Qiu et al. demonstrated field-free magnetization switching in 2022 [57]. We next discuss the antiferromagnetic spin Hall effect. The material in which this phenomenon was discovered, Mn_2Au, is a collinear antiferromagnet with a well-defined Néel vector. When current is applied, the Néel vector couples with spin–orbit interactions arising from the broken inversion symmetry of the crystal structure, producing an out-of-plane spin component. This was clarified by Song et al. in 2021 [58]. Based on this finding, the same authors also demonstrated field-free magnetization switching [58].

Finally, we discuss spin-current generation phenomena arising from altermagnetism. Altermagnetism is an emerging field of magnetism that has gained considerable attention in recent years. Broadly speaking, it combines characteristics of both ferromagnets and antiferromagnets. Although altermagnets belong to the antiferromagnetic family, they exhibit spin splitting similar to ferromagnets. This concept was proposed by Šmejkal et al. in 2022 [59, 60]. Conventional antiferromagnets such as NiO preserve time-reversal symmetry, resulting in degenerate spins. This is due to the high symmetry of the NiO crystal (rocksalt structure), where a translation operation following time reversal returns the spin state to its original configuration. In contrast, altermagnets such as RuO_2, although antiferromagnetic, have anisotropic crystal fields surrounding the Ru sublattice sites A and B. Thus, applying the same operations as for NiO does not restore the original spin state. The original state can instead be restored by a crystal rotation following time reversal. This combination of crystal structure and antiferromagnetic order allows the

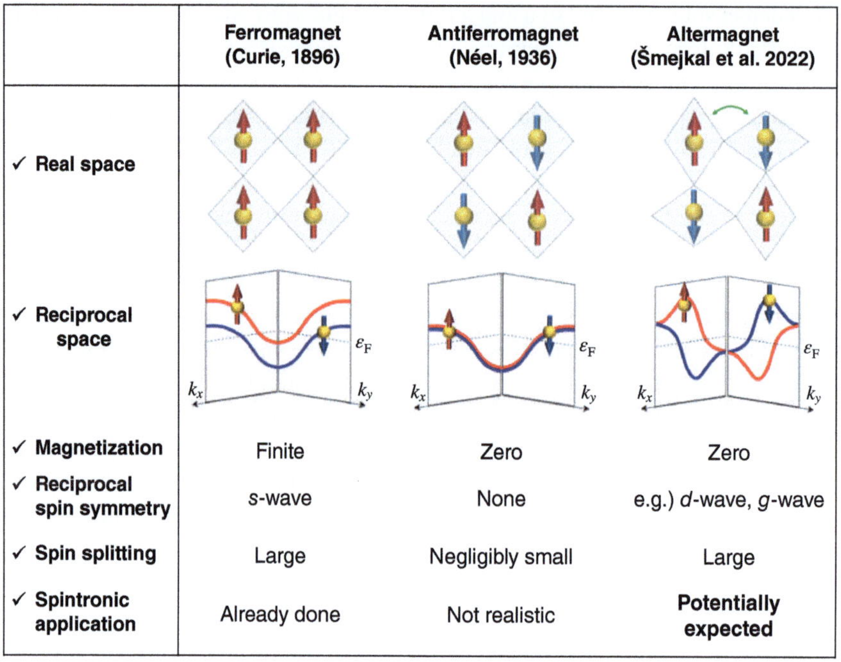

Figure 17.8 Magnetic family consisting of ferromagnet, antiferromagnet, altermagnet. Real-space spin density, band structure, and Fermi contour in momentum space, spin group, presence or absence of time-reversal symmetry, presence or absence of magnetic moment compensation, and presence or absence and shape of spin symmetry are shown. Red and blue colors correspond to up-spin and down-spin, respectively. While ferromagnets exhibit isotropic spin splitting due to conventional exchange coupling, altermagnets show anisotropic spin splitting depending on the crystal symmetry. In contrast, antiferromagnets have spin-degenerate states.

system to break time-reversal symmetry. Figure 17.8 compares conventional ferromagnets, antiferromagnets, and the altermagnets of interest. Because they exhibit anisotropic spin splitting, altermagnets can be regarded as momentum-dependent ferromagnets. The anisotropy of the spin splitting is determined by the crystal symmetry: RuO_2 (rutile structure) exhibits d-wave spin symmetry, whereas CrSb and MnTe (NiAs-type structure) exhibit g-wave spin symmetry. Importantly, although the net magnetization is zero due to the compensated magnetic moments typical of antiferromagnets, spin splitting occurs as in ferromagnets. Therefore, if the Néel vector of an altermagnet could be used as the information carrier in MRAM, the absence of net magnetization would prevent stray-field interference between bits, enabling high-density storage. Additionally, the terahertz-range spin-dynamic characteristics arising from the strong antiferromagnetic exchange coupling would allow ultrafast operation, making altermagnets promising candidates for next-generation MRAM. Various research efforts are ongoing in spintronics toward this goal.

Based on the spin splitting in altermagnets, spin-current generation is expected, particularly near the Fermi level. In 2021, González-Hernández et al. theoretically predicted a nonrelativistic spin-current generation phenomenon arising purely from the spin splitting of altermagnets, which does not require spin–orbit interaction [61].

This phenomenon, originating from spin-split bands, is referred to as the spin-splitter effect. It is particularly expected in metallic antiferromagnets such as d-wave altermagnetic RuO_2. The phenomenon is characterized by magnetic transport and a fourfold symmetric spin-split feature. Soon after the theoretical proposal, three independent groups reported experimental observations of the spin-splitter effect almost simultaneously [62–64]. Figure 17.9 shows results reported by Karube et al. When the x-direction is always defined as parallel to the applied electric field, the spin Hall effect generates a y-polarized spin component propagating along z-direction, which is detected as a spin current. Accordingly, Figure 17.9b shows a finite y-component in all cases. On the other hand, for RuO_2(101) (red plot), by properly aligning the spin-splitting direction with the electric-field (crystal) direction, anisotropic x- and z-polarized spin components (Figure 17.9a and c), which do not appear in the spin Hall effect, can be obtained. Karube et al. further utilized this z-polarized spin component to deterministically reverse the perpendicular magnetization of an adjacent Co layer without an external magnetic field (Figure 17.9d and 17.9e) [62]. Thus, the spin-split characteristics of altermagnets can also induce beneficial field-free magnetization switching for MRAM. As described earlier, spin-current-driven magnetization switching, which was previously difficult with conventional spin Hall effects,

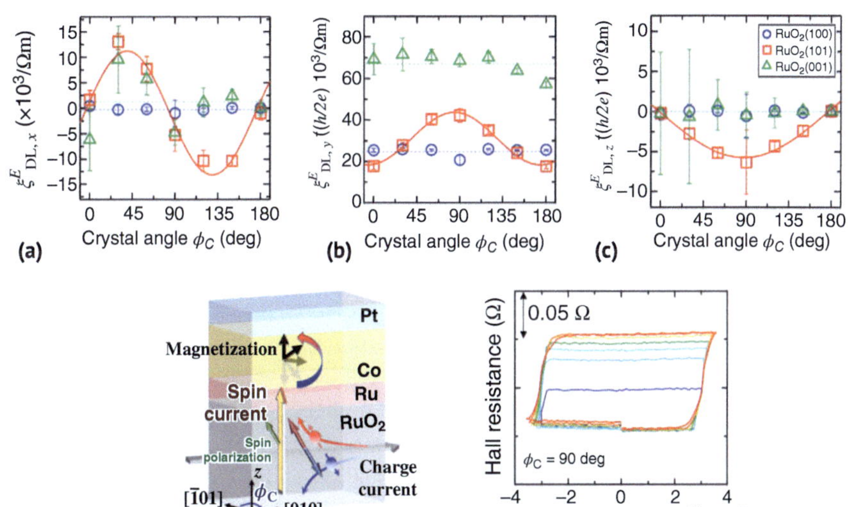

Figure 17.9 Crystal-angle ϕ_C dependence of DL-torque generation efficiency for spin currents with (a) x-polarization, (b) y-polarization, and (c) z-polarization in RuO_2 [62]. Here, ϕ_C denotes the relative angle with respect to a specific crystal direction, and the blue, red, and green plots correspond to the (100), (101), and (001) planes of RuO_2, respectively. The x-direction is always defined as parallel to the electric field. (d) Schematic of perpendicular magnetization switching of Co using z-polarized spin current from RuO_2 [62]. For the RuO_2 (101) plane, applying an electric field along the [010] direction generates a z-polarized component, enabling deterministic field-free magnetization switching. (e) Current-density dependence of the anomalous Hall resistance of Co [62]. Clear hysteresis corresponding to magnetization switching is observed. The color difference in the hysteresis curves represents different applied current values; as the charge current is monotonically increased, the region of magnetization switching expands, and the anomalous Hall resistance increases monotonically. This behavior originates from the multi-domain nature of Co.

has now been experimentally demonstrated based on various new insights. Looking ahead, entirely new phenomena are expected to enable magnetization reversal, further enhancing the performance of MRAM as a key device for information processing.

References

1 Baibich, M.N., Broto, J.M., Fert, A., Nguyen Van Dau, F., Petroff, F., Etienne, P., Creuzet, G., Friederich, A. and Chazelas, J. (1988) Giant magnetoresistance of (001)Fe/(001)Cr magnetic superlattices. *Phys. Rev. Lett.*, 61, 2472–5.

2 Binasch, G., Grünberg, P., Saurenbach, F. and Zinn, W. (1989) Enhanced magnetoresistance in layered magnetic structures with antiferromagnetic interlayer exchange. *Phys. Rev. B*, 39, 4828–30.

3 Parkin, S.S.P., More, N. and Roche, K.P. (1990) Oscillations in exchange coupling and magnetoresistance in metallic superlattice structures: Co/Ru, Co/Cr, and Fe/Cr. *Phys. Rev. Lett.*, 64, 2304–7.

4 Parkin, S.S.P. (1991) Systematic variation of the strength and oscillation period of indirect magnetic exchange coupling through the 3d, 4d, and 5d transition metals. *Phys. Rev. Lett.*, 67, 3598–601.

5 McGuire, T.R. and Potter, R.I. (1975) Anisotropic magnetoresistance in ferromagnetic 3d alloys. *IEEE Trans. Magn.*, 11, 1018–38.

6 Dieny, B., Speriosu, V.S., Parkin, S.S.P., Gurney, B.A., Wilhoit, D.R. and Mauri, D. (1991) Giant magnetoresistive in soft ferromagnetic multilayers. *Phys. Rev. B*, 43, 1297–300.

7 Daughton, J.M. (1999) GMR applications. *J. Magn. Magn. Mater.*, 192, 334–42.

8 Miyazaki, T. and Tezuka, N. (1995) Giant magnetic tunneling effect in Fe/Al_2O_3/Fe junctions. *J. Magn. Magn. Mater.*, 139, L231–4.

9 Moodera, J.S., Kinder, L.R., Wong, T.M., and Meservey, R. (1995) Large magnetoresistance at room temperature in ferromagnetic thin film tunnel junctions. *Phys. Rev. Lett.*, 74, 3273–6.

10 Mathon, J. and Umerski, A. (2001) Theory of tunneling magnetoresistance of an epitaxial Fe/MgO/Fe(001) junction. *Phys. Rev. B*, 63, 220403R.

11 Yuasa, S., Nagahama, T., Fukushima, A., Suzuki, Y., and Ando, K. (2004) Giant room-temperature magnetoresistance in single-crystal Fe/MgO/Fe magnetic tunnel junctions. *Nat. Mater.*, 3, 868–71.

12 Parkin, S.S.P., Kaiser, C., Panchula, A., Rice, P.M., Hughes, B., Samant, M. and Yang, S.-H. (2004) Giant tunnelling magnetoresistance at room temperature with MgO(100) tunnel barriers. *Nat. Mater.*, 3, 862–7.

13 Ikeda, S., Hayakawa, J., Ashizawa, Y., Lee, Y.M., Miura, K., Hasegawa, H., Tsunoda, M., Matsukura, F., and Ohno, H. (2008) Tunnel magnetoresistance of 604% at 300 K by suppression of Ta diffusion in CoFeB/MgO/CoFeB pseudo-spin-valves annealed at high temperature. *Appl. Phys. Lett.*, 93, 082508.

14 Yuasa, S. and Djayaprawira, D.D. (2007) Giant tunnel magnetoresistance in magnetic tunnel junctions with a crystalline MgO(001) barrier. *J. Phys. D: Appl. Phys.*, 40, R337–54.

15 Nguyen, V.D., Rao, S., Wostyn, K., and Couet, S. (2024) Recent progress in spin-orbit torque magnetic random-access memory. *npj Spintronics*, 2, 48.

16 Nozaki, T., Yamamoto, T., Miwa, S., Tsujikawa, M., Shirai, M., Yuasa, S. and Suzuki, Y. (2019) Recent progress in the voltage-controlled magnetic anisotropy effect and the challenges faced in developing voltage-torque MRAM. *Micromachines*, 10, 327.

17 Chen, B.J., Zeng, M., Khoo, K.H., Das, D., Fong, X., Fukami, S., Li, S., Zhao, W., Parkin, S.S.P., Piramanayagam, S.N. and Lim, S.T. (2023) Spintronic devices for high-density memory and neuromorphic computing – a review. *Mater. Today*, 70, 193–217.

18 Bhatti, S., Sbiaa, R., Hirohata, A., Ohno, H., Fukami, S. and Piramanayagam, S.N. (2017) Spintronics based random access memory: a review. *Mater. Today*, 20, 530–48.

19 Ryu, J., Lee, S., Lee, K. J. and Park, B.-G. (2020) Current-induced spin–orbit torques for spintronic applications. *Adv. Mater.*, 32, 1907148.

20 D'yakonov, M.I. and Perel, V.I. (1971) Possibility of orientating electron spins with current. *Sov. Phys. JETP Lett.*, 13, 467.

21 Kato, Y.K., Myers, R.C., Gossard, A.C., and Awschalom, D.D. (2004) Observation of the spin Hall effect in semiconductors. *Science*, 306, 1910–3.

22 Hirsch, J.E. (1999) Spin Hall effect. *Phys. Rev. Lett.*, 83, 1834–7.

23 Valenzuela, S.O. and Tinkham, M. (2006) Direct electronic measurement of the spin Hall effect. *Nature*, 442, 176–9.

24 Saitoh, E., Ueda, M., Miyajima, H. and Tatara, G. (2006) Conversion of spin current into charge current at room temperature: inverse spin-Hall effect. *Appl. Phys. Lett.*, 88, 182509.

25 Kimura, T., Otani, Y., Sato, T., Takahashi, S. and Maekawa, S. (2007) Room-temperature reversible spin Hall effect. *Phys. Rev. Lett.*, 98, 156601.

26 Hoffmann, A. (2013) Spin Hall effects in metals. *IEEE Trans. Magn.*, 49, 5172–93.

27 Sinova, J., Valenzuela, S.O., Wunderlich, J., Back, C.H. and Jungwirth, T. (2015) Spin Hall effects. *Rev. Mod. Phys.*, 87, 1213–60.

28 Niimi, Y., Morota, M., Wei, D.H., Deranlot, C., Basletic, M., Hamzic, A., Fert, A. and Otani, Y. (2011) Extrinsic spin Hall effect induced by iridium impurities in copper. *Phys. Rev. Lett.*, 106, 126601.

29 Niimi, Y., Kawanishi, Y., Wei, D.H., Deranlot, C., Yang, H.X., Chshiev, M., Valet, T., Fert, A. and Otani, Y. (2012) Giant spin Hall effect induced by skew scattering from bismuth impurities inside thin film CuBi alloys. *Phys. Rev. Lett.*, 109, 156602.

30 An, H., Kageyama, Y., Kanno, Y., Enishi, N. and Ando, K. (2016) Spin–torque generator engineered by natural oxidation of Cu. *Nat. Commun.*, 7, 13069.

31 Karube, S., Sato, T., Kajiwara, Y., Uchida, K., Takahashi, S., Maekawa, S. and Otani, Y. (2020) Enhancement of spin–charge current interconversion by oxidation of rhenium. *J. Magn. Magn. Mater.*, 516, 167298.

32 Demasius, K. U., Phung, T., Zhang, W., Hughes, B.P., Yang, S. H., Kellock, A., Han, W., Pushp, A. and Parkin, S.S.P. (2016) Enhanced spin–orbit torques by oxygen incorporation in tungsten films. *Nat. Commun.*, 7, 10644.

33 Murakami, S. (2005) Intrinsic spin Hall effect. *Adv. Solid State Phys.*, 45, 197–209.

34 Guo, G.-Y., Murakami, S., Chen, T. W. and Nagaosa, N. (2008) Intrinsic spin Hall effect in platinum: first-principles calculations. *Phys. Rev. Lett.*, 100, 096401.

35 Tanaka, T., Kontani, H., Naito, M., Naito, T., Hirashima, D.S., Yamada, K. and Inoue, J. (2008) Intrinsic spin Hall effect and orbital Hall effect in 4d and 5d transition metals. *Phys. Rev. B*, 77, 165117.

36 Hasan, M.Z. and Kane, C.L. (2010) Colloquium: topological insulators. *Rev. Mod. Phys.*, 82, 3045–67.

37 Manchon, A., Koo, H.C., Nitta, J., Frolov, S.M. and Duine, R.A. (2015) New perspectives for Rashba spin–orbit coupling. *Nat. Mater.*, 14, 871–82.

38 Edelstein, V.M. (1990) Spin polarization of conduction electrons induced by electric current in two-dimensional asymmetric electron systems. *Solid State Commun.*, 73, 233–5.

39 Ganichev, S.D., Ivchenko, E.L., Bel'kov, V.V., Tarasenko, S.A., Sollinger, M., Weiss, D., Wegscheider, W. and Prettl, W. (2002) Spin-galvanic effect. *Nature*, 417, 153–6.

40 Rojas Sánchez, J.C., Vila, L., Desfonds, G., Gambarelli, S., Attané, J.P., De Teresa, J.M., Magén, C. and Fert, A. (2013) Spin-to-charge conversion using Rashba coupling at the interface between non-magnetic materials. *Nat. Commun.*, 4, 2944.

41 Tserkovnyak, Y., Brataas, A. and Bauer, G.E.W. (2002) Spin pumping and magnetization dynamics in metallic multilayers. *Phys. Rev. B*, 66, 224403.

42 Mellnik, A.R., Lee, J.S., Richardella, A., Grab, J.L., Mintun, P.J., Fischer, M.H., Vaezi, A., Manchon, A., Kim, E.-A., Samarth, N. and Ralph, D.C. (2014) Spin-transfer torque generated by a topological insulator. *Nature*, 511, 449–51.

43 Rojas-Sánchez, J.-C., Oyarzún, S., Fu, Y., Marty, A., Vergnaud, C., Gambarelli, S., Vila, L., Jamet, M., Ohtsubo, Y., Taleb-Ibrahimi, A., Le Fèvre, P., Bertran, F., Reyren, N., George, J.-M. and Fert, A. (2016) Spin to charge conversion at room temperature by spin pumping into a new type of topological insulator: alpha-Sn films. *Phys. Rev. Lett.*, 116, 096602.

44 Karube, S., Kondou, K. and Otani, Y. (2016) Experimental observation of spin-to-charge current conversion at non-magnetic metal/Bi_2O_3 interfaces. *Appl. Phys. Express*, 9, 033001.

45 Tsai, H., Karube, S., Kondou, K., Yamaguchi, N., Ishii, F. and Otani, Y. (2018) Clear variation of spin splitting by changing electron distribution at non-magnetic metal/Bi_2O_3 interfaces. *Sci. Rep.*, 8, 5564.

46 Karube, S., Tezuka, N., Kohda, M. and Nitta, J. (2020) Anomalous spin-orbit field via the Rashba–Edelstein effect at the W/Pt interface. *Phys. Rev. Appl.*, 13, 024009.

47 Fukami, S., Anekawa, T., Zhang, C. and Ohno, H. (2016) A spin–orbit torque switching scheme with collinear magnetic easy axis and current configuration. *Nat. Nanotechnol.*, 11, 621–5.

48 Chernyshov, A., Overby, M., Liu, X., Furdyna, J.K., Lyanda-Geller, Y. and Rokhinson, L.P. (2009) Evidence for reversible control of magnetization in a ferromagnetic material by means of spin–orbit magnetic field. *Nat. Phys.*, 5, 656–9.

49 Liu, L., Lee, O.J., Chtchelkanova, A.Y. and Buhrman, R.A. (2012) Spin-torque switching with the giant spin Hall effect of tantalum. *Science*, 336, 555–8.

50 Fukami, S., Zhang, C., DuttaGupta, S., Kurenkov, A. and Ohno, H. (2016) Magnetization switching by spin–orbit torque in an antiferromagnet–ferromagnet bilayer system. *Nat. Mater.*, 15, 535–41.

51 Yu, G., Upadhyaya, P., Fan, Y., Alzate, J.G., Jiang, W., Wong, K.L., Takei, S., Bender, S.A., Chang, L.-T., Jiang, Y., Lang, M., Tang, J., Wang, Y., Tserkovnyak, Y., Khalili

Amiri, P. and Wang, K.L. (2014) Switching of perpendicular magnetization by spin–orbit torques in the absence of external magnetic fields. *Nat. Nanotechnol.*, 9, 548–54.
52 MacNeill, D., Stiehl, G.M., Guimaraes, M.H.D., Buhrman, R.A., Park, J. and Ralph, D.C. (2017) Control of spin–orbit torques through crystal symmetry in WTe_2/ferromagnet bilayers. *Nat. Phys.*, 13, 300–5.
53 Xie, Q., Lin, W., Sarkar, S., Shu, X., Chen, S., Liu, L., Zhao, T., Zhou, C., Wang, H., Zhou, J., Gradečak, S. and Chen, J. (2021) Field-free magnetization switching induced by the unconventional spin–orbit torque from WTe_2. *APL Mater.*, 9, 051114.
54 Taniguchi, T., Grollier, J. and Stiles, M.D. (2015) Spin-transfer torques generated by the anomalous Hall effect and anisotropic magnetoresistance. *Phys. Rev. Appl.*, 3, 044001.
55 Iihama, S., Taniguchi, T., Yakushiji, K., Fukushima, A., Shiota, Y., Tsunegi, S., Hiramatsu, R., Yuasa, S., Suzuki, Y. and Kubota, H. (2018) Spin-transfer torque induced by the spin anomalous Hall effect. *Nat. Electron.*, 1, 120–3.
56 Kimata, M., Chen, H., Kondou, K., Sugimoto, S., Muduli, P.K., Ikhlas, M., Omori, Y., Tomita, T., MacDonald, A.H., Nakatsuji, S. and Otani, Y. (2019) Magnetic and magnetic inverse spin Hall effects in a non-collinear antiferromagnet. *Nature*, 565, 627–30.
57 Hu, S., Shao, D.-F., Yang, H., Pan, C., Fu, Z., Tang, M., Yang, Y., Fan, W., Zhou, S., Tsymbal, E.Y. and Qiu, X. (2022) Efficient perpendicular magnetization switching by a magnetic spin Hall effect in a noncollinear antiferromagnet. *Nat. Commun.*, 13, 4447.
58 Chen, X., Shi, S., Shi, G., Fan, X., Song, C., Zhou, X., Bai, H., Liao, L., Zhou, Y., Zhang, H., Li, A., Chen, Y., Han, X., Jiang, S., Zhu, Z., Wu, H., Wang, X., Xue, D., Yang, H. and Pan, F. (2021) Observation of the antiferromagnetic spin Hall effect. *Nat. Mater.*, 20, 800–4.
59 Šmejkal, L., Sinova, J. and Jungwirth, T. (2022) Beyond conventional ferromagnetism and antiferromagnetism: a phase with nonrelativistic spin and crystal rotation symmetry. *Phys. Rev. X*, 12, 031042.
60 Šmejkal, L., Sinova, J. and Jungwirth, T. (2022) Emerging research landscape of altermagnetism. *Phys. Rev. X*, 12, 040501.
61 González-Hernández, R., Šmejkal, L., Výborný, K., Yahagi, Y., Sinova, J., Jungwirth, T. and Železný, J. (2021) Efficient electrical spin splitter based on nonrelativistic collinear antiferromagnetism. *Phys. Rev. Lett.*, 126, 127701.
62 Karube, S., Tanaka, T., Sugawara, D., Kadoguchi, N., Kohda, M. and Nitta, J. (2022) Observation of spin-splitter torque in collinear antiferromagnetic RuO_2. *Phys. Rev. Lett.*, 129, 137201.
63 Bose, A., Schreiber, N.J., Jain, R., Shao, D.-F., Nair, H.P., Sun, J., Zhang, X.S., Muller, D.A., Tsymbal, E.Y., Schlom, D.G. and Ralph, D.C. (2022) Tilted spin current generated by the collinear antiferromagnet ruthenium dioxide. *Nat. Electron.*, 5, 267–74.
64 Bai, H., Han, L., Feng, X.Y., Zhou, Y.J., Su, R.X., Wang, Q., Liao, L.Y., Zhu, W.X., Chen, X.Z., Pan, F., Fan, X. L. and Song, C. (2022) Observation of spin splitting torque in a collinear antiferromagnet RuO_2. *Phys. Rev. Lett.*, 128, 197202.

18

Triboelectric Materials for Sensor and Energy-harvesting Applications

Chuang Zhu[1] *and Xuqing Liu*[2]

[1] *College of Textiles, Donghua University, Shanghai, China*
[2] *School of Materials Science and Engineering, Northwestern Polytechnical University, Xi'an, Shaanxi, China*

18.1 Triboelectric Nanogenerator

A triboelectric nanogenerator (TENG) is an innovative energy-harvesting device that converts mechanical energy into electrical energy through the triboelectric effect, as illustrated in Figure 18.1 [1]. In essence, when two materials with different electron affinities come into contact and then separate, an electric charge is generated due to the imbalance of electrons [2]. This phenomenon can be harnessed to produce useful electrical power from a wide range of mechanical motions, such as vibrations [3], body movements [4–6], and even environmental phenomena like wind [7] and ocean waves [8–11]. TENG technology represents a promising avenue for sustainable and clean energy generation, especially as demands for portable, flexible, and self-powered electronic systems continue to grow. This introduction will delve into the definition of TENGs, discuss their key advantages, explain the fundamental principles behind their operation, and explore a variety of applications where they are making significant impacts.

A TENG is typically composed of two or more triboelectric layers, often polymers, metals, or other materials [12] that can transfer charges through friction or contact, as shown in Figure 18.2 [1]. The word "triboelectric" stems from the Greek "tribo," meaning "to rub," reflecting the central role of friction in initiating the charge transfer process. Although the triboelectric effect has been known for centuries (it's the same phenomenon responsible for static electricity when you rub a balloon on your hair), it was not until 2012 that researchers proposed the concept of using this effect at the nanoscale to harvest energy [13]. This breakthrough was crucial because it brought about a new class of nanogenerators that could be efficiently scaled down, integrated with other technologies, and tailored to specific applications [14].

One of the most notable advantages of TENGs is their simplicity. TENGs can often be manufactured using common and cost-effective materials, such as polytetrafluoroethylene [15], polyethylene terephthalate (PET) [16–17], or

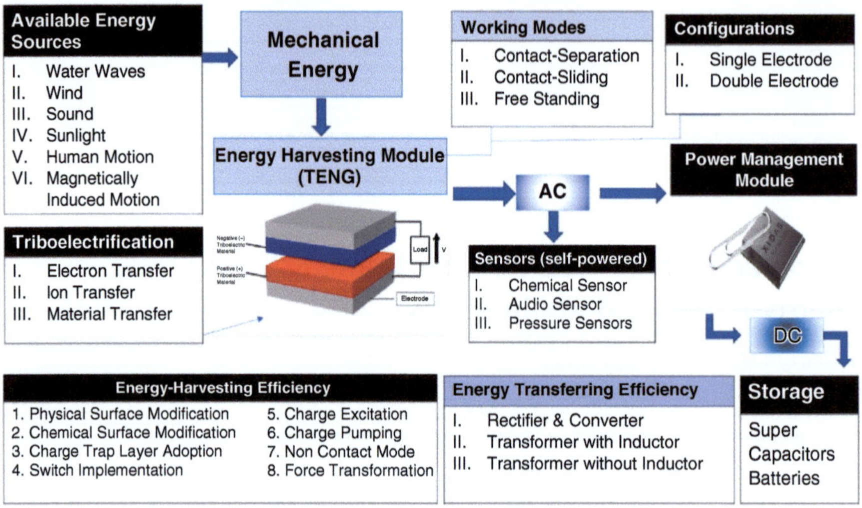

Figure 18.1 Schematic representation of the energy-harvesting system utilizing triboelectric nanogenerators (TENGs). *Source:* [1] with permission of Elsevier.

polydimethylsiloxane (PDMS) [18]. These materials are not only inexpensive but also widely available, making large-scale production feasible. Moreover, TENGs generally have a straightforward design: two layers of triboelectric materials, sometimes with patterned microstructures to enhance contact and increase the effective surface area [19]. This simplicity reduces both fabrication complexity and overall production costs. A second advantage lies in their high adaptability. TENGs can be easily scaled up or down, molded into different shapes, and integrated with various systems and devices without requiring drastic changes to the underlying technology [20]. For instance, flexible substrates can be used to create wearable TENGs that conform to the shape of the human body [21], or robust enclosures can be built to protect TENGs for outdoor deployment. Additionally, because triboelectric materials can be chosen from a broad range of options, ranging from everyday plastics to cutting-edge nanocomposites, researchers can tune the properties of the generator for higher electrical output or greater durability [22–23]. Furthermore, TENGs can function under a variety of conditions. Whether in humid or dry environments, TENGs continue to produce electricity as long as the fundamental requirement of contact and separation of two different materials remains intact. This makes them especially attractive for applications in remote or harsh settings where traditional energy sources or large-scale power infrastructure are not readily available. Since TENGs generate electricity from motion, they can also provide continuous energy harvesting as long as there is some form of repetitive movement or sporadic mechanical action [24–27].

At the heart of a TENG is the triboelectric effect, which involves two main processes: contact electrification and electrostatic induction. When two materials come into contact, electrons are transferred from one material to the other, depending on their respective electron affinities. This causes one material to become

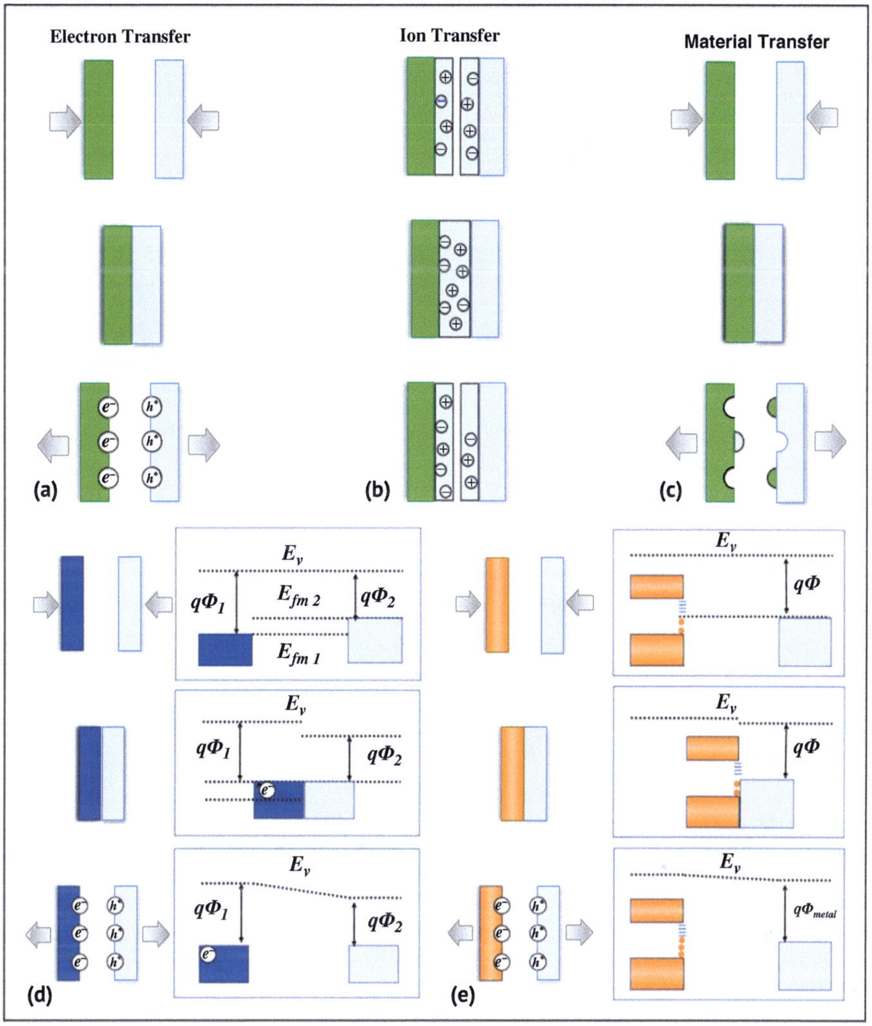

Figure 18.2 A comprehensive analysis of charge transfer phenomena in TENG systems: (a) the electron migration process within dielectric–dielectric interfaces, (b) ionic charge movement across insulating material boundaries, (c) material exchange dynamics at insulator interfaces, (d) the electronic band structure representation for metal–metal contact charging, and (e) the energy level diagram for metal–dielectric interfacial charging. *Source:* [1] with permission of Elsevier.

positively charged while the other becomes negatively charged. When the materials are separated, an electric potential difference is established between them. Electrodes connected to these materials capture the resulting electric current as the charges seek to rebalance [28]. There are four primary working modes commonly discussed in TENG research: vertical contact-separation, lateral sliding, single-electrode, and freestanding triboelectric-layer modes [29], as shown in Figure 18.3 [30]. In the vertical contact-separation mode, two materials approach each other perpendicularly and then separate, generating an alternating current. In the lateral sliding mode, the two surfaces slide against each other and then separate.

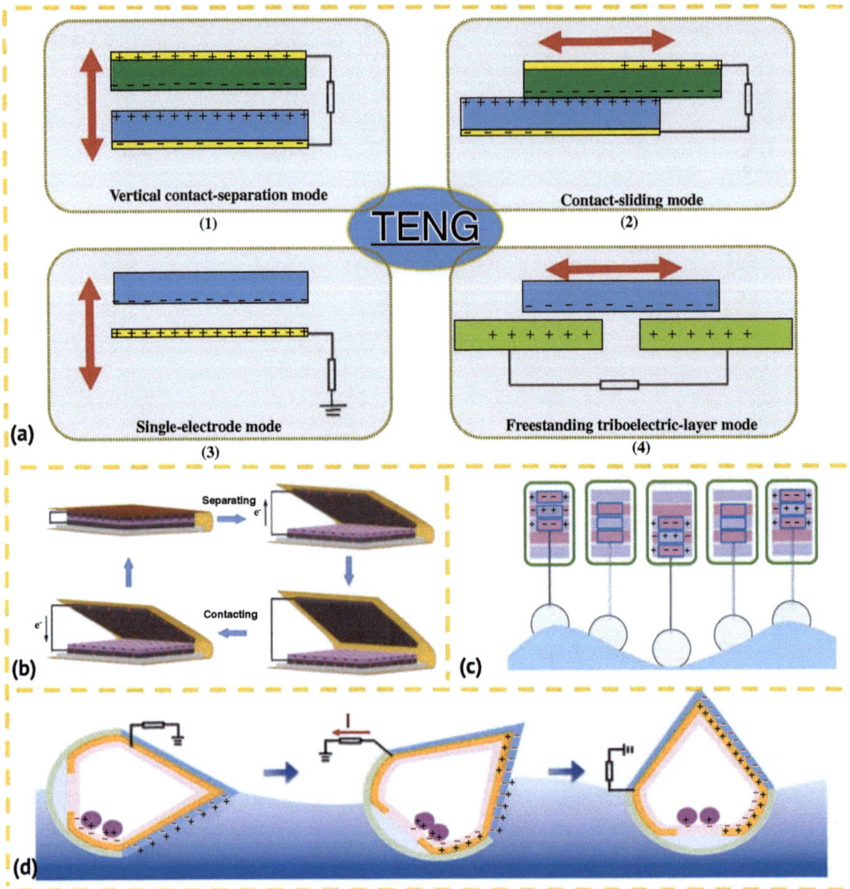

Figure 18.3 The schematic representation demonstrates various operational configurations of triboelectric nanogenerators: (a) four distinct working states of TENG devices, (b) operational mechanism of a single module in a folded multilayer TENG structure, (c) schematic illustration of the CMLS-TENG's functional principle, (d) working mechanism of TH-TENG showing transition from inclined to vertical orientation in marine environments. *Source:* [30] with permission of Elsevier.

The single-electrode mode uses only one triboelectric material, while the other material or object can be any conductive or nonconductive surface, simplifying the design. Lastly, in the freestanding mode, the triboelectric layer remains suspended or loosely attached, allowing charges to build up as it interacts with other surfaces. Each mode offers distinct advantages in terms of design, output power, and suitability for specific applications.

Because of their ability to scavenge energy from everyday motions, TENGs are widely explored in wearable electronics. A person's body can serve as a near-constant source of mechanical movement; walking, jogging, bending limbs, or even subtle gestures can all trigger contact and separation of triboelectric materials [31–32]. This opens the door to self-powered health monitors, smart textiles, and fitness trackers that no longer rely on frequent battery charging. For instance, TENGs integrated

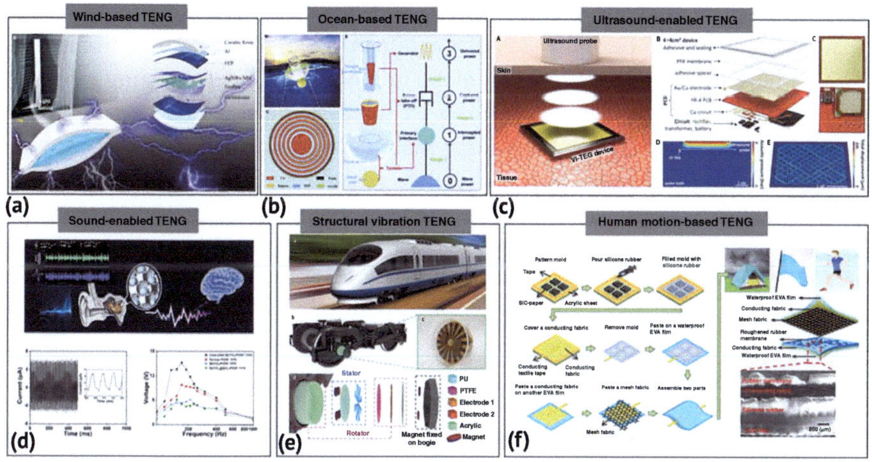

Figure 18.4 Innovative configurations of triboelectric nanogenerators for mechanical energy capture from diverse sources. (a) Aerodynamic energy-harvesting TENG design. (b) Marine wave energy conversion TENG configuration. (c) Ultrasonic wave-driven TENG architecture. (d) Acoustic energy-harvesting TENG system. (e) Vibration energy collection TENG prototype. (f) Biomechanical energy-harvesting TENG device. *Source:* [35] with permission of the American Chemical Society.

into shoe insoles can generate electricity with each step [33], potentially powering a pedometer or transmitting health data to a smartphone. In environmental sensing, TENGs present significant advantages as well. By capturing energy from environmental factors like wind, rain, or ocean waves, these devices can power remote sensors that monitor temperature, humidity, air quality, or other parameters without needing to replace or recharge batteries [34], as shown in Figure 18.4 [35]. This feature is particularly useful in fields such as agriculture, climate science, and wildlife observation, where devices often have to operate in isolated regions. Self-powered sensors reduce the need for wiring and manual battery replacements, thus lowering maintenance costs and environmental impact. Another emerging area is in the context of the Internet of Things (IoT). As more devices become connected and require power, TENGs can help alleviate the energy challenges associated with massive sensor networks. Placing TENGs in strategic locations on doors, floors, or any surface subject to motion can generate enough power for localized sensors or communication modules. This reduces dependency on battery-based systems and contributes to a more sustainable infrastructure overall [36]. Large-scale applications are also on the horizon. Researchers are investigating the use of TENGs in harvesting energy from ocean waves, rainfall, or even vehicle traffic on roads. By integrating triboelectric materials into road surfaces or along coastlines, it may be possible to capture a portion of the kinetic energy from passing cars or tidal movements. While challenges remain chiefly in ensuring device durability and optimizing power outputs, such endeavors highlight the versatility and scalability of TENG technology.

In summary, TENGs offer a novel and highly adaptable solution for harvesting mechanical energy and converting it into electricity. They rely on a principle of triboelectric charging that has long been known but only recently leveraged in a sophisticated manner for energy generation. With advantages that include low-cost materials, design flexibility, and the ability to operate in diverse environments, TENGs have carved out a promising niche in modern energy-harvesting research. Their applications range from self-powered wearable devices and environmental monitoring systems to potentially large-scale deployments capturing the kinetic energy of natural or human-made motions. As research continues to refine their efficiency, durability, and integration with emerging technologies, TENGs are poised to play an increasingly important role in the global shift toward cleaner, more sustainable energy solutions.

18.2 Fiber-based TENG

A fiber- or textile-based TENG is a TENG device in which one or more layers of triboelectric materials are engineered into yarns, fibers, or fully woven/knitted textiles. Unlike traditional rigid TENGs that employ flat sheets or plates, textile-based TENGs incorporate the triboelectric layers into flexible fibrous structures. In many cases, these fibers are coated or blended with conductive layers, polymeric films, or other materials designed to enhance the triboelectric effect [37]. By integrating fibers in an intertwined structure, such as weaving, knitting, or braiding, the device acquires the familiar properties of fabrics, namely, flexibility, breathability, and durability, while still being able to generate electrical power from mechanical motion, as illustrated in Figure 18.5 [38]. Textile-based TENGs can be composed entirely of triboelectric fibers or can incorporate triboelectric fibers into a standard textile substrate. In some designs, a single yarn itself can act as a TENG, combining a triboelectric component and an electrode layer in a coaxial configuration. In other instances, two different fiber types (e.g., positively and negatively charged fibers) are woven together, and their interactions during motion produce an electric potential difference [39–40]. As a result, fiber- and textile-based TENGs can be seamlessly integrated into everyday clothing, accessories, or industrial fabrics, making them a unique class of power-generating materials.

One of the most prominent advantages of fiber- and textile-based TENGs lies in their flexibility and comfort. Traditional TENGs, while effective in many scenarios, often rely on relatively rigid substrates. By contrast, textiles are inherently flexible and can be bent, stretched, or folded without losing their overall structural integrity [41–42]. This property makes them ideal for wearables or applications where conformability to curved surfaces is paramount. A textile-based TENG integrated into clothing can move with the human body, maintaining both comfort and performance over extended periods of use. Beyond flexibility, textile-based TENGs are lightweight and can maintain breathability, an important factor for both comfort and prolonged use. Conventional energy-harvesting devices may restrict airflow or add undesirable bulk, but fabrics can be engineered with open or mesh-like structures that allow for proper ventilation [43–44]. This is particularly beneficial in sports and health monitoring applications, where thermoregulation is crucial, as illustrated in Figure 18.6 [45]. Textiles are routinely subjected to washing,

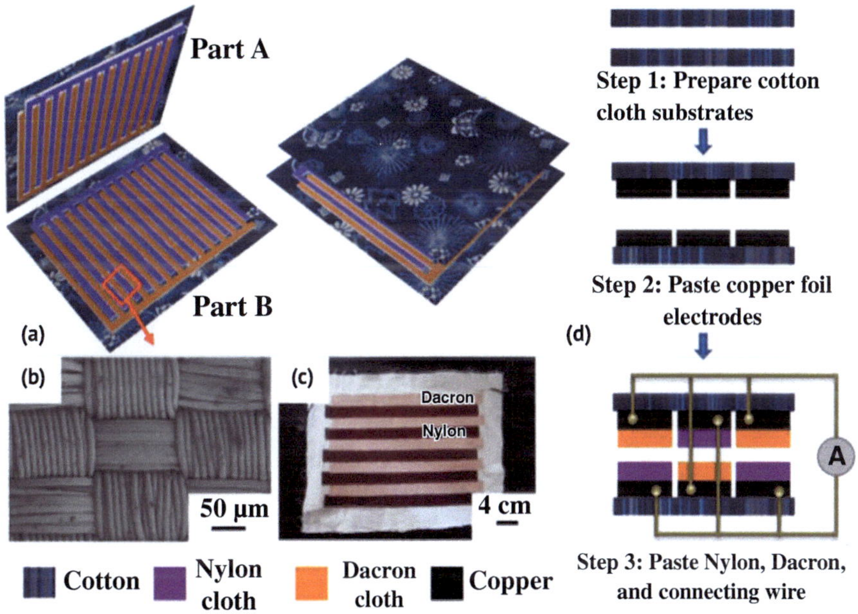

Figure 18.5 Visual representations of textile-based triboelectric nanogenerator: (a) conceptual illustration of fabric-integrated TENG featuring parallel-groove architecture, (b) microscopic visualization of woven nylon surface morphology through scanning electron microscopy, (c) physical demonstration of a TENG component module, (d) step-by-step manufacturing sequence for TENG assembly. *Source:* [38] with permission of Elsevier.

abrasion, and repeated folding or stretching. Consequently, many fiber- and textile-based TENGs are designed to resist damage from such everyday stresses. Researchers often coat fibers with protective layers or select robust polymer materials that safeguard the triboelectric properties during washing cycles [46–47]. This durability ensures that the TENG remains functional and maintains its performance over a product's typical lifespan, a key consideration for consumer acceptance and reliability. The textile industry is one of the largest manufacturing sectors globally, with well-established processes for spinning fibers, weaving or knitting fabrics, and finishing materials. By developing triboelectric fibers compatible with existing textile-manufacturing infrastructure, the large-scale production of TENG-embedded fabrics becomes more feasible. This opens up opportunities for cost-effective, high-volume manufacturing, positioning fiber-based TENGs as a potentially ubiquitous technology in the near future.

While fiber- and textile-based TENGs employ the same triboelectric effect as their rigid counterparts, several unique operational modes and design considerations arise due to the fabric-based architecture. The core mechanism remains the interplay of contact electrification and electrostatic induction. When two triboelectric materials with different electron affinities come into contact, electrons transfer from one material to the other. Upon separation, an electrostatic charge remains on the surfaces. Conductive or semiconductive layers integrated into the fibers act as electrodes, capturing the flow of electrons as charges seek to neutralize. This process can repeat every time the fabric experiences friction or contact separation, resulting in

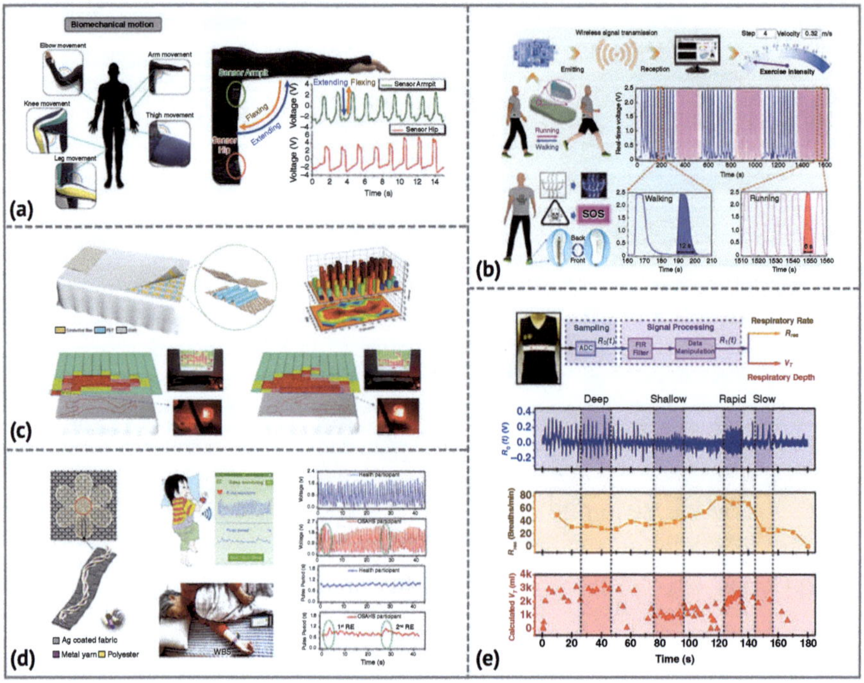

Figure 18.6 The application of t-TENGs spans from detecting human movements to enabling advanced health monitoring systems. (a) Shirts and trousers equipped with t-TENGs near bending joints for biomechanical motion analysis. (b) Smart shoes featuring t-TENGs in their soles to track human movements in real time. (c) A self-powered sleep monitor using a t-TENG-based pressure sensor array to analyze body posture, position, and pressure distribution. (d) A wristband with a t-TENG-integrated wireless system for recording pulse waves and diagnosing OSAHS. (e) A wearable respiratory monitor using a chest strap with t-TENGs to measure respiratory rate and depth. *Source:* [45] with permission of John Wiley & Sons.

a continuous or intermittent generation of electrical pulses. In textile-based structures, friction does not just occur between two large flat layers. Instead, multiple contact points arise where individual fibers rub against each other or against layers of material. This multitude of smaller contact points can enhance the overall triboelectric output if appropriately designed [48]. Moreover, motion in textiles such as bending, stretching, or compression encourages constant movement among fibers, translating into repeated contact-separation cycles. By optimizing the fiber weave, knit, or braiding pattern, it is possible to maximize these interactions, thereby improving energy-harvesting efficiency [49], as illustrated in Figure 18.7 [38]. The triboelectric output of fibers can be further optimized by surface modification techniques. For instance, chemical treatments, plasma processes, or nanostructured coatings can be applied to increase surface roughness or engineer specific electron affinities. This approach allows for finer tuning of the triboelectric properties without sacrificing other fabric characteristics like softness or breathability. Additionally, hierarchical structures—where nanostructures are introduced onto microfibers—can significantly increase the effective contact area and boost the

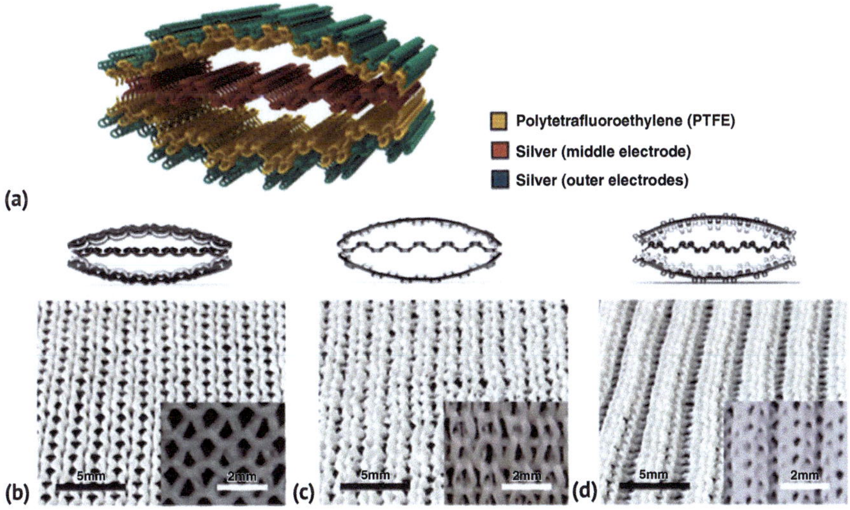

Figure 18.7 The textile-based triboelectric nanogenerator (T-TENG) incorporating three distinct fabric architectures: (a) conceptual representation of the T-TENG configuration, (b) single-layer interlooped textile design, (c) dual-layer knitted framework, and (d) ribbed fabric pattern with detailed microscopic views provided in the supplementary images. *Source:* [38] with permission of Elsevier.

device's electrical performance [50]. In fiber-based TENGs, electrodes can be either internal or external. Internal electrodes might be formed by embedding conductive materials (e.g., metal nanowires, carbon nanotubes, conductive polymers) directly within the fiber structure. External electrodes, on the other hand, can be achieved by coating fibers with conductive materials or weaving conductive yarns alongside triboelectric fibers. The choice of electrode integration strategy depends on target applications and requirements for mechanical robustness, electrical output, and washability.

Perhaps the most intuitive application for fiber-based TENGs is in wearable electronics. By integrating TENG fibers into clothing, shoes, wristbands, or other accessories, it becomes possible to harvest energy from everyday movements such as walking, running, or even subtle gestures. The electricity generated can power low-consumption electronics, like pedometers, heart rate monitors, or Bluetooth modules, thereby reducing or eliminating the need for battery replacements. Additionally, because the fabric remains comfortable and breathable, it can be incorporated into sportswear or medical garments that continuously track physiological signals [51]. Fiber- and textile-based TENGs lend themselves well to the development of self-powered sensors in fields ranging from healthcare to environmental monitoring. For example, a pressure sensor woven into a seat cover could detect whether a person is seated or not, using the TENG's electrical signals to power an alert system. In healthcare, garments with integrated TENG fibers might track posture, breathing rate, or movement patterns in patients with mobility challenges [52]. By harvesting energy from the very motion they are monitoring, these sensors eliminate or reduce the need for external power sources, making them more practical for continuous or remote operation. Another emerging area of interest is the integration of textile-based TENGs in soft robotics and human-machine interfaces. Soft robotic systems often require sensors that can detect bending, stretching, or external pressure without hindering the robot's flexible structure. Textile-based TENGs fit naturally into these applications, providing both tactile feedback and power generation. On the user interface side, interactive gloves or wearable controllers can use the triboelectric signals generated by finger movements to communicate commands to machines, offering a portable, self-powered control system [44, 53]. Beyond wearables, textile TENGs can be used in various household or public settings. Curtains that harvest energy from being drawn open and closed, carpets that generate power from foot traffic, or upholstery that reacts to sitting and standing movements are all possibilities. These applications turn common household or commercial products into micro-energy power stations, contributing to the broader concept of distributed energy generation, as illustrated in Figure 18.8 [54]. While the power output of a single carpet panel might be small, the cumulative effect across multiple surfaces could be meaningful for low-power electronics, sensor networks, or smart home systems.

Despite the notable promise of fiber- and textile-based TENGs, there are still several challenges that need to be addressed to ensure widespread adoption, as illustrated in Figure 18.9 [54]:

Efficiency and Power Density [55]: While TENGs can generate electricity from low-frequency and irregular movements, their overall energy output can be relatively small compared to more conventional energy sources. Ongoing research focuses on surface modifications, device architecture optimizations, and material innovations to increase power density.

Durability and Washability: Real-world usage requires the device to withstand repeated washing, sweat, environmental exposure, and mechanical wear. Researchers are exploring robust coatings and encapsulation techniques that preserve triboelectric properties without significantly degrading flexibility or breathability.

Scalability of Manufacturing [56]: Though textile manufacturing is well-established, embedding functional materials into fibers at a large scale can present technical hurdles. Ensuring consistent quality and performance across meters or kilometers of textile requires precisely controlled fabrication techniques and careful material selection.

Standardization and Testing: The nascent nature of this technology means there are no widely accepted standards for testing and benchmarking performance, particularly when it comes to wearables. Developing standardized protocols for durability, washability, and electrical output measurements will be crucial for commercial acceptance.

Integration with Electronics [57]: Seamlessly integrating the generated power into electronic circuits, including storage (e.g., supercapacitors or batteries), signal conditioning, and wireless transmission modules, remains an area of active exploration. Flexible, lightweight, and unobtrusive circuit solutions will likely be needed to complement the benefits of textile-based TENGs.

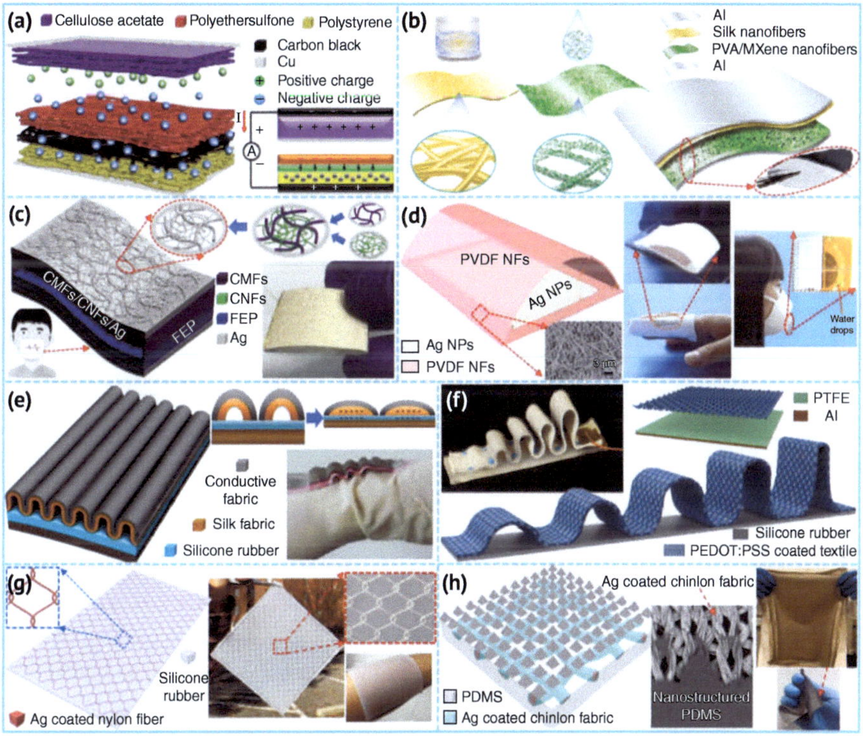

Figure 18.8 Textile-based triboelectric nanogenerators (TENGs) are developed using nanofiber networks or membrane structures incorporating textiles. (a) A TENG with multiple layers utilizes cellulose acetate nanofibers as the top layer and a composite of polyethersulfone, carbon black, and polystyrene nanofibers as the bottom layer. (b) A fully electrospun TENG pairs MXene/PVA composite nanofibers with silk fibroin as triboelectric materials. (c) A TENG based on hierarchical nanocellulose fibers transforms one-dimensional environmentally friendly cellulose microfibers/nanofibers into a two-dimensional CMF/CNF/Ag nanostructure. (d) A self-powered sensor with an arched design for monitoring respiration uses an electrospun PVDF nanofiber film paired with an electrode made from screen-printed Ag nanoparticles. (e) A TENG with a corrugated textile design features wavy silk conductive fabric on top and silicone rubber-coated conductive fabric underneath. (f) A strain sensor based on a TENG with varying arch heights is created by attaching a textile treated with PEDOT:PSS in an arch shape to a silicone rubber elastomer. (g) An electronic skin TENG incorporates stretchable yarn by embedding networks of conductive yarn into a silicone rubber elastomer. (h) A highly flexible, scalable textile-based TENG is made by applying a nanostructured PDMS film to a chinlon fabric coated with silver. *Source:* [54] with permission of John Wiley & Sons.

Looking ahead, advances in materials science, such as the development of novel polymers, conductive inks, and nanocomposites [58], are poised to enhance the performance of textile TENGs further. In parallel, improvements in high-throughput manufacturing methods could drive down costs and pave the way for mass-market products. With the growing interest in wearable technology, soft robotics, and the IoT, there is a strong impetus to push the capabilities of fiber-based TENGs, making them more efficient, durable, and seamlessly integrated into daily life.

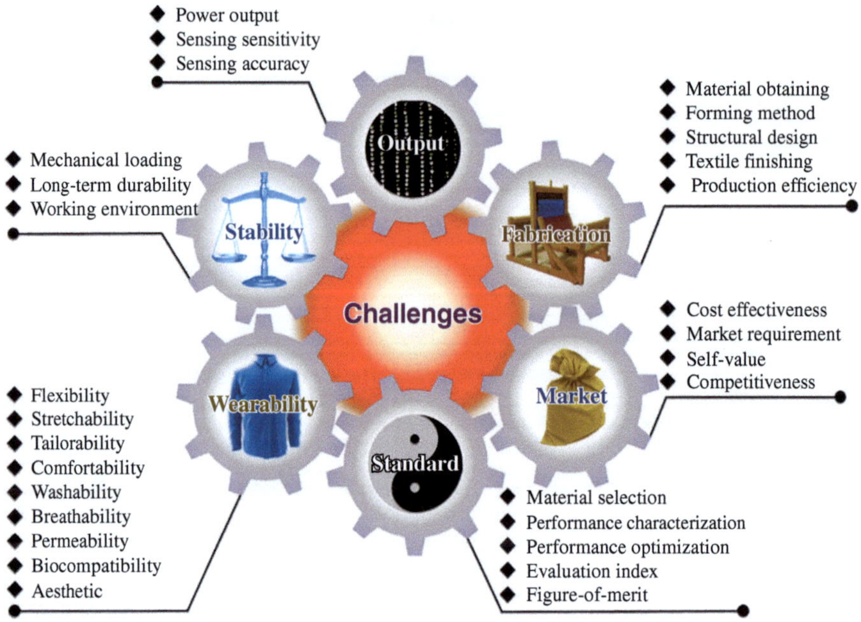

Figure 18.9 Several obstacles and challenges hinder the widespread industrial production and practical commercialization of textile-based triboelectric nanogenerators (TENGs). *Source:* [54] with permission of John Wiley & Sons.

Fiber- and textile-based TENGs represent a compelling evolution of TENG technology, bridging the gap between rigid energy-harvesting devices and the flexible, user-friendly textiles that dominate our everyday lives. By transforming ordinary fibers into power-generating elements through material selection, surface modification, and engineering design, these TENGs can harvest mechanical energy from a plethora of sources: human motion, environmental vibrations, and even routine activities like sitting on a chair or walking across a floor. The inherent advantages of flexibility, breathability, comfort, and large-scale manufacturability of textiles make them ideal candidates for next-generation wearable electronics, self-powered sensors, and distributed energy-harvesting solutions. Nevertheless, challenges remain in ensuring consistent performance, durability against real-world conditions, and straightforward integration with electronic components. Researchers worldwide are tackling these hurdles through innovative materials research, advanced manufacturing techniques, and novel device architectures, as illustrated in Figure 18.10 [45], steadily pushing the boundaries of what fiber-based TENGs can achieve [57]. As the technology matures, we can anticipate an array of new applications in healthcare monitoring, sports performance, smart homes, and beyond [59–60], ultimately transforming the textiles we wear and use into smart, self-powered systems that enhance our quality of life while minimizing environmental impact. With ongoing innovation, fiber- and textile-based TENGs are poised to play a pivotal role in the shift toward flexible, wearable, and sustainable energy solutions in the decades to come.

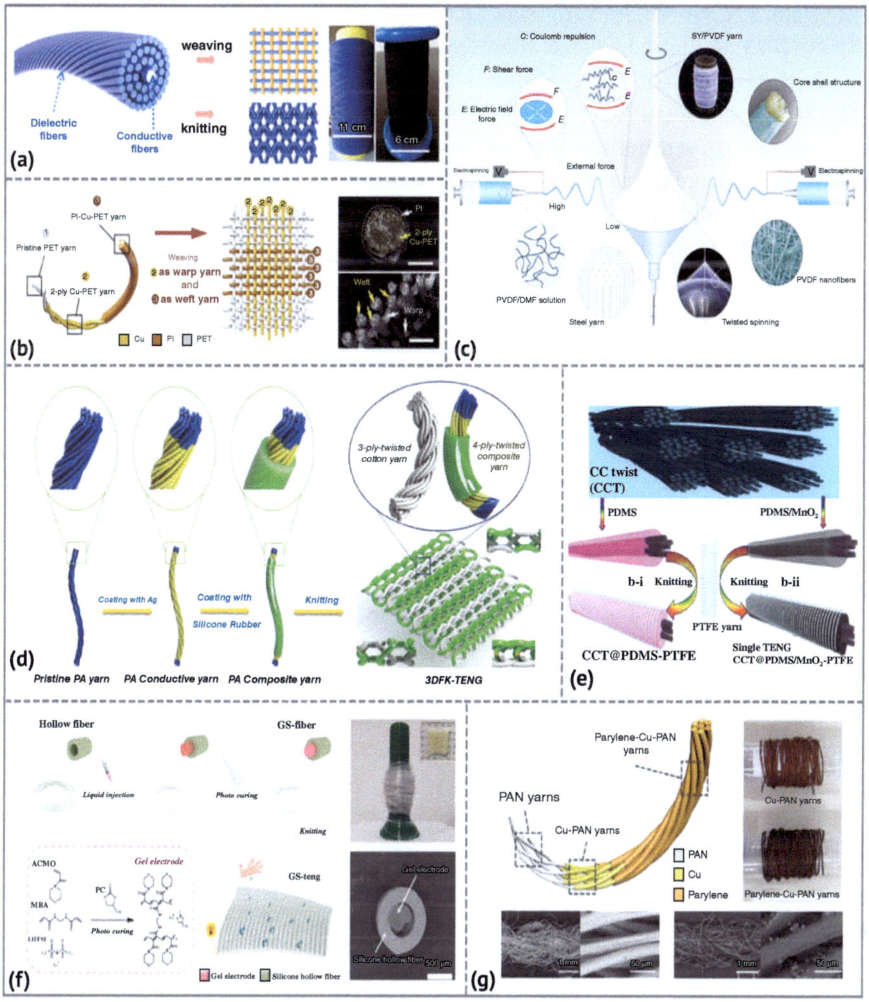

Figure 18.10 The development of the triboelectric layer involves the modification of fibers and yarns. (a) Diagram showing yarns designed for t-TENGs with a core–shell configuration. (b) Diagram depicting Cu–PET warp yarns and PI–Cu–PET weft yarns, accompanied by SEM images of their cross sections and woven patterns. (c) Diagram illustrating the yarn spinning process using electroassisted core spinning technology. (d) Production process of PA composite yarns. (e) Diagram demonstrating the round-tripping knitting technique for creating elastic yarns of CCT@PDMS and CCT@PDMS/MnO_2NW. (f) Manufacturing process of fibers based on organogel electrodes. (g) Diagrams and SEM images of Cu-PAN yarns and parylene-Cu-PAN yarns. *Source:* [45] with permission of John Wiley & Sons.

References

1 Iqbal, M.S., Lu, H., Khaladkar, S., Wu, X., Roy, S., Ullah, Z., Yan, Y., Zhang, Y. and Guo, B. (2024). Recent advances in triboelectric nanogenerators: Mechanism, rational designing and applications. *Mater. Today Energy*, 46, 101732.

2 Alam, S.N., Ghosh, A., Shrivastava, P., Shukla, U., Garg, K., Edara, A.C. and Sahoo, N. (2023). An introduction to triboelectric nanogenerators. *Nano-Struct. Nano-Objects*, 34, 100980.

3 Chen, J. and Wang, Z.L. (2017). Reviving vibration energy harvesting and self-powered sensing by a triboelectric nanogenerator. *Joule*, 1, 480.

4 Wang, L., Liu, W., Yan, Z., Wang, F. and Wang, X. (2021). Stretchable and shape-adaptable triboelectric nanogenerator based on biocompatible liquid electrolyte for biomechanical energy harvesting and wearable human–machine interaction. *Adv. Funct. Mater.*, 31, 2007221.

5 Wang, H., Han, M., Song, Y. and Zhang, H. (2021). Design, manufacturing and applications of wearable triboelectric nanogenerators. *Nano Energy*, 81, 105627.

6 Lai, Y.-C., Hsiao, Y. C., Wu, H. M. and Wang, Z.L. (2019). Waterproof fabric-based multifunctional triboelectric nanogenerator for universally harvesting energy from raindrops, wind, and human motions and as self-powered sensors. *Adv. Sci.*, 6, 1801883.

7 Ren, Z.W., Wang, Z.M., Liu, Z.R., Wang, L.F., Guo, H.Y., Li, L.L., Li, S.T., Chen, X.Y., Tang, W. and Wang, Z.L. (2020). Energy harvesting from breeze wind (0.7–6 $m \cdot s^{-1}$) using ultra-stretchable triboelectric nanogenerator. *Adv. Energy Mater.*, 10, 2001770.

8 Xia, K.Q., Fu, J.M. and Xu, Z.W. (2020). Multiple-frequency high-output triboelectric nanogenerator based on a water balloon for all-weather water wave energy harvesting. *Adv. Energy Mater.*, 10, 2000426.

9 Rodrigues, C., Nunes, D., Clemente, D., Mathias, N., Correia, J.M., Rosa-Santos, P., Taveira-Pinto, F., Morais, T., Pereira, A. and Ventura, J. (2020). Emerging triboelectric nanogenerators for ocean wave energy harvesting: state of the art and future perspectives. *Energy Environ. Sci.*, 13, 2657.

10 Liang, X., Jiang, T., Liu, G.X., Feng, Y.W., Zhang, C. and Wang, Z.L. (2020). Spherical triboelectric nanogenerator integrated with power management module for harvesting multidirectional water wave energy. *Energy Environ. Sci.*, 13, 277.

11 Liang, X., Jiang, T., Feng, Y.W., Lu, P.J., An, J. and Wang, Z.L. (2020). Triboelectric nanogenerator network integrated with charge excitation circuit for effective water wave energy harvesting. *Adv. Energy Mater.*, 10, 2002123.

12 Zhang, R.Y. and Olin, H. (2020). Material choices for triboelectric nanogenerators: a critical review. *EcoMat*, 2, e12062.

13 Fan, F.R., Tian, Z.Q. and Wang, Z.L. (2012). Flexible triboelectric generator. *Nano Energy*, 1, 328.

14 Shi, Z., Zhang, Y.H., Gu, J.W., Liu, B., Fu, H., Liang, H.Y. and Ji, J.H. (2024). Triboelectric nanogenerators: state of the art. *Sensors*, 24, 4298.

15 Maharjan, P., Toyabur, R.M. and Park, J.Y. (2018). A human locomotion inspired hybrid nanogenerator for wrist-wearable electronic device and sensor applications. *Nano Energy*, 46, 383.

16 Roy, S., Maji, P.K. and Goh, K.L. (2021). Sustainable design of flexible 3D aerogel from waste PET bottle for wastewater treatment to energy harvesting device. *Chem. Eng. J.*, 413, 127409.

17 Lai, W.L., Sharma, S., Roy, S., Maji, P.K., Sharma, B., Ramakrishna, S. and Goh, K.L. (2022). Roadmap to sustainable plastic waste management: a focused study on recycling PET for triboelectric nanogenerator production in Singapore and India. *Environ. Sci. Pollut. Res.*, 29, 51234.

18 Jiang, B., Long, Y., Pu, X., Hu, W.G. and Wang, Z.L. (2021). A stretchable, harsh condition-resistant and ambient-stable hydrogel and its applications in triboelectric nanogenerator. *Nano Energy*, 86, 106086.

19 Zu, G.Q., Wei, Y., Sun, C.Y. and Yang, X.J. (2022). Humidity-resistant, durable, wearable single-electrode triboelectric nanogenerator for mechanical energy harvesting. *J. Mater. Sci.*, 57, 2813.

20 Zhao, L.M., Li, H., Meng, J.P. and Li, Z. (2020). The recent advances in self-powered medical information sensors. *InfoMat*, 2, 212.

21 Ouyang, H., Tian, J.J., Sun, G.L., Zou, Y., Liu, Z., Li, H., Zhao, L.M., Shi, B.J., Fan, Y.B., Fan, Y.F., Wang, Z.L. and Li, Z. (2017). Self-powered pulse sensor for antidiastole of cardiovascular disease. *Adv. Mater.*, 29, 1703456.

22 Li, C.X., Zhu, Y.S., Sun, F.X., Jia, C.J., Zhao, T.M., Mao, Y.P. and Yang, H.D. (2022). Research progress on triboelectric nanogenerator for sports applications. *Energies*, 15, 5807.

23 Feng, X., Li, Q. and Wang, K. (2021). Waste plastic triboelectric nanogenerators using recycled plastic bags for power generation. *ACS Appl. Mater. Interfaces*, 13, 400.

24 Zhang, Z.Y., Hu, Z.Y., Wang, Y., Wang, Y.W., Zhang, Q.Q., Liu, D.H., Wang, H. and Xu, M.Y. (2022). Multi-tunnel triboelectric nanogenerator for scavenging mechanical energy in marine floating bodies. *J. Mar. Sci. Eng.*, 10, 455.

25 Zaw, N.Y.W., Yun, J., Goh, T.S., Kim, I., Kim, Y., Lee, J.S. and Kim, D. (2022). All-polymer waterproof triboelectric nanogenerator towards blue energy harvesting and self-powered human motion detection. *Energy*, 247, 123422.

26 Qu, Z.G., Huang, M.K., Chen, C.X., An, Y., Liu, H.Z., Zhang, Q.P., Wang, X.P., Liu, Y., Yin, W.L. and Li, X.F. (2022). Spherical triboelectric nanogenerator based on eccentric structure for omnidirectional low frequency water wave energy harvesting. *Adv. Funct. Mater.*, 32, 2202048.

27 Niu, Q.Q., Wei, H.F., Hsiao, B.S. and Zhang, Y.P. (2022). Biodegradable silk fibroin-based bio-piezoelectric/triboelectric nanogenerators as self-powered electronic devices. *Nano Energy*, 96, 107101.

28 Wang, Z.L. (2020). Triboelectric nanogenerator (TENG)—sparking an energy and sensor revolution. *Adv. Energy Mater.*, 10, 2000137.

29 Bu, X.F., Zhou, B.G., Li, J.C., Gao, C. and Guo, J.S. (2024). Orange peel-like triboelectric nanogenerators with multiscale micro-nano structure for energy harvesting and touch sensing applications. *Nano Energy*, 122, 109280.

30 Ye, J.C., He, C.-S., Gong, X. R., Zhang, H. H. and Li, X. (2025). Blue energy harvesting based on triboelectric nanogenerators (TENG): structural design, performance optimization, and application prospects. *J. Alloys Compd.*, 1014, 178710.

31 Bai, Z.Q., Xu, Y.L., Lee, C.K. and Guo, J.S. (2021). Autonomously adhesive, stretchable, and transparent solid-state polyionic triboelectric patch for wearable power source and tactile sensor. *Adv. Funct. Mater.*, 31, 2104365.

32 Shen, H., Lei, H., Gu, M.W., Miao, S., Gao, Z.Q., Sun, X.H., Sun, L.N., Chen, G.Y., Huang, H.B., Chen, L.G. and Wen, Z. (2022). A wearable electrowetting on dielectrics sensor for real-time human sweat monitor by triboelectric field regulation. *Adv. Funct. Mater.*, 32, 2204525.

33 Li, Y.Y., Zhang, Y.H., Yi, J., Peng, X., Cheng, R.W., Ning, C., Sheng, F.F., Wang, S., Dong, K. and Wang, Z.L. (2022). Large-scale fabrication of core-shell triboelectric braided fibers and power textiles for energy harvesting and plantar pressure monitoring. *EcoMat*, 4, e12191.

34 Liu, L., Wu, M., Zhao, W., Tao, J., Zhou, X. and Xiong, J. (2024). Progress of triboelectric nanogenerators with environmental adaptivity. *Adv. Funct. Mater.*, 34, 2308353.

35 Choi, D., Lee, Y., Lin, Z. H., Cho, S., Kim, M., Ao, C.K., Soh, S., Sohn, C., Jeong, C.K., Lee, J., Lee, M., Lee, S., Ryu, J., Parashar, P., Cho, Y., Ahn, J., Kim, I. D., Jiang, F., Lee, P.S., Khandelwal, G., Kim, S. J., Kim, H.S., Song, H. C., Kim, M., Nah, J., Kim, W., Menge, H.G., Park, Y.T., Xu, W., Hao, J., Park, H., Lee, J. H., Lee, D. M., Kim, S. W., Park, J.Y., Zhang, H., Zi, Y., Guo, R., Cheng, J., Yang, Z., Xie, Y., Lee, S., Chung, J., Oh, I. K., Kim, J. S., Cheng, T., Gao, Q., Cheng, G., Gu, G., Shim, M., Jung, J., Yun, C., Zhang, C., Liu, G., Chen, Y., Kim, S., Chen, X., Hu, J., Pu, X., Guo, Z.H., Wang, X., Chen, J., Xiao, X., Xie, X., Jarin, M., Zhang, H., Lai, Y.-C., He, T., Kim, H., Park, I., Ahn, J., Huynh, N.D., Yang, Y., Wang, Z.L., Baik, J.M. and Choi, D. (2023). Recent advances in triboelectric nanogenerators: from technological progress to commercial applications. *ACS Nano*, 17, 11087.

36 Jiang, C.M., Li, X.J., Ying, Y.B. and Ping, J.F. (2021). Fluorinated graphene-enabled durable triboelectric coating for water energy harvesting. *Small*, 17, 2007805.

37 Wang, J.X., He, J.M., Ma, L.L., Yao, Y.L., Zhu, X.D., Peng, L., Liu, X.R., Li, K.S. and Qu, M.N. (2021). A humidity-resistant, stretchable and wearable textile-based triboelectric nanogenerator for mechanical energy harvesting and multifunctional self-powered haptic sensing. *Chem. Eng. J.*, 423, 130200.

38 Paosangthong, W., Torah, R. and Beeby, S. (2019). Recent progress on textile-based triboelectric nanogenerators. *Nano Energy*, 55, 401.

39 Kwak, S.S., Kim, H., Seung, W., Kim, J., Hinchet, R. and Kim, S.-W. (2017). Fully stretchable textile triboelectric nanogenerator with knitted fabric structures. *ACS Nano*, 11, 10733.

40 Yang, Y., Sun, N., Wen, Z., Cheng, P., Zheng, H., Shao, H., Xia, Y., Chen, C., Lan, H., Xie, X., Zhou, C., Zhong, J., Sun, X. and Lee, S.-T. (2018). Liquid-metal-based super-stretchable and structure-designable triboelectric nanogenerator for wearable electronics. *ACS Nano*, 12, 2027.

41 Dassanayaka, D.G., Alves, T.M., Wanasekara, N.D., Dharmasena, I.G. and Ventura, J. (2022). Recent progresses in wearable triboelectric nanogenerators. *Adv. Funct. Mater.*, 32, 2205438.

42 Dong, K., Peng, X., Cheng, R.W., Ning, C.A., Jiang, Y., Zhang, Y.H. and Wang, Z.L. (2022). Advances in high-performance autonomous energy and self-powered sensing textiles with novel 3D fabric structures. *Adv Mater.*, 34, 2109355.

43 Feng, P.Y., Xia, Z.K., Sun, B.B., Jing, X., Li, H., Tao, X.M., Mi, H.Y. and Liu, Y.J. (2021). Enhancing the Performance of Fabric-Based Triboelectric Nanogenerators by Structural and Chemical Modification. *ACS Appl. Mater. Interfaces.*, 13, 16916.

44 Doganay, D., Cicek, M.O., Durukan, M.B., Altuntas, B., Agbahca, E., Coskun, S. and Unalan, H.E. (2021). Fabric based wearable triboelectric nanogenerators for human machine interface. *Nano Energy*, 89, 106412.

45 Zhao, Z. and Hu, Y. (2024). Textile triboelectric nanogenerator: Future smart wearable energy-integration technology. *Adv. Mater. Technol.*, 9, 2302012.

46 Ye, C., Yang, S., Ren, J., Dong, S.J., Cao, L.T., Pei, Y. and Ling, S.J. (2022). Electroassisted core-spun triboelectric nanogenerator fabrics for IntelliSense and artificial intelligence perception. *ACS Nano*, 16, 4415.

47 Jing, T.T., Xu, B.G. and Yang, Y.J. (2021). Organogel electrode based continuous fiber with large-scale production for stretchable triboelectric nanogenerator textiles. *Nano Energy*, 84, 105867.

48 Zhang, H., Zhang, D.Z., Wang, D.Y., Xu, Z.Y., Yang, Y. and Zhang, B. (2022). Flexible single-electrode triboelectric nanogenerator with MWCNT/PDMS composite film for environmental energy harvesting and human motion monitoring. *Rare Met.*, 41, 3117.

49 Xu, F., Dong, S.S., Liu, G.X., Pan, C.X., Guo, Z.H., Guo, W.B., Li, L.W., Liu, Y.P., Zhang, C., Pu, X. and Wang, Z.L. (2021). Scalable fabrication of stretchable and washable textile triboelectric nanogenerators as constant power sources for wearable electronics. *Nano Energy*, 88, 106247.

50 Zhang, C., Zhang, L., Bao, B., Ouyang, W.Y., Chen, W.Y., Li, Q.T. and Li, D.C. (2023). Customizing triboelectric nanogenerator on everyday clothes by screen-printing technology for biomechanical energy harvesting and human-interactive applications. *Adv. Mater. Technol.*, 8, 2201138.

51 Dong, K., Deng, J.A., Ding, W.B., Wang, A.C., Wang, P.H., Cheng, C.Y., Wang, Y.C., Jin, L.M., Gu, B.H., Sun, B.Z. and Wang, Z.L. (2018). Versatile core–sheath yarn for sustainable biomechanical energy harvesting and real-time human-interactive sensing. *Adv. Energy Mater.*, 8, 1801114.

52 Jiang, J., Wang, Q., Wang, B.Y., Dong, J.Q., Li, Z.L., Li, X.Y., Zi, Y.L., Li, S.T. and Wang, X.F. (2019). Direct lift-off and the piezo-phototronic study of InGaN/GaN heterostructure membrane. *Nano Energy*, 59, 545.

53 Ge, G., Lu, Y., Qu, X.Y., Zhao, W., Ren, Y.F., Wang, W.J., Wang, Q., Huang, W. and Dong, X.C. (2020). Muscle-inspired self-healing hydrogels for strain and temperature sensor. *ACS Nano*, 14, 218.

54 Dong, K., Peng, X. and Wang, Z.L. (2020). Fiber/fabric-based piezoelectric and triboelectric nanogenerators for flexible/stretchable and wearable electronics and artificial intelligence. *Adv. Mater.*, 32, 1902549.

55 Miao, Y., Zhou, M., Yi, J., Wang, Y., Tian, G., Zhang, H., Huang, W., Wang, W., Wu, R. and Ma, L. (2024). Woven fabric triboelectric nanogenerators for human-computer interaction and physical health monitoring. *Nano Res.*, 17, 5540.

56 Shen, Y., Chen, C., Chen, L., Shang, L., Wang, T., Wang, K., Zhao, Y., Hou, R., Cong, H., Jiang, G., Zhang, A., Ma, P. and Wang, Z.L. (2024). Mass-production of biomimetic fur knitted triboelectric fabric for smart home and healthcare. *Nano Energy*, 125, 109510.

57 Bayan, S., Pal, S. and Ray, S.K. (2022). Interface engineered silver nanoparticles decorated g-C3N4 nanosheets for textile based triboelectric nanogenerators as wearable power sources. *Nano Energy*, 94, 106928.

58 Neelakandan, S., Srither, S.R., Dhineshbabu, N.R., Maloji, S., Dahlsten, O., Balaji, R. and Singh, R. (2024). Recent advances in wearable textile-based triboelectric nanogenerators. *Nanomaterials*, 14, 1500.

59 Zhang, S., Rana, S.M.S., Bhatta, T., Pradhan, G.B., Sharma, S., Song, H., Jeong, S. and Park, J.Y. (2023). 3D printed smart glove with pyramidal MXene/Ecoflex composite-based toroidal triboelectric nanogenerators for wearable human-machine interaction applications. *Nano Energy*, 106, 108110.

60 Hao, Y., Zhang, Y., Mensah, A., Liao, S., Lv, P. and Wei, Q. (2023). Scalable, ultra-high stretchable and conductive fiber triboelectric nanogenerator for biomechanical sensing. *Nano Energy*, 109, 108291.

19

Multiscale Optimal Design of Smart Materials

Yasutomo Uetsuji

Department of Mechanical Engineering, Osaka Institute of Technology, Osaka, Japan

19.1 Introduction

Smart materials can change mechanical and electrical properties in response to external stimuli such as electric fields, magnetic fields, forces, and temperature. Representative materials include piezoelectric (PE) materials, electrostrictive materials, piezoelectric materials (PM), magnetostrictive materials, shape memory alloys, bimetals, electrorheological elastomers, and magnetorheological elastomers. They are used in various electromechanical devices as important elements of sensors, actuators, and energy harvesting in a wide range of fields, from automobiles to aerospace, home appliances, information and communications, and chemistry and medicine [1]. In this chapter, we focus on PE materials, PM, and their composite materials, and explain multiscale optimal design to improve their functional properties.

Starting with the discovery of the magnetostrictive effect in 1842, the PE effect in 1880, and the magnetoelectric (ME) effect in 1960, the coupling effect linking mechanics, electricity, and magnetism gained attention as a breakthrough for new devices. The PE effect is classified into a positive effect, which generates electric fields when a mechanical strain is externally applied, and an inverse effect, which generates strains when an electric field is externally applied. The inverse PE effect is a first-order electromechanical interaction and generates strains proportional to the applied electric field. On the other hand, the electrostrictive effect is a second-order electromechanical interaction and generates strains proportional to the square of the applied electric field. Similarly, the PM effect is a first-order and the magnetostrictive effect is a second-order magnetomechanical interaction and generates strains proportional to the first and second orders of the applied magnetic field, respectively. Furthermore, for example, when a magnetic field is externally applied to composite materials of PE and PM phases, the strains generated in the PM phase by the PM effect are transmitted to the PE phase, and the PE effect generates electric fields, resulting in the positive ME effect. On the other hand, when an electric field is externally applied, strains generated in the PE phase by the PE effect are transmitted to the PM phase, and magnetic fields are generated by the PM effect, resulting in the inverse ME effect.

In the case of simple materials without coupling effects, material properties can be improved by single crystallization, uniform organization, and uniform polarization. On the other hand, in the case of functional interference materials, the response to external loads involves multiple effects, and stress concentration also has a significant impact on the apparent functional properties, so it is important to bring out the synergistic effects of the heterogeneous structure and coupling properties. In other words, functional material design that proactively utilizes heterogeneous structures holds the potential to improve previously unknown properties.

Theories for dealing with heterogeneous structures include the classical mixture rule, Mori-Tanaka mean field theory [2], self-consistent theory [3], and asymptotic homogenization theory [4]. Among them, the asymptotic homogenization theory can derive governing equations that are mathematically consistent across scales for hierarchical structures with different scales and have high affinity with the finite element method. To date, multiscale finite element analysis based on the homogenization theory has been widely and actively applied to a wide range of multiphysics problems, from elasticity [5] to damage nonlinearity [6], elastoplasticity [7–9], viscoelasticity [10], heat conduction [11], piezoelectricity [12–18], ferroelectricity [19, 20], and electromagnetism [21–24] in composite materials. In this chapter, we apply multiscale finite element analysis to smart materials and discuss the optimal design of heterogeneous microstructures to maximize their macroscopic functional properties [13, 25–27]. This chapter is composed of five sections. Section 19.1 is the introduction, Section 19.2 summarizes the calculation method, Section 19.3 shows an example of the optimum design of polycrystalline PE material, Section 19.4 shows an example of the optimum design of ME composite material, and Section 19.5 describes the findings and future prospects.

19.2 Computational Methods

Figure 19.1 shows the overall scheme for multiscale optimal design for linearly coupled problems involving displacement, electric potential, and magnetic potential. The microstructure is, a PE polycrystal, a PM polycrystal body, or a polycrystalline composite, taking into account heterogeneity in polarization orientation. The macrostructure is a homogeneous body that reflects the microstructural characteristics, and mechanical, electrical, and magnetic boundary conditions are set to capture the physical response to external forces. The governing equations were derived by applying asymptotic homogenization theory to the scale coupling between the microstructure and the macrostructure (Appendix A). A multiscale simulation was developed to obtain approximate solutions using the finite element method for the derived microstructure and macrostructure equations. This multiscale simulation enables the following three numerical analyses: (1) Obtain homogenized physical properties of the macrostructure from various microstructural characteristics.

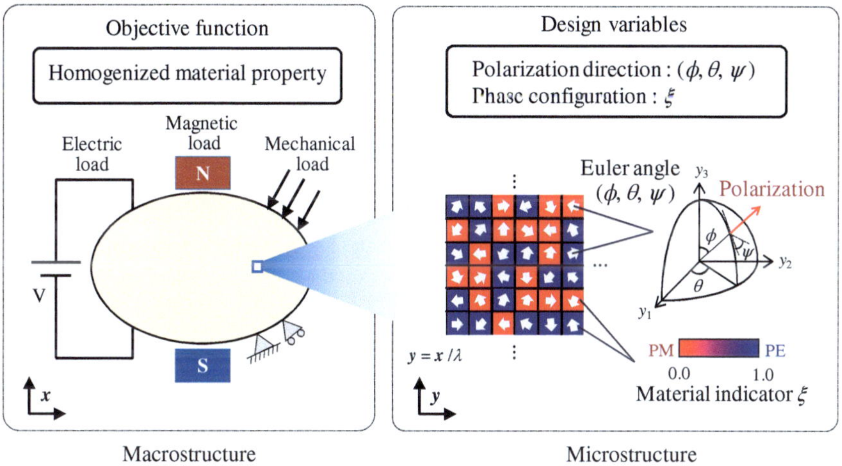

Figure 19.1 Multiscale modeling and optimization for polycrystalline PE materials, PM materials, and ME composites.

(2) Capture the physical response of the macrostructure to which homogenized physical properties are applied under mechanical, electrical, and magnetic boundary conditions. (3) Analyze the physical state of the heterogeneous microstructure from the physical response of the macrostructure.

For an optimization problem spanning from the microstructure to the macrostructure, the objective function was set to maximize the homogenized physical properties of the macrostructure, and the design variables were set to the characteristic parameters of the microstructure (Appendix B). The characteristic parameters considered were the electric and magnetic polarization directions defined by Euler angles, and the material indicator that determines PE or PM. The material indicator was set as a parameter that changes continuously using trigonometric functions from 0.0, which indicates a perfect state of PM, to 1.0, which indicates a perfect state of PE. The steepest descent method was applied to the previous multiscale optimization problem to search for the optimal solution.

19.3 Application to PE Polycrystals

Figure 19.2 shows the orientation dependence of the PE strain constants ${}^{\text{micro}}d_{333}$ and ${}^{\text{micro}}d_{311}$ in the micro-coordinate system for a single crystal of barium titanate $BaTiO_3$ (BTO) applied to each domain forming the microstructure. Here, the dependence on the angle ϕ between the micro-coordinate y_3 axis and the spontaneous polarization c axis is shown under the condition that $\phi = 45°$ and $\psi = 0°$ for the Euler angles (ϕ, θ, ψ) that determine the crystal orientation. From Figure 19.2, the PE strain constants of the BTO single crystal are the maximum values of ${}^{\text{micro}}d_{333} = 250.52\ \text{pC/N}$ and ${}^{\text{micro}}d_{333} = -229.33\ \text{pC/N}$ when the spontaneous polarization angle ϕ is 54.7°, that is, in the [111] orientation. The BTO single crystal generates large shear strain in the electric field in the a- and b-axis directions; that is, the

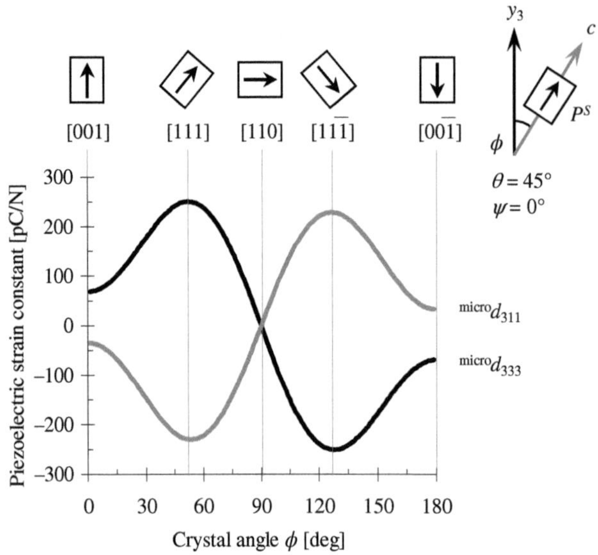

Figure 19.2 Dependence of PE strain constants on crystal orientation in BTO single crystal. *Source*: [13]/with permission of Elsevier.

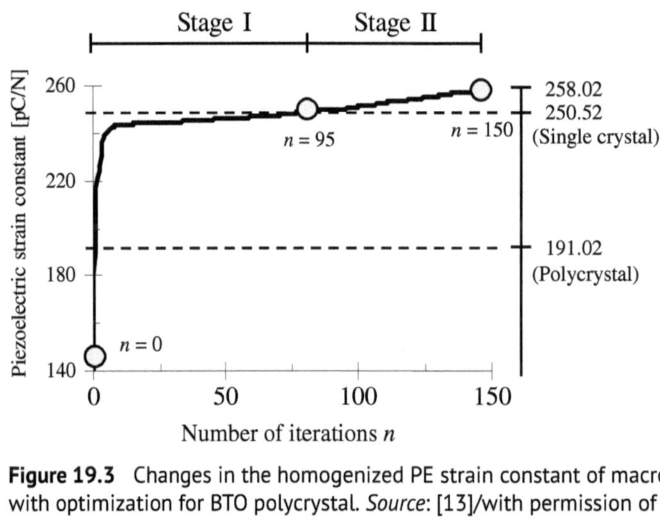

Figure 19.3 Changes in the homogenized PE strain constant of macrostructure $^{\text{macro}}d_{333}$ with optimization for BTO polycrystal. *Source*: [13]/with permission of Elsevier.

PE strain constants $^{\text{crystal}}d_{131}$ and $^{\text{crystal}}d_{223}$ in the crystal coordinate system are relatively large. As a result, when an off-axis electric field is applied, the shear strain is effectively generated in addition to the normal strain, resulting in the maximum PE response along the [111] direction rather than the c axis of the spontaneous polarization that is, the [001] direction.

The change in the homogenized PE strain constant of the macrostructure with optimization of the BTO polycrystal is shown in Figure 19.3, and the change in the crystal orientation distribution and the optimized microstructure is shown in Figure 19.4. As shown in Figure 19.4a, with an increase in the number of microstructure updates n, the Euler angle ϕ that governs the PE strain constant d_{333} of the single crystal first changed significantly from the initial state, where the crystal orientation was random. After that, it was confirmed that θ, which has no effect on d_{333}, tended to converge to three specific values in the single crystal. Focusing on this change in crystal orientation, the optimization process was roughly divided into Stage I ($n = 0$–95) and Stage II ($n = 96$–150). As shown in Figure 19.3, in Stage I, the initial state $n = 0$ gave 140.20 pC/N, and at $n = 95$, the $_{\text{macro}}d_{333}$ reached the maximum value of 250.52 pC/N for the single crystal shown in Figure 19.2. From Figure 19.4a, in stage I, the crystal orientations were ring-shaped in the (001) pole figure and [111] orientations ($\phi = 54.7°$, $\theta = 0°$–360°, $\psi = 45°$) in the inverse pole figure. Next, in stage II, at $n = 150$, the value of 258.02 pC/N was obtained, exceeding that of the single crystal, and the convergence condition was satisfied. Compared to the conventional bulk polycrystal, the PE strain constant was improved by 35%. The crystal orientation distribution showed some variation in the (001) pole figure, but a tendency for it to be localized to three specific orientations could be confirmed. In the spatial distribution of the optimized microstructure shown in Figure 19.4b, it can be seen that the three specific orientations are distributed in layers, arranged in the order of orientation A ($\theta = 0°$), orientation C ($\theta = 240°$), and orientation B ($\theta = 120°$) from the bottom layer of the microstructure.

The objective function, the PE strain constant $^{\text{macro}}d_{333}$, is defined as the perpendicular strain in the electric field direction generated per unit electric field E_3, and the perpendicular strain, is generated by mechanical effects $\left(s^E_{33kl}\sigma_{kl}\right)$ and electrical

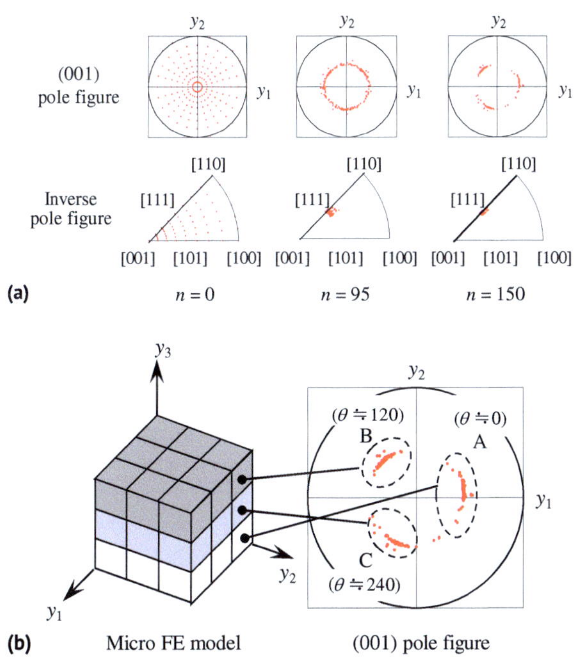

Figure 19.4 Change of the crystal orientation distribution and the optimized microstructure for BTO polycrystal. (a) Change of crystal orientation distribution in microstructure due to optimization. (b) Optimized microstructure. *Source*: [13]/with permission of Elsevier.

effects ($d_{m33}E_m$). In stage I in Figure 19.3, each domain was oriented in the [111] direction, and as shown in Figure 19.2 each domain reached its maximum value, and the strain due to the electrical effect was maximized, which is thought to increase the macroscopic PE strain constant. On the other hand, in stage II, the crystal orientations were localized to three regions and distributed in layers in the microstructure, which generated high microscopic stress due to the crystal orientation difference between the layers, and it is presumed that the strain due to the mechanical effect increased. Therefore, to more accurately verify the increase in the macroscopic PE strain constant due to the mechanical effect in stage II, we considered a simplified microcrystal assembly that ignored the variation in crystal orientations in the optimal solution. In other words, we considered a microcrystal assembly with a three-layer structure in which all domains are oriented in the [111] direction with $\phi = 54.7°$ and $\psi = 45°$, and only θ differs. Here, the Euler angles θ of each layer are θ_1, θ_2, and θ_3, respectively. Under the condition that $\theta_1 = 0°$, θ_2 and θ_3 were changed from 0° to 360°, and ${}^{\mathrm{macro}}d_{333}$ was evaluated. Next, a uniform unit electric field $E_3 = 1.0$ V/m was applied in the x_3 axis direction to the macrostructure shown in the figure under the boundary condition that did not constrain the expansion and contraction deformation in the x_1, x_2, and x_3 axis directions, and the micro-response associated with this was analyzed. In other words, the micro strain $({}^{\mathrm{micro}}\varepsilon_{33})$, which corresponds to the macro PE strain constant ${}^{\mathrm{macro}}d_{333}$, was decomposed into the strain due to mechanical effects $({}^{\mathrm{micro}}s^{E}_{33kl}\,{}^{\mathrm{micro}}\sigma_{kl})$ and the strain due to electrical effects $({}^{\mathrm{micro}}d_{m33}\,{}^{\mathrm{micro}}E_m)$ for the investigation.

Figure 19.5a shows the strain due to electrical effects, and Figure 19.5b illustrates the strain due to mechanical effects in a microcrystal assembly. Since there is little difference between the layers, the values calculated from the state quantities in the first layer are shown. The horizontal axis shows the Euler angles θ_2 and θ_3 of the second and third layers. As shown in Figure 19.5a, since all layers were oriented in the [111] direction, their PE strain constant was maximized, and it can be confirmed that the strain due to the electrical effect was always constant, regardless of θ_2 and θ_3. On the other hand, the strain due to mechanical effects shown in Figure 19.5b changed depending on the Euler angle and was at its maximum when $\theta_2 = 120°$, $\theta_3 = 240°$ and when $\theta_2 = 240°$, $\theta_3 = 120°$. In other words, it was verified that when each layer was oriented in the [111] direction and the difference in crystal orientation between the layers was 120°, the microstrain due to mechanical effects was maximized, and therefore the macro PE strain constant reached its maximum value. Furthermore, strain due to mechanical effects occurred due to the inhomogeneity of the crystal orientations in the microstructure and could not occur in a single crystal. Therefore, the optimized polycrystal exhibited PE properties that exceeded those of the single crystal.

19.4 Application to ME Composites

A multiscale optimization design was carried out for a composite material in which the PE phase was a BTO single crystal and the PM phase was a cobalt ferrite $CoFe_2O_4$ (CFO) single crystal. The objective function was the maximization of the homogenized macro-transverse ME constant, and the design variables were the

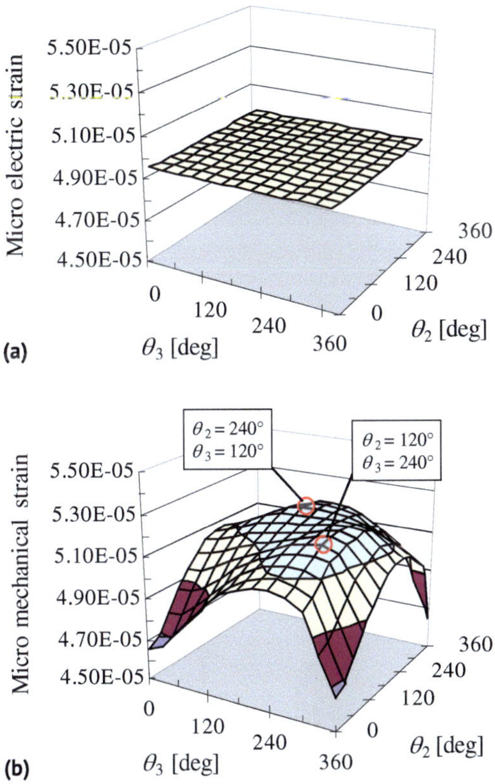

Figure 19.5 Dependence of the polarization directions of each layer on the strain generated in the optimized microstructure when a uniform electric field was applied to the macrostructure. (a) Electrical effect. (b) Mechanical effects. *Source*: [13]/with permission of Elsevier.

microstructural polarization orientations and the material indicators that determine the arrangement of the two phases. A regular division mesh was used for the microstructure model. The material indicator that changes continuously between the PE and PM phases and the polarization orientation (Euler angle) were independently assigned to each integration point in the element of the microstructure model. Figure 19.6 shows the numerical analysis results obtained from the multiscale optimization to maximize the transverse ME constant a_{31} of the BTO/CFO composite. The first vertical axis on the left shows the homogenized transverse ME constant $|a_{31}|$ of the macrostructure, normalized by the maximum value $|a_{31}| = 1.06$ nNs/VC in the existing laminate structure. The second vertical axis on the right shows the volume fraction x_{PE} of the PE phase calculated by binarizing the phase variables. The horizontal axis shows the number of times the microstructure was updated. The solid line in the figure indicates $|a_{31}|$, and the dotted line indicates x_{PE}. When the optimization started, $|a_{31}|$ increased rapidly and then increased gradually. After $n = 250$, there was no significant change, and it remained at a constant value at $n = 1000$. Next, to escape from the local solution, the design variables of the microstructure were cleaned at A in the figure. That is, the material variables were binarized, and similar orientations for electric and magnetic polarization were

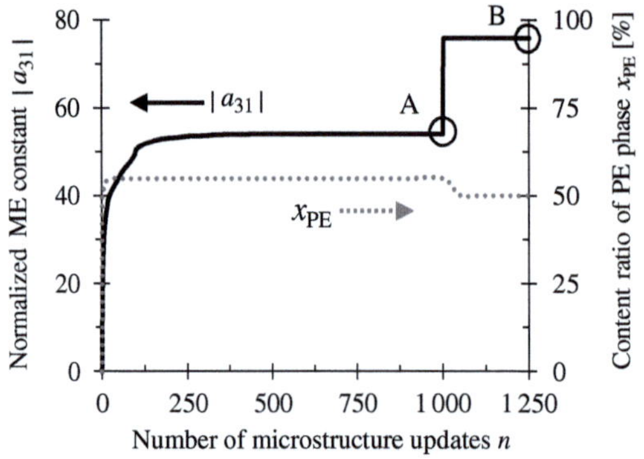

Figure 19.6 Change in macro-homogenized transverse ME constant $|a_{31}|$ with optimization for BTO/CFO composite.

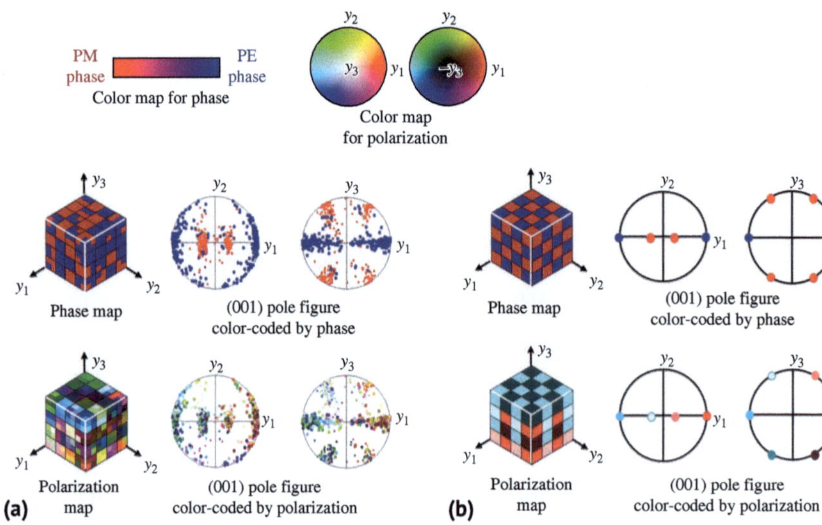

Figure 19.7 Changes in phase arrangement and crystal orientation distribution of microstructure with optimization of BTO/CFO composite. (a) n = 1 000. (b) n = 1 250. *Source*: [27]/with permission of Springer Nature.

grouped, and the orientations of the same group were replaced with their average value to simplify the microstructure. Cleaning caused $|a_{31}|$ to increase further and converged at $n = 1250$, as shown at B in the figure. On the other hand, the content of the PE phase changed significantly up to about $n = 30$ but then remained at $x_{PE} = 54.9\%$ and converged to $x_{PE} = 50.0\%$ after cleaning of the microstructure. The normalized ME constant of the optimized microstructure finally reached 71.6, a result far exceeding that of existing laminate structures. The x_{PE} at which the ME constant was maximized was 87.5% in the laminate structure; that is, the PE phase was rich, whereas in the optimal structure it was 50.0%.

Figure 19.7 shows the change in the microstructure in the optimization process to maximize the transverse ME constant of the BTO/CFO composite shown in Figure 19.6. Figure 19.6a shows the stationary state at $n = 1\,000$ before cleaning the microstructure (A in Figure 19.6), and Figure 19.6b shows the final state at $n = 1\,250$ (B in Figure 19.6). In each state, the phase arrangement and (001) pole figures in which polarization orientations in the microstructure were projected onto the y_2–y_3 and y_1–y_3 planes are shown, from the left. The phases were color-coded in succession based on the material indicator, from the PM phase in red to the PE phase in blue, as shown in the upper panel, and the phase arrangement and the pole figures of polarization orientations in the microstructures marked with these colors are shown in the upper figures. In addition, the polarization orientations in the pole figures were colored continuously using hexadecimal RGB, and the polarization orientation arrangement and pole figures in the microstructure marked with those colors are shown in the lower figures. First, we focus on the stationary state microstructure at $n = 1000$ shown in Figure 19.6a. As the optimization proceeded, the phase variables were almost binary, either 0.0 or 1.0, and the virtual intermediate phase naturally disappeared, revealing a microstructure composed of the PE and PM phases. Although it was difficult to find order in the arrangement of the two phases, a tendency could be seen for the PE and PM phases to be arranged alternately in all three directions. The polarization orientations of the PE phase, shown in blue, were polarized and oriented in the positive and negative directions of the y_1 axis in the pole figure and tended to be perpendicular to the y_3 axis. On the other hand, the polarization orientations of the PM phase, shown in red, were oriented in the positive and negative directions of the y_3 axis and were slightly tilted in the positive and negative directions of the y_1 axis. Next, focusing on the final state of $n = 1\,250$ shown in Figure 19.6b, a regular microstructure has been constructed in which the PE and PM phases were arranged alternately in all three directions. The polarization orientations of the PE phase were oriented in two directions, the positive and negative directions of the y_1 axis. In addition, the polarization orientations of the PM phase were oriented along the y_3 axis while tilting toward the y_1 axis, and considering the positive and negative directions of the coordinate axes, they were oriented in a total of four directions. If we look closely at the arrangement of the polarization orientations in the microstructure shown in the lower figures, it can be seen that the positive and negative signs of the polarization orientations were inverted in layers with respect to the y_3 axis.

To clarify the mechanism of the enhanced electric and magnetic effects in the optimized structure, the microstructural response of the BTO/CFO composite to the application of a magnetic field in the macrostructure was analyzed. The transverse ME constant a_{31} component is defined as the electric flux density generated in the third axis in response to the application of a magnetic field in the first axis. Therefore, a magnetic field $H_1 = 1.0$ A/m in the first axis was applied to the macrostructure, and the strain transfer in the microstructure and the generation of the electric flux density D_3 in the macrostructure were investigated. Specifically, we focused on the fact that the polarization orientations of the PE and PM phases in the optimized structure are regularly inverted and discussed the effect of this on strain transfer. As shown in Figure 19.7b, in the optimized structure, the polarization orientations of the PE phase were oriented in the positive and negative directions of the first axis, and the polarization orientations of the PM phase were oriented in the positive and negative directions of the third axis while tilting to the first axis, and the signs of these orientations in both phases were alternately reversed for the third axis.

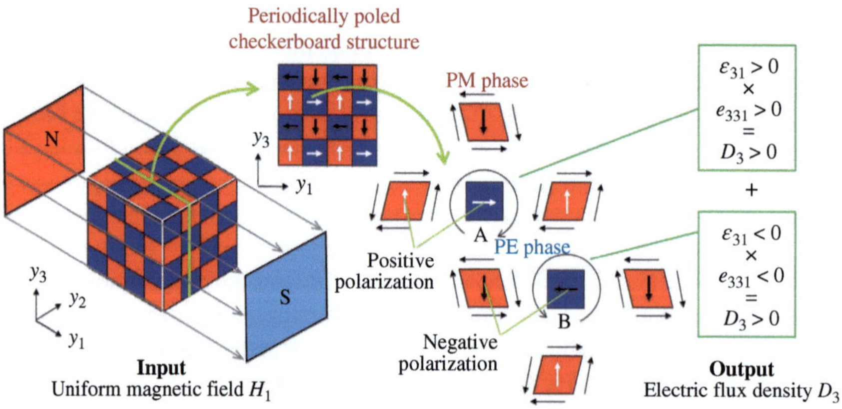

Figure 19.8 Schematic diagram of the deformation and electric flux density generation mechanism of the PM and PE phases in the optimized microstructure when a uniform magnetic field is applied to the macrostructure. *Source*: [27]/with permission of Springer Nature.

For simplification, the inclination of the polarization orientations of the FM phase to the first axis was ignored. The polarization orientations in the microstructure and the deformation of the PM and PE phases when an external magnetic field was applied are shown in Figure 19.8. The PE phase was shown in blue and the PM phase in red. The white arrow indicates the polarization orientation in the positive direction, and the black arrow indicates the polarization orientation in the negative direction.

Focusing on one PE phase with a polarization orientation oriented in the positive direction of the y_1 axis, the two adjacent PM phases on the left and right are oriented in the positive direction of the y_3 axis, and positive shear deformation occurs when the magnetic field is applied in the y_1 axis direction. In addition, the two adjacent PM phases on the top and bottom are oriented in the negative direction of the y_3 axis, and negative shear deformation occurs when the magnetic field is applied. The shear deformation of these four adjacent FM phases causes a clockwise twisting deformation around the FE phase oriented in the positive direction. Since the strain is transmitted from the PM phase to the PE phase with the sign reversed as a disturbance in the microstructure, negative shear strain is transmitted to the PE phase oriented in the positive direction. The polarization directions of the left and right and the top and bottom PM phases are opposite in sign, so that the torsional deformations do not cancel each other out but reinforce each other. As a result, a large shear strain is generated in the PE phase. In this positively oriented PE phase, the negative shear strain ε_{31} generates a negative electric flux density D_3 due to the positive PE effect e_{331}.

On the other hand, when we focus on the FE phase, whose polarization direction is oriented in the negative direction of the y_1 axis, the two adjacent PM phases on the left and right are oriented in the negative direction of the y_3 axis and generate negative shear strain when the magnetic field is applied. In addition, the two adjacent PM phases on the top and bottom are oriented in the positive direction of the y_3 axis and generate positive shear strain due to the magnetic field. The shear deformation of these four adjacent PM phases generates a counterclockwise twisting deformation around the negatively oriented PE phase. As mentioned earlier, the polarization directions of the left and right and the top and bottom PM phases are opposite in sign so that the torsional deformation is strengthened and a large shear strain is generated in the PE phase. In the negatively oriented PE phase, negative D_3 is generated by the negative PE effect e_{331} from positive ε_{31}. Therefore, shear strains of opposite signs are transmitted to the positively and negatively oriented PE phases, but the PE effects are also of opposite signs, so that electric flux densities of the same sign are ultimately generated. In other words, the electric flux densities generated in all PE phases are not canceled out, which increases the electric and magnetic properties. Furthermore, the polarization orientations of the adjacent phases above and below are inverted so that twist deformations of opposite signs are generated continuously and periodically around the positively and negatively oriented PE phases like a puzzle. In the optimized microstructure, the shear deformations of all PM phases efficiently contribute to strengthening the twist deformation of the adjacent PE phases without waste, and a mechanism was clarified that allows the electric flux densities generated in all PE phases to contribute to the expression of electric and magnetic properties without cancellation.

Taking the BTO/CFO composite as an example, the enhancement of the transverse ME effect by the heterostructure is summarized in Figure 19.9. The absolute value of the transverse ME constant $|a_{31}|$ is shown in the upper part, the phase arrangement in the middle part, and the polarization directions in the lower part. Here, the minimum unit cell of the periodically poled checkerboard structure was adopted for the microstructure model, and the analysis accuracy was improved by using a finer element division mesh than that of the optimization calculation. From the left, a multilayer structure, a one-sided checkerboard structure in which the PE and PM phases were alternately arranged in the first and second axes, and a three-sided checkerboard structure in which the PE and PM phases were alternately arranged in

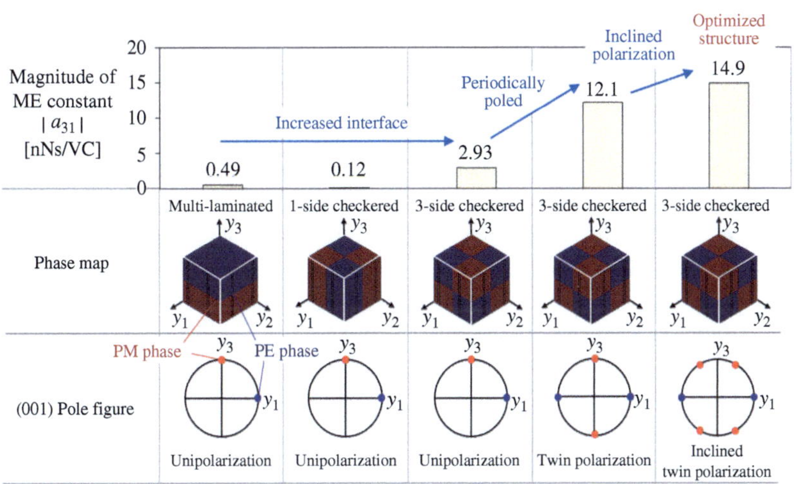

Figure 19.9 Comparison of the transverse ME constant with the change in the microstructure of BTO/CFO composite. *Source*: [27]/with permission of Springer Nature.

all three axes are shown. In these three microstructures, as shown in the (001) pole figure in the lower part, the polarization directions of the PE phase were oriented in the positive direction of the first axis, and the PM phase was oriented in the positive direction of the third axis; that is, each phase was oriented in a unidirectional state. In addition, for the three-sided checkerboard structure, we compared the properties of the twin-directional alignment state in which the PE phase was aligned in the positive and negative directions of the first axis and the PM phase was aligned in the positive and negative directions of the third axis, and the inclined twin-directional alignment state in which the polarization directions of the PM phase were inclined 30.4° to the first axis in the same alignment state. The inclined twin-directional alignment state was the optimized structure shown in Figure 19.7b.

In Figure 19.9, the PE phase content was standardized to 50.0% in all structures, and $|\alpha_{31}|$ was compared. In the conventional multilayer structure shown at the far left, $|\alpha_{31}|$ appeared, but it was small at 0.494 nNs/VC. In the one-sided checkerboard structure (the second from the left) and the three-sided checkerboard structure (the third), the area of the phase boundary, that is, the interface, increased, but the efficiency of strain transfer was poor, and the transverse ME constant hardly appeared. In the one-sided checkerboard structure, when the PE and PM phases were polarized in the same vertical or horizontal direction, vertically aligned nanocomposites exhibit excellent vertical ME constants, but this was not effective for the transverse ME constants. In the three-sided checkerboard structure, the interface was in the largest state, but the transverse ME constant did not appear in the unidirectional orientation state. In contrast, in the three-sided checkerboard structure, in the twin-directional orientation state in which the polarizations of the PE and PM phases were oriented in the positive and negative directions, $|\alpha_{31}|$ increased dramatically due to the polarization reversal effect, reaching 58.7 nNs/VC. Furthermore, in the inclined twin-directional orientation state in which the polarization directions of the PM phase were controlled, $|\alpha_{31}|$ increased by another 23%, reaching a maximum value of 75.9 nNs/VC.

19.5 Conclusion

As representative smart materials, we focused on PE polycrystals and ME composites and carried out multiscale optimal designs to maximize the macroscopic functional properties. For PE polycrystals, we discovered an optimal microstructure that exceeded that of the single crystal when the PE strain constant was used as the objective function. In the optimal microstructure, the strain due to the electrical effect was maximized, and the strain due to the mechanical effect was increased by the crystal orientation difference between the layers. On the other hand, for the ME composite, we discovered a periodically poled checkerboard structure that exhibited a transverse ME constant approximately 30 times higher than that of conventional laminate structures. In the periodically poled checkerboard structure, the shear deformation of all PM phases efficiently contributed to strengthening the torsional deformation of the adjacent PE phases, and the electric flux density generated in all PE phases contributed to the electric and magnetic properties without being offset.

In both of the aforementioned cases, the mechanism of the expression of excellent functional properties was due to heterogeneity in the microstructure, which overturns the conventional common sense in material development based on single crystallization, uniform organization, and uniform polarization. The key points in amplifying functional properties are the coupling effect and heterogeneity, and it is expected that smart materials that combine these points will lead to the discovery of even more unique microstructures. Meanwhile, while this study was an optimum search within the scope of continuum mechanics, it is hoped that constructing layered structures in PE polycrystals and periodically poled checkerboard structures in ME composites at the atomic level will lead to a breakthrough in ultra-high-performance smart materials.

Appendix A

The subject of this computation is the static problem of PE, PM polycrystals, and their composites with inhomogeneous polarization orientation distribution. These materials are handled with linear coupling constitutive equations, which relate the stress σ_{ij}, strain ε_{ij}, electric displacement D_i, electric field E_i, magnetic displacement B_i, and magnetic field H_i:

$$\sigma_{ij} = C_{ijmn}\varepsilon_{mn} - e_{nij}E_n - q_{nij}H_n \tag{A.1}$$

$$D_i = e_{imn}\varepsilon_{mn} + \kappa_{in}E_n + a_{in}H_n \tag{A.2}$$

$$B_i = q_{imn}\varepsilon_{mn} + a_{ni}E_n + \mu_{in}H_n \tag{A.3}$$

where C_{ijmn} is the fourth-order tensor for the elastic stiffness coefficient, κ_{ij} and $\mu_{ij}$are the second-order tensors for the electric permittivity and magnetic permeability, e_{mij} and q_{mij} are the third-order tensors for the PE and PM coupling coefficients, and a_{mn} is the second-order tensor for the ME coupling coefficient.

Asymptotic homogenization theory is employed for scale bridging between the macrostructure and the microstructures (Figure 19.1). The macrostructure is defined as a homogeneous body, and the microstructures are inhomogeneous structures composed of PE and/or PM crystal domains with independent polarization. It is assumed that the microstructures are sufficiently small compared to the macrostructure and have periodicity. The macrostructure occupies a region Ω, and each microstructure is defined as a region Y at every point in the macrostructure. Only one microstructure is considered in this computation. $\mathbf{x} = (x_1, x_2, x_3)^{\mathrm{T}}$ is employed for the coordinates for the macrostructure, while $\mathbf{y} = (y_1, y_2, y_3)^{\mathrm{T}}$ is used for the coordinates defined in every unit cell of the microstructure. The relation between $\mathbf{x}$ and $\mathbf{y}$ is as follows:

$$\mathbf{y} = \frac{1}{\lambda}\mathbf{x} \tag{A.4}$$

Here, λ indicates the scale ratio of the representative lengths between the macrostructure and the microstructure. The overall structure of interest is $\Omega^{\lambda} = \Omega \times \lambda Y$, which consists of the macrostructure and the microstructure.

Two assumptions are employed for scale bridging. The first is that the displacement u_i^λ, electric potential φ^λ, and magnetic potential ψ^λ for the overall structure are expressed by the following asymptotic expansions of the scale ratio:

$$u_i^\lambda = u_i^0(\mathbf{x}, \mathbf{y}) + \lambda u_i^1(\mathbf{x}, \mathbf{y}) + \cdots \tag{A.5}$$

$$\varphi^\lambda = \varphi^0(\mathbf{x}, \mathbf{y}) + \lambda \varphi^1(\mathbf{x}, \mathbf{y}) + \cdots \tag{A.6}$$

$$\psi^\lambda = \psi^0(\mathbf{x}, \mathbf{y}) + \lambda \psi^1(\mathbf{x}, \mathbf{y}) + \cdots \tag{A.7}$$

The superscript λ is a parameter for the overall structure and is related to both the macrostructure and the microstructure. $u_i^0(\mathbf{x}, \mathbf{y})$, $\varphi^0(\mathbf{x}, \mathbf{y})$, and $\psi^0(\mathbf{x}, \mathbf{y})$ are the displacement, electric potential, and magnetic potential for the macrostructure. These variables are finally independent of $\mathbf{y}$ and can be represented as $u_i^0(\mathbf{x})$, $\varphi^0(\mathbf{x})$, and $\psi^0(\mathbf{x}).u_i^1(\mathbf{x}, \mathbf{y})$, $\varphi^1(\mathbf{x}, \mathbf{y})$, and $\psi^1(\mathbf{x}, \mathbf{y})$ indicate perturbations to the displacement, electric potential, and magnetic potential in the inhomogeneous microstructure. They depend on the macrostructural and the microstructural coordinates.

The second assumption is that the microstructural perturbations $u_i^1(\mathbf{x}, \mathbf{y})$, $\varphi^1(\mathbf{x}, \mathbf{y})$, and $\psi^1(\mathbf{x}, \mathbf{y})$ change in proportion to the macrostructural mechanical strain $\varepsilon_{mn}^0(\mathbf{x})$, electric field $E_n^0(\mathbf{x})$, and magnetic field $H_n^0(\mathbf{x})$: that is,

$$u_i^1(\mathbf{x}, \mathbf{y}) = {}^{u\varepsilon}\chi_i^{mn}(\mathbf{x}, \mathbf{y})\,\varepsilon_{mn}^0(\mathbf{x}) - {}^{uE}\chi_i^n(\mathbf{x}, \mathbf{y})\,\mathrm{E}_n^0(\mathbf{x}) - {}^{uH}\chi_i^n(\mathbf{x}, \mathbf{y})\,H_n^0(x) \tag{A.8}$$

$$\varphi^1(\mathbf{x}, \mathbf{y}) = {}^{\varphi\varepsilon}\chi^{mn}(\mathbf{x}, \mathbf{y})\,\varepsilon_{mn}^0(\mathbf{x}) - {}^{\varphi E}\chi^n(\mathbf{x}, \mathbf{y})\,E_n^0(x) - {}^{\varphi H}\chi^n(\mathbf{x}, \mathbf{y})\,H_n^0(x) \tag{A.9}$$

$$\psi^1(\mathbf{x}, \mathbf{y}) = {}^{\psi\varepsilon}\chi^{mn}(\mathbf{x}, \mathbf{y})\,\varepsilon_{mn}^0(x) - {}^{\psi E}\chi^n(\mathbf{x}, \mathbf{y})\,E_n^0(x) - {}^{\psi H}\chi^n(\mathbf{x}, \mathbf{y})\,H_n^0(\mathbf{x}) \tag{A.10}$$

${}^{u\varepsilon}\chi_i^{mn}(\mathbf{x}, \mathbf{y})$, ${}^{uE}\chi_i^n(\mathbf{x}, \mathbf{y})$, and ${}^{uH}\chi_i^n(\mathbf{x}, \mathbf{y})$ indicate the microstructural perturbations of displacement per unit macrostructural mechanical strain, electric field, and magnetic field, respectively. ${}^{\varphi\varepsilon}\chi^{mn}(\mathbf{x}, \mathbf{y})$, ${}^{\varphi E}\chi^n(\mathbf{x}, \mathbf{y})$, and ${}^{\varphi H}\chi^n(\mathbf{x}, \mathbf{y})$ are the microstructural perturbations of the electric potential, and ${}^{\psi\varepsilon}\chi^{mn}(\mathbf{x}, \mathbf{y})$, ${}^{\psi E}\chi^n(\mathbf{x}, \mathbf{y})$, and ${}^{\psi H}\chi^n(\mathbf{x}, \mathbf{y})$ are the microstructural perturbations of the magnetic potential per unit macrostructural mechanical strain, electric field, and magnetic field, respectively. These proportionality constants are called the characteristic function tensors, and they satisfy Y-periodicity.

By applying these two assumptions, the macro-equations and micro-equations can be obtained as the conditions to establish the virtual work principle equations of the overall structure for any virtual displacement, electric potential, and magnetic potential [12]. The macro-equations result in three virtual work principle equations related to displacement, electric potential, and magnetic potential, and they determine the behavior of the macrostructure with respect to external loads. The macro-equations reflect the homogenized material properties described below, obtained from the microstructure. The micro-equations can be organized into nine equations related to the characteristic function tensors introduced in Eqs. (A.8)–(A.10). By solving the micro-equations under the periodic boundary conditions, the unknown characteristic function tensors can be obtained. From the obtained characteristic functions and material properties of each phase constituting the microstructure, the homogenized material properties of the macrostructure are derived as follows:

$$
\begin{aligned}
{}^{\text{macro}}C_{ijkl} = \frac{1}{|Y|}\int_Y \Bigg(&{}^{\text{micro}}C_{ijkl} + {}^{\text{micro}}C_{ijmn}\frac{\partial\, {}^{u\varepsilon}\chi_m^{kl}(\mathbf{x},\mathbf{y})}{\partial y_n} + \\
&{}^{\text{micro}}e_{mij}\frac{\partial\, {}^{\varphi\varepsilon}\chi^{kl}(\mathbf{x},\mathbf{y})}{\partial y_m} + {}^{\text{micro}}q_{mij}\frac{\partial\, {}^{\psi\varepsilon}\chi^{kl}(\mathbf{x},\mathbf{y})}{\partial y_m}\Bigg)\,\mathrm{d}Y
\end{aligned}
\tag{A.11}
$$

$$
\begin{aligned}
{}^{\text{macro}}\kappa_{ik} = -\frac{1}{|Y|}\int_Y \Bigg(&{}^{\text{micro}}\kappa_{ik} + {}^{\text{micro}}\kappa_{im}\frac{\partial\, {}^{\varphi E}\chi^{k}(\mathbf{x},\mathbf{y})}{\partial y_m} - {}^{\text{micro}}e_{imn}\frac{\partial\, {}^{uE}\chi_m^{k}(\mathbf{x},\mathbf{y})}{\partial y_n} \\
&+ {}^{\text{micro}}a_{im}\frac{\partial\, {}^{\psi E}\chi^{k}(\mathbf{x},\mathbf{y})}{\partial y_m}\Bigg)\,\mathrm{d}Y
\end{aligned}
\tag{A.12}
$$

$$
\begin{aligned}
{}^{\text{macro}}\mu_{ik} = -\frac{1}{|Y|}\int_Y \Bigg(&{}^{\text{micro}}\mu_{ik} + {}^{\text{micro}}\mu_{im}\frac{\partial\, {}^{\psi H}\chi^{k}(\mathbf{x},\mathbf{y})}{\partial y_m} - {}^{\text{micro}}q_{imn}\frac{\partial\, {}^{uH}\chi_m^{k}(x,\mathbf{y})}{\partial y_n} \\
&+ {}^{\text{micro}}a_{im}\frac{\partial\, {}^{\varphi H}\chi^{k}(\mathbf{x},\mathbf{y})}{\partial y_m}\Bigg)\,\mathrm{d}Y
\end{aligned}
\tag{A.13}
$$

$$
\begin{aligned}
{}^{\text{macro}}e_{kij} = \frac{1}{|Y|}\int_Y \Bigg(&{}^{\text{macro}}e_{kij} + {}^{\text{macro}}e_{mij}\frac{\partial\, {}^{\varphi E}\chi^{k}(\mathbf{x},\mathbf{y})}{\partial y_m} + {}^{\text{micro}}q_{mij}\frac{\partial\, {}^{\psi E}\chi^{k}(\mathbf{x},\mathbf{y})}{\partial y_m} \\
+ {}^{\text{micro}}C_{mnij}\frac{\partial\, {}^{uE}\chi_m^{k}(\mathbf{x},\mathbf{y})}{\partial y_n}\Bigg)\,\mathrm{d}Y &= \frac{1}{|Y|}\int_Y \Bigg({}^{\text{micro}}e_{kij} + {}^{\text{micro}}e_{kmn}\frac{\partial\, {}^{u\varepsilon}\chi_m^{ij}(\mathbf{x},\mathbf{y})}{\partial y_n} \\
&- {}^{\text{micro}}\kappa_{km}\frac{\partial\, {}^{\varphi\varepsilon}\chi^{ij}(\mathbf{x},\mathbf{y})}{\partial y_m} - {}^{\text{micro}}a_{km}\frac{\partial\, {}^{\psi\varepsilon}\chi^{ij}(\mathbf{x},\mathbf{y})}{\partial y_m}\Bigg)\,\mathrm{d}Y
\end{aligned}
\tag{A.14}
$$

$$
\begin{aligned}
{}^{\text{macro}}q_{kij} = \frac{1}{|Y|}\int_Y \Bigg(&{}^{\text{macro}}q_{kij} + {}^{\text{macro}}q_{mij}\frac{\partial\, {}^{\psi H}\chi^{k}(\mathbf{x},\mathbf{y})}{\partial y_m} + {}^{\text{micro}}e_{mij}\frac{\partial\, {}^{\varphi H}\chi^{k}(\mathbf{x},\mathbf{y})}{\partial y_m} \\
+ {}^{\text{micro}}C_{mnij}\frac{\partial\, {}^{uH}\chi_m^{k}(\mathbf{x},\mathbf{y})}{\partial y_n}\Bigg)\,\mathrm{d}Y &= \frac{1}{|Y|}\int_Y \Bigg({}^{\text{micro}}q_{kij} + {}^{\text{micro}}q_{kmn}\frac{\partial\, {}^{u\varepsilon}\chi_m^{ij}(\mathbf{x},\mathbf{y})}{\partial y_n} \\
&- {}^{\text{micro}}a_{km}\frac{\partial\, {}^{\varphi\varepsilon}\chi^{ij}(\mathbf{x},\mathbf{y})}{\partial y_m} - {}^{\text{micro}}\mu_{km}\frac{\partial\, {}^{\psi\varepsilon}\chi^{ij}(\mathbf{x},\mathbf{y})}{\partial y_m}\Bigg)\,\mathrm{d}Y
\end{aligned}
\tag{A.15}
$$

$$
\begin{aligned}
{}^{\text{macro}}a_{ik} = \frac{1}{|Y|}\int_Y \Bigg(&{}^{\text{micro}}a_{ik} + {}^{\text{micro}}a_{im}\frac{\partial\, {}^{\psi H}\chi^{k}(\mathbf{x},\mathbf{y})}{\partial y_m} - {}^{\text{micro}}e_{imn}\frac{\partial\, {}^{uH}\chi_m^{k}(\mathbf{x},\mathbf{y})}{\partial y_n} \\
+ {}^{\text{micro}}\kappa_{im}\frac{\partial\, {}^{\varphi H}\chi^{k}(\mathbf{x},\mathbf{y})}{\partial y_m}\Bigg)\,\mathrm{d}Y &= \frac{1}{|Y|}\int_Y \Bigg({}^{\text{micro}}a_{ik} + {}^{\text{micro}}a_{im}\frac{\partial\, {}^{\varphi E}\chi^{k}(\mathbf{x},\mathbf{y})}{\partial y_n} \\
&- {}^{\text{micro}}q_{imn}\frac{\partial\, {}^{uE}\chi_m^{k}(\mathbf{x},\mathbf{y})}{\partial y_n} + {}^{\text{micro}}\mu_{im}\frac{\partial\, {}^{\psi E}\chi^{k}(\mathbf{x},\mathbf{y})}{\partial y_m}\Bigg)\,\mathrm{d}Y
\end{aligned}
\tag{A.16}
$$

Here, the upper left subscript "macro" means that it is the homogenized material property of the macrostructure, and the upper left subscript "micro" means that it is the material property of the constituent phase of the microstructure. If the microstructure consists of a single phase and there is no inhomogeneity, then all characteristic function tensors are zero. In this case, the homogenized material properties of the macrostructure naturally match the material properties of the single phase of the microstructure. On the other hand, when the microstructure has nonuniformity, the homogenized material properties of the macrostructure are calculated as the volume average of the microstructure.

The obtained macro-equations and micro-equations were discretized by applying the three-dimensional finite element method. A hexahedral isoparametric linear element was adopted, and approximate solutions satisfying both equations were obtained.

Appendix B

Consider PE and PM single domains with uniform electric or magnetic polarization directions as constituent units of the microstructure of polycrystalline materials. Both the PE and PM domains are treated as orthogonal anisotropic bodies. That is, they have anisotropy in the spontaneous polarization direction and are isotropic with respect to the two axes orthogonal to the spontaneous polarization direction. The third axis of the crystal coordinate system, the [001] orientation, is set in the spontaneous polarization direction. The orientation of the PE and PM domains is the design variable of the microstructure. As shown in Figure 19.1, the [001] orientations of the PE and PM domains are defined by the Euler angles of the following equation:

$$\boldsymbol{E}_k = \left\{ \varphi_k \quad \theta_k \quad \psi_k \right\}^T \quad k = 1, 2, \cdots, n \tag{B.1}$$

where $\boldsymbol{E}_k$ denotes the Euler angle of the kth domain in the microstructure.

In the case of ME composites consisting of PE and PM domains, the arrangement of both domains is reflected in the microstructure by introducing a material indicator ξ_k, as shown in the following equation, and continuously changing the material properties between the PM domain and the PE domain:

$$^{\text{micro}}X_k = \frac{1 + \cos(\pi \xi_k)}{2} X^{PM} + \frac{1 + \cos\{\pi(\xi_k + 1)\}}{2} X^{PM} \tag{B.2}$$

where $^{\text{micro}}X_k$ indicates a material property such as elastic stiffness constant, electric permittivity, magnetic permeability, PE and PM coupling coefficients for the kth domain, and X^{PE} and X^{PM} denote the material properties of the PE and PM domains, respectively. For example, it is the PM domain at $\xi_k = 0.0$ and becomes the PE domain at $\xi_k = 1.0$, while it is a virtual intermediate phase at $\xi_k = 0.5$. For the convenience of optimization, the material properties can be continuously varied, but for the optimization ξ_k is spontaneously binarized when improving the ME constants. Therefore, no virtual intermediate phase in the microstructure is considered, and the material properties of microstructures composed only of the PE and PM domains are evaluated.

In case of polycrystalline PE materials, to search for the microstructure that maximizes the PE effect, we set the multiscale optimization problem as follows:

$$f(E_1, E_2, \cdots, E_n) - {}^{macro}d_{mij} \to \max \tag{B.3}$$

where ${}^{macro}d_{mij}$ is the homogenized PE strain constant of the given macrostructure. Similarly, in the case of polycrystalline composites, to search for the microstructure that maximizes the ME effect, we set the multiscale optimization problem as follows:

$$f(E_1, E_2, \cdots, E_n, \xi_1, \xi_2, \cdots, \xi_n) = {}^{macro}a_{ij} \to \max \tag{B.4}$$

where ${}^{macro}a_{ij}$ is the homogenized ME constant of the given macrostructure. The steepest-decent method is employed to find the optimal solution. The microstructure is a cube, and a regular division mesh using hexahedral primary elements with eight integration points is applied. The Euler angles and material indicator are set independently for the integral points within the element.

References

1 Narita, F. and Fox, M. (2018) A review on piezoelectric, magnetostrictive, and magnetoelectric materials and device technologies for energy harvesting applications. *Adv. Eng. Mater.*, 20, 1700743.

2 Mori, T. and Tanaka, K. (1973) Average stress in matrix and average elastic energy of materials with misfitting inclusions. *Acta Metall.*, 21, 571–4.

3 Hill, R. (1965) A self-consistent mechanics of composite materials. *J. Mech. Phys. Solids*, 13, 213–22.

4 Gudes, J.M. and Kikuchi, N. (1990) Preprocessing and postprocessing for materials based on the homogenization method with adaptive finite element methods. *Comput. Method Appl. Mech. Eng.*, 83, 143–98.

5 Yasuda, S., Teramoto, Y., Ogoe, S. and Uetsuji, Y. (2022) Fiber contribution on elastic properties of cellulose composites: A multiscale numerical study. *Polym. Compos.*, 43, 1656–64.

6 Takano, N., Uetsuji, Y., Kashiwagi, Y. and Zako, M. (1999) Hierarchical modelling of textile composite materials and structures by the homogenization method. *Model. Simul. Mater. Sci. Eng.*, 7, 207–31.

7 Uetsuji, Y., Yasuda, S. and Teramoto, Y. (2022) Effect of fibre aspect ratio and aggregation on nonlinear material property of random cellulose reinforced composites: a multiscale computational study. *Compos. Struct.*, 301, 116201.

8 Uetsuji, Y., Hamamoto, R., Luo, C., Tsuyuki, Y., Tsuchiya, K., Ikura, R. and Takashima, Y. (2023) Fiber morphology design of cellulose composites through multiscale simulation. *Int. J. Mech. Sci.*, 258, 108581.

9 Kinoshita, T., Chao, L., Ide, K., Aoki, K. and Uetsuji, Y. (2023) Mechanical characterization and design of hybrid reinforced polypropylene composites with talc fillers and cellulose fibers. *Compos. Part B Eng.*, 266, 110971.

10 Ohno, N., Wu, X. and Matsuda, T. (2000) Homogenized properties of elastic–viscoplastic composites with periodic internal structures. *Int. J. Mech. Sci.*, 42, 1519–36.

11 Uetsuji, Y., Yasuda, Y., Yamauchi, S., Matsushima, E., Adachi, M., Fuji, M. and Ito, H. (2021) Multiscale study on thermal insulating effect of a hollow silica-coated polycarbonate window for residential buildings. *Renew. Sustain. Energy Rev.*, 152, 111718.

12 Uetsuji, Y., Nakamura, Y., Ueda, S. and Nakamachi, E. (2004) Numerical investigation on ferroelectric properties of piezoelectric materials by a crystallographic homogenization method. *Model. Simul. Mater. Sci. Eng.*, 12, S303–17.

13 Uetsuji, Y., Horio, M. and Tsuchiya, K. (2008) Optimization of crystal microstructure in piezoelectric ceramics by multiscale finite element analysis. *Acta Mater.*, 56, 1991–2002.

14 Uetsuji, Y., Satou, Y., Nagakura, H., Nishioka, H., Kuramae, H. and Tsuchiya, K. (2008) Crystal morphology analysis of piezoelectric ceramics using electron backscatter diffraction method and its application to multiscale finite element analysis. *J. Comput. Sci. Technol.*, 2, 568–77.

15 Uetsuji, Y., Kimura, S., Kuramae, H., Tsuchiya, K. and Kamlah, M. (2012) Multiscale finite element simulations of piezoelectric materials based on two- and three-dimensional EBSD-measured microstructures. *J. Intell. Mater. Syst. Struct.*, 23, 563–73.

16 Uetsuji, Y., Kuramae, H. and Tsuchiya, K. (2013) Two-step homogenization simulation for multidomain and multigrain structures in piezoelectric materials. *Acta Mater.*, 61, 7304–12.

17 Yu, Y., Luo, C., Suto, T., Uetsuji, Y. and Narita, F. (2023) Fabrication, evaluation and multi-scale simulation of unidirectional carbon fiber reinforced piezoelectric composites for flexible motion sensor. *Small*, 20, 2307689.

18 Yamamoto, N., Machida, K., Ide, K., Aoki, K. and Uetsuji, Y. (2023) Enhancement effect of two-tiered heterostructure in piezoelectric filler-dispersed polymer composites. *Materialia*, 32, 101912.

19 Uetsuji, Y., Hata, T., Kuramae, H. and Tsuchiya, K. (2014) Homogenization modeling of domain switching in ferroelectric materials. *Acta Mech.*, 225, 2969–86.

20 Uetsuji, Y., Hata, T., Oka, T., Kuramae, H. and Tsuchiya, K. (2015) Multiscale simulation of domain switching behavior in polycrystalline ferroelectric materials. *Comput. Mater. Sci.*, 106, 100–10.

21 Uetsuji, Y., Wada, T. and Tsuchiya, K. (2016) Multiscale numerical investigation on effective physical properties of multiferroic $BaTiO_3$ / $CoFe_2O_4$ composites. *Int. J. Appl. Electromagn. Mech.*, 52, 1245–50.

22 Uetsuji, Y., Wada, T., Uekita, Y., Terada, J. and Tsuchiya, K. (2016) Computational and experimental investigation of the crystal orientation control effect on the electric permittivity and magnetic permeability of multiferroic composite materials. *Acta Mech.*, 228, 2879–93.

23 Uetsuji, Y., Wada, T. and Tsuchiya, K. (2017) Statistical investigation of homogenized physical properties of polycrystalline multiferroic composites materials. *Acta Mech.*, 230, 1387–401.

24 Uetsuji, Y., Wada, T. and Tsuchiya, K. (2019) Multiscale numerical study on origin of magnetoelectric effect in view of localization of microstructural strain field for multiferroic composite materials. *Comput. Mater. Sci.*, 158, 159–69.

25 Uetsuji, Y. and Wada, T. (2020) Computational study on microstructural optimization of magnetoelectric composite materials. *Comput. Mater. Sci.*, 172, 109365.

26 Sano, F. and Uetsuji, Y. (2020) Multiscale optimization of magnetoelectric composite materials – effect of material combination. *Compos. Struct.*, 290, 115500.

27 Uetsuji, Y., Sano, F. and Takeuchi, S. (2023) New proposal of heterostructure for magnetoelectric effect enhancement from multiscale optimization. *J. Mater. Sci.*, 58, 3625–40.

20

Application of Numerical Simulation on Devices using Smart Materials

Kotaro Mori

Department of Mechanical Systems Engineering, College of Engineering, Ibaraki University, Hitachi, Ibaraki, Japan

20.1 Introduction

Smart materials can autonomously respond to environmental changes over time. This section briefly introduces numerical simulations for device design using smart materials, such as magnetostrictive and piezoelectric materials.

20.1.1 Piezoelectric Materials

Piezoelectric materials generate an electric charge when subjected to mechanical stress and distortion when an electric field is applied. These materials can convert mechanical energy into electrical energy and vice versa [1]. This electromechanical coupling property makes them extremely useful in mechanical systems because they can serve as sensors and actuators. Piezoelectric materials are manufactured in various sizes and shapes, including plates, sheets, blocks, disks, hemispheres, tubes, and rings, and they can be used independently or in combination. Piezoelectric materials have a long history, with natural piezoelectric materials being studied as early as the 1880s. However, high-performance piezoelectric ceramics only emerged in the 1950s with the discovery of lead zirconate titanate (PZT). Since then, many piezoelectric materials have been studied, such as polyvinylidene fluoride (PVDF), zinc oxide (ZnO), and barium titanate ($BaTiO_3$). These materials exhibit diverse properties [2, 3]. For example, PZT and $BaTiO_3$ have a large electromechanical coupling coefficient and e constant, making them suitable for actuators. PVDF is flexible and has a large g constant, making it suitable for sensors. PZT remains the predominant piezoelectric ceramic and is widely used owing to its high output, compact size, and high power density. However, in recent years, lead-free piezoelectric materials and composites have been developed owing to the lead content of PZT. Piezoelectric composites are valuable because they allow a wide range of solid solutions with various additives, enabling manufacturers to customize properties for different applications [4].

20.1.2 Magnetostrictive Materials

Magnetostrictive materials deform in response to an external magnetic field and alter their magnetization in response to an applied force. These materials can convert mechanical energy into magnetic energy and vice versa [5]. Magnetostrictive materials, similar to piezoelectric materials, are widely used as sensors and actuators owing to their valuable properties, such as contactless actuation, fast response, and high energy density. The magnetostrictive effect was first described in the 19th century by English physicist James Joule. Following the discovery of the magnetostrictive effect, research into magnetostrictive materials continued, leading to a significant finding in the 1960s that pseudobinary iron-rare earth alloys exhibit substantial magnetostriction. In particular, Terfenol-D ($Tb_{0.3}Dy_{0.7}Fe_{1.93}$), an alloy of iron and the rare earth elements terbium and dysprosium, exhibits remarkable magnetostrictive properties [6, 7]. In addition, various magnetostrictive materials, such as Galfenol, Metaglas, and FeCo, have been developed, exhibiting various properties [8–11]. For example, Terfenol-D is recognized as a "giant" magnetostrictive material (over 1500 ppm), and Galfenol is known to have large magnetostriction (over 200 ppm) and good processability. Metaglas has good energy conversion efficiency per unit volume, and FeCo alloys are low-cost materials with properties similar to those of iron, such as high permeability, high strength, and high ductility. Magnetostrictive materials have properties similar to those of piezoelectric materials. However, their magnetic-field-driven response facilitates contactless actuation. In addition, without the need for wiring or electrodes, they offer the advantage of greater design flexibility. Notably, their properties significantly vary based on preload and the applied bias magnetic field [7, 9].

20.1.3 Composite Materials

Composite materials are fabricated by combining two or more base materials. Composites can achieve high strength, which is challenging to achieve with a single material, and excellent properties, such as specific rigidity, high-temperature strength, and heat shielding, by changing the ratio of the components and how they are combined. Various types of composite materials exist, such as laminated, fiber-dispersed, and particle-dispersed materials, each with significantly different properties, such as anisotropy. In addition, adopting intelligent materials in composite materials enhances material properties and adds functionality and responsiveness [12–14]. Furthermore, combining intelligent materials can create multifunctional materials with sensor, processor, and actuator functions. However, the properties of composite materials can significantly change based on the preparation conditions, even when using the same base materials. For example, the ferroelectric, piezoelectric, and pyroelectric properties of piezoelectric composites can be significantly altered by manipulating the polymer processing conditions, heat treatment, nanoconfinement, blending, nanocomposite formation, electrospinning, and the creation of porous structures [4, 14, 15].

20.1.4 Finite Element Method

The finite element method (FEM) is widely used for the numerical analysis of various physical phenomena, such as solid mechanics, fluid dynamics, heat transfer, material diffusion, and electromagnetism [16]. It effectively reproduces the mechanical behaviors of materials and structures, such as in structural analyses. The FEM has gained popularity owing to its user-friendly application, scalability, and adaptability to nonlinear behaviors [17]. However, this method requires careful implementation to ensure accurate results, particularly because of its high degree of freedom. Critical factors influencing its accuracy include element discretization (mesh) and the choice of interpolation functions. Considering the constitutive equations and their nonlinear characteristics is crucial when applying the FEM to smart materials. The nonlinear behavior of smart materials stems from their microscopic properties, which can be accurately modeled and incorporated into macroscopic elements by embedding appropriate interpolation functions within the constitutive equations.

20.2 Basic Equations

The basic equations for piezoelectric and magnetostrictive materials are presented here. Considering a rectangular Cartesian coordinate system O-$x_1x_2x_3$, the equilibrium equations can be written as follows [18]:

$$\sigma_{ji,j} = \rho u_{i,tt} \tag{20.1}$$

$$D_{i,i} = 0 \tag{20.2}$$

$$B_{i,i} = 0, \tag{20.3}$$

where σ_{ij}, D_i, B_i, ρ, and u_i are the components of the stress tensor, electric displacement vector, magnetic flux density vector, mass density, and elastic displacement, respectively. The comma followed by the index denotes partial differentiation with respect to the space coordinates x_i and time t, and the summation convention for the tensor indices is applied. The constitutive relationships of piezoelectric and magnetostrictive materials are expressed as follows:

$$\varepsilon_{ij} = s^E_{ijkl}\sigma_{kl} + d_{kij}E_k \tag{20.4}$$

$$D_i = d_{ikl}\sigma_{kl} + \epsilon^T_{ik}E_k, \tag{20.5}$$

and

$$\varepsilon_{ij} = s^H_{ijkl}\sigma_{kl} + d^m_{kij}H_k \tag{20.6}$$

$$B_i = d^m_{ikl}\sigma_{kl} + \mu_{ik}H_k, \tag{20.7}$$

where ε_{ij} is the component of the strain tensor, E_i is the electric field intensity vector, and s^E_{ijkl}, d_{kij}, and ϵ^T_{ij} are the elastic compliance, piezoelectric constant, and dielectric permittivity, respectively. H_i is the magnetic field intensity vector, and s^H_{ijkl}, d^m_{kij}, and μ_{ij} are the elastic compliance, piezomagnetic constant, and magnetic permittivity, respectively. The valid symmetry conditions for the material constants are as follows:

$$s^E_{ijkl} = s^E_{jikl} = s^E_{ijlk} = s^E_{klij},\ d_{kij} = d_{kji},\ \epsilon^T_{ij} = \epsilon^T_{ji}, \tag{20.8}$$

$$s^H_{ijkl} = s^H_{jikl} = s^H_{ijlk} = s^H_{klij},\ d^m_{kij} = d^m_{kji},\ \mu_{ij} = \mu_{ij}. \tag{20.9}$$

It should be noted that although the constitutive equations for piezoelectric and magnetostrictive materials have similar forms, the magnetostrictive phenomenon [19] is fundamentally different from the piezoelectric phenomenon. The strain, electric field, and magnetic field are derivable from the elastic displacement, electric potential, ϕ_i, and magnetic potential, φ_i, as follows:

$$\varepsilon_{ij} = \frac{1}{2}\left(u_{i,j} + u_{j,i}\right) \tag{20.10}$$

$$E_i = -\phi_{,i} \tag{20.11}$$

$$H_i = \varphi_{,i}. \tag{20.12}$$

20.3 Case Study

20.3.1 Piezoelectric Functionally Graded Composites

This section presents the fabrication and power-generation properties of piezoelectric functionally graded composites made of $BaTiO_3$ and epoxy resin. Piezoelectric composites were fabricated using epoxy and $BaTiO_3$ particles, and three types of samples with the same amount of $BaTiO_3$ particles were compared. The responses to vibration and temperature changes were analyzed [20].

20.3.1.1 Introduction

Piezoelectric materials have been used for energy harvesting to convert mechanical energy, such as vibrations and shocks, into electrical energy [21, 22]. However, more recently, studies have shown that barium titanate laminates can induce voltage upon temperature change [23]. This mechanism is believed to be caused by the temperature dependence of their material properties, deformation, and internal stress caused by thermal expansion. This mechanism differs from that by which thermoelectric and pyroelectric materials generate electricity. Energy harvesting owing to temperature changes expands the possibility of using thermal energy sources for piezoelectric materials, particularly in environments where vibration and heat are present, such as in factory plumbing systems and motors.

20.3.1.2 Numerical Procedure

20.3.1.2.1 Basic Equations

In this section, a temperature term is added to the material constitutive Eqs. (20.4) and (20.5) to account for temperature changes.

$$\varepsilon_{ij} = s^{ET}_{ijkl}\sigma_{kl} + d_{kij}E_k + \alpha^E_{ij}\Delta T \tag{20.13}$$

$$D_i = d_{ikl}\sigma_{kl} + \epsilon^{\sigma}_{ik}E_k + p^{\sigma}_i\Delta T, \tag{20.14}$$

where α_{ij} is the thermal expansion coefficient and p_i is the pyroelectric constant. Furthermore, assuming that the piezoelectric coefficient, linear expansion coefficient, and dielectric constant are temperature dependent, the constitutive Eqs. (20.13) and (20.14) can be rewritten as follows:

$$\varepsilon_{ij} = s^{ET}_{ijkl}\sigma_{kl} + d_{kij}\,(T)\,E_k + \alpha^E_{ij}(T)\Delta T \tag{20.15}$$

$$D_i = d_{ikl}\,(T)\,\sigma_{kl} + \epsilon^{\sigma}_{ik}\,(T)\,E_k + p^{\sigma}_i\Delta T. \tag{20.16}$$

Although numerous studies have determined the temperature dependence of physical properties through measurements or calculations, the present analysis was conducted with reference to reference [23] and actual measurements (Figure 20.1). For simplicity, the elastic compliance was assumed to be independent of the temperature, and the pyroelectric constant was ignored in the finite element analysis (FEA). With these introductions and assumptions, Eqs. (20.15) and (20.16) are expanded to include the thermal expansion and temperature dependence of the material properties.

20.3.1.2.2 Model

Single- and six-layer piezoelectric composite models were used in this analysis, as shown in Figure 20.2. Figure 20.3 shows the rectangular piezoelectric composite model, with dimensions l, w, and h, representing its length, width, and thickness, respectively. The piezoelectric material fraction is graded solely through the thickness, divided into six layers along the thickness direction, with each layer assumed to have constant material properties throughout its thickness. Let the coordinate

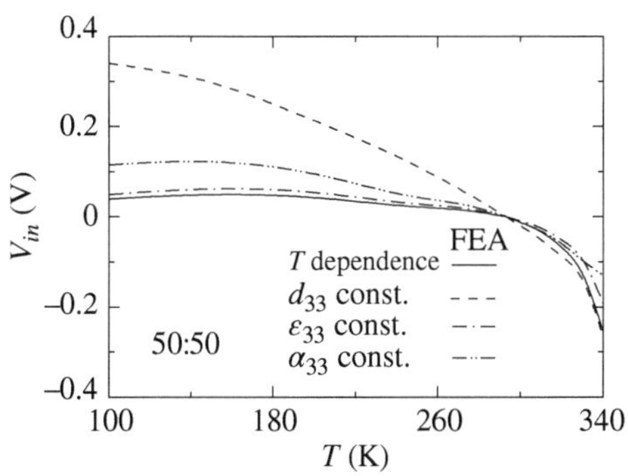

Figure 20.1 Measured coefficient of linear expansion versus temperature.

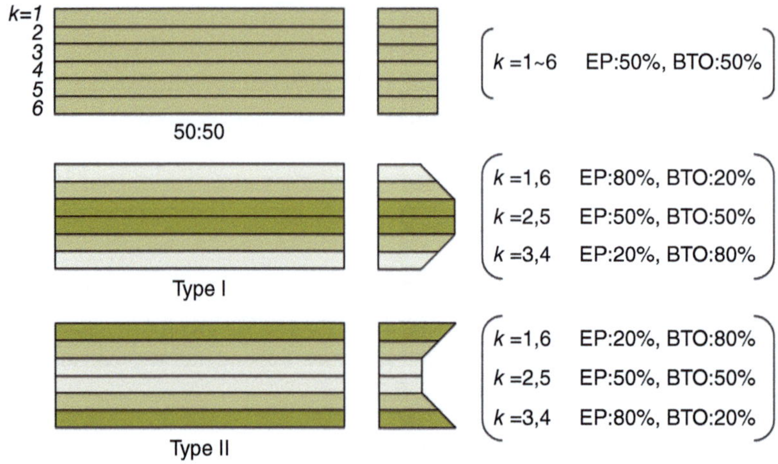

Figure 20.2 Laminated pattern of piezoelectric composites.

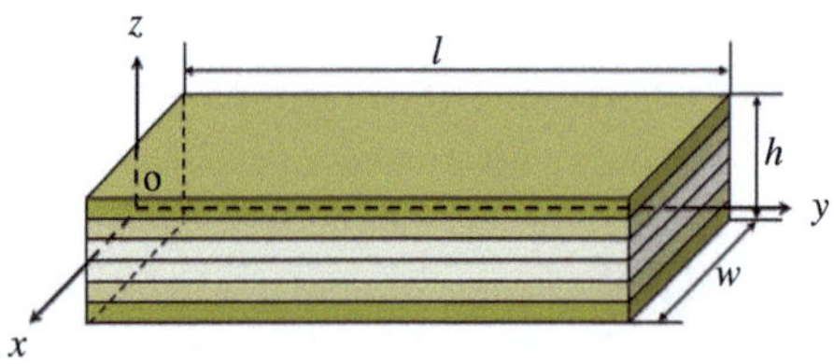

Figure 20.3 Illustration of piezoelectric composites.

axes $x = x_1$ and $y = x_2$ be aligned with the middle plane of the composite, and let the $z = x_3$ axis be perpendicular to this plane. In the single-layer material model, the physical properties are consistent across all layers. By contrast, in the six-layer model, that is, the functionally graded piezoelectric composite model, each layer has different physical properties. The material constants of the kth layer of the piezoelectric composite model are expressed as follows:

$$s_k = s^B V_k + s^R (1 - V_k) \tag{20.17}$$

$$\alpha_k = \alpha^B V_k + \alpha^R (1 - V_k) \tag{20.18}$$

$$\rho_k = \rho^B V_k + \rho^R (1 - V_k) \tag{20.19}$$

where the superscripts B and R represent $BaTiO_3$ and epoxy, respectively, and V_k represents the volume fraction calculated from the mass ratio of $BaTiO_3$ in the kth layer. The piezoelectric constants and permittivities are expressed as follows [24–26]:

$$d_k = d^B \frac{\epsilon^R}{\epsilon^B} \frac{15V_k}{(2 + 3V_k)(1 - V_k)} \tag{20.20}$$

$$\epsilon_k = \frac{\epsilon^B V_k ab + \epsilon^R (1 - V_k)}{V_k ab + (1 - V_k)}, \tag{20.21}$$

where parameters a and b are expressed as:

$$a = \frac{3\epsilon^R}{\epsilon^D + 2\epsilon^R} \tag{20.22}$$

$$b = 1 + \frac{3V_k\left(\epsilon^B - \epsilon^R\right)}{\epsilon^B + 2\epsilon^R}. \tag{20.23}$$

This study used single- and six-layer composite models with the same $BaTiO_3$ content ($BaTiO_3$:epoxy (BTO:EPO, mass ratio (wt.%)) = 50:50). Two types of six-layer models were designed for the functionally graded composites: Type I, in which the BTO content decreases from the midplane at $z = 0$ toward the outside in the thickness direction, and Type II, in which the BTO content increases from the midplane at $z = 0$ toward the outside in the thickness direction. Tables 20.1 and 20.2 list the material properties of $BaTiO_3$ and epoxy, respectively. The piezoelectric composites were analyzed as cantilevers and poled in the z-direction. The electrodes were not modeled because they were substantially thin compared with the piezoelectric composite; however, the voltages on the top and bottom surfaces of the piezoelectric composite were coupled in each plane. The mesh used 3D coupled-field solid elements. The number and size of the meshes were determined by verifying the convergence process of the solution before the simulation. First, the bending behavior of the piezoelectric composite cantilever owing to vibration at room temperature (293 K) was evaluated. A vibration $u_z\exp(i\omega t)$ equivalent to 1 G was applied to the fixed end, and the tip displacement and induced voltage were calculated. For simplicity, the damping ratio was not included in the calculation. Subsequently, the voltages induced in the piezoelectric composites under temperature changes were analyzed. The temperature was varied from room temperature to 100 and 340 K. Furthermore, the effect of the temperature-dependent material properties on the induced voltage was investigated.

Table 20.1 Material properties of $BaTiO_3$.

Elastic compliance ($\times 10^{-12}$ m^2/N)					Piezoelectric constant ($\times 10^{-12}$ m/V)			Permittivity ($\times 10^{-10}$ C/Vm)		Thermal expansion coefficient ($\times 10^{-6}$/K)	Density (kg/m^3)
s_{11}^{ET}	s_{33}^{ET}	s_{44}^{ET}	s_{12}^{ET}	s_{13}^{ET}	d_{31}	d_{33}	d_{15}	ϵ_{11}^{σ}	ϵ_{33}^{σ}	α^E	ρ
8.60	8.97	23.3	−2.49	−2.92	−27.5	141	270	143	150	15.7	6060

Table 20.2 Material properties of epoxy.

Elastic compliance ($\times 10^{-12}$ m^2/N)					Permittivity ($\times 10^{-12}$ C/Vm)		Thermal expansion coefficient ($\times 10^{-6}$/K)	Density (kg/m^3)
s_{11}^{ET}	s_{33}^{ET}	s_{44}^{ET}	s_{12}^{ET}	s_{13}^{ET}	ϵ_{11}^{σ}	ϵ_{33}^{σ}	α^E	ρ
265	265	709	−89.9	−89.9	35.4	35.4	60	1160

20.3.1.3 Experimental Procedure

This study used epoxy as the matrix material owing to its excellent flexibility, mechanical properties, moldability, and availability. The matrix and filler were, respectively, epoxy (Bisphenol-F) and $BaTiO_3$ particles with a diameter of 0.3 μm. Piezoelectric composites were fabricated as follows: first, the epoxy and $BaTiO_3$ particles were weighed in a precalculated ratio, and the $BaTiO_3$ particles were mixed with the epoxy. Subsequently, a curing agent was added to the mixture, and the mixture was heated at a relatively low temperature for defoaming. After defoaming, the mixture was shaped to achieve a uniform thickness and cured. The functionally graded composites were prepared by repeating this procedure using laminating layers. A specimen with dimensions of 12 mm × 30 mm × 0.9 mm was cut from the fabricated piezoelectric composite. Figure 20.4 shows an overview and a microscopic image of the cross section of the fabricated piezoelectric composite material. Although some agglomerates of $BaTiO_3$ particles were observed in specific areas, the particles were mostly uniformly dispersed within the layer. A similar trend was observed for the other specimens. Electrode deposition and polarization treatment were performed on the fabricated piezoelectric composite. For the electrode deposition, Au electrodes with a thickness of 30 nm were sputtered onto the top and bottom surfaces of the sample using a sputtering apparatus. Poling treatment was performed using a corona poling system at 8 kV and 373 K for 30 min to polarize the material in the thickness direction. Before testing, the material properties of the fabricated piezoelectric composites were measured at room temperature. The Young's modulus was determined by the cantilever deflection, and the piezoelectric coefficient and permittivity were measured using a piezo-d_{33} meter and an LCR meter, respectively. The Young's modulus E, piezoelectric constant d_{33}, and permittivity ϵ_{33} of each specimen were as follows: 3.57, 4.00, and 7.05 GPa; 10.8, 10.8, and 11.3 × 10^{-12} m/V; and 8.35, 18.5, and 29.4 × 10^{-9} C/Vm for 50:50, Type I, and Type II, respectively. The results confirmed that introducing a gradient distribution improved the material characteristics, even when the $BaTiO_3$ content was the same. Two types of

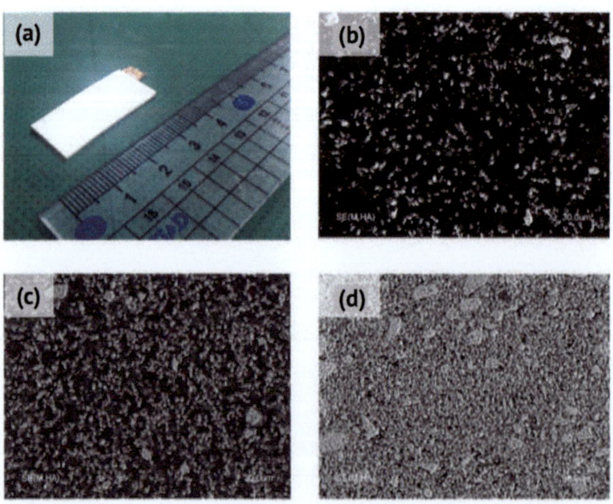

Figure 20.4 (a) Overall image and (b)–(d) cross-sectional SEM images of the fabricated piezoelectric composite. (b) BTO:EPO = 20:80, (c) 50:50, (d) 80:20.

tests were conducted: a vibration loading test and a temperature variation test. During the vibration load test, the fabricated piezoelectric composite was attached to a jig 5 mm from the end; that is, the length of the cantilever was 25 mm. In addition, the vibration conditions were adjusted such that the input frequency varied between 240 and 600 Hz, with an acceleration of 1 G at each frequency. During the test, the tip displacement and induced voltage of the cantilever were measured using a laser displacement meter and oscilloscope (input impedance of 1 MΩ), respectively. In the temperature change test, the piezoelectric composite was immersed in liquid nitrogen (77 K), and the induced voltage was measured using an oscilloscope.

20.3.1.4 Results and Discussion

Figure 20.5 shows the normalized displacement (u_{tip}/u_{y0}) versus input frequency (f) for the three piezoelectric composites. The plots represent the experimental results, and the lines denote the FEA results. Here, u_{y0} is the applied displacement amplitude (at $y = 0$). The measured resonant frequencies of the composites with BTO:EPO ratios of 50:50, Type I, and Type II were approximately 354, 320, and 342 Hz, respectively. The calculated resonant frequencies were approximately 363, 330, and 394 Hz, respectively. The experimental and FEA results of the piezoelectric composites with BTO:EPO ratios of 50:50 and Type I demonstrated similar tendencies, in contrast to those of the piezoelectric composite with Type II. The discrepancy in resonant frequency between the experiment and analysis can be attributed to the differences in size and density between the fabricated specimen and FEA model and the difference in apparent cantilever length owing to differences in clamping at the fixed end. This result demonstrates that the resonant frequency can be altered even for the same $BaTiO_3$ content and dimensions, extending design flexibility. Figure 20.6 shows the induced voltage of the piezoelectric composite (V_{in}) against the input frequency (f), presenting the test and FEA results. The highest induced voltage was observed for Type II and the lowest for Type I. These results indicate that the induced voltage was affected more by the degree of dispersion than by the total $BaTiO_3$ content. This is likely because, in the bending vibration, the largest deformation occurred in the outermost layers, resulting in a higher induced voltage proportional to the $BaTiO_3$ content.

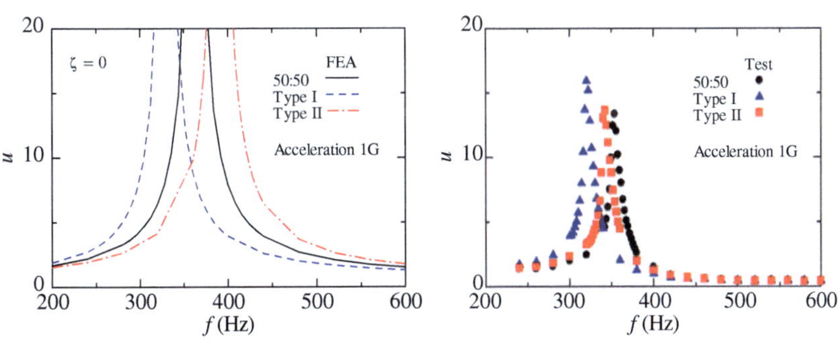

Figure 20.5 Normalized displacement versus frequency for piezoelectric composites: (left) FEA and (right) test.

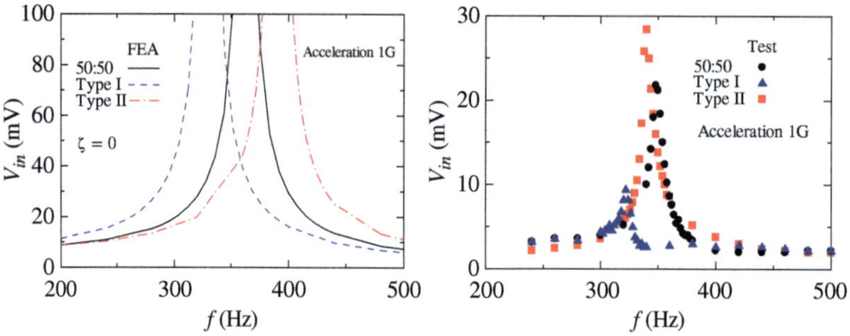

Figure 20.6 Induced voltage versus frequency for piezoelectric composites: (left) FEA and (right) test.

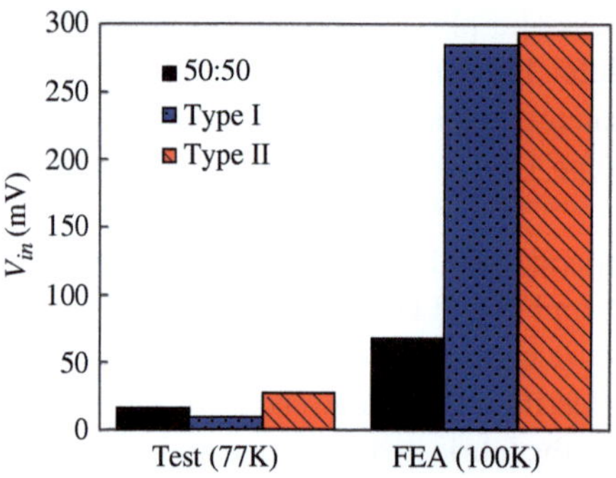

Figure 20.7 Induced voltage owing to the temperature change from room temperature to cryogenic temperature.

This study discusses the induced voltage with temperature changes and the effect of the temperature dependence of material properties on power-generation characteristics. Figure 20.7 shows the induced voltage (V_{in}) when the temperature of the piezoelectric composites is changed. Note that the results presented here are based on temperature variations relative to room temperature (293 K). The test results show only results from room temperature to cryogenic temperatures (77 K), while the analysis results are only up to 100 K. This is because liquid nitrogen was used in the experiment, and the temperature dependence of the material properties referenced in the analysis was evident only up to 100 K. The experimental and analytical results significantly differed. This discrepancy can be attributed to the difference in heat transfer rate and the simplification of the functionally graded material model. In the experiment, even if the specimen was placed in liquid nitrogen, there was a time lag before the specimen cooled down. By contrast, in the analysis, the temperature instantly changed to the input temperature. Furthermore, the induced voltage of piezoelectric materials is time dependent, and the amount of induced voltage is higher with rapid temperature changes. This study modeled the functionally graded material as a laminated material. Consequently, the stress gap at the interfaces between the layers, owing to temperature changes, significantly influenced the analysis results. Figure 20.8 shows the analysis results of the induced voltage for temperature variations other than 100 K. On the low-temperature side, Types I and

II exhibited larger values than 50:50. This is because the functionally graded material model in this analysis was created by laminating layers with different physical properties. Moreover, a stress gap occurred at the laminate interface owing to the difference in linear expansion coefficient, resulting in a larger maximum stress than that in the 50:50 model, which is a uniform material. For the high-temperature side, similar to that in the low-temperature side, Types I and II exhibited extremely large voltages compared with 50:50. This is likely because the linear expansion coefficient of the reference epoxy resin increased sharply at the boundary of 330 K, causing a large deformation of the entire test piece and the generation of a large stress gap at the laminate interface. The difference in trend from the vibration test is attributed to the difference between the bending and elongation deformations. Subsequently, the effects of the temperature dependence of the material properties on the induced voltage were investigated. To eliminate the influence of the laminated structure, a 50:50 model was used, and the temperature was varied from room temperature to 100 and 340 K. Four cases were considered: accounting for the temperature dependence of all the piezoelectric constants, dielectric constants, and linear expansion coefficients (T dependence), constant piezoelectric constants (d_{33} const.), constant dielectric constants (ϵ_{33} const.), and constant linear expansion coefficients (α const.). On the low-temperature side, the results for the cases where the physical properties were constant (no temperature dependence) were larger than those in the case considering all temperature dependencies. Notably, the temperature dependence of the piezoelectric constants had significant effects, with the induced voltage substantially increasing when these constants were kept unchanged. In addition, the temperature dependence of the dielectric constant and linear expansion coefficient had less significant effects than that of the piezoelectric constant. The peak values, except when the piezoelectric constants were constant, indicated that the temperature dependence of the piezoelectric constant significantly affected the induced voltage. Similarly, on the high-temperature side, the highest induced voltage was observed when the piezoelectric constant was held constant. In addition, the linear expansion coefficient had a significant effect; the induced voltage increased nearly linearly when the linear expansion coefficient was constant, whereas the increase was nonlinear in all results when the temperature dependence of the linear expansion coefficient was considered. When designing devices that rely on temperature changes, controlling the physical properties of the piezoelectric material is challenging, making material selection crucial. In such cases, selecting materials with a low temperature dependence of the piezoelectric constant is effective. Moreover, the temperature dependence of the dielectric constant is assumed to have a minimal effect on the induced voltage. Figure 20.9 shows the distributions of the normal and shear stresses at the fixed end along the thickness direction when the temperature was changed from room temperature to 100 K. The stress continuously varied in the 50:50 case, whereas large stress gaps were confirmed at interlaminar interfaces in Types I and II. Moreover, the range of the stress change was larger in Types I and II than in the 50:50 case, which likely affected the magnitude of the induced voltage owing to the temperature change, as shown in Figure 20.8. The shear stress distribution was analyzed. The maximum stress decreased in the order of Type I, Type II, and 50:50, with the largest interlaminar stress gap observed in Type II. Notably, stress gaps between the

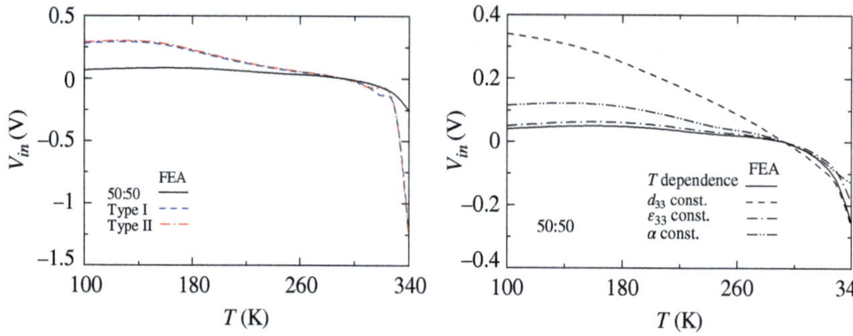

Figure 20.8 Calculated induced voltage under temperature change: (left) effect of modified laminated pattern and (right) material properties with temperature dependence.

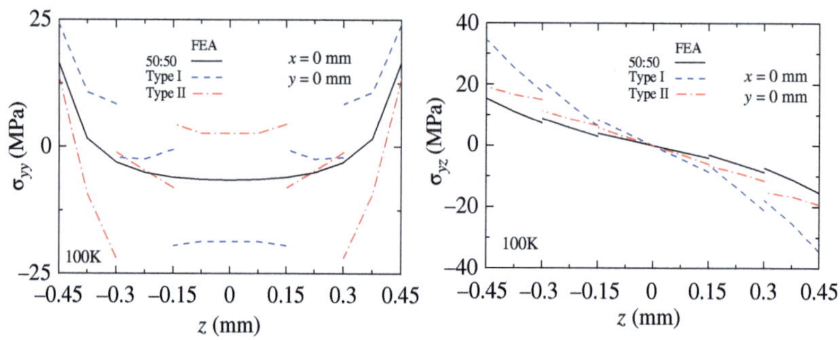

Figure 20.9 Stress distribution for the piezoelectric composite owing to temperature change from room temperature to 100 K: (left) normal stress and (right) shear stress.

layers were also observed in the 50:50 model. In theory, stress gaps should not occur because it is a single layer. However, because layers with the same material properties were modeled as a stack, a small stress gap was observed. Similar to the normal stress distribution, the 50:50 model yielded the lowest stress and stress gap.

20.3.1.5 Conclusion

This section summarizes the findings obtained from the fabrication and power-generation properties of the piezoelectric functionally graded composites made of $BaTiO_3$ and epoxy. The results confirmed that the piezoelectric functionally graded composites generated electricity from vibration and temperature changes. Temperature-dependent changes in the material constants, particularly the piezoelectric constants, significantly affected the induced voltage of the piezoelectric composite. This highlights an important factor to consider in material selection. In conclusion, piezoelectric functionally graded composites can be used to control internal stress and obtain induced voltages from various power sources. This may be valuable for designing energy-harvesting devices intended for environments with high vibrations and large temperature changes.

20.3.2 Magnetostrictive/Piezoelectric Laminated Composites

This section introduces magnetostrictive/piezoelectric laminated composites and analyzes the nonlinear behaviors of magnetostrictive and piezoelectric materials.

The behaviors of magnetostrictive/piezoelectric laminate composites under a load, electric field, and magnetic field are discussed [27–29].

20.3.2.1 Introduction

The most distinctive feature of magnetostrictive/piezoelectric composites is their magnetoelectric (ME) effect. The ME effect is defined as the induced polarization in an applied magnetic field (direct ME effect) or induced magnetization in an applied electric field (inverse ME effect). However, for magnetostrictive/piezoelectric composites, mechanical energy is also involved in the magnetic and electrical energy conversion; thus, one input can produce two outputs, expanding the scope of the device design. For example, a health monitoring function can be added to an actuator, or sensors and energy harvesting can be used simultaneously. This can reduce space and weight while enabling higher-level functions that systematically link multiple functions. However, this triad of energy transformations leads to a complex behavior [30, 31]. Therefore, understanding the internal electromagneto-mechanical fields is necessary for optimal ME composite design. Predicting intensified electromagneto-mechanical fields requires detailed finite element calculations.

20.3.2.2 Numerical Procedure

20.3.2.2.1 Basic Equations

The constitutive equation for ME materials can be written as follows:

$$\varepsilon_{ij} = s^{EH}_{ijkl}\sigma_{kl} + d^{H}_{kij}E_k + d'^{E}_{kij}H_k \tag{20.24}$$

$$D_i = d^{H}_{ikl}\sigma_{kl} + \epsilon^{TH}_{ik}E_k + \beta^{T}_{ik}H_k \tag{20.25}$$

$$B_i = d'^{E}_{kij}\sigma_{kl} + \beta^{T}_{ik}E_k + \mu^{TH}_{ik}H_k \tag{20.26}$$

where β^{T}_{ik} is ME susceptibility at constant stress. However, when ME materials are fabricated as laminates and there is no direct interaction between the materials, that is, $\beta^{T}_{ik} = 0$, they can be considered as separate constitutive equations for each material, as shown in Section 20.2.

This section introduces an analysis that considers the nonlinear behaviors of magnetostrictive and piezoelectric materials. This nonlinear behavior is caused by domain wall movement in piezoelectric materials and magnetic domain wall motion in magnetostrictive materials. This indicates that, in analyzing smart materials, microscopic factors are closely related to their macroscopic behavior. This study considered a model in which the polarization direction of the piezoelectric material is along the x-direction, and the axis of easy magnetization of the magnetostrictive material is along the z-direction, as shown in Figure 20.10.

First, the domain wall motion in piezoelectric materials was considered. The domain wall displacement in piezoelectric materials causes changes in the strain and electric dipole moment [32]. When an AC electric field ($E_x = E_0\exp(i\omega t)$) is applied in the x-direction to a grain in the xz-plane, the elastic coefficient s^{E}_{11}, piezoelectric constant d^{*}_{31}, and dielectric constant ϵ^{T}_{33} change owing to the displacement of the domain walls as follows [33]:

$$s^{E*}_{11} = s^{E}_{11} + \Delta s^{E}_{11} \tag{20.27}$$

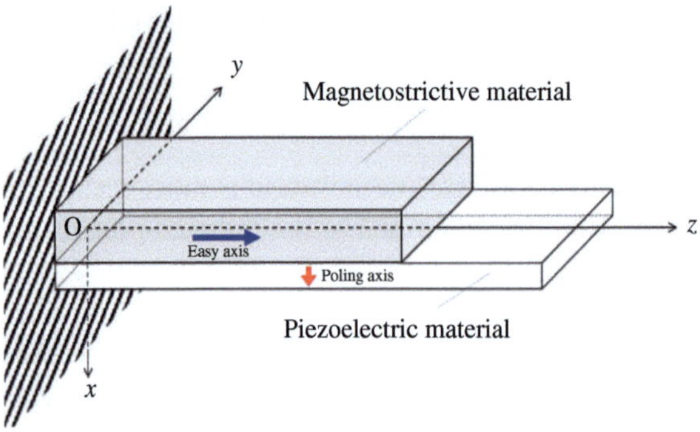

Figure 20.10 Illustration of magnetostrictive/piezoelectric laminate configuration.

$$d_{31}^{*} = d_{31} + \Delta d_{311} \tag{20.28}$$

$$\epsilon_{33}^{T*} = \epsilon_{33}^{T} + \Delta\epsilon_{33}^{T}, \tag{20.29}$$

where Δs_{11}^{E}, Δd_{31}^{*}, and $\Delta\epsilon_{33}^{T}$ are the increments associated with the domain wall motion. For a detailed explanation, refer to the following references [34]. The strain ε_{zz} and electric displacement D_x become

$$\varepsilon_{zz} = s_{13}^{E}\sigma_{xx} + s_{12}^{E}\sigma_{yy} + s_{11}^{E*}\sigma_{zz} + d_{31}^{*}E_x \tag{20.30}$$

$$D_x = d_{33}\sigma_{xx} + d_{31}^{*}\sigma_{yy} + d_{31}^{*}\sigma_{zz} + \epsilon_{33}^{T*}E_x. \tag{20.31}$$

Experimental studies on PZTs have shown that 45–70% of the dielectric and piezoelectric moduli values may originate from extrinsic contributions. The extrinsic dielectric constant $\Delta\epsilon_{33}^{T}$ was approximately estimated to be two-thirds of the measured value [35]. The following equation [34] describes $\Delta\epsilon_{33}^{T}$ in terms of the AC electric field amplitude E_0 and coercive electric field E_c:

$$\Delta\epsilon_{33}^{T} = \epsilon_{33}^{T}\frac{2E_0}{3E_c} \tag{20.32}$$

Based on the above, a model that considers the dependence of piezoelectric material properties on the AC electric field can be used to calculate the nonlinear dynamic bending behavior.

Subsequently, the magnetic domain wall motion in magnetostrictive materials is examined. The nonlinearity of magnetostriction versus magnetic filed curves arises from the rotation of magnetic domains. Magnetic domain switching leads to changes in the magnetoelastic constants, and the constants d'_{15}, d'_{31}, and d'_{33} for the Terfenol-D layer under a uniform magnetic field of magnetic induction $B_z = B_0$ are expressed as follows:

$$d'_{15} = d_{15}^{m} \tag{20.33}$$

$$d'_{31} = d_{31}^{m} + m_{31}H_z \tag{20.34}$$

$$d'_{33} = d_{33}^{m} + m_{33}H_z, \tag{20.35}$$

where d_{15}^m, d_{31}^m, and d_{33}^m are the piezomagnetic constants and m_{31} and m_{33} are the second-order magnetoelastic constants. When the length of the magnetostrictive material is significantly larger than its width and thickness, and the magnetic field is along the length direction (easy axis), the longitudinal (33) magnetostrictive deformation mode dominates. Therefore, it is assumed that only d'_{33} varies with the magnetic field H_z [36], and m_{31} equals zero. The constant m_{33} effectively predicts nonlinearity without the need for complex computations or additional parameters. However, it should be noted that the constants m vary depending on the materials and dimensions used. Although other researchers have proposed more complex models, caution is advised when incorporating them into the analyses [37, 38]. In addition, d_{15} changes when a magnetic field is applied in the x-direction [39].

The coupled-field solid elements in ANSYS were used for the analysis. Considering the symmetry of the model, only half of the cantilever width was modeled to reduce the calculation time. A total of 27 500 elements and 32 291 nodes were used in the magnetostrictive/piezoelectric cantilever model. It should be noted that before performing the simulations, a mesh sensitivity study was performed to ensure that the mesh was sufficiently fine. The fundamental equations for magnetostrictive materials are mathematically equivalent to those for piezoelectric materials. First, the strain, tip deflection, and induced voltage were calculated when a magnetic field was applied to the magnetostrictive layer. Subsequently, the tip deflection and induced magnetic field when an AC electric field was applied to the piezoelectric layer were calculated. Finally, the induced voltage and magnetic field were calculated when a repeated load was applied to the tip of the cantilever.

20.3.2.3 Experimental Procedure

A giant magnetostrictive/piezoelectric cantilever was fabricated by bonding Terfenol-D (15 mm × 5 mm × 3 mm) to PZT C-91 (20 mm × 5 mm × 0.5 mm). Tables 20.3 and 20.4 list the material properties of C-91 and Terfenol-D, respectively. The coercive electric field of C-91 was approximately 0.35 MV/m [34]. First, a magnetic field was applied in the z-direction, and the magnetostriction at the center of the surface of the Terfenol-D layer was evaluated using a strain gauge to determine the constant m_{33}. Subsequently, an experiment was performed in which an AC electric field was applied. Consider a Terfenol-D/PZT cantilever driven by an AC voltage with amplitude V_0, $V_0 \exp(i\omega t)$. An AC voltage was applied to the bottom surface of the PZT layer, while the top surface of the PZT layer was grounded. The displacement of the Terfenol-D/PZT cantilever under an AC voltage was measured using a laser displacement meter. The induced magnetic field in these laminates was obtained from the voltage generated in the search coil, measured using a data logger. The coil had an outside radius of 16.4 mm, an inside radius of 8.1 mm, and a length of 17 mm. It was wound from 80 μm diameter enameled copper wire and had 4810 turns. Finally, an experiment was performed in which a repeated load was applied to the tip of the cantilever. To generate a cyclic load P at the tip of the Terfenol-D/PZT cantilever, a cantilever load cell was used with its tip in contact with the tip of the Terfenol-D/PZT cantilever. The load ratio $R = P_{min}/P_{max}$ and cyclic frequency $f = \omega/2\pi$ were set to 0.1 and 50 Hz, respectively, and a cyclic load P was applied at the tip of the piezoelectric layer. The induced voltage at the bottom surface of the

PZT layer was measured using an oscilloscope. The top surface of the PZT layer was grounded. The induced magnetic field in the cantilevers was obtained from the generated voltage in the search coil, measured using a data logger.

20.3.2.4 Results and Discussion

Figure 20.11 shows the relationship between the applied magnetic field and strain of the magnetostrictive material. A nonlinear relationship between strain and magnetic field is observed, with the constant m_{33} being 3.3×10^{-12} m^2/A^2. Although the results are not presented here, they confirmed that m changes with the thickness of the magnetostrictive material. This is consistent with the second-order magnetoelastic constant expression $m_{33} = m + \sigma$, indicating that the internal stress depends on the thickness of the Terfenol-D layer. Figure 20.12 shows the analysis results of the tip deflection and induced voltage of the magnetostrictive/piezoelectric laminate material when a magnetic field is applied. It should be noted that a large electromagnet was used to apply the magnetic field, and measuring the deflection was

Table 20.3 Material properties of C-91.

Elastic compliance ($\times 10^{-12}$ m^2/N)					Piezoelectric constant ($\times 10^{-12}$ m/V)			Permittivity ($\times 10^{-10}$ C/Vm)		Density (kg/m^3)
s_{11}^E	s_{33}^E	s_{44}^E	s_{12}^E	s_{13}^E	d_{31}	d_{33}	d_{15}	ϵ_{11}^T	ϵ_{33}^T	ρ
17.1	18.6	41.4	−6.3	−7.3	−340	645	836	395	490	7800

Table 20.4 Material properties of Terfenol-D.

Elastic compliance ($\times 10^{-12}$ m^2/N)					Piezomagnetic constant ($\times 10^{-9}$ m/A)			Magnetic permittivity ($\times 10^{-6}$ H/m)		Density (kg/m^3)
s_{11}^H	s_{33}^H	s_{44}^H	s_{12}^H	s_{13}^H	$d^m{}_{31}$	$d^m{}_{33}$	$d^m{}_{15}$	μ_{11}	μ_{33}	ρ
17.9	17.9	26.3	−5.88	−5.88	−5.3	11	28	6.29	6.29	9250

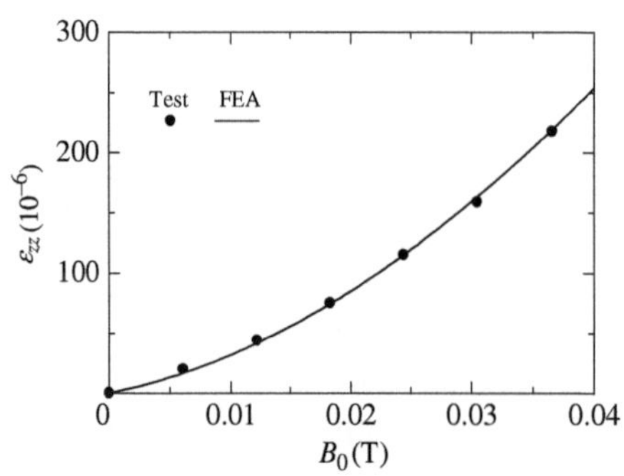

Figure 20.11 Strain versus magnetic field for Terfenol-D/PZT cantilever.

challenging owing to the measuring equipment; therefore, only analysis was performed. For the induced voltage, only the analytical results are presented owing to the minimal time variation in the applied magnetic field, making it challenging to measure. The deflection and induced voltage exhibit nonlinear behaviors, similar to the strain of the magnetostrictive material. Figure 20.13 shows the tip deflection and induced magnetic field under an AC electric field. The plots and lines represent the experimental and analytical results, respectively, and the black and red lines represent the tip deflection (left axis) and induced magnetic field (right axis), respectively. The tip deflection and induced magnetic field exhibit values nearly identical between the experimental and analytical results, reflecting the nonlinear behavior of domain switching. Finally, Figure 20.14 shows the results of the induced voltage and induced magnetic field under cyclic loading applied to the tip of the Terfenol-D/PZT cantilever. The experimental results are smaller than the analytical results, with minimal nonlinearity observed. In particular, the induced magnetic field was extremely small compared with that when an AC electric field was applied. This is attributed to the load applied at the cantilever tip, which did not contribute to the overall deformation of the Terfenol-D/PZT cantilever but rather caused localized deformation near the tip.

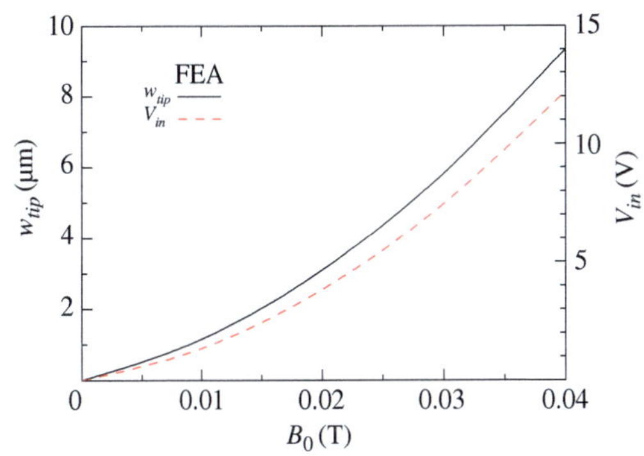

Figure 20.12 Tip deflection and induced voltage versus magnetic field for the Terfenol-D/PZT cantilever.

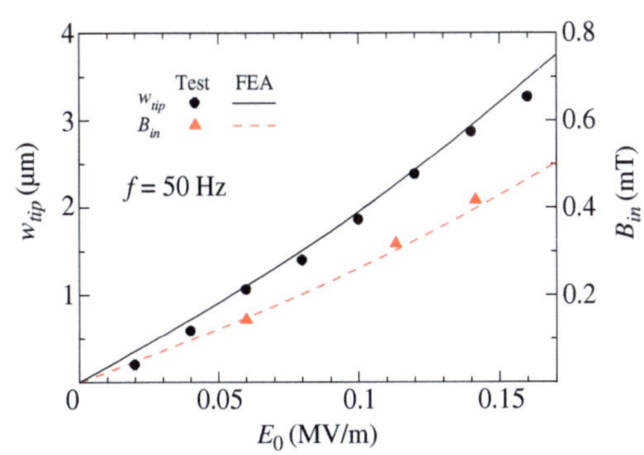

Figure 20.13 Tip deflection and induced magnetic field versus AC electric field amplitude for the Terfenol-D/PZT cantilever.

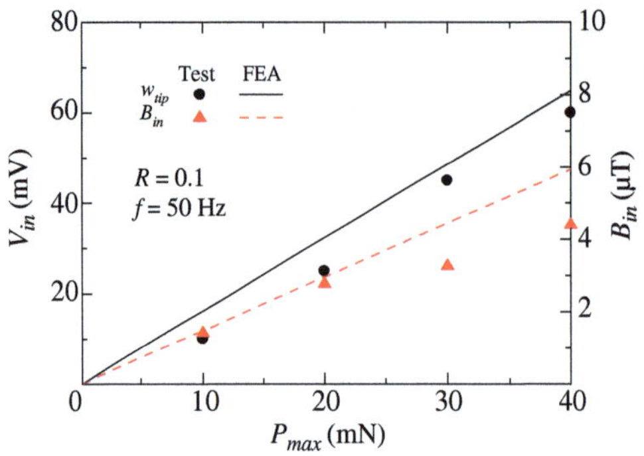

Figure 20.14 Induced voltage and induced magnetic field versus cyclic load for the Terfenol-D/PZT cantilever.

20.3.2.5 Conclusion

This section examines giant magnetostrictive/piezoelectric laminated composites. Magnetostrictive and piezoelectric materials exhibit nonlinear behaviors owing to their microcharacteristics; therefore, ingenuity is required when designing them macroscopically. However, giant magnetostrictive/piezoelectric laminated composites can obtain two outputs with one input, which allows actuators and sensors to add self-sensing/monitoring and energy-harvesting functions, while reducing weight and space.

20.4 Summary

Smart materials are expected to play an increasingly significant role in smart structures and Internet of Things applications, such as in energy harvesting and self-sensing. Therefore, achieving structural integrity and functional performance is crucial. This study explored the integration of only two materials. However, in practice, combining multiple materials involves optimizing multiple design parameters. FEA facilitates exploring various parameter combinations without creating a prototype, making it easy to achieve an optimal design. The scale at which smart materials are required is becoming smaller every year. In addition to the characteristics of macro- and mesoscale structures, the micro- and nanoscale properties and the interaction across these scales are increasingly explored. Therefore, considering the correlation between variables and the model to be used is crucial. Integrating FEA and machine learning offers the potential for efficient design.

References

1 Nye, J.F. (1957) Physical Properties of Crystals, The Clarendon Press, Oxford, UK.
2 Shrout, T.R. and Zhang, S.J. (2007) Lead-free piezoelectric ceramics: alternatives for PZT? *J. Electroceram.*, 19, 111–24.
3 Gomes, J., Nunes, J.S., Sencadas, V. and Lanceros-Mendez, S. (2010) Influence of the β-phase content and degree of crystallinity on the piezo- and ferroelectric properties of poly(vinylidene fluoride). *Smart Mater. Struct.*, 19, 065010.

4 Dunn, L. and Taya, M. (1993) Micromechanics predictions of the effective electroelastic moduli of piezoelectric composites. *Int. J. Solids Struct.*, 30, 161–75.

5 Engdahl, G. (2000) Handbook of Giant Magnetostrictive Materials, Academic Press, San Diego.

6 Clark, A.E., Teter, J.P. and McMasters, O.D. (1988) Magnetostriction "jumps" in twinned $Tb_{0.3}Dy_{0.7}Fe_{1.9}$. *J. Appl. Phys.*, 63, 3910–2.

7 Moffett, M.B., Clark, A.E., Wun-Fogle, M., Linberg, J., Teter, J.P. and McLaughlin, E.A. (1991) Characterization of Terfenol-D for magnetostrictive transducers. *J. Acoust. Soc. Am.*, 89, 1448–55.

8 Kellogg, R.A., Russell, A.M., Lograsso, T.A., Flatau, A.B., Clark, A.E., and Wun-Fogle, M. (2004) Tensile properties of magnetostrictive iron-gallium alloys. *Acta Mater.*, 52, 5043–50.

9 Atulasimha, J., Flatau, A.B. and Summers, E. (2007) Characterization and energy-based model of the magnetomechanical behavior of polycrystalline iron-gallium alloys. *Smart Mater. Struct.*, 16, 1265–76.

10 Yamamoto, M. (1942) Dynamical measurement of the *DE*-effect in iron-cobalt alloys. *J. Jpn. Inst. Met. Mater.*, 6, 581–92.

11 Yamazaki, T., Furuya, Y. and Nakao, W. (2019) Experimental evaluation of domain wall dynamics by Barkhausen noise analysis in $Fe_{30}Co_{70}$ magnetostrictive alloy wire. *J. Magn. Magn. Mater.*, 475, 240–8.

12 Dong, X., Qi, M., Guan, X. and Ou, J. (2011) Fabrication of $Tb_{0.3}Dy_{0.7}Fe_2$/epoxy composites: enhanced uniform magnetostrictive and mechanical properties using a dryprocess. *J. Magn. Magn. Mater.*, 323, 351–5.

13 Kurita, H., Keino, T., Senzaki, T. and Narita, F. (2022) Direct and inverse magnetostrictive properties of Fe-Co-V alloy particle-dispersed polyurethane matrix soft composite sheets. *Sens. Actuator A-Phys.*, 337, 113427.

14 Lin, P., Zhu, Y., Chen, Z., Fei, C., Chen, D., Zhang, S., Li, D., Yang, Y. and Chai, C. (2022) Design and fabrication of non-periodic 1-3 composite structure for ultrasonic transducer application. *Compos. Struct.*, 285, 115249.

15 Lin, C.-H. and Muliana, A. (2014) Polarization switching responses of 1-3 and 0-3 active composites. *Compos. Struct.*, 116, 535–51.

16 Zienkiewicz, O.C. (1977) The Finite Element Method, Third Edition. McGraw–Hill Book Company (UK) Limited.

17 Lee, H.-H. (2018) Finite Element Simulations with ANSYS Workbench 18. SDC publications.

18 Alshits, V.I., Darinskii, A.N. and Lothe, J. (1992) On the existence of surface waves in half-infinite anisotropic elastic media with piezoelectric and piezomagnetic properties. *Wave Motion*, 16, 265–83.

19 De Lacheisserie, E.T. (1993) Magnetostriction theory and applications of magnetoelasticity. CRC Press.

20 Mori, K., Narita, F., Wang, Z., Horibe, T. and Maejima, K. (2022) Thermoelectromechanical characteristics of piezoelectric composites under mechanical and thermal loading. *Adv. Eng. Mater.*, 24, 2101212.

21 Narita, F. and Fox, M. (2018) A review on piezoelectric, magnetostrictive, and magnetoelectric materials and device technologies for energy harvesting applications. *Adv. Eng. Mater.*, 20, 1700743.

22 Tan, T., Yan, Z., Zou, H., Ma, K., Liu, F., Zhao, L., Peng, Z. and Zhang, W. (2019) Renewable energy harvesting and absorbing via multi-scale metamaterial systems for Internet of things. *Appl. Energy*, 254, 113717.

23 Narita, F., Fox, M., Mori, K., Takeuchi, H., Kobayashi, T. and Omote, K. (2017) Potential of energy harvesting in barium titanate based laminates from room temperature to cryogenic/high temperatures: measurements and linking phase field and finite element simulations. *Smart Mater. Struct.*, 26, 115027.

24 Venkatragavaraj, E., Satish, B., Vinod, P.R. and Vijaya, M.S. (2001) Piezoelectric properties of ferroelectric PZT-polymer composites. *J. Phys. D: Appl. Phys.*, 34, 487–92.

25 Garnett, J.C.M. (1904) Colours in metal glasses and in metallic films. *Philos.* Trans. R. Soc. A, Phil. Trans. *A*, 203, 385–420.

26 Zhang, X., Le, M.-Q., Zahhaf, O., Capsal, J. F., Cottinet, P. J. and Petit, L. (2020) Enhancing dielectric and piezoelectric properties of micro-ZnO/PDMS composite-based dielectrophoresis. *Mater. Des.*, 192, 108783.

27 Shindo, Y., Mori, K. and Narita, F. (2010) Electromagneto-mechanical fields of giant magnetostrictive/piezoelectric laminates. *Acta Mech.*, 212, 253–61.

28 Mori, K., Shindo, Y. and Narita, F. (2012) Dynamic electromagneto-mechanical behavior of clamped-free giant magnetostrictive/piezoelectric laminates under AC electric fields. *Smart Mater. Struct.*, 21, 115003.

29 Mori, K., Shindo, Y., Narita, F. and Okura, S. (2016) Detection and response characteristics of giant magnetostrictive/piezoelectric laminated cantilevers under cyclic bending. *Mech. Adv. Mater. Struct.*, 23, 180–6.

30 Wang, T.-Z. and Zhou, Y. H. (2011) A theoretical study of nonlinear magnetoelectric effect in magnetostrictive-piezoelectric trilayer. *Compos. Struct.*, 93, 1485–92.

31 Nan, C.W., Li, M. and Huang, J.H. (2001) Calculations of giant magnetoelectric effects in ferroic composites of rare-earth-iron alloys and ferroelectric polymers. *Phys. Rev. B - Condens. Matter Mater. Phys.*, 63, 144415.

32 Li, S., Cao, W. and Cross, L.E. (1991) The extrinsic nature of nonlinear behavior observed in lead zirconate titanate ferroelectric ceramic. *J. Appl. Phys.*, 69, 7219–24.

33 Arlt, G. and Dederichs, H. (1980) Complex elastic, dielectric and piezoelectric constants by domain wall damping in ferroelectric ceramics. *Ferroelectrics*, 29 47–50.

34 Narita, F., Shindo, Y. and Mikami, M. (2005) Analytical and experimental study of nonlinear bending response and domain wall motion in piezoelectric laminated actuators under AC electric fields. *Acta Mater.*, 53, 4523–9.

35 Li, S., Bhalla, A.S., Newnham, R.E. and Cross, L.E. (1993) Quantitative evaluation of extrinsic contribution to piezoelectric coefficient d_{33} in ferroelectric PZT ceramics. *Mater. Lett.*, 17, 21–6.

36 Jia, Z., Liu, W., Zhang, Y., Wang, F. and Guo, D. (2006) A nonlinear magnetomechanical coupling model of giant magnetostrictive thin films at low magnetic fields. *Sens. Actuator A-Phys.*, 128, 158–64.

37 Wan, Y., Fang, D. and Hwang, K.-C. (2003) Non-linear constitutive relations for magnetostrictive materials. *International Journal of Non-Linear Mechanics*, 38, 1053–65.

38 Pei, Y. and Fang, D. (2010) A magnetoelastic model of nonlinear behaviors of Tb-Dy-Fe alloys based on domain rotation. *Int. J. Appl. Electromagn. Mech.*, 33, 883–9.

39 Shindo, Y., Narita, F., Mori, K. and Nakamura, T. (2009) Nonlinear bending response of giant magnetostrictive laminated actuators in magnetic fields. *J. Mech. Mater. Struct.*, 4, 941–9.

21

Future Outlook

Yu Jia

College of Engineering and Physical Sciences, Aston University, Birmingham, UK

21.1 Executive Summary and Market Projections

Energy harvesting holds the promise of providing decentralized self-sustaining power source. This makes energy harvesting a key enabling technology [1], serving many emerging digital technologies, including the Internet of Things (IoT), wearables, remote sensing, digital healthcare, and smart maintenance. As IoT expands toward the trillion-sensor universe [2], there is an increasing demand for this smart, decentralized, and self-sustaining power infrastructure to complement traditional battery power sources.

Energy harvesting has witnessed a robust expansion in both academic research and commercialization. The global energy-harvesting materials and systems market is projected to grow from $0.61 billion USD in 2025 to $0.94 billion USD by 2030 [3], with a compounded annual growth rate of 9.1%. Within the energy-harvesting sector, mechanical and thermal energy harvesting are projected to be the most dominant [4], led by piezoelectric, electromechanical, and thermoelectric transducers.

Looking ahead, there will be an ever-increasing trend of new IoT devices, wearables, and remote sensors that are integrated with vibration and/or thermal energy harvesters. New material design for these applications is also likely to couple with wider trends in miniaturization and sustainability. This has direct implications for favoring scalable and sustainable materials, such as the emergence of lead-free and micromachining-compatible piezoelectric material candidates [5] over the traditionally popular lead zirconate titanate (PZT) bulk material in the case of mechanical energy harvesters.

21.2 Technological Road Map and Evolution

The technological evolution of energy-harvesting transducers represents a critical pathway toward future applications that emphasize autonomous sensor networks, wearable electronics, and IoT implementations. There is a trend of convergence toward hybrid systems combining multiple transduction mechanisms. Advanced materials engineering emphasizes sustainability through lead-free compositions and reduced environmental impact. Manufacturing advances support scalable production through printing technologies and solution processing. This section takes stock of the current state and future trajectory of four top transduction mechanisms dominant in the mechanical and thermal energy-harvesting space: piezoelectric, triboelectric, electromagnetic, and thermoelectric.

21.2.1 Piezoelectric Materials

Piezoelectric transducers convert mechanical stress into electrical energy through the direct piezoelectric effect, where mechanical deformation generates an electrical charge. The fundamental performance metric is the piezoelectric coefficient d_{33}, which quantifies the charge generated per unit of applied mechanical stress, typically measured in picocoulombs per Newton (pC/N).

Recent developments demonstrate significant advancement in piezoelectric performance. Contemporary PVDF-based composites can achieve d_{33} values of 51.20 pC/N through melt-state energy implantation techniques [6], while shell-core PVDF/PMMA composite structures can attain 16 pC/N [7]. Ceramic systems show superior performance, with porous $BiFeO_3$-$BaTiO_3$ ceramics achieving 112–124 pC/N [8]. Lead-free alternatives such as $(Ba_{0.85}Ca_{0.15})(Zr_{0.1}Ti_{0.9})O_3$ with porous structures demonstrate 22–41% enhancement in d_{33} compared to dense counterparts [9].

Power-generation capabilities have demonstrated practical viability, with PVDF/BTO composites achieving power densities of 58.7 mW/m^2 under 10 N pressure [6]. Industrial applications show modest but viable outputs, with commercial piezoelectric sensors achieving continuous operation for low-power IoT applications [10].

Sustainability considerations center on lead-free alternatives to traditional PZT ceramics. Environmental regulations drive development toward bismuth-based systems and biocompatible materials such as PLA/BT composites with d_{33} values of 35 pC/N [11]. These materials offer comparable performance while eliminating toxic lead content.

Integrated circuit (IC) compatibility remains challenging due to high-voltage, low-current outputs. Future developments emphasize sustainability, miniaturization, and integration with CMOS technology for autonomous sensor networks.

21.2.2 Triboelectric Nanogenerators

Triboelectric nanogenerators (TENGs) exploit contact electrification between dissimilar materials to generate electrical energy through mechanical motion. The primary performance metrics include charge density (measured in $\mu C/m^2$), power density (W/m^2), and energy conversion efficiency.

Recent breakthroughs have achieved exceedingly promising performance levels, making them increasingly competitive against piezoelectric counterparts. Advanced TENG designs using PEI/P(VDF-TrFE-CFE) composites demonstrate charge densities of 9.23 mC/m^2 [12], while vacuum conditions enable charge densities up to 1003 $\mu C/m^2$ [13]. Bioinspired self-lubricating TENGs achieve power densities of 4.6 W/m^2 with 99% output retention after 60 000 cycles [14].

Material innovation focuses on enhancing charge retention and durability. TiO_2-MXene-incorporated polystyrene nanofiber charge-trapping layers achieve power densities of 25.2 $\mu W/cm^2$ [15]. Marine wave energy harvesters demonstrate power densities of 185.4 W/m^3 Hz with multitunnel-grating electrodes [16].

Sustainability advantages include material simplicity and processing efficiency. Most TENG systems utilize readily available polymers and metals, supporting cost-effective manufacturing while reducing environmental impact compared to traditional energy storage systems. IC integration benefits from voltage step-down and impedance-matching circuits. Future developments emphasize miniaturization and integration with energy storage systems for practical autonomous applications.

21.2.3 Electromagnetic Transducers

Electromagnetic induction harvesters convert mechanical motion into electrical energy through Faraday's law, utilizing relative motion between conductors and magnetic fields. Performance metrics include coupling coefficient, power density, and operational frequency bandwidth.

Contemporary electromagnetic harvesters achieve practical power outputs suitable for electronic applications in the range of 1's to 10's mW [17]. Hybrid systems combining electromagnetic and piezoelectric mechanisms demonstrate enhanced performance capabilities. Large-scale applications show significant potential for trackside sensor networks and industrial monitoring systems.

Sustainability considerations focus on permanent magnet materials and copper conductors. Rare earth magnets present supply chain challenges, driving research into alternative magnetic materials. Copper recycling infrastructure supports circular economy principles. IC compatibility represents a significant consideration, with output characteristics closely matching electronic system requirements. Future developments include wide-bandwidth and nonlinear operations and adaptive tuning for variable frequency environments [18].

21.2.4 Thermoelectric Energy Generators

Thermoelectric materials convert temperature gradients into electrical energy through the Seebeck effect. The fundamental performance metric is the dimensionless figure of merit ZT, defined as $ZT = \alpha^2\sigma T/\kappa$, where α is the Seebeck coefficient, σ is the electrical conductivity, T is the absolute temperature, and κ is the thermal conductivity.

Recent material developments demonstrate significant ZT improvements. ZnO nanostructures interfaced with graphene quantum dots achieve ZT values of 0.486 at 580 K [19]. $Bi_2Te_3/Bi_{0.5}Sb_{1.5}Te_3$ superlattice films reach $ZT = 1.44$ at 400 K [20].

Oxide-based materials such as $Ba_1/_3CoO_2$ demonstrate $ZT = 0.55$ at 600 °C with excellent stability [21].

Sustainability advantages include material abundance and processing simplicity. Earth-abundant materials increasingly replace rare elements, while waste heat recovery applications provide dual environmental benefits. IC compatibility benefits from moderate voltage and current outputs. Future developments emphasize flexible substrates and distributed architectures for wearable and IoT applications.

21.3 Next-generation Material Platforms

Advancements in energy-harvesting and sensor systems are progressing rapidly toward increased efficiency, multifunctionality, flexibility, and sustainability in devices. Table 21.1 highlights a few main classes, their core features, challenges, and application prospects.

21.3.1 Two-dimensional Materials

Two-dimensional (2D) materials, including graphene, transition metal dichalcogenides, and MXenes, are gaining traction due to their high surface-to-volume ratio, tunable electrical properties, and mechanical flexibility. These characteristics make them ideal for flexible and wearable energy harvesters such as triboelectric, piezoelectric, and photovoltaic generators [22, 23].

Table 21.1 Summary of emerging next-generation energy-harvesting material developments.

Material system	Key features and opportunities	Application focus	Challenges
2D materials	High surface area, flexibility, multimodal harvesting	Wearables, low-power sensors	Scalability, integration
MOFs/COFs	Tunable porosity/ functionality, hybrid integration	PENGs, TENGs, self-powered sensors	Stability, synthesis
Lead-free piezoelectrics	Nontoxic, robust, suitable for sustainable energy harvesting	Green wearables, soft robotics	Lower responsiveness, durability
Hybrid Perovskites	Multifunctional energy conversion, high PV efficiency	Solar cells, pyroelectric sensors	Stability, toxicity
Quantum materials	Quantum-limited sensitivity, new physical effects	Quantum sensors, imaging	Fabrication complexity
Electrets/hydrogels	Self-healing, robust, biointegration	Implantable devices, soft generators	Processing, retention
Organic–inorganic hybrids	Flexible, printable, biocompatible	Disposable/stretchable sensors	Lifetime performance

- **Multimodal Harvesting:** Capable of harnessing solar, thermal, and mechanical energies.
- **Integration:** Easily embedded in textiles and polymers for smart wearables.
- **Challenges:** Large-scale, cost-effective production and robust integration.

21.3.2 Framework Materials: Metal-organic Frameworks and Covalent Organic Frameworks

Metal-organic frameworks (MOFs) and covalent organic frameworks offer structural versatility and electrical and mechanical properties, beneficial for piezoelectric and TENGs. Recent work demonstrates their potential for self-powered sensors and wearable platforms, though stability and scalable synthesis remain obstacles [24].

21.3.3 Lead-free Piezoelectrics

Eco-friendly, lead-free piezoelectrics such as potassium sodium niobate (KNN) and barium titanate ($BaTiO_3$) are replacing traditional lead-based materials. These materials support green energy harvesting with robust mechanical compliance, though some performance limitations persist [25].

21.3.4 Hybrid Perovskite Materials

Hybrid perovskites exhibit high performance in photovoltaics, with power conversion efficiencies exceeding 25%. Multifunctional properties permit the coupling of piezoelectric, pyroelectric, and photovoltaic effects. However, issues around chemical stability and toxicity are active areas of research [26].

21.3.5 Quantum and Smart Materials

Quantum materials, notably those based on quantum dots, topological insulators, and tailored molecular structures, facilitate ultra-sensitive quantum sensing platforms for next-generation sensor devices, offering new standards in spatial/temporal resolution [27].

21.3.6 Electret, Hydrogel, and Luminescent Materials

- **Electrets**: Provide long-term charge retention for enhanced triboelectric devices.
- **Hydrogels**: Serve as ionic conductors and self-healing matrices for biointegrated sensors.
- **Luminescent Materials**: Enable multiplexed, low-energy optical-sensing platforms for environmental and biomedical systems [28].

21.3.7 Hybrid and Biocompatible Platforms

Organic–inorganic hybrids combine mechanical flexibility with tunable electronic properties, suitable for printable, disposable, and stretchable sensors. Biocompatible systems cater to implantable, environmentally benign applications and support sustainable device design [29].

21.4 Circular Futures: Sustainability, Market Considerations, and Emerging Applications

The intersection of sustainability, circular economy principles, and application expansion is redefining the trajectory of materials science and energy harvesting. Forward-thinking strategies are being implemented not only to enhance the performance of advanced materials and devices but to ensure their entire life cycle minimizes environmental impact while addressing future market needs.

21.4.1 Designing for Circularity and Closing Loops in Energy Waste

Material innovation prioritizes recyclability, toxicity reduction, and overall environmental compatibility. The adoption of lead-free piezoelectrics and biopolymer-based substrates is steadily replacing hazardous or hard-to-recycle alternatives. Modular product architectures—engineered for simplified disassembly andreprocessing—are reducing waste and supporting maintenance and upgrade cycles. Concurrently, life cycle assessments (LCAs) are increasingly used to evaluate and reduce environmental footprint at each development phase.

Energy-harvesting technologies present inherent advantages for circular economy integration because they convert ambient or wasted energy into useful electrical power. Thermoelectric modules recover energy from otherwise dissipated heat, while TENGs, manufactured from recycled or biodegradable feedstocks, demonstrate new approaches for both energy recovery and sustainable materials utilization. Agricultural and municipal waste streams are increasingly seen as potential inputs for device fabrication and active layers, reinforcing the concept of waste valorization.

21.4.2 Circular Business Models and Emerging Applications

The paradigm is shifting from linear product supply chains to service-oriented systems such as product-as-a-service (PaaS). Under these models, clients subscribe to device functionality—like energy generation or data monitoring—while ownership, maintenance, and recycling responsibilities stay with the provider, maximizing reuse and end-of-life recovery. Reverse logistics, enhanced by digital tracking and regulatory mechanisms such as extended producer responsibility (EPR), underpin these approaches. Some of the emerging applications that will leverage energy harvesting and sensors include:

- **Smart Cities:** Deployments of low-maintenance, energy-autonomous sensors for infrastructure health, environmental analytics, and utilities optimization.
- **Wearables and e-Textiles:** Flexible, energy-harvesting textiles and medical monitors designed for safe disposal or recovery, using bioderived or recyclable materials.
- **Decentralized Energy Systems:** Microgrids built from locally sourced or reused materials, employing autonomous sensors and harvesters to maximize resilience.

Table 21.2 Circularity, business, and policy considerations for energy-harvesting development.

Dimension	Key actions and trends	Examples/opportunities	Future challenges
Materials design	Renewable, recyclable, modular platforms	Lead-free ceramics, bioplastics	Material availability, performance
Circular business models	Move to service-based, life cycle-managed products	Energy-as-a-service, sensor networks	System integration, user adoption
Waste-to-energy	Convert waste/biomass to electricity or materials	TENGs from agrowaste, heat recovery systems	Scale, logistics, economics
Application expansion	Sustainable devices for new domains	Smart cities, microgrids, e-textiles	Interoperability, adoption
Policy and regulation	Standards for energy harvesters and ecodesign	EPR, carbon disclosure, green procurement	Regulatory harmonization, enforcement
Data-driven circularity	Digital tracking and optimization	AI for recycling, smart product passports	Privacy, digital infrastructure

- **Data-driven Circularity:** The application of artificial intelligence and IoT to optimize product life cycles, manage logistics, and enable predictive maintenance.

Meeting the dual imperatives of sustainability and technological advancement in materials design catalyzes both environmental and economic value. A truly circular approach will require not only technical innovation but also harmonized policies, new business partnerships, and transparent, data-driven management of resources. Table 21.2 summarizes the various dimensions to coherently consider for future development.

21.5 Conclusion and Strategic Recommendations

The integration of advanced materials and device architectures represents the decisive force propelling energy harvesting and sensing toward true multifunctionality, efficiency, and sustainability. The research frontier increasingly converges on the design of materials and systems that not only achieve enhanced performance metrics but also embody principles of resource conservation, recyclability, and responsible life cycle management.

Material innovation remains the core driver. Significant advancements in 2D materials, such as graphene and transition metal dichalcogenides, are enabling new paradigms in device miniaturization, flexibility, and hybrid energy conversion. Developments in MOFs and lead-free ceramics are providing tunable chemistries, augmented charge transport, and improved environmental profiles. The field's continued shift toward low-toxic, biodegradable, and modular compounds ensures that future devices will be better suited for integration into circular material streams.

Life cycle-informed design is increasingly foundational in materials research. The early incorporation of LCA tools into research practice allows technologists to evaluate and mitigate the environmental footprint at each development phase. Emphasis on design-for-recycling and energy-efficient fabrication—such as the adoption of additive manufacturing and scalable roll-to-roll processes—supports sustainable production at scale.

Scalable processing and deposition techniques, particularly those compatible with novel nanomaterials and hybrid architectures, are crucial for translating laboratory-scale innovations to real-world use. Methods such as solution processing, inkjet printing, and molecular self-assembly continue to bridge the gap between high-performance research materials and durable, manufacturable components.

Cross-disciplinary collaboration within the research ecosystem—across materials science, chemistry, digital engineering, and sustainability science—is growing increasingly indispensable. Cooperative frameworks and shared research platforms are accelerating the pace of discovery and integration, informing both the properties of emerging materials and the environmental implications of their widespread adoption.

Research priorities looking forward should center on the synthesis and characterization of next-generation compounds with inherent circularity, integration of these materials into flexible and multifunctional device platforms, and the development of real-time digital feedback systems that optimize device longevity and reclamation. Emphasizing these strategies will ensure that energy-harvesting and sensing technologies evolve not only in performance but in responsibility, positioning the scientific community as a leader in the global transition toward sustainable innovation.

References

1 Akinaga, H. (2020) Recent advances and future prospects in energy harvesting technologies. *Jpn. J. Appl. Phys.*, 59(11), 110201, http://doi.org/10.35848/1347-4065/abbfa0.

2 Kaminaga, S. (2018) Trillion sensors and MEMS. *Sens. Mater.*, 30(4), 723–31. http://dx.doi.org/10.18494/SAM.2018.1814.

3 Markets and Markets. (2025) Energy harvesting system market size, share & trends, 2025 to 2030. Market report: https://www.marketsandmarkets.com/Market-Reports/energy-harvesting-market-734.html [accessed on 21 July 2025]

4 Energy Harvesting Market. (2025) Energy Harvesting Market Growth - Trends & Forecast 2025 to 2035. Market report: https://www.futuremarketinsights.com/reports/global-energy-harvesting-market [accessed on 21 July 2025]

5 Wei, H., Wang, H., Xia, Y., Cui, D., Shi, Y., Dong, M., Liu, C., Ding, T., Zhang, J., Ma, Y., Wang, N., Wang, Z., Sun, Y., Wei, R. and Guo, Z. (2018) An overview of lead-free piezoelectric materials and devices. *J. Mater. Chem. C*, 6(46), 12446–67. https://doi.org/10.1039/C8TC04515A.

6 Huang, Z.-X., Li, L. W., Huang, Y. Z. Rao, W. X., Jiang, H.-W., Wang, J., Zhang, H. H., He, H. Z., and Qu, J. P. (2024) Self-poled piezoelectric polymer composites via melt-state energy implantation. *Nat. Commun.*, 15, 819. https://doi.org/10.1038/s41467-024-45184-4.

7 Li, X., Yuan, C., Gao, G., Zhou, H., Zhang, N., Yang, X., Liu, X. and Sun, H. (2024) All-organic shell–core PVDF/PMMA composite fiber films with enhanced piezoelectric performance for flexible sensing and energy harvesting. *J. Appl. Polym. Sci.*, 141(28), e55626. https://doi.org/10.1002/app.55626.

8 Narayan, B., Li, Z., Wang, B., Haugen, A.B., Hall, D., Khanbareh, H. and Roscow, J. (2024) Temperature-dependent ferroelectric properties and aging behavior of freeze-cast bismuth ferrite–barium titanate ceramics. *ACS Appl. Mater. Interfaces*, 16(15), 19283–97. https://pubs.acs.org/doi/10.1021/acsami.4c03002.

9 Li, Z., Roscow, J., Khanbareh, H., Davies, P.R., Han, G., Qin, J., Haswell, G., Wolverson, D. and Bowen, C. (2024) Porous structure enhances the longitudinal piezoelectric coefficient and electromechanical coupling coefficient of lead-free $(Ba_{0.85}Ca_{0.15})(Zr_{0.1}Ti_{0.9})O_3$. *Adv. Sci.*, 11, 2406255. https://doi.org/10.1002/advs.202406255.

10 Rigo, F., Migliorini, M. and Pozzebon, A. (2024) Piezoelectric sensors as energy harvesters for ultra low-power IoT applications. *Sensors*, 24(8), 2587. https://doi.org/10.3390/s24082587.

11 Puszczykowska, N., Rytlewski, P., Mirkowska, A., Augustyn, P. and Fiedurek, K. (2025) Fabrication, thermal, mechanical, and piezoelectric characterization of PLA/BT piezocomposites. *Smart Mater. Struct.*, 34(2), 025013. https://doi.org/10.1088/1361-665X/ad9971.

12 Liu, X., Zhao, Z., Zhang, B., Hu, Y., Qiao, W., Gao, Y., Wang, J., Guo, Z., Zhou, L., Wang, Z.L. and Wang, J. (2025) Superior charge density of triboelectric nanogenerator via trap engineering. *Adv. Funct. Mater.*, 35, 2416944. https://doi.org/10.1002/adfm.202416944.

13 Wang, J., Wu, C., Dai, Y., Zhao, Z., Wang, A., Zhang, T. and Wang, Z.L. (2017) Achieving ultrahigh triboelectric charge density for efficient energy harvesting. *Nat. Commun.*, 8(1), 88. https://doi.org/10.1038/s41467-017-00131-4.

14 Chen, A., Zeng, Q., Tan, L., Wang, T., Xu, F., Wang, J., Tao, X., Yang, Y. and Wang, X. (2024) Charge accumulation effect enabled by a bioinspired self-lubricating triboelectric nanogenerator for both high average power density and long durability. *Adv. Funct. Mater.*, 34, 2405698. https://doi.org/10.1002/adfm.202405698.

15 Venkatesan, M., Chandrasekar, J., Hsu, Y. C., Sun, T. W., Li, P. Y., King, X. T., Chung, M. A., Chung, R. J., Lee, W. Y., Zhou, Y., Lin, J. H. and Kuo, C. C. (2024) Rationally improved surface charge density of triboelectric nanogenerator with TiO_2-MXene/polystyrene nanofiber charge trapping layer for biomechanical sensing and wound healing application. *Adv. Sci.*, 11, 2404019. https://doi.org/10.1002/advs.202404019.

16 Wang, Y., Du, H., Yang, H., Xi, Z., Zhao, C., Qian, Z., Chuai, X., Peng, X., Yu, H., Zhang, Y., Li, X., Hu, G., Wang, H. and Xu, M. (2024) A rolling-mode triboelectric nanogenerator with multi-tunnel grating electrodes and opposite-charge-enhancement for wave energy harvesting. *Nat. Commun.*, 15, 6834. https://doi.org/10.1038/s41467-024-51245-5.

17 Beeby, S.P. and O'Donnell, T. (2009). Electromagnetic Energy Harvesting. In: Priya, S. and Inman, D.J. (eds.) Energy Harvesting Technologies, Springer, Boston, MA. https://doi.org/10.1007/978-0-387-76464-1_5.

18 Jia, Y. (2020) Review of nonlinear vibration energy harvesting: duffing, bi-stability, parametric, stochastic and others. *J. Intell. Mater. Syst. Struct.*, 31(7), 921. https://doi.org/10.1177/1045389X20905989.

19 Choi, M., An, J., Lee, H., Jang, H., Park, J.H., Cho, D., Song, J.Y., Kim, S.M., Oh, M.W., Shin, H. and Jeon, S. (2024) High figure-of-merit for ZnO nanostructures by interfacing lowly-oxidized graphene quantum dots. *Nat Commun.*, 15, 1996. https://doi.org/10.1038/s41467-024-46182-2.

20 Park, N.-W., Lee, W. Y., Yoon, Y. S., Kim, G.-S., Yoon, Y. G., and Lee, S. K. (2019) Achieving out-of-plane thermoelectric figure of merit $ZT = 1.44$ in a p-type $Bi_2Te_3/Bi_{0.5}Sb_{1.5}Te_3$ superlattice film with low interfacial resistance. *ACS Appl. Mater. Interfaces*, 11(41), 38247–54. https://doi.org/10.1021/acsami.9b11042.

21 Zhang, X., Zhang, Y., Wu, L., Tsuruta, A., Mikami, M., Cho, H.J. and Ohta, H. (2022) $Ba_{1/3}CoO_2$: a thermoelectric oxide showing a reliable ZT of ~0.55 at 600 °C in air. *ACS Appl. Mater. Interfaces.*, 14(29), 33355–60. https://doi.org/10.1021/acsami.2c08555.

22 Khandelwal, G., Agarwal, S., Jaiswal, A., Tiwari, S., Tripathi, S., Kumar, A. and Kim, S. J. (2023) Recent developments in 2D materials for energy harvesting applications. *J. Phys. Energy*, 5(3), 032001. https://doi.org/10.1088/2515-7655/acc7c8.

23 Kumar, A. and Mandal, D. (2024) 2D-pyroelectric materials for waste thermal energy harvesting and beyond. *Adv. Energy Mater.*, 14, 2403319. https://doi.org/10.1002/aenm.202403319.

24 Priyadarshini, A., Divya, S., Swain, J., Das, N., Swain, S., Hajra, S., Panda, S., Samantaray, R., Belal, M., Kaja, K.R., Kumar, N., Kim, H.J., Oh, T.H., Vivekananthan, V. and Sahu, R. (2025) Advancements in framework materials for enhanced energy harvesting. *Nanoscale*, 17(4), 1790–811. https://doi.org/10.1039/d4nr04570j.

25 Akshayveer, A., Buroni, F.C., Melnik, R., Rodriguez-Tembleque, L. and Saez, A. (2025) Environment-friendly technologies with lead-free piezoelectric materials: a review of recent developments, applications, and modelling approaches. *Mater. Sci.*, https://doi.org/10.48550/arXiv.2502.20250.

26 Pandey, R., Vats, G., Yun, J., Bowen, C.R., Ho-Baillie, A.W.Y., Seidel, J., Butler, K.T. and Seok, S.I. (2019) Mutual insight on ferroelectrics and hybrid halide perovskites: a platform for future multifunctional energy conversion. *Adv Mater.*, 31(43), e1807376. https://doi.org/10.1002/adma.201807376.

27 Yu, C.J., von Kugelgen, S., Laorenza, D.W. and Freedman, D.E. (2021) A molecular approach to quantum sensing. *ACS Cent. Sci.*, 7(5), 712–23. https://doi.org/10.1021/acscentsci.0c00737.

28 Wu, X., Li, G. and Lee, D.W. (2015) A hydrogel-based energy harvester with broad bandwidth driven by ambient vibrations, *2015 Transducers – 2015 18th International Conference on Solid-State Sensors, Actuators and Microsystems (TRANSDUCERS)*, Anchorage, AK, USA, 938–41. https://doi.org/10.1109/TRANSDUCERS.2015.7181078.

29 Wang, S., Kang, Y., Wang, L., Zhang, H., Wang, Y. and Wang, Y. (2013) Organic/inorganic hybrid sensors: a review. *Sens. Actuator B Chem.*, 182, 467–81. https://doi.org/10.1016/j.snb.2013.03.042.

Index

Note: Page numbers in *italics* and **bold** refers to figures and tables respectively.

f

n